INTRODUCTORY ALGEBRA

THIRD EDITION

J. P. Wood

University of Houston—Downtown College

CHARLES E. MERRILL PUBLISHING COMPANY
A Bell and Howell Company
Columbus   London   Toronto   Sydney

Published by
Charles E. Merrill Publishing Company
A Bell & Howell Company
Columbus, Ohio 43216

This book was set in Times Roman.
The production editors were Ann Mirels and Linda M. Johnstone.
The cover was prepared by Will Chenoweth.

International Standard Book Number: 0-675-08511-X
Library of Congress Catalog Card Number: 76-44128
3 4 5 6-81 80 79 78 77

Printed in the United States of America

FOR
RACHEL
AND
JEFFREY
WILSON

PREFACE

This book takes as it point of departure the student's elementary experience with numbers. Hence, no background other than basic arithmetic is required. Algebra is explored from its beginnings as generalized arithmetic and is developed along intuitive lines that are tied to familiar concepts, guiding the student, at a carefully cadenced pace, toward an understanding of the basic techniques of beginning and intermediate algebra. The chapters are arranged in modular formats: Each chapter is a self-contained treatment of a central topic, and each ends with a chapter test that serves also as a pretest for the succeeding chapter.

The intuitive treatment, particularly evident in the first five chapters, is couched in language that is readily comprehensible by the beginning student. The general weakness endemic to mathematics students at all levels in computation with exponents and radicals has motivated a simplified but emphatic treatment of these topics in the chapter preceding quadratic equations. In the later chapters, section references are liberally supplied so that the student may recall the elementary illustrations of the basic operations covered earlier.

All topics covered in the text are accompanied by an ample selection of completely worked-out examples and copious exercises. Stated problems, emphasized throughout the text, focus on everyday affairs rather than on mathematical puzzles, and all measurements in the problems utilize the metric system. If more practice is needed, a workbook entitled *Student Guide to Introductory Algebra* is available. The workbook is keyed to the text and provides supplementary problems.

In this edition manipulative skills occupy center stage, beginning in Chapter 1 with the immediate introduction of operations with signed numbers. The sharp focus of the earlier editions on operational techniques has been intensified in three ways: (1) additional topics and exercises on basic operations with fractions, equations, the metric system, and stated problems have been added; (2) the use of set notation and language has been greatly reduced and appears only in topics in which sets are particularly useful, such as the sets of ordered pairs defined by linear equations; and (3) the theoretical aspects of algebra have been removed from the body of text and placed in an Appendix.

Another feature of this edition is an expanded treatment of quadratic equations, functions and graphs, and inequalities. The last-named topic appears in a separate chapter on first-degree inequalities comprised of inequalities in one and two variables, absolute-value inequalities, and systems of linear inequalities. Because more advanced students frequently exhibit unspoiled innocence of manipulative skill with inequalities, an intuitive introduction to the properties of inequalities with additional stress on techniques of simplification and on illustrative and practical examples replaces a formal treatment.

Above all else, this book seeks effective communication with the student. Consequently, the language of the text begins mathematically at ground zero and is constantly adapted to the student's level of progress. The decidedly elementary language of the first three chapters, designed to reassure the student and to build self-confidence, is coupled with a sequence of topics arranged to capture and to hold his or her interest. Specifically, no topic is introduced until its need and usefulness can be immediately evidenced. Symbolic language is introduced slowly, always with emphasis on the convenience and common sense which sufficiently justify its use. The commutative and associative laws, the identity elements, the distributive axiom, and the language of sets are not presented as original Commandments but as responses to specific needs.

Finally, the book has been written in the hope that by starting at humble beginnings and leading with deliberate simplicity from the relatively naive to the relatively sophisticated, *Introductory Algebra* will help formerly disaffected students to discover the elegance, the fun, and, indeed, the handiness of mathematics.

I am greatly indebted to my colleagues Dr. Betty Hinman and Professor Betty Bollinger for comments and suggestions in the preparation of the manuscript. I also wish to express my appreciation to Dr. L. A. Colquitt of Texas Christian University, Professor H. E. Hall of DeKalb Community College, Dr. David Outcalt of the University of California at Santa Barbara, and Professor Mary Jo Knobelsdorf of the University of Houston Central Campus for their comments.

J. P. Wood
Houston, Texas

CONTENTS

Chapter 10 Radicals; Complex Numbers   359

Chapter 11 Quadratic Equations in One Variable   393

Chapter 12 Logarithms   435

1

THE INTEGERS; OPERATIONS WITH SIGNED NUMBERS

The best way to begin a study of algebra is to start with an understanding of how algebra itself began. It started quite simply as a general or symbolic method for representing the processes of arithmetic. Thus if 3×2 means "three times two" and 3×6 means "three times six," then if a represents a number, $3 \times a$ means "three times a."

By letting the letters of the alphabet represent numbers, we can write, *symbolically*, the operations of arithmetic. For example, if a is any number and b is any other number, then $a \times b$ means "a times b" and is a concise way of representing the *product* of the two numbers.

In like manner, $a + b$ means the *sum* of two numbers; $a - b$ means the *difference* of two numbers, and a/b means the *quotient* of two numbers provided that b is not equal to zero. The beginning student can avoid many errors if he will firmly keep in mind this basic fact: *algebra began as generalized arithmetic*.

Example:

Since

$$\frac{2+4}{2} = \frac{2}{2} + \frac{4}{2} = 1 + 2 = 3$$

then

$$\frac{a+b}{a} = \frac{a}{a} + \frac{b}{a} = 1 + \frac{b}{a}$$

(provided that a is not equal to zero).

Consequently, the letters in elementary algebra represent the numbers of arithmetic, and in operations with the numbers of arithmetic, we have the proper beginnings of algebra.

1.1
The Natural Numbers

The simplest (and in all likelihood the first) numbers that the human race devised in the attempt to classify different quantities are the so-called *counting numbers*, or *natural numbers*. This means the sequence of numbers used in counting:

$$1, \overset{\cdot\cdot}{2}, 3, 4, 5, 6, 7, 8, 9 \ldots$$

In mathematics, the three dots placed at the end are used to indicate that this sequence of numbers has no ending, or, as we say, that it continues indefinitely. Please observe that the natural numbers include *only* those numbers which we use to count the discrete (individually distinct) objects in any sort of collection, such as the number of ships in a convoy, or gazelles in a herd, or students in a classroom. One could not have, for example, one-third of a ship in a convoy nor one-half of a gazelle in a herd. In other words, although $\frac{1}{2}$ and $\frac{1}{3}$ are numbers, they are certainly not *counting* (or natural) numbers. All of the natural numbers are a part of a collection, or set, of numbers called *integers*. The succession of the natural numbers in order of size can be represented on a straight line.

If we mark off a straight line in uniform graduations, the natural numbers serve to count off the successive marks in order. Since a straight line can be extended indefinitely, and since the natural numbers continue indefinitely, we have a continuing succession of points and numbers. If the first graduation is considered merely as a starting point, then the label "4" on a point denotes that it is the *fourth* graduation to the *right* of the starting point. The number zero is used to denote this starting point (or origin), and the natural numbers identify the succession of uniform graduations to the right of zero. This is shown in Figure 1.1.

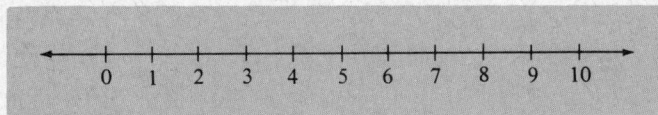

Figure 1.1

The point on this line designated by the number 10 can be considered to be the *graph* of that number on the *number line*, and it means that the point so labeled lies 10 units to the *right of zero*. This scheme outlines a beginning for a relation of numbers to points on a line, and from this beginning, we shall devise a geometric representation of the numbers of arithmetic. The attempt to locate points on a line and to designate by appropriate numbers their distances from zero leads to the development, part of which is retraced here, of the *real number system*. Thus the natural numbers form a part of this real number system; that is, every natural number is a real number, just as every Ford is an automobile. Please note, however, that not every automobile is a Ford, nor is every real number a natural number. Later we shall consider other types of real numbers. For the greater part of this course we shall be concerned specifically with the algebra of real numbers, and until otherwise stated, the word *number* will be used to mean *real numbers only*.

The natural numbers clearly do not identify the location of every point on the number line. For example, since the number line (also called the continuum) is infinite in length, what numbers would identify successive uniform graduations to the *left* of zero?

1.2
The Negative Integers

If we wish to identify by number successive graduations on the number line to the left of zero, how should we proceed? (See Figure 1.2)

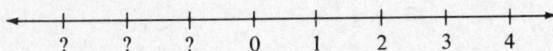

Figure 1.2

Obviously, the point which lies one unit to the *left* of zero is the same distance from zero as the point which lies one unit to the *right*, but how do we signify the opposite direction? As our concept of numbers and operations with them developed, similar bothersome questions arose. Just as there is a difference in direction to be accounted for in our present problem, there are many differences of a similar nature involved with the use of numbers. Examples are the difference between having ten dollars if one is free of debt and owing ten dollars if one is broke; or the difference between a temperature of 60° above zero and 60° below zero. All of these examples involve quantities, but they also all involve a reversal in direction, a reversal in finances, a reversal in temperature. From the necessity of dealing with situations like these, the negative numbers were born.

The two opposite signs ($+$ and $-$) give us a simple device for designating opposite directions. Thus the point one unit to the left of zero on the number line is designated by -1, and the point two units to the left, by -2. In similar fashion, successive graduations to the left of zero are designated by the *negatives* of the natural numbers. This set of numbers is called the set of *negative integers* (Figure 1.3). The natural numbers, which identify successive graduations to the right of the origin, are the opposites of these, or the *positive integers*. The negative integers are written with the negative sign affixed; for this reason the positive integers (or natural numbers) may be indicated by the absence of a sign. When such a number has no sign affixed to it, then it is *always understood to be positive*. Thus the point which is three units to the right of zero on the number line corresponds to the number $+3$ or simply 3.

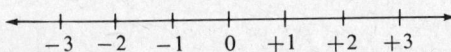

Figure 1.3

These two sets of numbers, the set of positive integers and the set of negative integers, are both infinite and are both part of the real number system.

The set of positive integers (the natural numbers) may be written in this way:

$$\{1, 2, 3, 4, 5, 6, \ldots\}$$

Similarly the set of negative integers may be written:

$$\{-1, -2, -3, -4, -5, -6, \ldots\}$$

The phrase "a set of numbers" is used here in the same sense that we would use the word *set* to refer to a set of tools or a set of encyclopedias. As our investigation of real numbers proceeds, we shall take a closer look at the usefulness of the set concept in mathematics. For the present we shall employ the word *set* as a convenient way to refer to certain collections of numbers included in the real number system.

Thus the positive and negative integers correspond respectively to uniform graduations on the number line to the right and left of zero, as shown in Figure 1.4. But

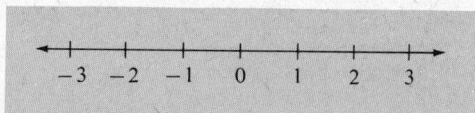

Figure 1.4

what about the integer zero? Since zero obviously cannot lie to the right or left of itself, *zero is neither positive nor negative.* We shall see in succeeding discussions that zero is the only real number that is not so classified.

1.3
The Sets N and J

The collection of all the numbers which have thus far been matched one-to-one with the starting point, zero, and with successive graduations to the right and left of zero on the number line is called the *set of integers.* Since the integers are exclusively those numbers which identify the uniform graduations (which are each one unit in length) on the number line, the set of integers necessarily includes all of the natural numbers, the negatives of all natural numbers, and the number zero. Moreover, the set of integers includes *only* those numbers.

The integers are the natural numbers, the negatives of the natural numbers, and zero.

Numbers in a particular set are said to "belong to" the set and are called the "*elements*" or the "*members*" of the set. We can identify a specific set of numbers by simply listing its elements, setting off each separate element by commas. To signify that we mean *all* the elements of the set, we enclose the listed elements in *braces.*

Thus the set of natural numbers (positive integers) was previously written

$$\{1, 2, 3, 4, 5, \ldots\}$$

Using the same notation we may represent the set of integers in the following way:

$$\{\ldots -5, -4, -3, -2, -1, 0, 1, 2, 3, 4, 5, \ldots\}$$

Remember that the three dots on either end indicate that this set of numbers continues indefinitely; e.g., there is no end to the positive integers nor any beginning to the negative integers.

Although *all* of the natural numbers are included in the set of integers, it should be noted carefully that the *reverse* of that statement is not true. For example, we can say, correctly, that every horse is an animal, but we cannot truthfully *reverse* the statement and say that every animal is a horse. In like manner we can say, correctly, that every natural number is an integer, but the reverse of the statement is clearly not true.

In preceding paragraphs we have referred to the set of natural numbers by two other names: the set of counting numbers and the set of positive integers. We can simplify all of this nomenclature considerably by selecting a particular *symbol* to represent the set of natural numbers and then referring to that set exclusively by the symbol selected. It is common in mathematics to refer to the set of natural numbers by the capital letter N. We shall adopt that symbol to represent this set throughout this text.

$$N = \{1, 2, 3, 4, 5, \ldots\}$$

This conventional agreement will simplify all references to the set of natural numbers. Adopting this convention is an example of the fact that the symbolic language used in mathematics is prompted by convenience and common sense and certainly is not intended to obscure the subject by a fog of mysterious hieroglyphics.

Using this same convenient tactic, we shall henceforth refer to the set of integers by the capital letter J:

$$J = \{\ldots -5, -4, -3, -2, -1, 0, 1, 2, 3, 4, 5, \ldots\}$$

The two sets of numbers, N and J, as well as their relationship to each other can be pictorially represented by using the regions enclosed by geometric figures (circles, rectangles, squares, triangles, etc.) to represent particular sets. For example, we can draw a circle and label the region enclosed by it with the letter J, which means that the enclosed region (containing infinitely many points) is intended to represent the infinitely many numbers in the set of integers. (See Figure 1.5.)

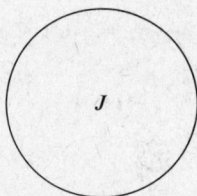

Figure 1.5

The size of the circle has no relationship to the "size" of the set: the circle is meant only to suggest a pictorial representation, or diagram, of the set.

Now suppose that we use the region enclosed by a triangle to represent the set N and then draw the diagram to show the relationship of N to J. This can be done in the following manner:

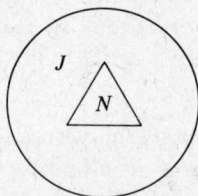

Figure 1.6

The fact that the triangle labeled N is *completely contained within* the circle labeled J shows that every natural number is contained in the set of integers. At the same time, the "J circle," by including points *outside* the "N triangle," shows that there are some integers which are *not* natural numbers. The diagram thus gives us an instant and organized view of the two sets, N and J.

Diagrams of this type are called Venn diagrams because they were widely used by a man named John Venn. They can be helpful in devising a composite view of the entire real number system, not only because they provide a simple schematic drawing ("blueprint") of different sets which make up the system, but also because they can be made to depict quite clearly the relationships of the different sets to each other. This latter point is illustrated by the information implicit in the diagram of the two sets just discussed. Because such diagrams and the language of sets will be employed throughout our discussions of the real numbers, they should be considered carefully here and retained for future reference.

Thus far we have classified two sets of numbers that are among the numbers encountered in arithmetic. Later we shall look at other types of numbers which also belong to the real number system, numbers such as $\frac{3}{4}$, $-\frac{5}{8}$, .4, 2.6, $\sqrt{3}$, and $\sqrt{5}$.

While all of those numbers are encountered in arithmetic, note that *none* of them is an *integer*. As stated earlier, the integers are exclusively those numbers that identify the points marking the unit graduations to the right and left of zero on the number line. We shall find that the set of real numbers includes all of the numbers needed to label by number *every point* on the number line.

Now let us take a list of real numbers and beside each one write the following notations: *n.n.* if the number is a natural number; *int.* if the number is an integer; and *neither* if the number is not included in either of the two sets *N* or *J*.

Examples:

(1) 286 *int., n.n.*   (2) $\frac{2}{3}$ *neither*   (3) .5 *neither*

(4) 4.25 *neither*   (5) −860 *int.*   (6) 5618 *int., n.n.*

(7) $-\frac{5}{9}$ *neither*   (8) 7.5 *neither*   (9) 11 *int., n.n.*

Exercise 1:

Classify each number listed below in the indicated manner: *n.n.* if the number belongs to *N*; *int.* if the number belongs to *J*; and *neither* if the number is not an element of *N* or *J*.

(1) $5\frac{1}{2}$ (2) −17   (3) $\frac{7}{8}$

(4) 784   (5) 9.3   (6) 12

(7) 2.65   (8) $-\frac{3}{4}$ (9) −3840

(10) 0   (11) .86   (12) −.6

1.4
Absolute Value

Frequently in mathematics we find that, for a given number, we need to express *its distance from zero only*, without regard to its *direction* from zero. Since every real number gives both distance and direction, another way of writing these numbers, to express distance only, was needed. The method devised was the use of two vertical bars, | |. When a real number is enclosed in vertical bars, its distance from zero is designated with no regard for the direction; such a symbol is read "the absolute value" of the number enclosed.

In this manner the symbol |6| means "how many units *distance* lie between 0 and 6?" The distance is 6 units, as you can see in Figure 1.7, and |6| = 6. This is read

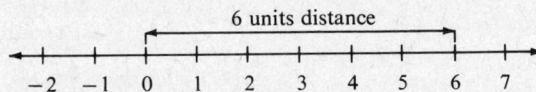

Figure 1.7

"the absolute value of 6 is 6." The symbol |−6| means "how many units *distance* lie between 0 and −6?" The distance is 6 units, and |−6| = 6. This is read "the absolute value of −6 is 6." See Figure 1.8.

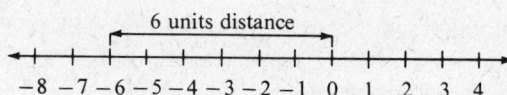

Figure 1.8

In similar fashion:

$$|3| = 3 \qquad\qquad |10| = 10 \qquad\qquad \left|4\frac{1}{2}\right| = 4\frac{1}{2}$$

$$|-3| = 3 \qquad\qquad |-10| = 10 \qquad\qquad \left|-4\frac{1}{2}\right| = 4\frac{1}{2}$$

Since the reversal of a sign means the reversal of a given situation we can define the meaning of absolute values in the symbolic language of algebra in the following manner.

Let a be *any* real number, positive, negative, or zero. Then

$$|a| = a \text{ if } a \text{ is } positive \text{ or } zero$$
$$|a| = -a \text{ if } a \text{ is } negative.$$

For example, if $a = -3$, then $|-3| = -(-3) = 3$. This result, $-(-3) = 3$, can be seen intuitively from the number line: if the number -3 designates the point three units to the *left* of zero, then the reversal of the situation, e.g., $-(-3)$, would indicate the number representing the point three units to the *right* of zero, which is the number 3.

If we have the following:

$$|8 - 5| = |3| = 3$$

we can show the meaning on the number line in the following way:

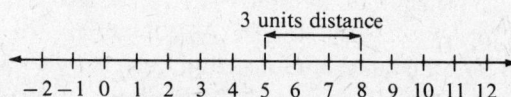

Figure 1.9

Therefore $|8 - 5| = |3| = 3$ actually measures the distance between $+5$ and $+8$ on the number line (See Figure 1.31.)

Compare the following:

$$|6| = |6 - 0| = 6, \text{ the distance between 6 and 0}$$

$$|-6| = |-6 - 0| = 6, \text{ the distance between } -6 \text{ and 0}$$

$$|8 - 5| = 3, \text{ the distance between 8 and 5}$$

To state this *symbolically* we can say that if *a* is a number and *b* is another number, $|a - b|$ measures the *distance only* between *a* and *b*. Following are examples of problems involving absolute values of numbers:

(1) $|9| = 9$

(2) $|-7| = 7$

(3) $|9| + |-7| = 9 + 7 = 16$

(4) $|9| - |-7| = 9 - 7 = 2$

(5) $|9| \times |-7| = 9 \times 7 = 63$

(6) $\sqrt{9} = |3| = 3$

(7) $\sqrt{9} = |-3| = 3$

(8) $|7 - 2| - |5 - 1| = |5| - |4| = 5 - 4 = 1$

Exercise 2:

Compute the value of the following:

(1) $|9| = ?$

(2) $|32| = ?$

(3) $|-14| = ?$

(4) $|-7| = ?$

(5) $|128| = ?$

(6) $|-316| = ?$

(7) $|12| = ?$

(8) $|-25| = ?$

(9) $|13 - 5| = ?$

(10) $|130 - 20| = ?$

(11) $|5| \times |4| = ?$

(12) $|-5| \times |4| = ?$

(13) $|5| \times |-4| = ?$

(14) $|-5| \times |-4| = ?$

(15) $|3| + |4| = ?$

(16) $|5| + |-2| = ?$

(17) $|-10| + |-3| = ?$

(18) $|8 - 2| + |-4| = ?$

(19) $|6 - 2| + |10 - 3| = ?$

(20) $\sqrt{25} = |5| = ?$

(21) $\sqrt{25} = |-5| = ?$

(22) $|11 - 3| - |4 - 2| = ?$

OPERATIONS WITH SIGNED NUMBERS

The four fundamental operations that we perform with numbers are the following:

Division Multiplication Subtraction Addition

When one's comprehension of the number system progresses beyond the positive numbers, it becomes necessary to review these fundamental operations as they apply to *all* real numbers, to the negative numbers and zero as well as to the positive numbers.

For the present we shall proceed intuitively and naively assume that the fundamental operations for all real numbers obey the same laws as for the first real numbers we learned. Intuition is not an infallible guide (far from it!), and it is certainly no substitute for mathematical proof, but it is still the natural avenue along which the mind travels to seek out new ideas. We are concerned here, therefore, not with a revision of previous knowledge, but merely with an extension of it.

1.5
Addition

As previously stated, we proceed on the assumption that addition means the same thing with all real numbers that it means in Grade One. Looking back over our earliest experiences with this process, we recall that, when we were told to add 3 to 5, we wrote down the following:

$$\begin{array}{r} 3 \\ 5 \\ \hline 8 \end{array}$$

We know now that all of these numbers are positive; so we could write

$$\begin{array}{r} +3 \\ +5 \\ \hline +8 \end{array}$$

and mean exactly the same thing.

Now let us analyze that example. Exactly what did we do in the above addition? We added two numbers which had the *same sign* (both $+$), and to do that, we took the arithmetic sum of the numbers and gave the result the *same sign*. That, in effect, was what happened every time we added two numbers which had the same sign.

$$\begin{array}{rrrr} +6 & +8 & +9 & +3 \\ +5 & +10 & +6 & +7 \\ \hline +11 & +18 & +15 & +10 \end{array}$$

We shall now assume that addition of real numbers is a consistent process (it is) and that it means the same thing all the time (it does). Therefore, to *add* two numbers having the *same sign*, we always do the *same thing*: that is we *take the arithmetic sum of the numbers and give to the result their common sign*.

> If $+3$ plus $+5$ equals $+8$
> then -3 plus -5 equals -8

This process makes excellent sense if one thinks of it in terms of credits and debits. If I have 3 dollars ($+3$) and I gain 5 dollars more ($+5$), then when the credits are added, the sum of my financial situation is that I am 8 dollars *ahead* ($+8$ dollars). On the other hand, assume that I am broke and that I owe Jim 3 dollars (-3), and I also owe Joe 5 dollars (-5); then when the debits are added, I am surely the sum of 8 dollars in *arrears* (-8 dollars).

Addition of numbers having like signs:

				$+4$	$+8$
$+3$	$+8$	$+10$	$+27$	$+6$	$+9$
$+4$	$+11$	$+5$	$+13$	$+3$	$+10$
$+7$	$+19$	$+15$	$+40$	$+13$	$+27$
				-6	-801
-32	-8	-1064	-27	-4	-920
-47	-11	-523	-13	-3	-1038
-79	-19	-1587	-40	-13	-2759

Summary

To add numbers having the same sign, take the arithmetic sum of the numbers and give the result their common sign.

Exercise 3:

Add the following numbers:

(1) $+6$
       $+9$

(2) -6
       -9

(3) $+10$
       $+5$

(4) -10
       -5

(5)   8
      7

(6) -21
       -3

(7)   24
      16

(8) -13
       -10

(9)   10
      13
       8

(10) -10
       -13
       -8

(11) 6
      14
       5

(12) -8
       -13
       -1

(13) 20
      32
      40

(14) -13
       -14
       -36

(15) 84
       6
      32

(16) -27
       -13
       -4

(17)	37	(18)	-62	(19)	425	(20)	-375
	92		-48		760		-610
	85		-79		418		-848

Since every real number (except zero) is either positive or negative, it follows that any time two nonzero numbers are added, one of two things is bound to happen. Either the signs of the two numbers will be alike or else the signs will be opposite. We have considered what happens when the signs are alike.

Now let us consider the problem of adding two numbers which have opposite signs. If we *add* -8 to $+5$, what should the result be? We could translate this into debits and credits and put the same question this way: If I have exactly 5 dollars ($+5$) and I owe 8 dollars (-8), how should I describe my financial situation? The situation is clearly that I am 3 dollars in *arrears* (-3). Consequently, -8 and $+5$ equals -3.

Here common sense dictates that we take the arithmetic *difference* of the numbers and give the result the sign of the number having the *greater* absolute value. Since the indebtedness is greater than the cash on hand, we quite obviously wind up in the minus column.

In like manner:

$$-10 \text{ and } +4 \text{ equals } -6$$
$$-7 \text{ and } +2 \text{ equals } -5$$
$$+10 \text{ and } -12 \text{ equals } -2$$
$$+10 \text{ and } -5 \text{ equals } +5$$
$$-2 \text{ and } +6 \text{ equals } +4$$

Summary

To add two numbers having unlike signs, take the arithmetic difference of the numbers and give the result the sign of the number having the greater absolute value.

Examples:

$+12$	$+12$	-12	-12
-4	-14	$+6$	$+17$
$+8$	-2	-6	$+5$

We should observe that real numbers, in the operation of addition, exhibit a most important property: *the sum of two numbers is the same regardless of the order in which they are added.* To illustrate this property, note the following examples:

1. $+4 + 5 = 9$
   $+5 + 4 = 9$ therefore $+4 + 5 = +5 + 4$

2. $-8 - 3 = -11$
   $-3 - 8 = -11$ therefore $-8 - 3 = -3 - 8$

3. $+6 - 4 = 2$
$-4 + 6 = 2$ therefore $+6 - 4 = -4 + 6$

4. $-10 + 2 = -8$
$+2 - 10 = -8$ therefore $-10 + 2 = +2 - 10$

These are all examples of a property of real numbers called the *commutative law of addition*. We assume that this is true for all real numbers and state the fact symbolically, with *a* and *b* representing real numbers, as follows:

> **The Commutative Law of Addition**
>
> $a + b = b + a$

A second important property of real numbers in the operation of addition can be illustrated by adding any three numbers. Consider the following:

$$4 + 5 + 2 = ?$$

We might begin by adding the *first two* numbers and continue by adding the last one to the sum of the first two:

$$4 + 5 + 2 = (4 + 5) + 2 = 9 + 2 = 11$$

But we might also add the *last two* numbers and then add the first one to the sum of the last two:

$$4 + 5 + 2 = 4 + (5 + 2) = 4 + 7 = 11$$

Repeating these operations in order to compare them, we have the following:

$$4 + 5 + 2 = (4 + 5) + 2 = 9 + 2 = 11$$

$$4 + 5 + 2 = 4 + (5 + 2) = 4 + 7 = 11$$

The sum of the three numbers is clearly the same regardless of the manner in which the numbers are paired together. Note that we did not rearrange the *order* of the numbers: we added them in *different pairs*. The fact that the sum remains the same for the different pairings is called the *associative law of addition* and may be stated symbolically, with *a*, *b*, and *c* representing real numbers, as follows:

> **The Associative Law of Addition**
>
> $(a + b) + c = a + (b + c)$

Examples:

Add the following numbers.

(1) $\begin{array}{r} -900 \\ 243 \\ 525 \\ \hline -132 \end{array}$ Using the associative law, add the two *positive* numbers first; thus we have $\begin{array}{r} -900 \\ 768 \\ \hline -132 \end{array}$

(2) $\begin{array}{r} 318 \\ -204 \\ 563 \\ \hline 677 \end{array}$ Using the commutative law, mentally *rearrange* the numbers and add the two *positive* numbers first: $\begin{array}{r} 318 \\ 563 \\ -204 \\ \hline \end{array}$ or $\begin{array}{r} 881 \\ -204 \\ \hline 677 \end{array}$

Exercise 4:

Add the following:

(1) $\begin{array}{r} +6 \\ -3 \\ \hline \end{array}$ (2) $\begin{array}{r} -6 \\ +3 \\ \hline \end{array}$ (3) $\begin{array}{r} +8 \\ -10 \\ \hline \end{array}$ (4) $\begin{array}{r} -8 \\ +10 \\ \hline \end{array}$ (5) $\begin{array}{r} -15 \\ +2 \\ \hline \end{array}$

(6) $\begin{array}{r} +15 \\ -2 \\ \hline \end{array}$ (7) $\begin{array}{r} +18 \\ -6 \\ \hline \end{array}$ (8) $\begin{array}{r} -18 \\ +6 \\ \hline \end{array}$ (9) $\begin{array}{r} -3 \\ -4 \\ \hline \end{array}$ (10) $\begin{array}{r} +8 \\ +6 \\ \hline \end{array}$

(11) $\begin{array}{r} +7 \\ -3 \\ \hline \end{array}$ (12) $\begin{array}{r} -3 \\ +7 \\ \hline \end{array}$ (13) $\begin{array}{r} -10 \\ +5 \\ \hline \end{array}$ (14) $\begin{array}{r} +5 \\ -10 \\ \hline \end{array}$ (15) $\begin{array}{r} -1170 \\ -465 \\ \hline \end{array}$

(16) $\begin{array}{r} +11 \\ +4 \\ \hline \end{array}$ (17) $\begin{array}{r} -11 \\ +4 \\ \hline \end{array}$ (18) $\begin{array}{r} +11 \\ -4 \\ \hline \end{array}$ (19) $\begin{array}{r} -8 \\ 6 \\ \hline \end{array}$ (20) $\begin{array}{r} 1008 \\ -429 \\ \hline \end{array}$

(21) $\begin{array}{r} -18 \\ 12 \\ \hline \end{array}$ (22) $\begin{array}{r} 257 \\ -273 \\ \hline \end{array}$ (23) $\begin{array}{r} -184 \\ -26 \\ \hline \end{array}$ (24) $\begin{array}{r} 26 \\ 13 \\ \hline \end{array}$ (25) $\begin{array}{r} +10 \\ +2 \\ -6 \\ \hline \end{array}$

(26) $\begin{array}{r} +2 \\ +10 \\ -6 \\ \hline \end{array}$ (27) $\begin{array}{r} -6 \\ +10 \\ +2 \\ \hline \end{array}$ (28) $\begin{array}{r} +10 \\ -6 \\ +2 \\ \hline \end{array}$ (29) $\begin{array}{r} +10 \\ -2 \\ -6 \\ \hline \end{array}$ (30) $\begin{array}{r} -10 \\ +2 \\ -6 \\ \hline \end{array}$

(31) $\begin{array}{r} +10 \\ -2 \\ +6 \\ \hline \end{array}$ (32) $\begin{array}{r} -10 \\ -2 \\ -6 \\ \hline \end{array}$ (33) $\begin{array}{r} 1053 \\ 260 \\ 687 \\ \hline \end{array}$ (34) $\begin{array}{r} 624 \\ -350 \\ 230 \\ \hline \end{array}$ (35) $\begin{array}{r} -800 \\ 286 \\ 309 \\ \hline \end{array}$

1.6
Subtraction

Like addition, subtraction is the same process in basic algebra that it is in arithmetic. Subtraction, however, is a one-way street, and we must know clearly what is being subtracted from what.

If we are told, for instance, to subtract 3 from 5, the number that is being subtracted is 3, and it is called the *subtrahend*. The other number, 5, is called the *minuend*, probably because we had to call it something to distinguish it from the subtrahend. In every subtraction problem, the *subtrahend* must be clearly understood.

Now, back to Grade One. Assume that we are told to subtract 3 from 5. As we now know, this means that we subtract +3 from +5. But when you were instructed to subtract 3 from 5, you did *not* write down +3 and +5. Instead you wrote the following:

$$\begin{matrix} 5 \\ -3 \\ \hline \end{matrix} \qquad \left(\begin{matrix} \text{which} \\ \text{means} \end{matrix} \quad \begin{matrix} +5 \\ -3 \\ \hline \end{matrix}\right)$$

Now exactly what did you do? The 5 is still +5 but the 3 has become −3. In other words, *you reversed the sign of the subtrahend.*

You did this because you were told that was what subtraction meant. Subtraction does mean just that: *Reverse the sign of the subtrahend* and then follow the rules for addition.

To subtract +3 from +5

1. Reverse the sign of the subtrahend and write
$$\begin{matrix} +5 \\ -3 \\ \hline \end{matrix}$$

2. Now, *add* the numbers:
$$\begin{matrix} +5 \\ -3 \\ \hline +2 \end{matrix}$$

Note that we have arrived at a sound conclusion: that when +3 is *subtracted from* +5 the result is +2. Therefore if subtraction is a consistent operation (it is), and if it means the same thing all the time (it does), we can subtract any number from another by doing the same thing: *reversing* the sign of the subtrahend and then *adding* the numbers.

Examples:

(1) Subtract +2 from +7

$$\begin{array}{r} +7 \\ -2 \\ \hline +5 \end{array}$$

(2) Subtract +3.76 from +10

$$\begin{array}{r} +10.00 \\ -3.76 \\ \hline +6.24 \end{array}$$

(3) Subtract −3 from +5

$$\begin{array}{r} +5 \\ +3 \\ \hline +8 \end{array}$$

(4) Subtract −4 from −9

$$\begin{array}{r} -9 \\ +4 \\ \hline -5 \end{array}$$

(5) Subtract −500 from −1132

$$\begin{array}{r} -1132 \\ +500 \\ \hline -632 \end{array}$$

(6) Subtract 4 from −6

$$\begin{array}{r} -6 \\ -4 \\ \hline -10 \end{array}$$

(7) Subtract 700 from −1306

$$\begin{array}{r} -1306 \\ -700 \\ \hline -2006 \end{array}$$

To subtract one number from another, reverse the sign of the subtrahend and then follow the rules for addition.

In the addition of real numbers we observed that the *sum* of two numbers was the same regardless of the order of the numbers, and this property is called the commutative law of addition.

Examples:

Add +4 to −6

$$\begin{array}{r} +4 \\ -6 \\ \hline -2 \end{array}$$

Add −6 to +4

$$\begin{array}{r} -6 \\ +4 \\ \hline -2 \end{array}$$

Here we should note that this law does *not* apply to subtraction.

Examples:

$$
\begin{array}{r}
-6 \\
-4 \\
\hline
-10
\end{array}
$$

Subtract +4 from −6

$$
\begin{array}{r}
+4 \\
+6 \\
\hline
+10
\end{array}
$$

Subtract −6 from +4

Since +10 and −10 are clearly not the same number, subtraction is not commutative.

Exercise 5:

In the following problems subtract the lower number from the upper number:

(1) $\begin{array}{r} +10 \\ +4 \\ \hline \end{array}$
(2) $\begin{array}{r} +17 \\ +8 \\ \hline \end{array}$
(3) $\begin{array}{r} +25 \\ +3 \\ \hline \end{array}$
(4) $\begin{array}{r} +8 \\ +2 \\ \hline \end{array}$
(5) $\begin{array}{r} +11 \\ +7 \\ \hline \end{array}$

(6) $\begin{array}{r} +8 \\ +5 \\ \hline \end{array}$
(7) $\begin{array}{r} +8 \\ -5 \\ \hline \end{array}$
(8) $\begin{array}{r} -8 \\ +5 \\ \hline \end{array}$
(9) $\begin{array}{r} -8 \\ -5 \\ \hline \end{array}$
(10) $\begin{array}{r} +16 \\ -2 \\ \hline \end{array}$

(11) $\begin{array}{r} +18 \\ +4 \\ \hline \end{array}$
(12) $\begin{array}{r} -12 \\ +3 \\ \hline \end{array}$
(13) $\begin{array}{r} -17 \\ -10 \\ \hline \end{array}$
(14) $\begin{array}{r} 24 \\ -3 \\ \hline \end{array}$
(15) $\begin{array}{r} -12 \\ 5 \\ \hline \end{array}$

(16) $\begin{array}{r} -2 \\ -4 \\ \hline \end{array}$
(17) $\begin{array}{r} -5 \\ 10 \\ \hline \end{array}$
(18) $\begin{array}{r} 6 \\ 15 \\ \hline \end{array}$
(19) $\begin{array}{r} 7 \\ -15 \\ \hline \end{array}$
(20) $\begin{array}{r} -3 \\ 21 \\ \hline \end{array}$

(21) $\begin{array}{r} 9 \\ 18 \\ \hline \end{array}$
(22) $\begin{array}{r} -7 \\ -12 \\ \hline \end{array}$
(23) $\begin{array}{r} 4 \\ 9 \\ \hline \end{array}$
(24) $\begin{array}{r} -16 \\ 27 \\ \hline \end{array}$
(25) $\begin{array}{r} -2 \\ 8 \\ \hline \end{array}$

(26) $\begin{array}{r} 9 \\ -18 \\ \hline \end{array}$
(27) $\begin{array}{r} -3 \\ -14 \\ \hline \end{array}$
(28) $\begin{array}{r} -17 \\ 23 \\ \hline \end{array}$
(29) $\begin{array}{r} 18 \\ 6 \\ \hline \end{array}$
(30) $\begin{array}{r} 23 \\ -6 \\ \hline \end{array}$

(31) $\begin{array}{r} -820 \\ 1306 \\ \hline \end{array}$
(32) $\begin{array}{r} 1600 \\ 973 \\ \hline \end{array}$
(33) $\begin{array}{r} -2 \\ -16 \\ \hline \end{array}$
(34) $\begin{array}{r} 47 \\ 133 \\ \hline \end{array}$
(35) $\begin{array}{r} 248 \\ -52 \\ \hline \end{array}$

1.7
Multiplication

Multiplication began as a method of rapid addition. It developed into quite an art from there, but in order to decide anything as basic as how to multiply signed numbers together, we might as well go back to the beginning. In seeking the unknown, there is nothing so sensible as starting with the well known.

The expression 6×2 means "add six twos." If we do this

$$
\begin{array}{c}
2 \\
2 \\
2 \\
2 \\
2 \\
2 \\
\hline
12
\end{array}
$$

we find that the result is 12; so $6 \times 2 = 12$.

Like the other fundamental operations, multiplication is indicated by a specific symbol, the most familiar of which is "x." For algebraic purposes, this symbol is not very handy. The symbol x is a letter of the alphabet, and if we propose to use letters to represent numbers, we would find it confusing, when we hand-write mathematical exercises, to distinguish between x as a symbol for multiplication and x as a symbol for a number. So here and now, we are going to discard the letter x as a symbol of multiplication and use something else instead. In the "something else" we have a choice of two widely used symbols for multiplication: one of them is a *dot*, $6 \cdot 2$; and the other is *parentheses*, $(6)(2)$.

Therefore, the familiar "6×2" will be written in one of the following ways:

$$
\left.
\begin{array}{r}
6 \cdot 2 \\
\text{or } (6)(2) \\
\text{or } \quad 6(2)
\end{array}
\right\} = 12
$$

As noted earlier, $(6)(2)$ means to add six twos. As noted in Chapter 1 of this book, the symbols 6 and 2 mean $+6$ and $+2$. Consequently $(6)(2)$ really means to "add six *positive* twos."

$$
\begin{array}{c}
+2 \\
+2 \\
+2 \\
+2 \\
+2 \\
+2 \\
\hline
+12
\end{array}
$$

So when we say $(6)(2) = 12$, we are actually saying that $(+6)(+2) = +12$.

$$(6)(2) = 12$$
$$\text{means } (+6)(+2) = +12$$

There is no mystery, then, about multiplying positive numbers together; we have been doing *that* for years.

$$(5)(3) = 15 \text{ means } (+5)(+3) = +15$$
$$(6)(7) = 42 \text{ means } (+6)(+7) = +42$$

Before we proceed further in the multiplication of signed numbers, we need to look at one of the laws of multiplication for real numbers.

If (6)(2) means to add six positive twos, then (2)(6) must mean to add two positive sixes.

$$
\begin{array}{cc}
+2 & +6 \\
+2 & +6 \\
+2 & \overline{+12} \\
+2 & \\
+2 & \\
+2 & \\
\overline{+12} &
\end{array}
$$

We note the following:

$$(+6)(+2) = +12$$
$$(+2)(+6) = +12$$

or

$$(6)(2) = 12$$
$$(2)(6) = 12$$

Therefore

$$(6)(2) = (2)(6)$$

This situation repeats itself endlessly.

$$\left.\begin{array}{l}(3)(5) = 15 \\ (5)(3) = 15\end{array}\right\} \quad (3)(5) = (5)(3)$$

$$\left.\begin{array}{l}(7)(2) = 14 \\ (2)(7) = 14\end{array}\right\} \quad (7)(2) = (2)(7)$$

$$\left.\begin{array}{l}(5)(6) = 30 \\ (6)(5) = 30\end{array}\right\} \quad (5)(6) = (6)(5)$$

In the multiplication of two numbers, it appears that the result (called the *product*) is independent of the order of the multiplication. We shall assume that this is true for all real numbers and state the fact algebraically; if a and b are any two numbers, then

$$(a)(b) = (b)(a)$$

This is called the *commutative law of multiplication*.

Please note that *only* multiplication and addition are commutative. In division, for example, $6 \div 3$ is *not* equal to $3 \div 6$. In symbolic language, with a and b representing real numbers, we have the following:

> **The Commutative Law of Multiplication**
> $$(a)(b) = (b)(a)$$

Taking into account that both multiplication and addition are commutative, let us again proceed to investigate the multiplication of signed numbers. If $(+6)(+2)$ always means to add six *positive* twos, then what does $(+6)(-2)$ mean? It must mean to add six *negative* twos. If we do this

$$\begin{array}{r} -2 \\ -2 \\ -2 \\ -2 \\ -2 \\ -2 \\ \hline -12 \end{array}$$

we get -12 as a result. Then $(6)(-2) = (+6)(-2) = -12$.

$$\text{Since} \qquad (+6)(-2) = -12$$

$$\text{then} \qquad (-2)(+6) = -12$$

by the commutative law of multiplication.

Let us consider $(2)(-6)$, which is the same as $(+2)(-6)$. This means to add two *negative* sixes.

$$\begin{array}{r} -6 \\ -6 \\ \hline -12 \end{array}$$

$$\text{Therefore} \qquad (2)(-6) = -12$$

$$\text{and} \qquad (-6)(2) = -12$$

by the commutative law of multiplication.

Now let us line up the above results:

$$\boxed{\begin{aligned} (+6)(-2) &= -12 \\ (-2)(+6) &= -12 \\ (+2)(-6) &= -12 \\ (-6)(+2) &= -12 \end{aligned}}$$

In *every* case above, we have multiplied together two numbers having *unlike* signs and in *every* case the product was *negative*. By using similar examples, we could continue to multiply positive and negative numbers together and the products would continue to be negative. Thus we shall conclude that *when two numbers having unlike signs are multiplied together, the product is always negative.*

Examples:

$$(+6)(-3) = -18 \qquad (+3)(-7) = -21 \qquad (+7)(-7) = -49$$
$$(-5)(+10) = -50 \qquad (-9)(+8) = -72 \qquad (-4)(+9) = -36$$

Thus far in our consideration of multiplication with signed numbers, we have arrived at these requirements:

1. Multiplication is commutative.

$$(a)(b) = (b)(a)$$

2. When two positive numbers are multiplied together, the product is positive.

$$(+6)(+2) = +12$$
$$(+2)(+6) = +12$$

3. When two numbers having *unlike* signs are multiplied together, the product is *always negative.*

$$(-6)(+2) = -12$$
$$(+6)(-2) = -12$$

Now let us consider this situation:

$$(-6)(-2) = ?$$

We shall begin by aligning this product with the others and taking another look:

$$
\begin{array}{l}
(+6)(+2) = +12 \\
(-6)(-2) = ? \\
(+6)(-2) = -12 \\
(-6)(+2) = -12 \\
(-2)(+6) = -12 \\
(+2)(-6) = -12
\end{array}
$$

Take a *good* look! If you were going to try an educated guess here, what would your best guess be? And why?

When we confront the product $(-6)(-2)$, we are ourselves confronted by the fact that intuition, unfortunately, has its limitations. We could worry around with this product and come up with some plausible-sounding arguments such as the following:

1. Looking back over other examples of multiplication with signed numbers that have already been considered, we might observe that, in every case, the product of the absolute values of the numbers has equaled the absolute value of the product; e.g.,

$$(6)(-2) = -12 \text{ and } |-12| = 12$$

$$\begin{aligned} |6| &= 6 \\ |-2| &= 2 \end{aligned} \quad \text{and } (6)(2) = 12$$

   This could prompt us to try the same thing with the product $(-6)(-2)$:

$$\begin{aligned} |-6| &= 6 \\ |-2| &= 2 \end{aligned} \quad \text{and } (6)(2) = 12$$

   and we might conclude that $(-6)(-2)$ should equal some kind of "12."
2. In the real number system there are only two kinds of 12's; namely, $+12$ and -12. Thus we could easily be led to conjecture that $(-6)(-2)$ must equal $+12$ or -12. But this unprofound conclusion could likely have been reached by pure guesswork in the first place.
3. Finally, we might just happen to take the view that

$$\text{if} \quad \underline{(+6)(-2)} = \underline{-12}$$

$$\text{then} \quad \underline{(-6)(-2)}$$

   should give the *opposite* result, which would be $\underline{+12}$.
   If we happened to take this view, then we would happen to be *right*, but the correctness of our conclusion would owe considerably more to luck than to logic.

The stark, simple truth is that $(-6)(-2)$ *is* $+12$, but there is no simple, intuitive way to demonstrate this fact at this time. In the following chapter, we shall use an axiom called the distributive axiom to demonstrate the fact that the product of two negative numbers must be a positive number, but the proof of this must be deferred to a discussion of the laws of algebra. For the present, therefore, we shall accept as a matter of faith that

$$(-6)(-2) = +12$$

Using the commutative law of multiplication we can now line up all possible combinations of products involving $+6$ or -6 and $+2$ or -2.

$$\boxed{\begin{aligned} (+6)(+2) &= +12 \\ (+2)(+6) &= +12 \\ (-6)(-2) &= +12 \\ (-2)(-6) &= +12 \end{aligned}}$$

$$(+6)(-2) = -12$$
$$(-2)(+6) = -12$$
$$(+2)(-6) = -12$$
$$(-6)(+2) = -12$$

Looking over these results carefully we can see that the multiplication of signed numbers resolves itself into two fairly simple situations.

The product of two numbers having *like signs* is *always positive*.

The product of two numbers having *opposite signs* is *always negative*.

Now let us look at a situation in which we have three numbers multiplied together instead of two. Suppose we have

$$(2)(3)(4) = ?$$

We could first multiply the (2) and (3) together, and then multiply the result by (4).

$$(2)(3)(4) = \underline{(2)(3)}(4) = (6)(4) = 24$$

But we could *also* multiply the (3) and (4) together first and then multiply the result by (2).

$$(2)(3)(4) = (2)\underline{(3)(4)} = (2)(12) = 24$$

Thus we have the following:

$$\underline{(2)(3)}(4) = (6)(4) = 24$$
$$(2)\underline{(3)(4)} = (2)(12) = 24$$

Therefore

$$\underline{(2)(3)}(4) = (2)\underline{(3)(4)}$$

or

$$(2\cdot3)(4) = (2)(3\cdot4)$$

This is an example of the *associative law* of multiplication. With *a*, *b*, and *c* representing real numbers, this law can be written symbolically in the following way:

The Associative Law of Multiplication

$$(a\cdot b)(c) = (a)(b\cdot c)$$

When numbers are multiplied together, they are called *factors*. According to the above law, we may pair factors together in different ways. Note that the associative law has nothing to do with the *order* of the factors but only with the *pairing* of them. But in the product $(a)(b)(c)$, $a \cdot b = b \cdot a$ by the commutative law. Therefore $(a)(b)(c) = (b)(a)(c) = (b \cdot a)(c) = (b)(a \cdot c)$. The meaning of all this is that in multiplying numbers together, *we use each factor once only and we may use the factors in any order and in any pairing we please.*

Examples:

$$(2)(3)(4) = 24 \qquad (3)(4)(2) = 24$$
$$(4)(3)(2) = 24 \qquad (3 \cdot 4)(2) = 24$$
$$(3)(2)(4) = 24 \qquad (3)(4 \cdot 2) = 24$$

Following are some examples of multiplication with signed numbers:

(1) $(5)(5) = 25$ (2) $(-5)(-5) = 25$ (3) $(-5)(5) = -25$
(4) $(-2)(-3) = 6$ (5) $(-2)(3) = -6$ (6) $(-2)(-2)(-2) = -8$
(7) $(-3)(5)(2) = -30$ (8) $(-1)(1)(-1) = 1$ (9) $(-2)(3)(-1)(4) = 24$

Exercise 6:

Multiply the following:

(1) $(-8)(-3)$ (2) $(-4)(-5)$
(3) $(+6)(+8)$ (4) $(+10)(+9)$
(5) $(+8)(-3)$ (6) $(-8)(+3)$
(7) $(-4)(+5)$ (8) $(-5)(+4)$
(9) $(4)(9)$ (10) $(-4)(9)$
(11) $(4)(-9)$ (12) $(-4)(-9)$
(13) $(2)(3)(6)$ (14) $(-2)(3)(6)$
(15) $(-2)(-3)(6)$ (16) $(-2)(-3)(-6)$
(17) $(3)(-2)(-6)$ (18) $(6)(-2)(-3)$
(19) $(-3)(6)(-2)$ (20) $(-6)(2)(-3)$
(21) $(-3)(-3)$ (22) $(-3)(-3)(-3)$
(23) $(2)(2)$ (24) $(-2)(-2)$
(25) $(-2)(-2)(-2)$ (26) $(7)(-1)(-4)(-1)$
(27) $(-1)(-1)$ (28) $(-1)(-1)(-1)$
(29) $(-1)(-1)(-1)(-1)$ (30) $(-3)(4)(2)(-1)$

1.8
Division

In the statement $12/4 = 3$, the number 12 is the *dividend;* the number 4 is the *divisor;* and the result, 3, is the *quotient.* In every division problem from here to eternity a quotient will be correct *only if* the quotient multiplied by the divisor gives the dividend.

Thus

$$\frac{+12}{+4} = +3 \text{ } because \text{ } (+3)(+4) = +12$$

$$\frac{+35}{+7} = +5 \text{ } because \text{ } (+5)(+7) = +35$$

Suppose we had

$$\frac{-12}{-4} = ?$$

In the above problem, the *quotient* (?) multiplied by the *divisor* (-4) must equal the *dividend* (-12). In other words ($?$)(-4) $= -12$. The only possible number for (?) is $+3$ because ($+3$)(-4) $= -12$. Therefore

$$\frac{-12}{-4} = +3 \text{ } because \text{ } (+3)(-4) = -12$$

$$\frac{-35}{-7} = +5 \text{ } because \text{ } (+5)(-7) = -35$$

Repeating the results in the above examples, we have the following:

$$\frac{+12}{+4} = +3 \longrightarrow (+3)(+4) = +12$$

$$\frac{+35}{+7} = +5 \longrightarrow (+5)(+7) = +35$$

$$\frac{-12}{-4} = +3 \longrightarrow (+3)(-4) = -12$$

$$\frac{-35}{-7} = +5 \longrightarrow (+5)(-7) = -35$$

In every example, we divided two numbers which had *like* signs and in every case the quotient was positive. From this we draw the following conclusion:

The quotient of two numbers having *like signs* is always a *positive* number.

Now let us consider the quotient of two numbers having unlike signs. Suppose that we had the following problems:

1. $\dfrac{+12}{-4} = ?$ (where (?)(-4) must equal $+12$)

2. $\dfrac{-12}{+4} = ?$ (where (?)($+4$) must equal -12)

In Example 1 the question is what number multiplied by -4 will equal $+12$? The only possible answer is -3. Therefore

$$\frac{+12}{-4} = -3 \; because \; (-3)(-4) = +12$$

In Example 2 the question is what number multiplied by $+4$ will equal -12? The only possible answer is -3. Therefore

$$\frac{-12}{+4} = -3 \; because \; (-3)(+4) = -12$$

In both of the examples, we were dividing two numbers with *unlike* signs, and in both cases the correct result was *negative*.

From these examples we draw the following conclusion:

> **The quotient of two numbers having *unlike signs* is always a *negative* number.**

Examples:

$$\frac{+18}{+6} = +3 \qquad \frac{-18}{-6} = +3 \qquad \frac{-18}{+6} = -3 \qquad \frac{+18}{-6} = -3$$

$$\frac{24}{4} = 6 \qquad \frac{-24}{-4} = 6 \qquad \frac{-24}{4} = -6 \qquad \frac{24}{-4} = -6$$

In these examples, one can check each result simply by showing that the quotient multiplied by the divisor gives the dividend.

Exercise 7:

Perform the division and check each answer.

(1) $\dfrac{+18}{+9}$ (2) $\dfrac{-18}{-9}$ (3) $\dfrac{-18}{+9}$ (4) $\dfrac{+18}{-9}$ (5) $\dfrac{27}{-3}$

(6) $\dfrac{-27}{3}$ (7) $\dfrac{27}{3}$ (8) $\dfrac{-27}{-3}$ (9) $\dfrac{-56}{-8}$ (10) $\dfrac{-32}{4}$

(11) $\dfrac{14}{-7}$ (12) $\dfrac{-121}{-11}$ (13) $\dfrac{36}{-4}$ (14) $\dfrac{-49}{-7}$ (15) $\dfrac{63}{9}$

(16) $\dfrac{-25}{5}$ (17) $\dfrac{54}{9}$ (18) $\dfrac{-52}{13}$ (19) $\dfrac{75}{-25}$ (20) $\dfrac{-100}{-10}$

(21) $\dfrac{-144}{4}$ (22) $\dfrac{-125}{-5}$ (23) $\dfrac{45}{-9}$ (24) $\dfrac{-42}{7}$ (25) $\dfrac{30}{6}$

(26) $\dfrac{-16}{8}$ (27) $\dfrac{-3}{3}$ (28) $\dfrac{-7}{7}$ (29) $\dfrac{5}{-5}$ (30) $\dfrac{-8}{-8}$

(31) $\dfrac{-81}{3}$ (32) $\dfrac{-32}{-16}$

1.9
Operations with Zero

Zero has some characteristics that set it apart from all of the other integers. It is the only one of the integers, for instance, that is neither positive nor negative, which is another way of saying that it has the same meaning with either sign. On the number line, $+0$ and -0 identify exactly the same point. To add zero to any other number or to subtract it is essentially the same operation. For example,

$$4 + 0 = 4$$

and

$$4 - 0 = 4$$

Stated symbolically, if a is any real number, then

$$a + 0 = a$$

> **There is an element $0 \in R$ such that for every $a \in R, a + 0 = a$.**

Because of this unique property, which no other real number has, zero is aptly called "the additive identity," meaning that when you add zero to any number, the number remains identically the same.

Zero is enormously helpful in that it enables us to symbolize numbers and operations which would be impossible to represent otherwise.

Example 1:

(1)   The number 2 means "2 units."
(2)   The number 20 means "*no* units and 2 tens."
(3)   The number 200 means "*no* units, *no* tens, and 2 hundreds."

Therefore zero provides us with a concrete way of saying "none of these." Without it in the positional notation of our number system, we would be hard put to distinguish between 2 and 20 and 200.

Example 2:

If we wish to take 6 units away from 6 units, we would obviously have no units left. But it would be cumbersome and slightly maddening if we had no symbol with which to express this simple fact. Imagine writing

$$6 - 6 = \text{no units left!}$$

Instead, we write

$$6 - 6 = 0$$

which says it neatly, symbolically, and correctly.

Example 3:

Zero provides us with an intelligible way of writing decimals.
  (1)   .1 means "one tenth."
  (2)   .01 means "*no* tenths and one hundredth."
  (3)   .003 means "*no* tenths, *no* hundredths, and three thousandths."
  (4)   .101 means "one tenth, *no* hundreths, and one thousandth."

Zero, then, is really a sort of miracle; it is the actual earthly manifestation of a quantity that isn't there! In adding it to or subtracting it from another number, as we have already noted, no change takes place.

In multiplication and division, however, this marvelously concrete ghost can cause havoc unless one is extremely careful in noting its behavior.

Multiplication by Zero   Let us consider multiplication by zero and look at the product

$$(6)(0) = ?$$

Since multiplication means repeated addition, (6)(0) means to add six zeros. If we do this

$$
\begin{array}{c}
0 \\
0 \\
0 \\
0 \\
0 \\
0 \\
\hline
0
\end{array}
$$

the sum is clearly zero. Therefore (6)(0) = 0, and by the commutative law of multiplication, (0)(6) = 0. In like manner

$$(3)(0) \text{ means } 0 + 0 + 0 = 0$$

Therefore

$$(3)(0) = 0 \text{ and } (0)(3) = 0$$

Suppose we added a thousand zeros? The sum of them all would still be zero, and consequently $(1000)(0) = 0$, and $(0)(1000) = 0$.

What is $\left(\frac{1}{2}\right)(0)$? This simply means $\frac{1}{2}$ of 0, which is the same as 0 divided by 2.

$$\left(\frac{1}{2}\right)(0) = \frac{0}{2}$$

And

$$\frac{0}{2} = 0 \text{ because } (0)(2) = 0$$

so

$$\left(\frac{1}{2}\right)(0) = 0 \text{ and } (0)\left(\frac{1}{2}\right) = 0$$

From these examples we can surmise a fundamental fact about zero and its multiplication by any number: *the product is always zero!*

Examples:

$(6)(0) = 0$	$(0)(b) = 0$	$(0)(a) = 0$
$(0)(6) = 0$	$(10{,}866)(0) = 0$	$(b)(0) = 0$
$(565)(0) = 0$	$(a)(0) = 0$	$\left(\frac{1}{3}\right)(0) = 0$
$(-4)(0) = 0$	$(-2)(8)(0) = 0$	$(a)(b)(0) = 0$

Division by Zero   When we tackle division by zero, we run headlong into the nonexistent. This can be illustrated quite simply by remembering that division was born as multiplication in reverse.

For example, we say

$$\frac{12}{4} = 3 \text{ only because } (3)(4) = 12$$

or

$$\frac{35}{7} = 5 \text{ only because } (5)(7) = 35$$

Now, consider this:

$$\frac{12}{0} = ?$$

When we say $12/0 = ?$ we must guarantee that $(?)(0)$ will equal 12. In other words, what number $(?)$ *multiplied* by 0 will give 12? **There is no such number.** No matter what number we might try, *when we multiply by 0*, we cannot get 12 because the product is always zero.

Examples:

$$\frac{6}{0} = ? \qquad ? \text{ is meaningless because } (?)(0) \text{ } cannot \text{ equal } 6$$

$$\frac{100}{0} = ? \qquad ? \text{ is meaningless because } (?)(0) \text{ } cannot \text{ equal } 100$$

$$\frac{-35}{0} = ? \qquad ? \text{ is meaningless because } (?)(0) \text{ } cannot \text{ equal } -35.$$

All of the above symbols $(6/0, 100/0, -35/0)$ have no more meaning mathematically than $?/!$ or $*/\$$, because none of them represent any number. In attempting to describe this situation, the only sensible thing we can do is to point out that the process of dividing by zero produces a meaningless result. Consequently, *division by zero is not defined*. As the above examples show, it is meaningless, and in all operations with numbers we rule out division by zero forever. Our heretofore convenient ghost has retreated completely into its ghostdom and has taken the division-by-zero process into nonexistence with it.

When we encounter this situation, it is customary to retreat rapidly using one of the following notations: if a is any *nonzero* number, then

$$\frac{a}{0} \text{ is meaningless}$$

or

$$\frac{a}{0} \text{ is undefined}$$

The Indeterminate Form, $\frac{0}{0}$ In every preceding example of division by zero, including the summary in symbolic language, we have examined only quotients in which the *dividend* did *not also equal zero*. Hence we emphasized in the preceding summary that $\frac{a}{0}$ is *undefined* provided that a is any *nonzero* number. If a *is* equal to zero, we would have $\frac{0}{0}$. Now, what is this? Actually it could be any number you like! For example

$$\frac{0}{0} = 782,564 \ \textit{because} \ \underset{\text{quotient}}{(782,564)} \quad \underset{\text{divisor}}{(0)} = \underset{\text{dividend}}{0}$$

$$\frac{0}{0} = 13\frac{1}{2} \ \textit{because} \ \left(13\frac{1}{2}\right)(0) = 0$$

$$\frac{0}{0} = .68593 \ \textit{because} \ (.68593)(0) = 0$$

This could obviously go on forever. We say that 0/0 is *indeterminate;* since it could be any number whatsoever, it cannot, in itself, identify a specific number.

The foregoing discussion has been exclusively concerned with division by zero, i.e., with zero as a *divisor*. Note carefully that no other real number is ruled out as a divisor. For example, the number $\frac{0}{7}$ has 7 as a divisor, and the quotient can be computed and proved as any other quotient which represents a real number:

$$\frac{0}{7} = 0 \ \text{ because } \ (0)(7) = 0$$

In other words, if 0 is the dividend, and any number *except zero* is the divisor, we have a quotient equal to the real number zero. In like manner:

$$\frac{0}{4} = 0 \qquad\qquad \frac{0}{106} = 0 \qquad\qquad \frac{0}{-25} = 0$$

To summarize, if a represents any *nonzero* real number, then the situations we encounter with any division of real numbers involving zero can be shown symbolically as follows:

$$\frac{0}{a} = 0 \ \text{ because } \ (0)(a) = 0$$

$\frac{a}{0}$ is *undefined* because the quotient represents *no* number at all

$\frac{0}{0}$ is *indeterminate* because the quotient represents *any* number

The Additive Inverse We have previously concluded that when a number is multiplied by zero, the product is always equal to zero. Here we should observe further that it can be proved that the *sum* of two numbers will equal zero only if the two numbers are the *same* in absolute value but are *opposite* in sign. In this situation each number is said to be the *additive inverse* of the other. If we add, for example, the numbers 6 and −6, we could write

$$6 + -6$$

but this expression is somewhat awkward. The meaning of this expression is clearer if we separate the plus and minus signs by parentheses and write instead

$$6 + (-6)$$

This expression, $6 + (-6)$, means the same thing as $6 + 1(-6)$ and is simplified by first multiplying the $+1$ and the -6:

$$6 + (-6) = 6 + 1(-6) = 6 - 6 = 0$$

Thus the *additive inverse* of 6 is -6 because their *sum* is zero.

On the number line, for any distance to the *right* of zero designated by a number such as a, the *same* distance to the *left* of zero would be designated by the number $-a$. In general, we assume for every positive real number, a, there exists another real number, $-a$, such that $a + (-a) = 0$. In this situation, $-a$ is called the *additive inverse* of a.

Examples:

Number	Additive Inverse	Sum
-5	5	$-5 + 5 = 0$
3	-3	$3 + (-3) = 0$
$-\dfrac{2}{7}$	$\dfrac{2}{7}$	$-\dfrac{2}{7} + \dfrac{2}{7} = 0$
$\dfrac{3}{4}$	$-\dfrac{3}{4}$	$\dfrac{3}{4} + \left(-\dfrac{3}{4}\right) = 0$
$-.6$	$.6$	$-.6 + .6 = 0$

Exercise 8:

Perform the indicated operation when possible. If an operation is undefined, or indeterminate, state that fact.

(1) $3 + 0$

(2) $-4 - 0$ $\quad -4$

(3) $-5 + 0$

(4) $0 - 6$ $\quad -6$

(5) $10 + (-10)$

(6) $0 + \dfrac{7}{8}$ $\quad 7/8$

(7) $\dfrac{4}{9} + \left(-\dfrac{4}{9}\right)$

(8) $(7)(0)$ $\quad 0$

(9) $7 \div 0$

(10) $0 \div 7$ $\quad 0$

(11) $(-17)(0)$

(12) $-17 \div 0$ $\quad$ and

(13) $(3)(-6)(0)$

(14) $\dfrac{5}{6} + \left(-\dfrac{5}{6}\right)$ $\quad 0$

(15) $8 \div (-2)$

(16) $(9)(-2)$ $\quad -18$

(17) $-56 \div 7$

(18) $24 \div 0$ $\quad$ und

(19) $0 \div 24$ (20) $-18 \div 0$ und   (21) $0 \div (-18)$

(22) $24 \div (-6)$ -4   (23) $-18 \div 3$ (24) $0 \div 0$ und

(25) $65 \div 0$ (26) $-8 + 8$ 0   (27) $(-8)(8)$

(28) $-8 \div 8$ -1   (29) $8 + (-8)$ (30) $-15 \div (-5)$ 3

(31) $0 \div (-6)$ (32) $\dfrac{7}{8} + \left(-\dfrac{7}{8}\right)$ 0   (33) $-6 \div 0$

(34) $(-15)(-5)$ 75   (35) $-15 + (-5)$ (36) $-7 \div (-7)$ 1

Chapter Test 1

I. Represent each of the following sets by listing several representative elements in an appropriate manner:
   1. N = the set of natural numbers
   2. J = the set of integers

II. Classify each number listed below in the following manner: *n.n.* if the number belongs to N; *int.* if the number belongs to J; and *neither* if the number is not an element of N or J.

   1. 32   2. $\dfrac{3}{4}$ 3. -125

   4. 5   5. $-8\dfrac{3}{5}$ 6. 0

   7. .5   8. -11 9. 6

III. Compute the value of the following.
   1. $|-8|$ 2. $|12|$
   3. $|-8| + |12|$ 4. $|8| + |-12|$
   5. $|-8| \times |12|$ 6. $|-8| \times |-12|$
   7. $|12 - 8|$ 8. $|12 - 8| + |-12|$

IV. Add the following:
   1. $\begin{array}{r} -9 \\ +18 \end{array}$ 2. $\begin{array}{r} -12 \\ -6 \end{array}$ 3. $\begin{array}{r} -283 \\ -67 \end{array}$ 4. $\begin{array}{r} 3460 \\ -2393 \end{array}$ 5. $\begin{array}{r} -41 \\ +24 \\ -3 \end{array}$

V. Subtract the lower number from the upper number:
   1. $\begin{array}{r} 12 \\ +12 \end{array}$ 2. $\begin{array}{r} -12 \\ +12 \end{array}$ 3. $\begin{array}{r} -12 \\ -12 \end{array}$ 4. $\begin{array}{r} 12 \\ -12 \end{array}$ 5. $\begin{array}{r} -156 \\ +46 \end{array}$ 6. $\begin{array}{r} -1383 \\ -417 \end{array}$

VI. Perform the indicated operations. If an operation is undefined or indeterminate, state that fact.
   1. $(4)(-8)(2)$ 2. $(-4)(2)(-8)$
   3. $(-8)(0)(3)$ 4. $(0)(17)$

5. $-18 \div (-9)$ 6. $18 \div (-9)$
7. $0 \div (-3)$ 8. $14 \div 0$
9. $-42 \div (6)$ 10. $(-2)(-3)(-5)$
11. $8 + (-8)$ 12. $(-2)(35)(0)$
13. $-30 \div -6$ 14. $(-4)(3)(-6)$

VII. Complete the following statements:

1. $(2)(7) = (7)(2)$ is an example of the _____ property of multiplication.
2. $(3 \cdot 5)(4) = (3)(5 \cdot 4)$ is an example of the _____ property of multiplication.
3. $8 + 5 = 5 + 8$ is an example of the _____ property of addition.
4. $(4 + 5) + 6 = 4 + (5 + 6)$ is an example of the _____ property of addition.
5. If zero is added to a number, the sum is _____.
6. Zero is called the additive _____.
7. If zero is multiplied by a number, the product is _____.
8. If zero is divided by a nonzero number, the quotient is _____.
9. If a nonzero number is divided by zero the quotient is _____.
10. If zero is divided by zero, the quotient is _____.
11. The additive inverse of -5 is _____.
12. The sum of any number and its additive inverse is _____.

2

OPERATIONS WITH ALGEBRAIC EXPRESSIONS

2.1
How Algebra Began

The techniques of algebraic computation evolved from a simple and very useful idea. The idea is that, in referring to some unknown number, it is more convenient to use a short symbol or label to denote the number than it is to retain the cumbersome phrase "an unknown number." In other words, we give the number a name so that we can refer to it and distinguish it from other numbers conveniently.

Rather than invent a collection of new symbols to use as names for different numbers, we found it handy to borrow the letters of the alphabet and use them for names. This happy choice also provided a conveniently wide range of available labels for numbers because the letters may be Latin or Greek, script or italic, or upper case or lower case, etc.

If we wish, for instance, to refer to the length and width of a rectangle, we proceed in the following manner. Every rectangle *has* length and width, and they are certainly not necessarily the same (Figure 2.1). Since we are talking about two possibly *different* numbers, we would assign them two different names. We let the letter *l* represent the

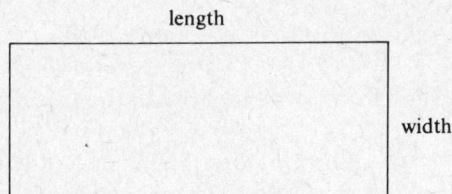

length

width

Figure 2.1

length of the rectangle; this means that *l* becomes a stand-in for the actual length. We then choose another letter, such as *w*, to represent the width of the rectangle. The rectangle is then labeled as in Figure 2.2.

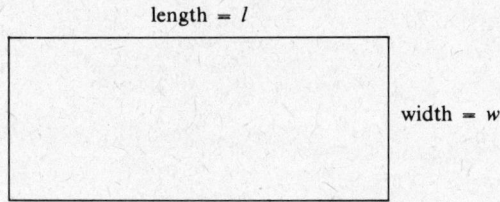

length = l

width = w

Figure 2.2

Now suppose that we wish to represent the area of any such rectangle: We write

> The area of a rectangle = the length of the rectangle times the width of the rectangle

Or we could write, using our convenient stand-ins

> The area of a rectangle = $(l)(w)$

Better yet, if we agree to let the letter A stand in for the area of the rectangle, then we could write

> $$A = (l)(w)$$

Notice that in the last statement we have represented the area of a rectangle very clearly and very compactly simply by letting letters stand in for numbers. The expression

$$A = (l)(w)$$

is obviously simpler to write *and* easier to read than

> The area of a rectangle = the length of the
> rectangle times the width of the rectangle

After this useful device of using letters for numbers was adopted, it immediately became necessary to figure out a way to do arithmetic with letters rather than with actual numbers. And that is exactly what basic algebra is all about: how to perform arithmetic computations with *literal* numbers.

To show why ordinary arithmetic with literal numbers soon becomes necessary, let us consider the perimeter of the rectangle whose length is l and whose width is w (Figure 2.3). The perimeter means the total distance around the rectangle; that distance is some number, which we shall call P. Since the figure is a rectangle, the opposite sides must be of equal length.

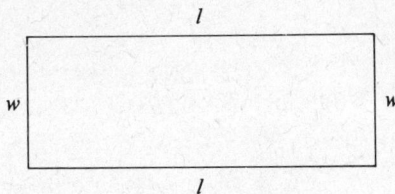

Figure 2.3

To compute the perimeter, P, we merely add the lengths of all four sides. Thus

$$P = l + w + l + w$$

Since addition is commutative, the sum of the middle numbers, $w + l$, is equal to $l + w$; so we could also say

$$P = l + l + w + w$$

Now the question arises, what is $l + l$? And what is $w + w$? How does one *add l* and l or w and w?

To answer these questions, we go back to arithmetic and ask the same kind of question: How do we add *any* two like numbers?

For example:

$$5 + 5 = 10$$

but we could also say

$$5 + 5 = 2(5)$$

In like manner:

$$4 + 4 = 2(4)$$
$$7 + 7 = 2(7)$$

These suggest that

$$l + l = 2(l)$$
$$w + w = 2(w)$$
$$x + x = 2(x)$$

When we mean the quantity $2(4)$, we cannot write 24, because the figure 24 *already* has another meaning; that is,

$$24 = 2(10) + 4$$

But when we mean the quantity $2(l)$, we could simplify the notation by writing $2l$, because the number $2l$ does *not* already have another meaning. Consequently, to make the notation simpler, we agree that 2(*l*) will be written 2*l*. In this way, we represent the multiplication of *literal* numbers with *no sign at all*.

Examples:

(1) $6 + 6 = 2(6)$
     $b + b = 2(b) = 2b$
(2) $4 + 4 + 4 = 3(4)$
     $x + x + x = 3(x) = 3x$
(3) $7 + 7 + 7 + 7 = 4(7)$
     $y + y + y + y = 4(y) = 4y$

Conversely:

(1) $2b = b + b$
(2) $3x = x + x + x$
(3) $4y = y + y + y + y$

2.2
The Symbols for the Fundamental Operations

Before we continue into arithmetic with literal numbers, let us look at the symbols for the four fundamental operations as they are used both in arithmetic and in algebra.

1. *Addition*

    3 plus 4 means $3 + 4$
    a plus *b* means $a + b$

2. *Subtraction*

    8 minus 2 means $8 - 2$
    a minus *b* means $a - b$

3. *Multiplication*

    4 times 5 means $4 \cdot 5$
        or $(4)(5)$
        or $4(5)$
    4 times *x* means $4 \cdot x$
        or $(4)(x)$
        or $4(x)$
        or $4x$
    a times *b* means $a \cdot b$
        or $(a)(b)$
        or $a(b)$
        or ab

4. *Division*

$$8 \text{ divided by } 2 \text{ means } 8 \div 2$$
$$\text{or } \frac{8}{2}$$
$$\text{or } 2\,\overline{\smash{)}8}$$
$$a \text{ divided by } b \text{ means } a \div b$$
$$\text{or } \frac{a}{b}$$
$$\text{or } b\,\overline{\smash{)}a}$$

Note that the parallel in symbolism between arithmetic and algebraic processes is exact, except in the matter of multiplication.

2.3
The Language of Algebra

In algebra, since we deal both with numerals and with letters, it becomes exceedingly important to be able to refer to any part of a quantity clearly and exactly. As we have observed previously:

$$7 + 7 = 2(7)$$

suggests that

$$x + x = 2(x) = 2x$$

Moreover,

$$-7 - 7 = 2(-7)$$

suggests that

$$-x - x = 2(-x) = -2x$$

The algebraic quantities shown above ($2x$ and $-2x$) are called *terms*.

A *term* is a quantity completely set off from other quantities (to the right or left) by either a *plus* or a *minus sign*.

In the term $2x$, the 2 and the x are called factors.

Factors are quantities which are *multiplied* together.

A single *term* may contain many *factors;* and a *factor* may contain more than one *term*.

Examples:

$$3x \quad \text{one term}$$
$$\text{two factors } (3), (x)$$
$$-2ab \quad \text{one term}$$
$$\text{three factors } (-2), (a), (b)$$
$$5xyz \quad \text{one term}$$
$$\text{four factors } (5), (x), (y), (z)$$
$$3a(b - c) \quad \text{one term}$$
$$\text{three factors } (3), (a), (b - c)$$

The different factors which make up a term are called *coefficients*. In the term $2x$, the number 2 is the *coefficient* of x, and x is the *coefficient* of 2. Because multiplication is commutative, $2x$ and $x2$ mean exactly the same thing. We frequently need to refer to the *numerical* coefficient of a term: in the term, $-5xy$, the numerical coefficient is -5. If no numeral appears in front of a term, then the numerical coefficient is always understood to be the number *one*.

$$x \text{ means } 1x$$
$$ab \text{ means } 1ab$$
$$xyz \text{ means } 1xyz$$
$$-y \text{ means } -1y$$
$$-ab \text{ means } -1ab$$

A literal number or numeral, or a quantity containing a combination of literal numbers and/or numerals, joined together by a finite number of the fundamental operations is called an *algebraic expression*.

Frequently the factors which make up a term are the same numbers. When this happens, rather than write the same factors over and over, we have devised a more convenient way of saying the same thing. The device is the use of a number, called an *exponent*, which is written slightly above and to the right of the repeating factor, usually in smaller type. The exponent tells how many times this same factor is repeated. Thus:

1. xx is written x^2 and means "x used as a *factor* two times."
2. yyy is written y^3 and means "y used as a *factor* three times."

Conversely:

1. x^4 means "x used as a *factor* four times." Then x^4 means $xxxx$.
2. x^3 means "x used as a *factor* three times." Then $x^3 = xxx$.
3. $x^5 = xxxxx$
4. $z^3 = zzz$
5. $y^6 = yyyyyy$

6. $4^2 = (4)(4)$
7. $8^3 = (8)(8)(8)$

In like manner:

1. $3x^2y = 3xxy$
2. $5a^2b^3 = 5aabbb$
3. $-7mn^4 = -7mnnnn$

When a number has no exponent, the exponent is understood to be *one*: x and x^1 are the same number.

Please note carefully the difference between numerical *coefficients* and *exponents*.

$$3x = x + x + x$$
$$x^3 = xxx$$

Exercise 9:

Use exponents to simplify the following terms:

(1) $5xx = 5x^2$ (2) $-3xyy$ (3) $6xxyyy$
(4) $-3aaabb$ (5) $7aabbbcccc$ (6) $-2xxxxyyyyy$
(7) $4aaaabbb$ (8) $-5xyyzzzzz$

Algebraic expressions may consist of one term only, or they may contain more than one term. *On this single fact rest all of the apparent differences in operations with the numbers of arithmetic and the numbers of algebra.* In algebra, quantities may be classified by the *number of terms* which they contain.

1. Any expression made up of exactly *one term* is a *monomial*.
2. Any expression which contains *more than one term* is called a *multinomial*.

Note that the number of factors in a term has nothing to do with whether an expression is a monomial or a multinomial.

Examples:

Monomials	Multinomials
7	$3 + 7$
$-\dfrac{8}{x}$	$8 - 2 + 3$

$$2x \qquad\qquad x + \frac{2}{y}$$

$$-3ab \qquad\qquad 2x^2 - 3x + 4$$

$$2a^2c^3 \qquad\qquad 4x^3 - 7xy + 2y^2 - 8y$$

$$-6x^2y^3z^4 \qquad\qquad \frac{3}{x} + \frac{4}{y} - 2x^2$$

Exercise 10:

Give the number of terms in each of the following expressions, and classify each as a monomial or a multinomial:

(1) $3x^2 - 4y + 2$ 　　　　　　　(2) $-10x^2y^3z$

(3) $4abc^2 - 7x^2y$ 　　　　　　 (4) $-4m^2 + mn - 2n^2$

(5) $a + b + c + d$ 　　　　　　　(6) $abcd$

(7) $3x^2 + y^2 - x + 3$ 　　　　　 (8) $256z^2b^3c$

(9) $a^2 - 2ab - b^2$ 　　　　　　 (10) $\dfrac{a}{b}$

(11) $\dfrac{a}{b} - c$ 　　　　　　　　(12) $\dfrac{x}{y} - \dfrac{4}{x} + 3$

Memorize the following statements:

1. A term is a quantity completely set off from other quantities by either a plus or a minus sign.
2. Factors are quantities which are multiplied together.
3. Every algebraic expression is either a monomial or a multinomial.

2.4
Addition

In algebra as well as in arithmetic, addition means the summing up of quantities and then the combining of terms which are exactly the same kinds of things. Unless the quantities to be added *are* the same kinds of things, we cannot actually combine them. The following examples are intended to give you a brief survey of addition from Grade One to the present. Look them over *carefully*, and be sure that you understand exactly what addition really means.

(1)   2 cows + 3 cows = 5 cows   *but* 2 cows + 3 pigs = 2 cows + 3 pigs
      (Note that the answer is not 5;
      it is 5 cows.)

(2) $\dfrac{2}{7} + \dfrac{3}{7} = \dfrac{5}{7}$ *but* $\dfrac{2}{7} + \dfrac{3}{8} = \dfrac{2}{7} + \dfrac{3}{8}$

(3) $2x + 3x = 5x$ *but* $2x + 3y = 2x + 3y$

(4) $2y^2 + 3y^2 = 5y^2$ *but* $2y^2 + 3y = 2y^2 + 3y$

(5) $2ab + 3ab = 5ab$ *but* $2ab + 3ab^2 = 2ab + 3ab^2$

Take a *long* look at the foregoing examples! Note that in the examples on the left side we were adding terms that were *all* cows, or *all* sevenths, or *all* x's, or *all* y^2's, or all ab's, and therefore we could *combine* the terms. This is not true in the examples on the right, and we could *not* combine the terms.

Since, in addition, we can only combine quantities which are exactly the same kinds of things, we can combine only terms which are exactly the same kinds of terms. When terms are exactly alike, except that their numerical coefficients may be different numbers, we get the sum of them by simply adding their numerical coefficients. You cannot combine terms that are *not* exactly alike any more than you can combine cows and pigs.

Examples:

(1)   Add $3x^2 + 2x + 5$ and $6x^2 - 4x - 8$
      $3x^2 + 2x + 5 + 6x^2 - 4x - 8$

Use the commutative law of addition and rearrange these terms in the following manner:

$3x^2 + 6x^2 + 2x - 4x + 5 - 8$

Now combine the like terms:

$9x^2 - 2x - 3$

In this problem the result, $9x^2 - 2x - 3$, cannot be made any simpler because all like terms have been combined. Because addition is commutative, the terms could be written in a different *order:* e.g., $-2x + 9x^2 - 3$ or $-3 - 2x + 9x^2$ or $9x^2 - 3 - 2x$ are all the same number as $9x^2 - 2x - 3$.

(2)   Add $3b^2$ and $7b^2$ and $10b^2$
      $3b^2 + 7b^2 + 10b^2 = 20b^2$

(3)   Add $4a^2 - 7a^2b^2 + 3b^2$ and $2a^2b^2 + 4b^2 - 3$
$$4a^2 - 7a^2b^2 + 3b^2 + 2a^2b^2 + 4b^2 - 3$$
$$4a^2 - 7a^2b^2 + 2a^2b^2 + 3b^2 + 4b^2 - 3$$
$$\underline{4a^2 - 5a^2b^2 + 7b^2 - 3}$$

Exercise 11:

Add the following:

(1)   3 dogs $-$ 4 cats, 7 dogs $+$ 8 cats, 9 dogs $-$ 10 cats

(2) $-3x,\ 5x,\ 2x,\ 6x$

(3) $5a^2,\ 4a,\ 2x^2,\ 5$

(4) $-6x,\ 3xy,\ 2x,\ 8xy$

(5) $-4x^2,\ 3y,\ 4x^2,\ 5y$

(6) $3a,\ 2ab,\ 7a$

(7) $-5x^2y^2,\ 3x^2,\ 4x^2y^2,\ 2y^2$

(8) $2a - 3b,\ 4a - 6b$

(9) $5x + y^2,\ 6x - 3y$

(10) $3x + xy,\ 7x - xy$

(11) $3ab - 8,\ 6a - 3$

(12) $5x + 2y,\ 4x + 3y,\ 2x - 5y$

(13) $3a^2 + 2b,\ 7a - 4b,\ a^2 - 6b$

(14) $2a^2 + 3ab - 5b^2,\ 4a^2 - 2ab + 6b^2$

(15) $4x^2 - 4xy + 2y^2,\ 3x^2 - 8xy - 7y^2$

(16) $2a - b,\ a^2 - 3b,\ 5a - b^2$

(17) $4x - 6,\ 3x^2 - 7,\ 5x^2 - 7x - 2$

(18) $5m^2 - 3n^2 + 4,\ 7m^2 - 8n^2 - 10$

(19) $3a - 4,\ 7a^2 + 5a,\ 6a - 8$

(20) $3x^2yz - 4xy,\ 2x^2yz + 3y^2,\ 4x^2yz - 2xy$

(21) $7a^2 - 4a^2b^2 + 5ab,\ 2a^2 - 3a^2b,\ 6a^2b^2 + 8ab$

(22) $x^2y - 4x^2 - xy^2,\ 5x^2 - 5xy^2,\ 7x^2y + 2x^2$

2.5
Subtraction

In considering the subtraction of literal numbers, we should take another look at one fundamental difference in the literal numbers of algebra and the numerals of arithmetic. The difference is that all of the numbers of elementary arithmetic can be written as monomials. For example, if we have $4 + 2$ (*two terms*), we can always write instead, 6 (*one term*). But *if* we have $a + b$ (*two terms*), we cannot reduce this to one term because we cannot combine a's and b's.

Arithmetic	Algebra

$$3 + 5 - 1 = \underline{7}$$
$$2a + b - c = \underline{2a + b - c}$$

$$13 - 5 = \underline{8}$$
$$6x - y = \underline{6x - y}$$

$$5 + 3 - 4 + 6 = \underline{10} \qquad 2x^2 - x + 3y^2 - y = \underline{2x^2 - x + 3y^2 - y}$$

In subtracting one number from another, we must always *reverse the sign of the subtrahend* and then follow the rules for addition. How does one reverse the sign of a multinomial? We can answer that by writing an arithmetic number as a multinomial and then reversing the sign.

$$6 = 6$$
$$4 + 2 = 6$$
$$8 + 3 - 5 = 6$$

Anytime, anywhere that we want to write 6, we can write it as $4 + 2$ or as $8 + 3 - 5$, because these quantities *always* mean 6 also. Obviously, to reverse the sign of *anything* that means 6, we must somehow come up with a quantity that means -6.

Number	Number with sign reversed
6	-6
$4 + 2$	$-4 - 2 \, (= -6)$
$8 + 3 - 5$	$-8 - 3 + 5 \, (= -6)$

Consequently, to reverse the sign of any multinomial, we must *reverse the sign of every term in the multinomial*.

Example 1:

(1) *Subtract $a^2 - 5a + 4$ from $6a^2 + 2a - 3$.*
(2) The *subtrahend* is $a^2 - 5a + 4$.
(3) The subtrahend with the sign *reversed* is $-a^2 + 5a - 4$.
(4) Having reversed the sign of the subtrahend, we *add* it to the minuend.

$$-a^2 + 5a - 4 + 6a^2 + 2a - 3 = 5a^2 + 7a - 7$$

Example 2:

Subtract $3x^2 - 5x + 2$ from $7x^2 + 6x + 8$
$$-3x^2 + 5x - 2 + 7x^2 + 6x + 8 = 4x^2 + 11x + 6$$

Exercise 12:

In the following problems *subtract* the *first* number from the *second:*

$(1)\quad 5a^2 - 2ab + b^2,\ 2a^2 - 5ab + 6b^2$

$(2)\quad -3x^2 + y^2,\ 4x^2 + 7y^2$

$(3)\quad 7a + ab + 2b,\ 3a - 6ab$

$(4)\quad x^2 - x^2y^2 + 2xy,\ 4x^2y + 5xy - 6$

$(5)\quad -3x^2 - 5x + 7,\ 4x^2 + 2xy - 6x$

$(6)\quad 3a^2 + 2a - 6,\ 8a^2 - 2ab - 2a$

$(7)\quad m^3 - n^3,\ m^3 - 2m^2n$

$(8)\quad 3x^2 - 8,\ 3x^2 + 8$

Exercise 13:

In the following problems, subtract the *second* number from the *first:*

$(1)\quad 3m - 6n,\ -4m^2 + 5n$

$(2)\quad 6x^2y^2 - 7xy + x^2 - y^2,\ -5x^2y^2 + 2xy - 4x^2 + 3y^2$

$(3)\quad m + n,\ -a - b$

$(4)\quad -2a + b,\ -2a + b$

$(5)\quad 6a^2 - 4a + 3,\ -7a + 5$

$(6)\quad 4x^2y^2 - 7x^2 + 8y^2,\ -3xy + 7x^2 + 8y^2$

$(7)\quad -4a^2 - 3a + 4ab,\ -2a^2 + 4a - 8b$

$(8)\quad -4x^2 + 3xy + 3y^2,\ -2x^2 - 3xy - 8y^2$

$(9)\quad 2a^2b - 3ab^2,\ 4ab^2 + 3a^2b^2$

$(10)\quad a^3 - a^2b + ab^2,\ b^3 - a^2b + ab^2$

2.6
Multiplication

Every number is either a monomial or a multinomial. Therefore anytime that you are multiplying two numbers, there are only *three situations* that can ever happen. They are the following:

1. *A monomial times a monomial*

   $(2)(3)$

   $(4x)(-5x)$

   $(-6y^2)(3xy)$

   $(-3a)(-4ab)$

2. *A monomial times a multinomial*

   $(2)(3 + 4)$

   $(a)(a + b)$

   $(2x)(3x^2 - 4x + 2)$

   $(-5x^2y)(4x - 7y + xy)$

3. *A multinomial times a multinomial*

$$(4 + 5)(3 + 2)$$
$$(a + b)(a - b)$$
$$(x - y)(x^2 - xy + y^2)$$
$$(a^2 - 5a + 3)(a^3 + 2a^2 - 4a + 2)$$

We shall examine these three situations separately. First we shall describe the general process of multiplication, and then we shall check it against the three different situations.

The Process of Multiplication

To multiply *any* two factors together, multiply *each term* in the first factor separately by *each term* in the second factor. Then combine similar terms, if any, produced by the separate products.

If you will only believe this, and remember it, and then make intelligent use of it, there are no two numbers in this world that you cannot multiply together correctly. Notice, however, that the above instructions will always be meaningless to anyone who does not know clearly what a *term* is, and who cannot distinguish between a *term* and a *factor*. In all of basic algebra, nothing is more vital than this: that you train your mind to react instantly to the tremendous difference between a *term* and a *factor*.

Case I: *A Monomial Times a Monomial*

In multiplying numbers we may arrange the factors in any order we please because of the commutative law. In literal numbers, we traditionally multiply the numerals together first and then the literal numbers in alphabetical order. This is not absolutely necessary, but it *is* convenient.

Examples:

(1) $(2)(4) = 2 \cdot 4 = 8$
(2) $(2x)(4) = 2 \cdot x \cdot 4 = 2 \cdot 4 \cdot x = 8x$
(3) $(2x)(4y) = 2 \cdot x \cdot 4 \cdot y = 2 \cdot 4 \cdot x \cdot y = 8xy$
(4) $(-3a)(2b) = -3 \cdot a \cdot 2 \cdot b = -3 \cdot 2 \cdot a \cdot b = -6ab$

In every case in the above examples we were multiplying a *monomial* times a *monomial*. In multiplication of this type we simply wind up with a series of factors which we arrange in a convenient order.

Examples:

$$(1) \quad (-3a)(4b) = -3 \cdot a \cdot 4 \cdot b$$
$$= -3 \cdot 4 \cdot a \cdot b$$
$$= -12ab$$

$$(2) \quad (-2x)(-5y) = -2 \cdot x \cdot -5 \cdot y$$
$$= -2 \cdot -5 \cdot x \cdot y$$
$$= 10xy$$

When we have *repeated* factors multiplied together, we can see by observation that the exponents of the repeated factors are actually *added*.

Examples:

$$(1) \quad (x^3)(x^2) = (xxx)(xx)$$
$$= x \cdot x \cdot x \cdot x \cdot x$$
$$= xxxxx$$
$$= x^5$$
Note: $(x^3)(x^2) = x^{3+2} = x^5$

$$(2) \quad (y^2)(y^4) = (yy)(yyyy)$$
$$= y \cdot y \cdot y \cdot y \cdot y \cdot y$$
$$= yyyyyy$$
$$= y^6$$
Note: $(y^2)(y^4) = y^{2+4} = y^6$

$$(3) \quad (a^5)(a^2) = (aaaaa)(aa)$$
$$= a \cdot a \cdot a \cdot a \cdot a \cdot a \cdot a$$
$$= aaaaaaa$$
$$= a^7$$
Note: $(a^5)(a^2) = a^{5+2} = a^7$

The foregoing examples are illustrations of everything that can happen when we multiply a monomial times a monomial. Following are some general examples of this.

Examples:

$$(1) \quad (-4x^2y)(-3x^2y^2) = -4 \cdot x^2 \cdot y \cdot -3 \cdot x^2 \cdot y^2$$
$$= -4 \cdot -3 \cdot x^2 \cdot x^2 \cdot y \cdot y^2$$
$$= \underline{12x^4y^3}$$

$$(2) \quad (-6x^5y^2)(2x^2y^3) = -6 \cdot x^5 \cdot y^2 \cdot 2 \cdot x^2 \cdot y^3$$
$$= -6 \cdot 2 \cdot x^5 \cdot x^2 \cdot y^2 \cdot y^3$$
$$= \underline{-12x^7y^5}$$

$$(3) \quad (3x)(-2xy)(4x^2y^2) = 3 \cdot x \cdot -2 \cdot x \cdot y \cdot 4 \cdot x^2 \cdot y^2$$
$$= 3 \cdot -2 \cdot 4 \cdot x \cdot x \cdot x^2 \cdot y \cdot y^2$$
$$= \underline{-24x^4y^3}$$

$$(4) \quad (-3x^2)^2 = (-3x^2)(-3x^2)$$
$$= -3 \cdot -3 \cdot x^2 \cdot x^2$$
$$= \underline{9x^4}$$

$$(5) \quad (-2a^2)^3 = (-2a^2)(-2a^2)(-2a^2)$$
$$= -2 \cdot -2 \cdot -2 \cdot a^2 \cdot a^2 \cdot a^2$$
$$= \underline{-8a^6}$$

$$(6) \quad (7x^3y^4)^2 = (7x^3y^4)(7x^3y^4)$$
$$= 7 \cdot 7 \cdot x^3 \cdot x^3 \cdot y^4 \cdot y^4$$
$$= \underline{49x^6y^8}$$

Exercise 14:

Multiply the following:

(1) $(-3)(-2a^2b)$

(2) $(-3a^2)(-2a^2b)$

(3) $(-3a^2b^2)(-2a^2b)$

(4) $(4x^2)(-3x^2y^2)$

(5) $(-2ab)(6a^3)$

(6) $(4x^2y)(x^3)$

(7) $(2a^3b)(-ab)$

(8) $(a^2b^2)(a^3b)$

(9) $(-a^2b^2)(a^3b)$

(10) $(3x^2y)(-2xy)(5x^3)$

(11) $(2a^2b)(-3ab^2)(-4a^3b^3)$

(12) $(4xyz)(-4x^2y)(-4y^2z^2)$

(13) $(-2bc)(-3ab)(-4cd)$

(14) $(6x^2y^2z)(-2x^3yz^3)$

(15) $(3a^2b^2c)(4ab)(6b^2c^2)$

(16) $(-2xy)(-3x^2y)(-x^2y^2)$

(17) $(-4x^2)(-2xy^2)(-x^2y^2z^2)$

(18) $(-5ab)(-2a^2)(-a^2b^2c^2)$

(19) $(4a^3)(4a^3)$

(20) $(-4a^3)(-4a^3)$

(21) $(-x^2y)(-x^2y)(-x^2y)$

(22) $(3x)^2$

(23) $(3x)^3$

(24) $(-3x)^3$

(25) $(-2a^3)^2$

(26) $(-2a^3)^3$

(27) $(3x^2y)^3$

(28) $(-3x^2y)^3$

(29) $(3a^3b^2)^2$

(30) $(-3a^3b^2)^2$

(31) $(-3a^3b^2)^3$

(32) $(-2x^3)^4$

Case II: *A Monomial Times a Multinomial*

If we have (4)(6), we know that the correct answer is 24. As we noted earlier, the quantity 6 can be written as a multinomial.

$$6 = 4 + 2$$
$$6 = 8 + 3 - 5$$

But no matter *how* we choose to write the quantity 6, any time that we multiply (6) by (4), the *only* correct answer is 24.

If we have, for instance,

$$(4)(4 + 2)$$

or

$$(4)(8 + 3 - 5)$$

we are multiplying (4) by a quantity that *always* means (6), and the only possible correct answer is 24.

Now, how do we multiply these quantities together so that we will wind up with 24? In the example

$$(4)(4 + 2)$$

the first factor (4) is a monomial, but the second factor (4 + 2) is composed of two *terms* and is a multinomial. Let us use the general description of the process of multiplication and multiply *each term* in the first factor, (+4), *separately* by *each term* in the second factor, (4 + 2), and then combine similar terms.

$$4(4 + 2) = 4(4) + 4(2)$$
$$= 16 + 8$$
$$= 24$$

By doing just that we have arrived at the correct result, which we knew in advance to be 24. In like manner:

$$(4)(6) = (4)(8 + 3 - 5)$$
$$= 4(8) + 4(3) + 4(-5)$$
$$= 32 + 12 - 20$$
$$= 44 - 20$$
$$= 24$$

The foregoing examples of multiplication illustrate one of the fundamental axioms of the algebra of real numbers. It is called the *distributive axiom*, and it can be stated symbolically, with *a*, *b*, and *c* representing real numbers, as follows:

> **The Distributive Axiom**
> $$a(b + c) = ab + ac$$

We say that multiplication is *distributive over addition*. Actually, without the distributive axiom, multiplication and addition would be two independent operations. The distributive axiom serves to connect multiplication and addition and because of that it powerfully influences algebra.

Examples:

(1) $(3)(5) = (3)(4 + 1) = 3(4) + 3(1)$
$$= 12 + 3$$
$$= 15$$

(2) $(3)(5) = (3)(4 + 7 - 6) = 3(4) + 3(7) + 3(-6)$
$$= 12 + 21 - 18$$
$$= 33 - 18$$
$$= 15$$

(3) $(-3)(5) = (-3)(4 + 7 - 6) = -3(4) - 3(7) - 3(-6)$
$$= -12 - 21 + 18$$
$$= -33 + 18$$
$$= -15$$

(4) $2x(x + y) = 2x(x) + 2x(y)$
$$= 2x^2 + 2xy$$

(5) $-3a(2a - 4b) = -3a(2a) - 3a(-4b)$
$$= -6a^2 + 12ab$$

(6) $(2x^2 + 3y)y = 2x^2(y) + 3y(y)$
$$= 2x^2y + 3y^2$$

(7) $3x^2(2x^2 - 5xy + y^2) = 3x^2(2x^2) + 3x^2(-5xy) + 3x^2(y^2)$
$$= 6x^4 - 15x^3y + 3x^2y^2$$

(8) $-2ab(3a^2 - 4ab - 2b^2) = -2ab(3a^2) - 2ab(-4ab) - 2ab(-2b^2)$
$$= -6a^3b + 8a^2b^2 + 4ab^3$$

Observe that in every one of the above examples we are following the distributive axiom, which means that we are multiplying *each term* in the first factor *separately* by *each term* in the second factor and then combining similar terms, if any, in the separate products. We are doing this for the very sensible reason that, as has been illustrated, it gives us the right answer.

The distributive axiom also serves to illustrate the fact that the product of two negative numbers must be a positive number.

Example:

We know that $(-5)(4) = -20$
Since $4 = 6 - 2$
then $(-5)(6 - 2) = -20$

By the distributive axiom $(-5)(6 - 2) = -5(6) - 5(-2)$
$$= -30 \pm ?10$$

Since we know the correct result is -20, the last term above must be $+10$; i.e., $-5(-2)$ must equal $+10$.

Consequently $(-5)(6 - 2) = -5(6) - 5(-2)$
$$= -30 + 10$$
$$= -20$$

Exercise 15:

Multiply the following:

(1) $(2a)(3a^2 - 4a)$ (2) $(-3xy)(x^2 - 2xy + y^2)$
(3) $6x(3x^2 - 4xy + 2y)$ (4) $-4a(a^2 - 2ab - b)$
(5) $-5x(x^3 - 2x^2 + 3x - 4)$ (6) $4(3 + 2)$
(7) $-4(3 + 2)$ (8) $(3 - 2)4$
(9) $-4(3 - 2)$ (10) $4a(3a + 2b)$
(11) $(3a + 2b)(-4a)$ (12) $4a(3a - 2b)$
(13) $-4a(3a - 2b)$ (14) $2x^2yz(x^2 - y^2 + 3z^2)$
(15) $5a^2(6a - 3a^2 + 4)$ (16) $a(b + c + d)$
(17) $(b + c - d)a$ (18) $m(x + y)$
(19) $(x + y)a$ (20) $2s^2(3s^2 - 5s + 6)$
(21) $3x^2y(2x + 4y - 1)$ (22) $-3a^2(2a^2 - 3b^2 - 4a + 2b)$
(23) $-2rs(3r^2 - 2rs + r^2s - 1)$ (24) $(6m^2 - 2n^2 + 4mn)2m^3$
(25) $-5x(3x - 4y - 2z - 3)$ (26) $ab(3a^2b - 4ab^2 + 2a^2b^2)$

Case III: *A Multinomial Times a Multinomial*

Multiplying a multinomial by a multinomial is the most complicated thing that can happen in the multiplication of algebraic expressions, but the situation is simplified considerably when we remember that all such multiplication means the *same thing* all the time. You actually cannot avoid the right answer in any such multiplication if you will just remember the distributive axiom: that is, *always* multiply *each term* in one factor *separately* by *each term* in the other factor.

Let us go back to the simple (and obvious) fact that

$$(4)(6) = 24$$

We can rewrite 4 and 6 in the following manner:

$$4 = 3 + 1$$
$$6 = 4 + 2$$

Therefore $(4)(6) = (3 + 1)(4 + 2) = 24$.

By the *distributive axiom*

$$(3 + 1)(4 + 2) = 3(4 + 2) + 1(4 + 2)$$
$$= 3(4) + 3(2) + 1(4) + 1(2)$$
$$= 12 + 6 + 4 + 2$$
$$= 24$$

Expressing this in symbolic language

$$(a + b)(c + d) = a(c + d) + b(c + d)$$
$$= ac + ad + bc + bd$$

By following this axiom, we can multiply any and all multinomials together.

Examples:

(1) $(3)(4) = (2 + 1)(3 + 1)$
$$= 2(3 + 1) + 1(3 + 1)$$
$$= 2(3) + 2(1) + 1(3) + 1(1)$$
$$= 6 + 2 + 3 + 1$$
$$= 12$$

(2) $(a + b)(a + b) = a(a + b) + b(a + b)$
$$= a(a) + a(b) + b(a) + b(b)$$
$$= a^2 + ab + ba + b^2$$

In the third term we have ba. But $ba = ab$ by the commutative law:

$$(a + b)(a + b) = a^2 + ab + ab + b^2$$

The two middle terms ab and ab are exactly alike and can be combined:

$$[ab + ab = 2ab]$$
$$(a + b)(a + b) = a^2 + 2ab + b^2$$

(3) $(x + y)(x - y) = x(x - y) + y(x - y)$
$$= x(x) + x(-y) + y(x) + y(-y)$$
$$= x^2 - xy + xy - y^2$$
$$= x^2 - y^2$$

(4) $(2x + 3)(x - 1) = 2x(x - 1) + 3(x - 1)$
$$= 2x(x) + 2x(-1) + 3(x) + 3(-1)$$
$$= 2x^2 - 2x + 3x - 3$$
$$= 2x^2 + x - 3$$

(5) $(m + n)(m^2 - mn + n^2) = m(m^2 - mn + n^2) + n(m^2 - mn + n^2)$
$$= m(m^2) + m(-mn) + m(n^2) + n(m^2) +$$
$$n(-mn) + n(n^2)$$
$$= m^3 - m^2n + mn^2 + m^2n - mn^2 + n^3$$
$$= m^3 + n^3$$

(6) $(3a - b)(2a + 4b) = 3a(2a + 4b) - b(2a + 4b)$
$$= 3a(2a) + 3a(4b) - b(2a) - b(4b)$$
$$= 6a^2 + 12ab - 2ab - 4b^2$$
$$= 6a^2 + 10ab - 4b^2$$

(7) $(x - y)^2 = (x - y)(x - y) = x(x - y) - y(x - y)$
$$= x(x) + x(-y) - y(x) - y(-y)$$
$$= x^2 - xy - yx + y^2$$
$$= x^2 - xy - xy + y^2$$
$$= x^2 - 2xy + y^2$$

Exercise 16:

Multiply the following:

(1) $(4x - 3)(3x + 1)$ (2) $(2x - 4)(7x + 1)$
(3) $(3y - 4)(2y - 5)$ (4) $(2a + 7)(a + 4)$
(5) $(5a + 2b)(6a - b)$ (6) $(2a + 4)(2a - 4)$
(7) $(5x + 3)(5x - 3)$ (8) $(a + b)(a - b)$
(9) $(2x + 3y)(2x - 3y)$ (10) $(3x - y)(2x^2 + 6x - y)$
(11) $(x + y)(x^2 - xy + y^2)$ (12) $(a + b)(a^2 - ab + b^2)$

(13) $(a + 2b)(a + 2b)$
  or $(a + 2b)^2$

(14) $(m + n)(m + n)$
  or $(m + n)^2$

(15) $(2x + y)(2x + y)$
  or $(2x + y)^2$

(16) $(x + y + z)(x + y + z)$
  or $(x + y + z)^2$

(17) $(a + b + c)^2$

(18) $(a^2 - 3a - 4)(2a^2 + 3a - 2)$

(19) $(3m^2 - 2m + 1)(2m^2 + 5m - 8)$

(20) $(x^2 - 5x + 2)(3x^2 - x - 4)$

(21) $(3x - 2y)(3x + 2y)$

(22) $(3x - 2y)^2$

(23) $(5a + 2b)^2$

(24) $(7x - 1)^2$

(25) $(3x^2 + 4)^2$

(26) $(2x - y + z)^2$

(27) $(2x + y + 3)(2x + y - 3)$

(28) $(4a - b + 1)(4a - b - 1)$

We may of course have more than *two* multinomials multiplied together. Suppose we had $(a + b)^3$. The exponent, 3, means that the number $a + b$ must be *used as a factor three times*. Therefore $(a + b)^3$ means $(a + b)(a + b)(a + b)$.

$$\begin{aligned}
(a + b)^3 &= (a + b)(a + b)(a + b) \\
&= \underline{(a + b)(a + b)}(a + b) \\
&= (a^2 + ab + ab + b^2)(a + b) \\
&= (a^2 + 2ab + b^2)(a + b) \\
&= a^2(a + b) + 2ab(a + b) + b^2(a + b) \\
&= a^3 + a^2b + 2a^2b + 2ab^2 + ab^2 + b^3 \\
(a + b)^3 &= a^3 + 3a^2b + 3ab^2 + b^3
\end{aligned}$$

You should now be convinced for all time, if you believe the distributive axiom, that $(a + b)^3$ **does not mean** $a^3 + b^3$!

Exercise 17:

Show that
$$(x + y)^3 = x^3 + 3x^2y + 3xy^2 + y^3$$

Exercise 18:

Multiply the following:

(1) $(m + n)^3$

(2) $(r + 2s)^3$

(3) $(a - b)^3$

(4) $(x - y)^3$

(5) $(2x - 3y)^3$

(6) $(3a - 1)^3$

(7) $(4x + 3)^3$

(8) $(5y - 2)^3$

The following set of problems represents a general review of multiplication. Remember that the factors may be arranged in any *order* (the commutative law) or any *pairing* (the associative law) and that multiplication is *distributive* over addition.

Exercise 19:

Multiply the following:

(1)	$(2ab)^2$	(2)	$(2a + b)^2$
(3)	$(2 + a + b)^2$	(4)	$(-3xy)^3$
(5)	$-3(x^2 + y)^2$	(6)	$-5x^2(2x - 7y^2)$
(7)	$(4x - 7)(2x - 3)$	(8)	$(x - 4)(3x^2 - 5x + 20)$
(9)	$3a(7a^2 - 5a + 4)$	(10)	$(-rs)^3(-2rs)$
(11)	$(r + s)^3$	(12)	$(3x - 1)(9x^2 + 3x + 1)$
(13)	$(-4xy)(-3x^2y^2)^2$	(14)	$(2ab)^2(5ab)$
(15)	$(2ab)(5a + b)$	(16)	$(2a + b)(5a + b)$
(17)	$(3x^2y^2)(-2xy)$	(18)	$(-2xy)(3x^3 + y^2)$
(19)	$(4x^2y^3)(-7xy)^2$	(20)	$(4x^2y^3)(-7x + y)$
(21)	$(2a + 1)(2a - 1)$	(22)	$(4x + 5y)(4x - 5y)$
(23)	$(2x + 1)^3$	(24)	$(4a - 3)^2$
(25)	$(2a)(a + b)^2$	(26)	$(x + y)(x - y)(x^2 + y^2)$

2.7
Grouping and the Order of Operations

In applying the distributive axiom to a problem such as

$$4(4 + 2)$$

we arrived at the statement

$$4(4 + 2) = 4(4) + 4(2)$$

where, in the expression on the right side of the equal mark, the multiplication was done *first*, and then the products were added to get the correct answer.

$$4(4) + 4(2)$$
$$16 + 8$$
$$24$$

Situations like this will occur frequently, and we must now consider the *order* of different operations (multiplication, addition, etc.) when we have more than one of them in the same problem. The human race agreed long ago to write *all* mathematical operations so that the following order should be observed:

Parentheses	*first*
Exponents	*second*
Multiplication ⎤	*next*, in the order in which
Division ⎦	they occur reading left to right
Addition ⎤	
Subtraction ⎦	*last*

This order is observed in all of mathematics; it is particularly important in computer mathematics, and it must be memorized. A good gimmick for remembering the order is the following phrase:

Please	Parentheses (parentheses are implied above and below any fraction bar)
Excuse	Exponents
My	Multiplication
Dear	Division
Aunt	Addition
Sally	Subtraction

As has been amply illustrated, the use of parentheses is most efficient for indicating multiplication of multinomials. In fact, because of the order of operations, some such device is absolutely necessary. For example, if we have the following:

1. $4 + 3(5)$
2. $(4 + 3)(5)$ or $(4 + 3)5$

we have two entirely different computations.

In problem 1, $4 + 3(5)$, we must do the multiplication *first* and then the addition.

$$4 + 3(5) = 4 + 15 = 19$$

In problem 2, $(4 + 3)(5) = (5)(4 + 3)$ by the commutative law, and $5(4 + 3) = 5(4) + 5(3)$ by the distributive axiom.

$$5(4) + 5(3) = 20 + 15$$
$$= 35$$

Therefore:

1. $4 + 3(5) = 19$
2. $(4 + 3)(5) = 35$

In like manner:

$$a + b(c) = a + bc$$
$$(a + b)(c) = ac + bc$$

We will frequently have multiplication within multiplication and to show this we use brackets [ ] and braces { } as well as parentheses.

Consider the following:

$$2a - 3(a + b)$$

We must do the multiplication *first*.

$$2a - 3(a + b) = 2a - 3a - 3b$$

Then we perform the addition.

$$2a - 3a - 3b = \underline{-a - 3b}$$

Suppose we had this situation:

$$4x - (x - 4)$$

When there is no coefficient in front of a parenthesis, it is *always understood to be the number one*. Therefore

---

$4x - (x - 4)$ *means* $4x - 1(x - 4)$

We do the multiplication *first*

$$4x - 1(x - 4) = 4x - x + 4$$

Then we perform the addition

$$4x - x + 4 = \underline{3x + 4}$$

Now consider this situation:

$$3a - 4[2(a - b) - (a + 2)] + 6a$$

Everything inside the brackets is to be multiplied by -4, but we shall first work *inside* the brackets and do the multiplication indicated by the parentheses inside. *We leave everything else alone.*

$$3a - 4[2(a - b) - 1(a + 2)] + 6a$$
$$\downarrow \quad \downarrow \quad \downarrow \qquad\qquad\qquad \downarrow \quad \downarrow$$
$$3a - 4[2a - 2b - a - 2] \quad + 6a$$

Now combine the like terms *inside* the brackets. Leave everything else alone.

$$3a - 4[a - 2b - 2] + 6a$$

Now multiply every term in the brackets by -4. Leave everything else alone.

$$3a - 4a + 8b + 8 + 6a$$

Finally, do the indicated addition by combining like terms.

$$5a + 8b + 8$$

The number above is the final, simplest, and *correct* result. Here are the processes repeated in the order that they should be performed.

$$3a - 4[2(a - b) - (a + 2)] + 6a =$$
$$3a - 4[2(a - b) - 1(a + 2)] + 6a =$$
$$3a - 4[2a - 2b - a - 2] + 6a =$$
$$3a - 4[a - 2b - 2] + 6a =$$
$$3a - 4a + 8b + 8 + 6a =$$
$$5a + 8b + 8$$

In removing signs of grouping, one should follow three simple procedures to guarantee the correct result.

1. Follow the order of operations.
2. Always work from the *inside* out.
3. Have the good sense to slow down and perform only one operation at a time.

Exercise 20:

Simplify the following:

(1) $3 + 2(8)$
(2) $(3 + 2)(8)$
(3) $2x - 3x^2(y)$
(4) $(2x - 3x^2)(y)$
(5) $x + 4(x - 1)$
(6) $(x + 4)(x - 1)$
(7) $2a + b(a - 2b)$

(8) $(2a + b)(a - 2b)$
(9) $3x + 4(2x - 5)$
(10) $(3x + 4)(2x - 5)$
(11) $4a - (2 - 3a) + 7$
(12) $5x + 2 - (3x - 4)$
(13) $a + 2(b + 6) - b - (2a + b)$
(14) $2x - 3[2(x - y) + 4x] - 6y$
(15) $3x^2 - 4x + 5 - (x^2 - 6x + 2)$
(16) $a^2 - 7a + 10 - (3a^2 + 2a - 4)$
(17) $4x + 3(x + 1) - (2x - 5) + 6x$
(18) $3x - 2[3(x + 1) - (2x + 3)] + 2x$
(19) $a - [3(a + 1) - 2(a - 4)] - 6a + 2$
(20) $3a - 2\{4a - [5(a - 1) - 2(a - 3)] - 6a\} - 5a$

2.8
Division

In the division of algebraic expressions we shall limit our attention to special kinds of monomials and multinomials called *polynomials*.

Polynomials Following is a collection of monomials and multinomials (zero is excluded from all denominators):

1. $4x - \dfrac{3}{x^2} + \sqrt{x} + 6$　　　　　　2. $5x^3 - 5x^2 + 2x - 4$

3. $3x^3$　　　　　　　　　　　　　　4. $5\sqrt{xyz}$

5. $7a^3 - 2a^2 + 3\sqrt{a} + 2$　　　　　6. $\dfrac{7x^2}{y}$

7. $\dfrac{4}{a} + 3b - 2$　　　　　　　　　8. $4b^3 - \sqrt{3}b^2 + 7b - 4$

Of this group, *only* 2, 3 and 8 are *polynomials*. Now what is a polynomial?

1. A polynomial may be either a monomial or a multinomial.

2. In polynomials *all exponents* of *literal* numbers (x, y, z, etc.) must be non-negative *integers*. This is another way of saying that there are no literal numbers in the *denominator* of any term of the polynomial, and that there are no roots of *literal* numbers such as $\sqrt{x}$ or $\sqrt[3]{x}$ or $\sqrt[4]{y}$.

3. In a *real* polynomial, all of the numerical coefficients are *real* numbers.

Summary

A real polynomial is any monomial or multinomial having only nonnegative integers as exponents of the *literal* numbers and only real numbers as numerical coefficients.

Exercise 21:

Tell which of the following are polynomials:

(1) $7x^4 - 3x^3 - \sqrt{5}x^2 + x - 6$

(2) $\dfrac{5}{x^2} - 3x^4 + \dfrac{2}{x} + 10$ (x cannot equal 0)

(3) $8y^5 - 1$

(4) $3y^2$

(5) $5\sqrt[6]{x}$

(6) $5a^2 - 2a^4 + 3\sqrt{a} + 2$

(7) $x^7 - 2x^3 + 3$

(8) $y^5 - 3y^4 - 2y^3 + y + \sqrt{y} - 5$

(9) $\dfrac{3}{x^2} - \dfrac{1}{x} + 6$ (x cannot equal 0)

(10) $\dfrac{3a^2}{b}$ (b cannot equal 0)

In the division processes which we shall consider, we shall include only rational polynomials. Unless otherwise specified, these numbers are referred to simply as polynomials.

Every division problem involves a dividend and a divisor. We state here the meaning of division as it applies to *all* situations: in a division problem *each separate term of the dividend is divided by the entire divisor*. This is always true regardless of whether the numbers involved are monomials or multinomials. To see *why* it is true, we can look at some examples from arithmetic.

1. $\dfrac{8}{4}$ (Divide *each term* of the dividend by the *entire divisor*)

$\dfrac{8}{4} = 2$

2. $\dfrac{4+2}{2}$ (Divide *each term* of the dividend by the *entire divisor*)

$$\frac{4+2}{2} = \frac{4}{2} + \frac{2}{2} = 2 + 1 = 3$$

or $\dfrac{4+2}{2} = \dfrac{6}{2} = 3$

3. $\dfrac{18-6}{3}$ (Divide *each term* of the dividend by the *entire divisor*)

$$\frac{18-6}{3} = \frac{18}{3} - \frac{6}{3} = 6 - 2 = 4$$

or $\dfrac{18-6}{3} = \dfrac{12}{3} = 4$

As Examples 2 and 3 clearly show, we perform division in this way because it gives us the right answer. In symbolic language

$$\frac{a+b}{c} = \frac{a}{c} + \frac{b}{c}$$ (provided that c is not zero)

The three situations which can arise in division are the following:

  I. *The Division of a Monomial by a Monomial*

    (a) $\dfrac{10}{2}$

    (b) $\dfrac{6x^2}{-2x}$ (x is not equal to 0)

    (c) $\dfrac{-10a^2b}{-2a}$ (a is not equal to 0)

 II. *The Division of a Multinomial by a Monomial*

    (a) $\dfrac{4+2}{2}$

    (b) $\dfrac{a+b}{b}$ (b is not equal to 0)

    (c) $\dfrac{6x^2 - 2x + 12x^3}{2x}$ (x is not equal to 0)

III. *The Division of a Monomial or a Multinomial by a Multinomial*

    (a) $\dfrac{x^2 - 5x + 6}{x - 2}$

    (b) $\dfrac{x^3}{x - 1}$ (Note: Zero is excluded from all denominators)

    (c) $\dfrac{5x^3 - 6x^2 + 2x - 4}{x^2 - 2x + 3}$

I.   *The Division of a Monomial by a Monomial*

In this situation we have only *one term* in the dividend and only *one term* in the divisor. This is analogous to simple arithmetic and amounts to reducing a fraction to its lowest terms.

Examples:

(1)   If we have $\dfrac{30}{12}$ we simplify the fraction by dividing out all common *factors*.

Zero is excluded from all denominators.

$$\frac{30}{12} = \frac{(6)(5)}{(6)(2)} = \frac{6}{6} \cdot \frac{5}{2} = 1 \cdot \frac{5}{2} = \frac{5}{2}$$

$$or \quad \frac{30}{12} = \frac{(\cancel{6})(5)}{(\cancel{6})(2)} = \frac{1 \cdot 5}{1 \cdot 2} = \frac{5}{2}$$

(2) $$\frac{24}{9} = \frac{(8)(3)}{(3)(3)} = \frac{8}{3} \cdot \frac{3}{3} = \frac{8}{3} \cdot 1 = \frac{8}{3}$$

$$or \quad \frac{24}{9} = \frac{(8)(\cancel{3})}{(3)(\cancel{3})} = \frac{8 \cdot 1}{3 \cdot 1} = \frac{8}{3}$$

(3) $$\frac{6a}{3a} = \frac{(2)(\cancel{3})(\cancel{a})}{(\cancel{3})(\cancel{a})} = \frac{2 \cdot 1 \cdot 1}{1 \cdot 1} = \frac{2}{1} = 2$$

(4) $$\frac{6a^2}{3a} = \frac{(2)(\cancel{3})(\cancel{a})(a)}{(\cancel{3})(\cancel{a})} = \frac{2 \cdot 1 \cdot 1 \cdot a}{1 \cdot 1} = \frac{2a}{1} = 2a$$

(5) $$\frac{3a}{3ab} = \frac{(\cancel{3})(\cancel{a})}{(\cancel{3})(\cancel{a})(b)} = \frac{1 \cdot 1}{1 \cdot 1 \cdot b} = \frac{1}{b}$$

When we are dividing monomials with *repeated factors*, we still divide out the common factors. In the following examples, x cannot equal 0:

1. $$\frac{x^4}{x^2} = \frac{(\cancel{x})(\cancel{x})(x)(x)}{(\cancel{x})(\cancel{x})} = x^2$$

2. $\dfrac{x^5}{x^2} = \dfrac{\overset{1}{(\cancel{x})}\overset{1}{(\cancel{x})}(x)(x)(x)}{\underset{1}{(\cancel{x})}\underset{1}{(\cancel{x})}} = x^3$

3. $\dfrac{x^4}{x^7} = \dfrac{\overset{1}{(\cancel{x})}\overset{1}{(\cancel{x})}\overset{1}{(\cancel{x})}\overset{1}{(\cancel{x})}}{\underset{1}{(\cancel{x})}\underset{1}{(\cancel{x})}\underset{1}{(\cancel{x})}\underset{1}{(\cancel{x})}(x)(x)(x)} = \dfrac{1}{x^3}$

You may have observed, in the discussion of these operations, that we are saying "divide out the common factors." You are probably addicted to the word "cancel." If you must use this word, please remember that, in simplifying fractions, it applies to *factors only*, and *never, never, NEVER to terms!* In fact a most important, a most basic, and a most necessary rule of all algebraic operations is the following:

NEVER CANCEL TERMS

unless, of course, you really wish to appear to be someone who never actually understood arithmetic. If you don't believe that is golden advice, then write it down; look at it; and *use your head*.

Take a good look at these two fractions:

1. $\dfrac{(4)(5)}{4}$

2. $\dfrac{4+5}{4}$

In fraction 1 we have two *factors* in the numerator. In fraction 2 we have two *terms* in the numerator. In *both* fractions we have the factor (4) in the denominator; i.e., $4 = (4)(1)$.

$$
\boxed{\begin{array}{c}
\text{Fraction 1 means} \\[4pt]
\dfrac{(4)(5)}{(4)} = \dfrac{20}{4} = 5 \\[10pt]
or \quad \dfrac{\overset{1}{(\cancel{4})}(5)}{\underset{1}{(\cancel{4})}} = \dfrac{5}{1} = 5
\end{array}}
$$

$$
\boxed{\begin{array}{c}
\text{Fraction 2 means} \\[4pt]
\dfrac{4+5}{4} = \dfrac{9}{4}
\end{array}}
$$

Please note that, in the numerator of fraction 2, the quantity $4 + 5$ is never going to mean anything in this world but 9.

If you had "canceled" the 4's in fraction 2, you would have gotten an idiotic result. Remember, we never cancel numbers because they are the same *numbers*, only because they are the same *factors*.

Dividing a monomial by a monomial simply means, therefore, that we are going to find all of the common *factors* both in the numerator and in the denominator and divide them out.

Examples:

Zero is excluded from all denominators.

(1) $\dfrac{6x^2y^3}{2xy^2} = \dfrac{(\overset{1}{\cancel{2}})(3)(\overset{1}{\cancel{2}})(x)(\overset{1}{\cancel{x}})(\overset{1}{\cancel{y}})(y)}{(\underset{1}{\cancel{2}})(\underset{1}{\cancel{x}})(\underset{1}{\cancel{y}})(\underset{1}{\cancel{y}})}$

$= \dfrac{3xy}{1}$

$= 3xy$

(2) $\dfrac{15a^3b}{-3ab^3} = \dfrac{(\overset{-1}{\cancel{3}})(5)(\overset{1}{\cancel{a}})(a)(a)(\overset{1}{\cancel{b}})}{(-\cancel{3})(\underset{1}{\cancel{a}})(\underset{1}{\cancel{b}})(b)(b)}$

$= \dfrac{-5a^2}{b^2}$ or $-\dfrac{5a^2}{b^2}$

(3) $\dfrac{-6a^2}{-6a^3} = \dfrac{(-\overset{1}{\cancel{6}})(\overset{1}{\cancel{a}})(\overset{1}{\cancel{a}})}{(-\underset{1}{\cancel{6}})(\underset{1}{\cancel{a}})(\underset{1}{\cancel{a}})(a)}$

$= \dfrac{1}{a}$

(4) $\dfrac{-a^3}{a^2} = \dfrac{(-a)(\overset{1}{\cancel{a}})(\overset{1}{\cancel{a}})}{(\underset{1}{\cancel{a}})(\underset{1}{\cancel{a}})}$

$= \dfrac{-a}{1}$ or $-a$

Exercise 22:

Divide the following (zero is excluded from all denominators):

(1) $\dfrac{a^3}{a}$ (2) $\dfrac{a}{a^3}$ (3) $\dfrac{-x^5}{x^2}$ (4) $\dfrac{b^4}{-b^2}$

(5) $\dfrac{y^2}{y^3}$ (6) $\dfrac{-a^3}{-a^5}$ (7) $\dfrac{12a^4}{3a^2}$ (8) $\dfrac{-35x^2}{7x^3}$

(9) $\dfrac{-8x^3}{-2x^3}$ (10) $\dfrac{3x^2y}{3xy^2}$ (11) $\dfrac{-16a}{4a^2}$ (12) $\dfrac{3ab}{3a^2b^2}$

(13) $\dfrac{6x^2y}{3x}$ (14) $\dfrac{-14a^2bc}{2ab}$ (15) $\dfrac{-25x^3y^2}{5xy^2}$ (16) $\dfrac{26x^2y^4}{-13x^3y^3}$

(17) $\dfrac{-10x^2y}{2x^4y^5}$ (18) $\dfrac{44a^3b}{11ab^4}$ (19) $\dfrac{24x^2y^3z}{9xyz^3}$ (20) $\dfrac{15a^4b^2c}{6ab^5c^3}$

(21) $\dfrac{x}{x^2y}$ (22) $\dfrac{x^2y}{x}$ (23) $\dfrac{-3ab}{6a^2b^2}$ (24) $\dfrac{6a^2b^2}{-3ab}$

(25) $\dfrac{45xy^3}{18x^4y^2}$ (26) $\dfrac{3ab^2c}{48a^4b^4c^3}$

II. *Division of a Multinomial by a Monomial*

In dividing a multinomial by a monomial, you cannot evade the *correct* result if you always follow the same process: divide *each separate term* of the *dividend* by the *entire divisor*. Since the dividend in this situation is a multinomial, it will always have more than one term.

Examples:

(1) $\dfrac{6+12}{3}$ (Divide each term of the dividend by the entire divisor)

$$\frac{6+12}{3} = \frac{6}{3} + \frac{12}{3} = 2 + 4 = 6$$

or $$\frac{6+12}{3} = \frac{18}{3} = 6$$

(2) $\dfrac{25-10}{5}$ (Divide each term of the dividend by the entire divisor)

$$\frac{25-10}{5} = \frac{25}{5} - \frac{10}{5} = 5 - 2 = 3$$

or $$\frac{25-10}{5} = \frac{15}{5} = 3$$

(3) $\dfrac{a+b}{a}$ (Divide each term of the dividend by the entire divisor)

$$\frac{a+b}{a} = \frac{a}{a} + \frac{b}{a} = 1 + \frac{b}{a}$$ (*a* cannot equal zero)

(4) $\dfrac{x^3 - 4x^2}{x}$ (Divide each term of the dividend by the entire divisor)

$$\frac{x^3 - 4x^2}{x} = \frac{x^3}{x} - \frac{4x^2}{x} = x^2 - 4x \qquad (x \text{ cannot equal zero})$$

(5) $\dfrac{6a^3 - 12a^2 - 3a}{3a}$ $\left(\begin{array}{l}\text{Divide each term of the dividend by the entire} \\ \text{divisor}\end{array}\right)$

$$\frac{6a^3 - 12a^2 - 3a}{3a} = \frac{6a^3}{3a} - \frac{12a^2}{3a} - \frac{3a}{3a}$$

$$= 2a^2 - 4a - 1 \qquad (a \text{ cannot equal zero})$$

(6) $\dfrac{3x^2y - 15xy^2 + 6xy}{9x^2y^2} =$ $\left(\begin{array}{l}\text{Divide each term of the dividend by the} \\ \text{entire divisor}\end{array}\right)$

$$\frac{3x^2y}{9x^2y^2} - \frac{15xy^2}{9x^2y^2} + \frac{6xy}{9x^2y^2} =$$

$$\frac{\cancel{3}\cancel{x}\cancel{x}\cancel{y}}{\cancel{(3)}(3)\cancel{x}\cancel{x}\cancel{y}y} - \frac{\cancel{(3)}(5)\cancel{x}\cancel{y}\cancel{y}}{\cancel{(3)}(3)\cancel{x}x\cancel{y}\cancel{y}} + \frac{\cancel{(3)}(2)\cancel{x}\cancel{y}}{\cancel{(3)}(3)\cancel{x}x\cancel{y}y} =$$

$$\frac{1}{3y} - \frac{5}{3x} + \frac{2}{3xy} \qquad (x \text{ and } y \text{ cannot equal zero})$$

Exercise 23:

Divide the following (zero is excluded from all denominators):

(1) $\dfrac{24 - 6}{3}$ (2) $\dfrac{18 - x}{3}$ (3) $\dfrac{a + 12}{4}$

(4) $\dfrac{x + 5}{5}$ (5) $\dfrac{b - 6}{6}$ (6) $\dfrac{5x + 10}{10}$

(7) $\dfrac{2 - 4a}{2}$ (8) $\dfrac{a - b}{a}$ (9) $\dfrac{m + n}{n}$

(10) $\dfrac{x - y}{x}$ (11) $\dfrac{5x^2 - 10x}{5x}$ (12) $\dfrac{4x^2 - 8x}{2x}$

(13) $\dfrac{12a^3 - 8a^2}{4a}$ (14) $\dfrac{15b^3 - 3b}{3b}$ (15) $\dfrac{12x^2y - 8xy^2}{4x^2y^2}$

(16) $\dfrac{3a^2b^3 - 6ab^2}{3a^2b}$ (17) $\dfrac{45x^3y^4 - 5x^2y^2}{5x^2y^2}$ (18) $\dfrac{14ab - 3a^2b^2}{7ab^3}$

(19) $\dfrac{15a^2b^2 - 9a^2b + 3ab}{3ab}$ (20) $\dfrac{12x^3 - 8x^2y^2 + 4y^2}{4xy}$

(21) $\dfrac{25m^3 + 10m^2n^2 + 15n^3}{20m^2n^2}$

(22) $\dfrac{14x^2y^2 - 21x - 7xy}{-7xy}$

(23) $\dfrac{24r^2t - 12rt^2 + 6r^2t^3}{-3r^3t^2}$

(24) $\dfrac{18a^4b - 6ab^4 + 3a^3b^3}{-3a^3b^3}$

III. *Division of a Monomial or a Multinomial by a Multinomial*

This is the most complicated computation that can happen in the division of polynomials, but again, all it means is that we follow the same routine for division all the time: that is, we must always divide each term of the dividend by the entire divisor. In the following example $x - 2$ cannot equal 0.

> If we have
> $$\frac{x^2 - 5x + 6}{x - 2}$$
> it *means*
> $$\frac{x^2}{x - 2} - \frac{5x}{x - 2} + \frac{6}{x - 2}$$

In the above example we have clearly divided each term of the dividend by the entire divisor, and therefore we must be right. Thus

$$\frac{x^2 - 5x + 6}{x - 2} = \frac{x^2}{x - 2} - \frac{5x}{x - 2} + \frac{6}{x - 2}$$

There is absolutely no question about the *meaning* of the above problem; the only question is, how on earth do you *do* it?

You stand here, in relation to algebraic processes, about where you stood in Grade Four when instead of dividing numbers like 792 by 2 (using simple "short" division) you were suddenly faced with the problem of dividing 792 by a number such as 22.

> $$\begin{array}{r} 396 \\ 2\overline{)792} \end{array} \qquad \begin{array}{r} ?? \\ 22\overline{)792} \end{array}$$

If you recall, it was necessary then to shift gears mentally and switch to a process called "long division." That is exactly what has happened here: *when the divisor is a multinomial we must switch to long division.*

How does one do long division with multinomials? Oddly enough, it is done the same way it was done in arithmetic and for the *same* reasons, although the actual process might not have been obvious to you at the time.

Let us consider what really happened when you used long division to divide 792 by 22. Because of the wonderful *positional notation* we use in our number system, the process of dividing 792 by 22 becomes quite automatic, but at the same time, the basic facts of what is going on tend to get buried in the automation.

Here is the automatic process:

$$
\begin{array}{r}
36 \\
22 \overline{)\ 792} \\
-66 \\
\hline
+132 \\
-132 \\
\hline
0
\end{array}
$$

This process worked out neatly, but only because our number system, being base 10, is actually written in *powers of 10*. In other words, every numeral of arithmetic is simply a *polynomial written in powers of 10*, and then condensed in form by positional notation.

Why does the collection of numerals 792 mean *seven hundred ninety-two?* Because:

---

792 really means $7(10)^2 + 9(10)^1 + 2$.

$$792 = 7(10)^2 + 9(10)^1 + 2$$

Notice that the number on the right is a polynomial in powers of 10.

$$792 = 7(100) + 9(10) + 2$$
$$792 = 700 + 90 + 2$$

22 really means $2(10)^1 + 2$.

$$22 = 2(10)^1 + 2$$
$$22 = 20 + 2$$

Consequently, 792 can be written $700 + 90 + 2$ and 22 can be written $20 + 2$. Thus

$$22 \,\overline{\big)\,792} \quad means \quad 20 + 2 \,\overline{\big)\,700 + 90 + 2}$$

$$
\begin{array}{r}
36 \\
22 \,\overline{\big)\,792} \quad or \\
-66 \\
\hline
+132 \\
-132 \\
\hline
0
\end{array}
\qquad
\begin{array}{r}
35 + 1 \\
20 + 2 \,\overline{\big)\,700 + 90 + 2} \\
-700 - 70 \\
\hline
+ 20 + 2 \\
- 20 - 2
\end{array}
$$

$$\frac{792}{22} = 36 \quad or \quad \frac{700 + 90 + 2}{20 + 2} = 35 + 1 = 36$$

When we are dividing *algebraic* polynomials, we have no positional notation available, and we are stuck with the *polynomial* form of the number. Therefore the division process must follow the pattern *on the right* in the above illustration.

Compare:

$$\frac{700 + 90 + 2}{20 + 2} \qquad \text{and} \qquad \frac{x^2 - 5x + 6}{x - 2}$$

$$\downarrow \qquad\qquad\qquad\qquad\qquad \downarrow$$

1.
$$
\begin{array}{r}
35 \\
20 + 2 \,\overline{\big)\,700 + 90 + 2} \\
700 + 70 \\
\hline
\end{array}
\qquad
\begin{array}{r}
x \\
x - 2 \,\overline{\big)\,x^2 - 5x + 6} \\
x^2 - 2x \\
\hline
\end{array}
$$

2.
$$
\begin{array}{r}
35 \\
20 + 2 \,\overline{\big)\,700 + 90 + 2} \\
-700 - 70 \\
\hline
+ 20
\end{array}
\qquad
\begin{array}{r}
x \\
x - 2 \,\overline{\big)\,x^2 - 5x + 6} \\
-x^2 + 2x \\
\hline
- 3x
\end{array}
$$

3.
$$
\begin{array}{r}
35 + 1 \\
20 + 2 \,\overline{\big)\,700 + 90 + 2} \\
-700 - 70 \\
\hline
20 + 2 \\
20 + 2 \\
\hline
\end{array}
\qquad
\begin{array}{r}
x - 3 \\
x - 2 \,\overline{\big)\,x^2 - 5x + 6} \\
-x^2 + 2x \\
\hline
-3x + 6 \\
- 3x + 6 \\
\hline
\end{array}
$$

$$\begin{array}{r} 35 + 1 \\ 20 + 2 \overline{\smash{\big)}\ 700 + 90 + 2} \\ \underline{-700 - 70} \\ 20 + 2 \\ \underline{-20 - 2} \\ 0 \end{array}$$

$$\begin{array}{r} x - 3 \\ x - 2 \overline{\smash{\big)}\ x^2 - 5x + 6} \\ \underline{-x^2 + 2x} \\ -3x + 6 \\ \underline{+3x - 6} \\ 0 \end{array}$$

If you have looked over the foregoing examples carefully (and thoughtfully), you may have observed that in each of the two problems, the terms of each polynomial were arranged in *descending* order of the exponents. This is exceedingly important to remember.

Because of the commutative law of addition, we may arrange the terms of a polynomial in any order we please.

Examples:

(1) $4 + 5 + 3 = 5 + 4 + 3$
$= 5 + 3 + 4$
$= 3 + 5 + 4$

(2) $x^2 + 5x + 6 = 5x + x^2 + 6$
$= 5x + 6 + x^2$
$= 6 + 5x + x^2$

In performing long division with polynomials, both dividend and divisor are arranged in descending order of the exponents. For example:

$$\frac{3x + 2 + x^3 - 2x^2}{-2 + x} \quad \text{will be written} \quad \frac{x^3 - 2x^2 + 3x + 2}{x - 2}$$

Now let us review what we did in each step of *both problems* in the foregoing examples.

1.   Arrange the terms in descending order of the exponents; then divide the *first term* in the dividend by the *first term* in the divisor.

2.   Write this result on the quotient line and then multiply it by the *entire divisor*.

3.   Write this product below the dividend and then subtract it *from* the dividend: i.e., change the signs and add. Now bring down the rest of the dividend.

4.   Step 3 gives us a *new* dividend and we *repeat* the whole process from the beginning: i.e., divide the *first term* in the *new* dividend by the *first term* in the divisor; write the result on the quotient line; etc.

This is the method by which all polynomials are divided *when the divisor is a multinomial.*

Example:

$(2x - 1$ cannot equal 0.)

$$
\text{(1)} \quad 2x - 1 \;\big|\; \overline{2x^3 - 5x^2 + 8x - 5} \quad \overset{x^2}{}
$$

(1) $\begin{array}{r} x^2 \\ 2x - 1 \,\big|\, \overline{2x^3 - 5x^2 + 8x - 5} \end{array}$

(2) $\begin{array}{r} x^2 \\ 2x - 1 \,\big|\, \overline{2x^3 - 5x^2 + 8x - 5} \\ \underline{2x^3 - x^2 } \end{array}$

(3) $\begin{array}{r} x^2 \\ 2x - 1 \,\big|\, \overline{2x^3 - 5x^2 + 8x - 5} \\ \underline{-2x^3 + x^2 } \\ - 4x^2 + 8x - 5 \end{array}$ (New dividend: now repeat the process)

(4) $\begin{array}{r} x^2 - 2x \\ 2x - 1 \,\big|\, \overline{2x^3 - 5x^2 + 8x - 5} \\ \underline{-2x^3 + x^2 } \\ - 4x^2 + 8x - 5 \\ \underline{- 4x^2 + 2x } \end{array}$

(5) $\begin{array}{r} x^2 - 2x \\ 2x - 1 \,\big|\, \overline{2x^3 - 5x^2 + 8x - 5} \\ \underline{-2x^3 + x^2 } \\ - 4x^2 + 8x - 5 \\ \underline{+ 4x^2 - 2x } \\ 6x - 5 \end{array}$ (New dividened: now repeat the process)

(6) $\begin{array}{r} x^2 - 2x + 3 \\ 2x - 1 \,\big|\, \overline{2x^3 - 5x^2 + 8x - 5} \\ \underline{-2x^3 + x^2 } \\ - 4x^2 + 8x - 5 \\ \underline{+ 4x^2 - 2x } \\ + 6x - 5 \\ \underline{- 6x + 3 } \\ - 2 = \text{Remainder} \end{array}$

Note: What does the nonzero remainder in this last example mean? Does it mean that a mistake was made or that it is wrong to divide $2x - 1$ into $2x^3 - 5x^2 + 8x - 5$? Of course not! An arithmetic analogy should clarify this situation: for example, the problem $38 \div 5$ has a *quotient* of 7 and a *remainder* of 3, which means that

$$\frac{38}{5} = 7 + \frac{3}{5}$$

Thus the problem $(2x^3 - 5x^2 + 8x - 5) \div (2x - 1)$ has a *quotient* of $x^2 - 2x + 3$ and a *remainder* of -2, which means that

$$\frac{2x^3 - 5x^2 + 8x - 5}{2x - 1} = x^2 - 2x + 3 + \frac{-2}{2x - 1}$$

Exercise 24:

Divide the following (zero is excluded from all denominators):

(1) $(x^2 - 6x + 8) \div (x - 4)$

(2) $(6y^2 - y - 2) \div (3y - 2)$

(3) $(a^2 - 2ab + b^2) \div (a - b)$

(4) $(x^2 - 10x + x^3 + 8) \div (x - 2)$

(5) $(6y^3 - 13y^2 + 21y - 10) \div (3y - 2)$

(6) $(4x^3 - 12x^2 - x + 6) \div (2x - 1)$

(7) $(12a^3 + 17a^2 - 3a - 7) \div (2 + 3a)$

(8) $(x^2 - 16) \div (x - 4)$

(9) $(x^3 + 8) \div (2 + x)$

(10) $(y^3 - 1) \div (y - 1)$

(11) $(x^4 - 3x^3 + 2x^2 - 5x + 2) \div (x - 1)$

(12) $(4x^3 - 5x + x^2 + 3x^5 - x^4) \div (2 + x)$

(13) $(2x^3 - 11x^2 + 16x - 6) \div (2x - 3)$

(14) $(17a^3 + 9a^2 + 6a^4 + 12a + 7) \div (2a + 5)$

(15) $(8x^4 - 6x^3 - 11x^2 - 5x + 2) \div (4x - 1)$

(16) $(6a^5 - 8a^4 + 5a^3 - 19a^2 + 2) \div (3a - 1)$

(17) $(x^5 - 1) \div (x - 1)$

(18) $(a^5 + 1) \div (a + 1)$

In the preceding examples of division, all dividends and all divisors were multinomials. But the long division process is required only when the *divisor* is a multinomial, regardless of the number of terms in the dividend. Thus if we had a monomial divided by a multinomial, we would have to use long division to find the quotient. The following example illustrates this situation:

$$\frac{a^3}{a + 2} \qquad (a + 2 \text{ cannot equal } 0)$$

Using long division, we would proceed as follows:

$$
\begin{array}{r}
a^2 - 2a + 4 \\
a + 2 \overline{\smash{)}\, a^3 } \\
\underline{a^3 + 2a^2 } \\
- 2a^2 \\
\underline{- 2a^2 - 4a } \\
4a \\
\underline{4a + 8} \\
- 8 = \text{Remainder}
\end{array}
$$

The following exercise in division contains miscellaneous problems designed to illustrate all of the situations that have been discussed. It would be wise to look over all of the problems first and decide in advance which of the problems will require long division; e.g., note carefully which problems have *multinomials* in the *divisors*.

Exercise 25:

Divide the following polynomials (zero is excluded from all denominators):

(1) $\dfrac{5x^2 - 25x^2y + 10y^2}{5x^2}$

(2) $\dfrac{-3x^3y^2z^4}{2xy^2}$

(3) $(6x^3 + x^2 + 4x - 2) \div (3x + 2)$

(4) $\dfrac{8x^2 - 2xy - 15y^2}{2x - 3y}$

(5) $\dfrac{24a^3b - 4a^2b^2 - 6ab^3}{-2a^3b^3}$

(6) $\dfrac{26c^3d^4}{39c^2d}$

(7) $\dfrac{x^3}{x + 1}$

(8) $\dfrac{x^3}{x^2 + 1}$

(9) $\dfrac{4x^2y^2 - 7x^3 + 5xy}{20x^2y^2}$

(10) $\dfrac{-28a^4b^5c^7}{-49a^3b^2c}$

(11) $\dfrac{16y^4 + 96y^3 + 216y^2 + 72y + 81}{2y + 3}$

(12) $\dfrac{a^2 + 4}{a + 2}$

(13) $\dfrac{a^2 - 4}{a + 2}$

(14) $\dfrac{-3a + 2b - 5c}{-3a}$

(15) $\dfrac{-54x^5yz^7}{12x^2y^5z}$

(16) $\dfrac{a - 2}{6a^4}$

(17) $\dfrac{6a^4}{a - 2}$

(18) $\dfrac{x^3 - 27}{x - 3}$

(19) $\dfrac{a^3 + 1}{a + 1}$

(20) $\dfrac{b^5 - 32}{b - 2}$

Chapter Test 2

  I. Define the following:
1. monomial
2. multinomial
3. polynomial

 II. Fill in the blanks in the following statements about the polynomial

$$2x^4 - 3x^3y^3 + 4x^2y - 5xy^4 + 7$$

1. There are _____ terms in the polynomial.
2. There are _____ variables in the polynomial.
3. The numerical coefficient of the second term is _____.
4. The coefficient of y in the third term is _____.
5. In the fourth term the exponent of x is _____, and the exponent of y is _____.

III. Add the following numbers:
1. $-6a^2, 4ab, 8a^2, ab$
2. $3x^2 - 6xy + 2y^2, 4xy - 2y^2$
3. $a^3 + 3a^2b^2 - 4b^2 + 2, 6a^2b^2 + 4b^2 + 8ab$

IV. Subtract the second number from the first.
1. $4x - 7, 3x + 8$
2. $-2x^3 - 3x + 2, -8x^3 + 2x^2 + 2$
3. $-5a^2 + 2ab + 6b^2, -2a^2 - 4ab + 2b^2$

 V. Multiply the following:

1. $(-2x^3y)^2$
2. $(3ab)(-2ab^2)$
3. $(a + 3)(a + 3)$
4. $(a + 3)(a - 3)$
5. $(-3xy)(-2x^2)$
6. $2x^2(3x - 4)$
7. $(3a^2b)^3$
8. $(5ab^2)^2$
9. $(5a + b^2)^2$
10. $(2x + 3)(5x - 2)$
11. $(x - 1)(x^2 + x + 1)$
12. $(2a - 3)(5a - 1)$
13. $(2x + y)^3$
14. $(2xy)^3$

VI. Divide the following (zero is excluded from all denominators):

1. $\dfrac{-8a^3b}{-2a}$
2. $\dfrac{a^2 - 4}{a}$
3. $\dfrac{a^2 - 4}{a - 2}$
4. $\dfrac{6x^2y}{3xy^2}$
5. $\dfrac{2a + 4b}{2a}$
6. $\dfrac{6x^2y - 8xy^2 + 2xy}{2xy}$
7. $\dfrac{x^3 - 2x^2 + x - 2}{x - 1}$
8. $\dfrac{6a^2 - a - 12}{2a - 3}$
9. $\dfrac{a^3 - 1}{a}$
10. $\dfrac{a^3 - 1}{a - 1}$
11. $\dfrac{a^3}{a - 1}$

VII. Simplify the following:
1. $x - 2(3x - 1)$
2. $(x - 2)(3x - 1)$
3. $4a - (3a - 4) + 2$
4. $2x - [3(x + 1) - 4] + 5x$
5. $5x^2 - 2[4(x - 1) - x(3x + 2)] - (8x + 1)$
6. $2a + 3b - 2\{4[a - (2a + b) + 3b] - a\} + 6b$

3
SPECIAL
PRODUCTS
AND
FACTORING

3.1
The Meaning and Importance of Special Products and Factoring

The two topics, special products and factoring, which will be investigated in this chapter embody two fairly simple ideas. The student should be warned, however, that his success or failure in this and other mathematics courses will depend in part on how well he masters these two skills. The special products and factoring patterns of algebra are actually universal patterns which are encountered throughout the whole of the first year and a half of college mathematics. Therefore, the student who does not learn these processes thoroughly would be well advised to make no further attempt to learn algebra or any related course, such as trigonometry, analytic geometry, or calculus. Any such attempt would be futile — as futile as trying to work long division in arithmetic with no knowledge of the multiplication tables.

The phrase "special products" means just that. There are some products in mathematics which occur so frequently that we set them aside from all others, classify them, and *memorize them*. You have, in fact, already memorized a large number of special products such as the following:

$$(8)(9) = 72$$
$$(3)(8) = 24$$
$$(7)(5) = 35$$

The equality sign in mathematics has many important ramifications, and one of them we need to review carefully right now: *every statement of equality is always true in reverse.*

Since $(7)(5) = 35$

then $35 = (7)(5)$

If the staggering importance of that statement escapes you, try reversing the following true statements and see what miserable untruths can result:

   1.   Every Buick is an automobile.
   2.   Cows are four-legged animals.
   3.   All monks are men.

This characteristic of the eternal reversibility of an equal mark is called the *symmetric property* of the equality relation. It has become quite important because our second topic, factoring, is merely the art of *writing a special product in reverse*. For example, to *factor* the number 21, we write

$$21 = (3)(7)$$

Please note that you could do this *only if* you had previously memorized the product $(3)(7) = 21$. Skill in factoring, therefore, necessarily depends on how skillfully you have previously memorized the special products. It would be an easy matter to list ten or twenty reasons why it is important to memorize the special products, but this one is more important than all the others: any attempt at factoring is useless without them.

Examples:

Special products.

   (1) $(5)(3) = 15$
   (2) $(7)(5) = 35$
   (3) $(a + b)(a + b) = a^2 + 2ab + b^2$

Examples:

Factoring.

   (1) $15 = (5)(3)$
   (2) $35 = (7)(5)$
   (3) $a^2 + 2ab + b^2 = (a + b)(a + b)$

At this point you may be uncomfortably aware that there were a great many special products to be memorized in arithmetic, usually sixty of them. In algebra we shall consider only five.

Two types of multinomials play an important part in the five special products; they are called *binomials and trinomials*.

   1.   *A binomial is a multinomial which contains exactly two terms.*

Examples:

Binomials.

$a + b$ $a^2 - 1$

$x^2 - 3y$ $4 + 2$

$4abc - 2xyz$ $\dfrac{x}{y} - 4$ (y cannot equal 0)

2. *A trinomial is a multinomial which contains exactly three terms.*

Examples:

Trinomials.

$$6x^2 - 5x + 2 \qquad\qquad \frac{x^2}{4} + \frac{2x}{5} + 3$$

$$3 + 4 - 7 \qquad\qquad \frac{a}{b^3} + \frac{c}{d^2} - 4 \qquad (b \text{ and } d \text{ cannot equal } 0)$$

$$7x^2yz + 3x^2y^2 - 4y^2z^2 \qquad\qquad y^7 - 10y^2 + y^4$$

We shall encounter both of these types of multinomials in the first special product which we shall consider, the square of a binomial.

3.2
The Square of a Binomial

Since the process of squaring any number consists of multiplying the number by itself, this special product arises when any binomial is multiplied by itself. To see exactly what does happen, we shall square a few binomials and then take a thoughtful look at the final results.

Examples:

$$\begin{aligned}
(1) \quad (a + b)^2 &= (a + b)(a + b) \\
&= a(a + b) + b(a + b) \\
&= a^2 + ab + ab + b^2 \\
&= a^2 + 2ab + b^2
\end{aligned}$$

$$\begin{aligned}
(2) \quad (x + y)^2 &= (x + y)(x + y) \\
&= x(x + y) + y(x + y) \\
&= x^2 + xy + xy + y^2 \\
&= x^2 + 2xy + y^2
\end{aligned}$$

$$\begin{aligned}
(3) \quad (a - b)^2 &= (a - b)(a - b) \\
&= a(a - b) - b(a - b) \\
&= a^2 - ab - ab + b^2 \\
&= a^2 - 2ab + b^2
\end{aligned}$$

$$\begin{aligned}
(4) \quad (x - y)^2 &= (x - y)(x - y) \\
&= x(x - y) - y(x - y) \\
&= x^2 - xy - xy + y^2 \\
&= x^2 - 2xy + y^2
\end{aligned}$$

Repeating the foregoing examples in a condensed form, we have the following:

$$
\begin{aligned}
&(1) \quad (a+b)^2 = a^2 + 2ab + b^2 \\
&(2) \quad (x+y)^2 = x^2 + 2xy + y^2 \\
&(3) \quad (a-b)^2 = a^2 - 2ab + b^2 \\
&(4) \quad (x-y)^2 = x^2 - 2xy + y^2
\end{aligned}
$$

If we examine the above products carefully, some striking similarities in the results soon become evident.

First, and most evident, *every* result is a *trinomial*. You could make a great beginning in special products by saying over to yourself a few hundred times, "the square of any *binomial* always produces a *trinomial*." While it is true that in the squares of certain binomials the resulting trinomials can sometimes be simplified by combining like terms, the significant thing is that the product always starts out as a *trinomial*.

Having absorbed that all-important fact, let us analyze the repeating pattern which occurs in every result in the boxed examples.

1. The *first term* in every trinomial is the *square* of the *first term* in the binomial.

$$ (\underline{a} - b)^2 = \underline{a^2} - 2ab + b^2 \qquad (a)(a) = a^2 $$

2. The *last term* in every trinomial is the *square* of the *last term* in the binomial.

$$ (a - \underline{b})^2 = a^2 - 2ab + \underline{b^2} \qquad (-b)(-b) = b^2 $$

3. The *remaining term* in every *trinomial* is equal to *twice* the *product* of the *two terms* in the *binomial*.

$$ (a - b)^2 = a^2 - \underline{2ab} + b^2 $$

1st term in binomial $= a$.
2nd term in binomial $= -b$.
Product of a and $-b = -ab$; *twice* that product is $-\underline{2ab}$.

This same pattern is repeated every time any binomial is squared; the pattern is absolutely automatic. Even if we had a nonsensical binomial, the *square* of it would faithfully follow that pattern.

$$ (\triangle + \square)^2 = \triangle^2 + 2\triangle\square + \square^2 $$
$$ (? - !)^2 = ?^2 - 2?! + !^2 $$

The above examples look like pure nonsense, but if the symbols represent numbers

(*any numbers*), the results are perfectly correct. By memorizing this pattern, therefore, we should be able to square any binomial on inspection. Study *carefully* the
following examples of squaring binomials.

Examples:

(1) $(4 + 2)^2 = 4^2 + 2(4)(2) + 2^2$
$= 16 + 16 + 4$
$= \underline{36}$

or $(4 + 2)^2 = 6^2 = \underline{36}$

(2) $(3 - 2)^2 = 3^2 + 2(3)(-2) + (-2)^2$
$= 9 - 12 + 4$
$= \underline{1}$

or $(3 - 2)^2 = 1^2 = \underline{1}$

(3) $(2x + 3)^2 = (2x)^2 + 2(2x)(3) + 3^2$
$= \underline{4x^2 + 12x + 9}$

(4) $(3y - 4)^2 = (3y)^2 + 2(3y)(-4) + (-4)^2$
$= \underline{9y^2 - 24y + 16}$

(5) $(x^2 - 2y^3)^2 = (x^2)^2 + 2(x^2)(-2y^3) + (-2y^3)^2$
$= \underline{x^4 - 4x^2y^3 + 4y^6}$

Exercise 26:

Square the following binomials by inspection:

(1) $(m + n)^2$	(2) $(m - n)^2$	(3) $(2m + n)^2$
(4) $(2m - n)^2$	(5) $(x + 1)^2$	(6) $(x - 1)^2$
(7) $(2x + y)^2$	(8) $(2x - y)^2$	(9) $(3a + 1)^2$
(10) $(4b - 3)^2$	(11) $(2a - 3b)^2$	(12) $(5x - 2y)^2$
(13) $(x^2 - 3)^2$	(14) $(2a^2 + 5)^2$	(15) $(7x - 4y)^2$
(16) $(y^4 + 1)^2$	(17) $(3b^4 - 2)^2$	(18) $(4 + 5)^2$
(19) $(6 - 3)^2$	(20) $(3ax - 4)^2$	(21) $(2ab + 7)^2$
(22) $(x^2 + 5)^2$	(23) $(x^2 - 5)^2$	(24) $(x^3 - 3y^4)^2$
(25) $(2a^4 - 7b^5)^2$	(26) $(5x^5 + 2y^3)^2$	

3.3
Recognizing and Factoring a Trinomial Perfect Square

A number is called a perfect square only if it is the product of two *identical* factors.

Examples:

Perfect squares.

$$49 = (7)(7) \text{ or } 7^2$$
$$64 = (8)(8) \text{ or } 8^2$$
$$a^2 + 2ab + b^2 = (a + b)(a + b) \text{ or } (a + b)^2$$
$$x^2 + 2xy + y^2 = (x + y)(x + y) \text{ or } (x + y)^2$$

A trinomial perfect square then is simply a trinomial which is the product of two identical factors. As we observed repeatedly in the preceding section, this is the kind of number produced every time a binomial is squared. We could conclude, therefore, with perfectly sound common sense, that *every trinomial perfect square must be the square of some binomial.*

$$\text{Since} \quad (a + b)^2 = a^2 + 2ab + b^2$$
$$\text{then} \quad a^2 + 2ab + b^2 = (a + b)^2$$

This process of reversing a special product, as we noted earlier, is called *factoring.* Not all trinomials are perfect squares any more than all numbers are perfect squares. Since trinomial perfect squares must be the squares of binomials we must think backward as to how such a number is produced in the first place.

$$\boxed{\begin{array}{l} a^2 + 2ab + b^2 = (a + b)^2 \\ a^2 - 2ab + b^2 = (a - b)^2 \end{array}}$$

By taking a good look at the above examples we can figure out what must be true of every trinomial that *is* a perfect square.

In both of the above examples, the *first* and *last* terms of each trinomial are obtained by squaring the first and last terms of each binomial; therefore, *in any trinomial perfect square two of the three terms must themselves be perfect squares.*

The *remaining term* in each of the trinomial perfect squares ($2ab$ and $-2ab$) must always equal (except for the sign) twice the product of the square roots of the two perfect squares. The following numbers are both trinomial perfect squares:

$$a^2 + 2ab + b^2$$
$$a^2 - 2ab + b^2$$

1. In both numbers, a^2 and b^2 are perfect squares.
2. $\sqrt{a^2} = a$ and $\sqrt{b^2} = b$*

---

*To simplify this discussion the literal numbers, a and b, are arbitrarily assumed to be positive so that $\sqrt{a^2} = a$ and $\sqrt{b^2} = b$. It can be shown by a slightly more complex argument that the factoring pattern would be the same regardless of the signs of a and b. Meanwhile, the assumption of positive numbers in no way affects the validity of the factoring pattern.

3. ab = the product of the square roots of the two perfect squares.
4. $2ab$ = *twice* the product of the square roots of the two perfect squares.

Therefore, in order for any trinomial to be a perfect square, two things must always be true.

In every trinomial perfect square:

1. **Two of the three terms must be perfect squares.**
2. **The absolute value of the remaining term must equal the absolute value of twice the product of the square roots of the two perfect squares.**

The above test will tell you unerringly whether or not any trinomial is a perfect square. Remember that if a trinomial *is* a perfect square it must be the square of a binomial.

Study the following examples and answer the questions.

Examples:

(1) $x^2 - 6x + 9 = (x - 3)^2$ or $(x - 3)(x - 3)$

(2) $a^2 - 10a + 25 = (a - 5)^2$ or $(a - 5)(a - 5)$

(3) $16a^2 + 8a + 1 = (4a + 1)^2$ or $(4a + 1)(4a + 1)$

(4) $x^2 - 4x + 4 = (x - 2)^2$ or $(x - 2)(x - 2)$

(5) $x^2 - 4x - 4$ is *not* a perfect square. Why?

(6) $x^2 + 6x + 9 = (x + 3)^2$ or $(x + 3)(x + 3)$

(7) $x^2 + 3x + 9$ is *not* a perfect square. Why?

(8) $x^2 + 36 + 12x$ $(36 + 12x = 12x + 36$ by the commutative law of addition)
$$x^2 + 36 + 12x = x^2 + 12x + 36 = (x + 6)^2 \text{ or } (x + 6)(x + 6)$$

(9) $a^2 + 2a + 4$ is *not* a perfect square. Why?

(10) $4x^4 - 12x^2 + 9 = (2x^2 - 3)^2$ or $(2x^2 - 3)(2x^2 - 3)$

(11) $36a^6 + 60a^3 + 25 = (6a^3 + 5)^2$ or $(6a^3 + 5)(6a^3 + 5)$

Exercise 27:

Determine which of the following are trinomial perfect squares, and factor those which are perfect squares:

(1) $x^2 - 8x + 16$ (2) $a^2 + 5a + 25$

(3) $x^2 + x + \dfrac{1}{4}$ (4) $y^2 + 36 + 12y$

(5) $x^2 - 2x + 1$ (6) $m^2 + 9 + 3m$

(7) $4x^2 + 20x + 25$

(8) $9a^2 - 24a + 16$

(9) $16y^2 + 1 + 4y$

(10) $4a^2 - 12ab + 9b^2$

(11) $16x^2 + 40x + 25$

(12) $9y^2 + 4 - 12y$

(13) $4a^2 + 16ab + 16b^2$

(14) $25x^2 - 40x - 16$

(15) $25x^2 - 20xy + 4y^2$

(16) $36a^2 + 12a + 1$

(17) $49y^2 - 28y + 4$

(18) $m^4 + 64 + 16m^2$

(19) $y^4 - 2y^2 + 1$

(20) $4a^6 + 12a^3 + 9$

(21) $a^4 - 4a^2 + 16$

(22) $25x^4 - 10x^2 + 1$

(23) $9y^4 - 12x^2y^2 + 4x^4$

(24) $49x^2 + 25y^2 + 70xy$

(25) $9a^2 + 30ab + 25b^2$

3.4
The Product of Two Binomials Having the Same Literal Numbers

This special product, that of two binomials having the same literal numbers, produces a special sort of result. The phrase "the *same literal number*" means the same variable with the same exponent.

Compare the following products:

1. $(2x + 1)(3y - 2)$
2. $(2x + 1)(3x^2 - 2)$
3. $(2x + 1)(3x - 2)$

The above examples are all products of two binomials. But in Examples 1 and 2 the literal numbers in the two binomials are *not* the same. In Example 3 the same literal number appears in both binomials. Now let us compare the results of the multiplication of these same three examples.

1. $(2x + 1)(3y - 2) = \underline{6xy - 4x + 3y - 2}$
2. $(2x + 1)(3x^2 - 2) = \underline{6x^3 + 3x^2 - 4x - 2}$
3. $(2x + 1)(3x - 2) = 6x^2 - 4x + 3x - 2 = \underline{6x^2 - x - 2}$

In Examples 1 and 2 the final products contain *four* terms, but in Example 3 two of the terms could be combined and the final result is a *trinomial*. With one exception, the product of two binomials having the *same* literal numbers will always be a trinomial. The exception occurs when two such binomials represent the sum and the difference of the same two numbers, a special case that will be excluded here and discussed in Section 3.9.

Examples:

(1) $(3x + 1)(2x - 5) = 6x^2 - 15x + 2x - 5$
$$= 6x^2 - 13x - 5$$

(2) $(2a + b)(5a + 2b) = 10a^2 + 4ab + 5ab + 2b^2$
$$= 10a^2 + 9ab + 2b^2$$

(3) $(y^2 - 3)(y^2 - 4) = y^4 - 4y^2 - 3y^2 + 12$
$$= y^4 - 7y^2 + 12$$

(4) $(x^3 + 2)(x^3 - 3) = x^6 - 3x^3 + 2x^3 - 6$
$$= x^6 - x^3 - 6$$

Not only are these products trinomials, but there is also a very special relationship between the exponents of the literal symbols in each trinomial: in every result, the *greatest* exponent of any literal symbol is always *twice as large* as the next greatest exponent of that same literal symbol.

By knowing in advance that the answer to such a product is a trinomial with a special relationship between any two exponents of the same literal symbol, we can compute such products automatically by using the following pattern.

Example I:

$(x + 2)(x + 3) = ?$

(1) To get the *first term* in the trinomial, multiply together the *first terms* in each binomial.

$$(\underline{x} + 2)(\underline{x} + 3) \qquad (x)(x) = x^2 \qquad \text{first term}$$

(2) To get the *last term* in the trinomial, multiply together the *last terms* in each binomial.

$$(x + \underline{2})(x + \underline{3}) \qquad (2)(3) = 6 \qquad \text{last term}$$

(3) To get the *middle term* in the trinomial, in the binomials multiply together the two terms farthest apart and the two terms closest together; then add the results.

$$(x + 2)(x + 3) \qquad \begin{array}{l} (x)(3) = 3x \\ (2)(x) = 2x \end{array}$$
$$2x + 3x = 5x \qquad \text{middle term}$$

(4)   Therefore

$$(x + 2)(x + 3) = \underbrace{x^2}_{\text{first term}} + \underbrace{5x}_{\text{middle term}} + \underbrace{6}_{\text{last term}}$$

Example II:

$$(\underline{a} + 3)(\underline{a} - 5) = a^2 + (-5a + 3a) + (3)(-5)$$
$$= a^2 - 2a - 15$$

Example III:

$$(\underline{2x} - 3)(\underline{x} - 1) = 2x^2 - 5x + 3$$

In symbolic language:

$$(\underline{ax} + b)(\underline{cx} + d) = acx^2 + adx + bcx + bd$$
$$= acx^2 + (ad + bc)x + bd$$

Exercise 28:

Multiply the following binomials by inspection:

(1) $(x + 1)(x + 3)$ (2) $(a + 2)(a + 3)$

(3) $(y + 5)(y + 2)$ (4) $(b + 4)(b + 5)$

(5) $(x + 2)(x + 3)$ (6) $(x + 2)(x - 3)$

(7) $(x - 2)(x + 3)$ (8) $(x - 2)(x - 3)$

(9) $(2x + 1)(x + 4)$ (10) $(2x + 1)(x - 4)$

(11) $(2x - 1)(x + 4)$ (12) $(2x - 1)(x - 4)$

(13) $(3a - 4)(2a + 1)$ (14) $(5y - 6)(y + 2)$

(15) $(2b - 1)(b - 3)$ (16) $(3x - 7)(2x + 5)$

(17) $(3y + 8)(y + 4)$ (18) $(x^2 + 4)(x^2 + 2)$

(19) $(x^2 - 4)(x^2 - 2)$ (20) $(2a^2 + 3)(3a^2 - 1)$

(21) $(y^2 - 6)(4y^2 - 5)$ (22) $(x^3 + 5)(x^3 + 1)$

(23) $(2x^3 - 3)(5x^3 - 1)$ (24) $(y^4 + 2)(y^4 - 8)$

(25) $(3x - 7y)(2x - 5y)$ (26) $(4a + 5b)(5a - 2b)$

(27) $(3x - 7y)(2x - 3y)$ (28) $(8y + 3)(9y + 1)$
(29) $(5x + 4)(3x + 7)$ (30) $(7a - 5b)(2a - 3b)$

3.5
Factoring Trinomials of the Type $ax^2 + bx + c$

Trinomials of the type $ax^2 + bx + c$ are those in which a, b, and c are stand-ins for fixed numbers (called *constants*) and x is a literal symbol called an *unknown* or a *variable*. The significant thing about trinomials of this type is that the greatest exponent of the variable is *always twice as large* as the next greatest exponent of the same variable. And since addition is commutative, we shall agree to arrange all terms in descending order of the exponents.

Examples:

(1) $x^2 + 6 + 5x = x^2 + 5x + 6$
(2) $3b^2 - 5 + 2b = 3b^2 + 2b - 5$
(3) $4a^3 + 5a^6 - 2 = 5a^6 + 4a^3 - 2$

If you will look back over the results of Exercise 28, you will find that *every product* was a trinomial of this particular type. To get this kind of product we multiplied together, in every case, two binomials having the same literal numbers. Therefore we could, with commendable good sense, reason backwards and predict that *if* such a trinomial has factors, *the factors will have to be two binomials each having the same literal numbers.*

$$\boxed{\begin{array}{l} \text{Since } (x + 2)(x + 3) = x^2 + 5x + 6 \\ \text{then } x^2 + 5x + 6 = (x + 2)(x + 3) \end{array}}$$

When we attempt to factor a trinomial of the type $ax^2 + bx + c$, we have no guarantee that such factors exist. But we can start out in the *certain knowledge* that if such factors exist *those factors must be two binomials*, *each having the same literal numbers*. With this knowledge, we proceed on a simple trial-and-error basis.

Examples:

(1) $x^2 + 5x + 6 = (? + ?)(? + ?)$
Factors must be two binomials having the *same* literal numbers.
$$x^2 + 5x + 6 = (x + ?)(x + ?)$$
x is the same literal number in both and $(x)(x) = x^2$, which fits the first term of the trinomial.

$$x^2 + 5x + 6 = (x + ?)(x + ?)$$

The missing numbers above must have a product equal to 6: $(?)(?) = 6$. These *could* be:

$$(6)(1) = 6 \text{ or } (-6)(-1) = 6$$
$$(3)(2) = 6 \text{ or } (-3)(-2) = 6$$
$$x^2 + 5x + 6 = (x + 6)(x + 1)?$$

No: $(6)(1) = 6$, but $(x + 6)(x + 1) = x^2 + \underline{7x} + 6$.

$$x^2 + 5x + 6 = (x + 3)(x + 2)?$$

Yes: $(3)(2) = 6$, and $(x + 3)(x + 2) = \underline{x^2 + 5x + 6}$.

Note that the *negative* factors of 6 would not fit because the middle term would be negative.

$$(x - 6)(x - 1) = x^2 \underline{- 7x} + 6$$
$$(x - 3)(x - 2) = x^2 \underline{- 5x} + 6$$

(2) $\qquad a^2 + 6a + 8 = (a + ?)(a + ?) \qquad$ (since $a \cdot a = a^2$)

The missing numbers must have a product equal to 8:

$$(8)(1) = 8 \text{ or } (-8)(-1) = 8$$
$$(4)(2) = 8 \text{ or } (-4)(-2) = 8$$

$$a^2 + 6a + 8 = (a + 8)(a + 1)?$$

No: $(8)(1) = 8$, but $(a + 8)(a + 1) = a^2 + \underline{9a} + 8$

$$a^2 + 6a + 8 = (a + 4)(a + 2)?$$

Yes: $(4)(2) = 8$ and $(a + 4)(a + 2) = \underline{a^2 + 6a + 8}$

Repeating the two examples above we have the following:

$$\boxed{\begin{aligned} x^2 + 5x + 6 &= (x + 3)(x + 2) \\ a^2 + 6a + 8 &= (a + 4)(a + 2) \end{aligned}}$$

Here we could exhibit some native intelligence and note carefully that when *all* of the terms of the *trinomial* are *positive*, then *all* of the terms of the two *binomial factors* are *also positive*.

Exercise 29:

Find the factors.

(1) $x^2 + 8x + 15 = (\quad + \quad)(\quad + \quad)$

(2) $x^2 + 3x + 2 = (\quad + \quad)(\quad + \quad)$

(3) $a^2 + 7a + 12 =$

(4) $y^2 + 5y + 4 =$

(5) $b^2 + 9b + 18 =$

Now consider the following examples:

$$x^2 - 5x + 6 = (x - 3)(x - 2)$$
$$a^2 - 6a + 8 = (a - 4)(a - 2)$$

Here we should observe that the *first* and *last* terms of each trinomial are *positive*, but the *middle term* of each one is *negative*. Now take a good look at the binomial factors. In this situation, the *last term* in each binomial factor will *always* be *negative*.

Exercise 30:

Find the factors.

(1) $x^2 - 7x + 12 = ($ $-$ $)($ $-$ $)$
(2) $a^2 - 9a + 20 = ($ $-$ $)($ $-$ $)$
(3) $b^2 - 7b + 12 =$
(4) $y^2 - 8y + 15 =$
(5) $x^2 - 11x + 30 =$

Finally, we might have this situation:

$$x^2 - x - 6 = (x - 3)(x + 2)$$
$$x^2 + x - 6 = (x + 3)(x - 2)$$

Here the *last term* in each trinomial is *negative*. This always means that the last terms in each of the binomial factors must have *opposite* signs, so that the product of the two numbers will be negative. The sign of the *middle term* in the trinomial tells us which one should be positive and which one negative.

Exercise 31:

Find the factors.

(1) $a^2 - a - 12 = ($ $+$ $)($ $-$ $)$
(2) $b^2 - 2b - 8 = ($ $+$ $)($ $-$ $)$
(3) $x^2 + 2x - 15 = ($ $+$ $)($ $-$ $)$
(4) $a^2 + 3a - 10 =$
(5) $y^2 - 3y - 28 =$

Exercise 32:

Factor the following trinomials:

(1) $x^2 + 12x + 32$ (2) $a^2 + 4a + 3$
(3) $b^2 + 9b + 14$ (4) $x^2 - 11x + 24$

(5) $a^2 - 7a + 10$ (6) $b^2 - 7b + 6$

(7) $x^2 + 9x + 20$ (8) $x^2 - 9x + 20$

(9) $x^2 - x - 20$ (10) $x^2 + x - 20$

(11) $a^2 + 6a + 9$ (12) $x^2 - 14x + 49$

(13) $a^2 - 7a - 18$ (14) $b^2 - b - 56$

(15) $c^2 + 8c - 20$ (16) $x^2 - 2xy - 15y^2$

(17) $a^2 - 10ab + 16b^2$ (18) $x^2 + 13xy + 30y^2$

(19) $a^2 - 4ab - 21b^2$ (20) $a^4 - 5a^2 - 6$

(21) $x^4 + x^2 - 30$ (22) $a^4 + a^2 - 6$

(23) $x^6 + 6x^3 + 8$ (24) $y^6 - 12y^3 + 35$

(25) $x^4 - 6x^2y^2 - 16y^4$ (26) $x^4 - 10x^2y + 21y^2$

(27) $a^2 + 11ab + 28b^2$ (28) $x^6 - 3x^3y - 28y^2$

(29) $y^6 + 8y^3 + 12$ (30) $a^4 - 3a^2b^2 - 10b^4$

In a polynomial, the numerical coefficient of the term having the greatest degree is called the "leading coefficient." When we arrange polynomials in descending order of the terms by degrees, the leading coefficient is simply the first one. In $3x^2 - 4x + 1$ the *leading coefficient* is 3. In $x^5 - 4x^3 + x^2 - 5x + 2$ the *leading coefficient* is 1.

In every trinomial which we have attempted to factor, the leading coefficient has always been 1. When the leading coefficient is not 1, the problem is more complicated and will often require more trials (and errors) before the correct factors are found. Like every other skill that human beings develop, skill in factoring comes from *practice*. As you factor more and more trinomials, you will find it increasingly easier to find the right combinations. Unfortunately, there is no other way to develop skill in factoring. After you have memorized the special product involved in a particular type of factoring, the "thinking backwards" from the product to the factors is fairly simple, but the amount of practice required is prodigious. In fact, there are no special intellectual gifts required to learn all of basic algebra; only special *effort* is required.

Note carefully and *repeatedly* the following examples of factoring trinomials by trial and error.

Examples:

(1) $2x^2 + 5x + 3 = (? + ?)(? + ?)$

     $2x^2 + 5x + 3 = (2x + ?)(x + ?)$

     $2x^2 + 5x + 3 = (2x + 1)(x + 3)$?

       No, because $(2x + 1)(x + 3) = 2x^2 + \underline{7x} + 3$

     $2x^2 + 5x + 3 = (2x + 3)(x + 1)$?

       Yes, because $(2x + 3)(x + 1) = \underline{2x^2 + 5x + 3}$

(2)　$4x^2 - 22x + 10 = (? - ?)(? - ?)$
　　　$4x^2 - 22x + 10 = (2x - ?)(2x - ?)$
　　　$4x^2 - 22x + 10 = (2x - 5)(2x - 2)?$
　　　　No, because $(2x - 5)(2x - 2) = 4x^2 - \underline{14x} + 10$
　　　$4x^2 - 22x + 10 = (4x - 5)(x - 2)?$
　　　　No, because $(4x - 5)(x - 2) = 4x^2 - \underline{13x} + 10$
　　　$4x^2 - 22x + 10 = (4x - 2)(x - 5)?$
　　　　Yes, because $(4x - 2)(x - 5) = \underline{4x^2 - 22x + 10}$

(3)　$5x^2 + 11x - 12 = (? + ?)(? - ?)$
　　　$5x^2 + 11x - 12 = (5x + ?)(x - ?)$
　　　$5x^2 + 11x - 12 = (5x + 6)(x - 2)?$
　　　　No, because $(5x + 6)(x - 2) = 5x^2 - \underline{4x} - 12$
　　　$5x^2 + 11x - 12 = (5x + 2)(x - 6)?$
　　　　No, because $(5x + 2)(x - 6) = 5x^2 - \underline{28x} - 12$
　　　$5x^2 + 11x - 12 = (5x + 4)(x - 3)?$
　　　　No, because $(5x + 4)(x - 3) = 5x^2 - \underline{11x} - 12$
　　　$5x^2 + 11x - 12 = (5x - 4)(x + 3)?$
　　　　Yes, because $(5x - 4)(x + 3) = \underline{5x^2 + 11x - 12}$

Exercise 33:

Factor the following trinomials:

(1)	$2x^2 + 7x + 3$	(2)	$6a^2 + 7a + 2$
(3)	$5y^2 - 14y + 8$	(4)	$4x^2 - 23x + 15$
(5)	$6x^2 + 11x - 10$	(6)	$10a^2 + 7a - 12$
(7)	$6x^2 - 7x - 24$	(8)	$5a^2 - 13a - 28$
(9)	$18y^2 - 21y + 5$	(10)	$6y^2 - 13y + 6$
(11)	$8x^2 + 34x + 21$	(12)	$5a^2 + 7ab - 6b^2$
(13)	$6x^2 - 7xy - 5y^2$	(14)	$6a^2 + 19ab + 15b^2$
(15)	$21x^2 - 31xy + 4y^2$	(16)	$6x^4 - 13x^2 - 5$
(17)	$8a^4 + 42a^2 + 27$	(18)	$6x^4 - 19x^2 + 15$
(19)	$5a^6 - 27a^3 - 18$	(20)	$14x^6 + 13x^3 + 3$
(21)	$6a^2 + 5a - 4$	(22)	$16b^2 + 14b + 3$
(23)	$3x^2 - 11xy + 6y^2$	(24)	$8a^2 + 14ab - 15b^2$
(25)	$12y^6 - 11y^3 + 2$	(26)	$12a^2 + 4a - 5$
(27)	$12x^2 - 7x - 12$	(28)	$5b^6 - 7b^3 - 6$
(29)	$35x^4 - 32x^2 + 5$	(30)	$6a^8 + 25a^4 + 24$

3.6
Prime Numbers; Set and Subset

A positive integer which is different from the integer one is called a *prime number* if it has no positive integral factors other than itself and one.

Examples of prime numbers:

$$2, 7, 13, 19, 41, \ldots$$

Mathematicians have busied themselves for thousands of years studying the succession of prime numbers in the real number system. If the occurrence of primes follows any sort of formula, no one has yet been able to find out exactly what that formula is.

First let us define a prime number and then look at a few of them. Remember that the set of natural numbers, N, is the infinite set of positive integers which begins with the number one:

$$N = \{1, 2, 3, 4, 5, \ldots\}$$

A prime number is any natural number different from one which has no positive integral factors other than itself and one.

Thus if p is a prime number, and S is the set of positive integral factors of p, then $S = \{1, p\}$.

The succession of prime numbers in the real number system begins with the numbers

$$2, 3, 5, 7, 11, 13, 17, 19, 23, 29, 31, 37, \ldots$$

The three dots of course indicate that this set of numbers goes on forever. Euclid proved that there are an infinite number of primes.

Any natural number, such as 10, which is greater than the number one and not a prime, can always be written as the product of two or more primes and is called a *composite number*.

$$10 = (2)(5) \longrightarrow \text{product of two primes}$$
$$30 = (2)(3)(5) \longrightarrow \text{product of three primes}$$

The fundamental theorem of arithmetic states that *every composite number can be expressed as a product of prime factors in one and only one way*, apart from the arrangement of the factors.

Suppose that Tom, Dick, and Harry were asked to find the *prime* factors of 24. According to the above theorem, no matter *how* each one starts out, they should all wind up with *exactly* the same factors, although the factors need not be arranged in the same order.

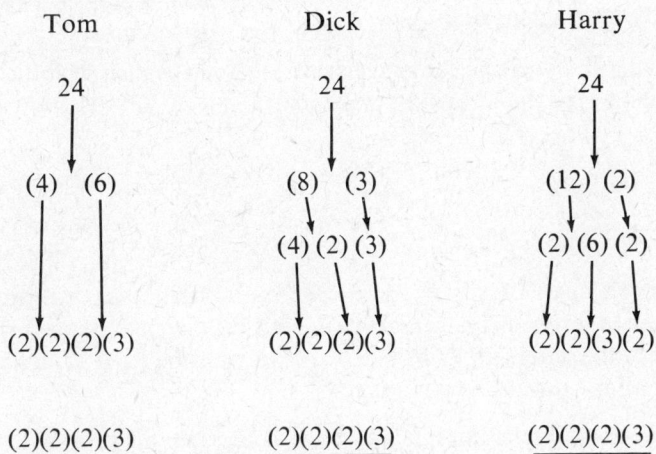

Tom	Dick	Harry
24	24	24
(4) (6)	(8) (3)	(12) (2)
	(4) (2) (3)	(2) (6) (2)
(2)(2)(2)(3)	(2)(2)(2)(3)	(2)(2)(3)(2)
(2)(2)(2)(3)	(2)(2)(2)(3)	(2)(2)(2)(3)

We shall henceforth use the letter P exclusively to designate the set of primes. Figure 3.1 shows the addition of the set of primes to our diagram in Section 1.3.

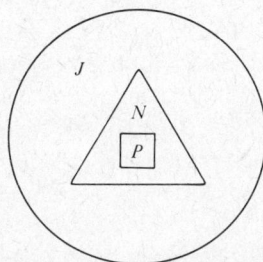

Figure 3.1

Set and Subset In Figure 3.1, observe that the set P is contained within the set of natural numbers, N; that is, every element of P is also an element of N. We describe this situation by saying that P is a subset of N. Similarly, N is a subset of the set of integers, J.

In general,

The set **A** is a *subset* of the set **B** if every element of **A** is also an element of **B**.

When A is a subset of B, we write

$$A \subseteq B$$

The symbol $\subseteq$ means "is a subset of" or "is contained in."

By the above definition, any set A is a subset of itself since every element of A is certainly an element of A.

Hence, in symbolic language we can describe the relationships of the three sets N, J, and P in the following manner:

$$P \subseteq N$$
$$P \subseteq J$$
$$N \subseteq J$$

All the elements of sets P, J, and N are real numbers, and each of the three sets is a subset of the set of real numbers, a set that we shall henceforth call R. Throughout this text the letters P, J, N, and R will always be used to designate these sets. These designations should be memorized.

Note carefully: While the sets P, N, and J are all subsets of R, these three sets by no means encompass *all* of the set R. Just as we might say that the three sets of Fords, Chevrolets, and Buicks are subsets of the set of all automobiles, there are still many automobiles that are neither Fords nor Chevrolets nor Buicks. Later we shall encounter other elements of R that do not belong to P, N, or J.

As observed in Section 1.2, we use the word *set* in mathematics exactly as we do when we speak of a set of dishes, a set of golf clubs, or a set of chessmen. A set is simply a collection of objects, and the objects could be dishes, chessmen, numbers, points, golf clubs, lines, fish, people, or any other objects on earth we find convenient to classify in collections. The set concept is so basic to our way of thinking that we use it unconsciously in such phrases as "a herd of cattle," "a flock of birds," "a school of fish," "the senior class," "the ship's complement," or "the church congregation." The set concept is also basic in mathematics, and it provides us with a means of organizing mathematical relationships in convenient compartments.

To study the behavior of numbers intelligently, it inevitably becomes necessary to classify the numbers in sets. Therefore we have devised a sort of language of sets so that we can all convey the same meaning in the simplest possible way.

We distinguish different sets by giving them different names, usually the capital letters of the alphabet. In a similar manner we refer to the elements of sets symbolically by using the small letters of the alphabet. Finally, we use the symbol $\in$ to mean "belongs to" or "is an element of" or "is a member of."

Examples:

(1) If A is a set and a is an element of the set A, we write

$$a \in A$$

If a is *not* an element of the set A, we write

$$a \notin A$$

(2) N is the set of natural numbers; thus

$$2 \in N$$

and

$$\frac{1}{3} \notin N$$

(3)   If F is the set of positive integral divisors of 12, then

$$
\begin{array}{ll}
1 \in F & \quad 4 \in F \\
2 \in F & \quad 6 \in F \\
3 \in F & \quad 12 \in F
\end{array}
$$

but

$$5 \notin F \quad \text{or} \quad 7 \notin F$$

We say that a set B is *well-defined* if, given any element b, it is possible to decide that one and only one of the following is true:

$$b \in B$$

or

$$b \notin B$$

Two ways in which we designate sets are the following:

1.   *A simple statement.*

$$N = \text{the set of natural numbers}$$

2.   *A listing of the elements of the set, always enclosed in braces.*

$$N = \{1, 2, 3, 4, 5, \ldots\}$$

The three dots show that the set N is an infinite set. Note that when we speak of listing the elements of the set, we are actually referring to listing the names of the elements of the set.

Examples of sets:

(1) $T =$ the set of United States presidents serving between 1932 and 1944
or $T = \{$Franklin Delano Roosevelt$\}$

(2) $C =$ the set of continents on the planet Earth
or $C = \{$North America, South America, Australia, Africa, Europe, Asia, Antarctica$\}$

(3) A = the set of positive integral divisors of 12
     or $A = \{1, 2, 3, 4, 6, 12\}$

(4) J = the set of integers
     or $J = \{\ldots -4, -3, -2, -1, 0, 1, 2, 3, 4, \ldots\}$

The Null Set The preceding examples illustrate that a set may contain one element or many elements or infinitely many elements. Indeed, a set may be "empty" and contain no elements at all, such as the set of five-legged cows or the set of integers between 5 and 6. An empty set is called a null set, specifically *the* null set, because all sets having no elements are the same set. The symbol for the null set is $\varnothing$, and it is written without braces.

Examples of the null set:

(1) T = the set of 800-pound gnats = $\varnothing$
(2) M = the set of primes divisible by 4 = $\varnothing$

The null set is a subset of every set. Since the null set has no elements at all, it certainly has no elements that do not belong to every other set. If we wished to list all possible different subsets of a given set, then we would have to include the null set as one of them.

Equal Sets Note that in the examples of the null set the equal sign was used to show that the given sets were equal to the null set. As another example of equal sets, consider the set of positive integral divisors of 4. This set contains three elements and can be identified by listing those elements. Hence, we would write

The set of positive integral divisors of 4 = $\{1, 2, 4\}$

Two sets A and B are equal if and only if they contain exactly the same elements. This means that every element of A is an element of $B (A \subseteq B)$ and every element of B is an element of A $(B \subseteq A)$. The elements may be repeated in one and not the other, or they may be written in a different order, but they must be the *same* elements.

The Symbols $\in$ and $\subseteq$ In set notation it is important to distinguish between the symbols $\in$ and $\subseteq$. The symbol $\in$ always identifies an *element* (or elements) of a set, but it is never used to denote a *subset* of a given set. On the other hand, the symbol $\subseteq$ always identifies a *subset* of a given set but never an element of a set. In the same manner that we use the symbol $\notin$ to mean "is *not* an element of" a set, we use the symbol $\nsubseteq$ to mean "is *not* a subset of" a given set.

One special significance of the braces used to enclose the elements of sets becomes evident when one must distinguish between sets and elements of sets. The symbol 6 represents "the number six"; but the symbol $\{6\}$ represents "the *set* whose only element is the number 6."

If $A = \{-1, 0, 4\}$ we would write

$$4 \in A \qquad \text{4 is an element of } A$$
$$\{4\} \subseteq A \qquad \{4\} \text{ is a subset of } A$$

In referring to elements or to subsets of A, we *never* write $4 \subseteq A$ or $\{4\} \in A$. In like manner:

$$-1 \in A \qquad \{-1\} \subseteq A$$
$$-1, 4 \in A \qquad \{-1, 4\} \subseteq A$$
$$2 \notin A \qquad \{-1, 2\} \nsubseteq A$$

Examples:

$$E = \text{the set of positive even integers}$$

(1) $6 \in E$	(2) $\{4\} \subseteq E$	(3) $\{2, 4, 6\} \subseteq E$
(4) $-12 \notin E$	(5) $8, 10 \in E$	(6) $\{8, 10\} \subseteq E$
(7) $\{6, 11\} \nsubseteq E$	(8) $7 \notin E$	

Exercise 34:

(1) Express each number as the product of prime factors. If the number is already prime, state the fact.

   Example: $42 = (7)(6) = (7)(3)(2)$
                $41 \quad$ prime

   (a) 72   (b) 28   (c) 61   (d) 100
   (e) 52   (f) 60   (g) 39   (h) 53

(2) Define each of the following sets by listing the elements:

   (a) $D = $ the set of positive integral divisors of 18
   (b) $E = $ the set of positive even integers
   (c) $G = $ the set of integers between -3 and 5
   (d) $K = $ the set of primes between 40 and 60

(3) Using the symbol $\in$, assign each of the following numbers to all of the sets P, N, J, R to which the number belongs:

   Examples: $5 \in P, N, J, R$
                $-4 \in J, R$

   (a) 36   (b) 13   (c) -20 (d) 0
   (e) 47   (f) 1   (g) 79   (h) -65

(4) If $E = $ the set of positive even integers, put the correct symbol $\in, \notin, \subseteq,$ or $\nsubseteq$ in the blanks provided in the following statements:

(a) 13 ____ E (b) 26 ____ E (c) 10, 12, 14 ____ E
(d) {4} ____ E (e) 6, 10 ____ E (f) {4, 8, 16} ____ E
(g) 17 ____ E (h) {17} ____ E (i) -32 ____ E
(j) 18,432 ____ E (k) {2, 3, 4} ____ E (l) 0 ____ E

(5) Answer the following questions yes or no:

(a) Is $P \subseteq N$?   (b) Is $P \subseteq R$?   (c) If $J \subseteq J$?
(d) Is $J \subseteq N$?   (e) Is $P \subseteq J$?   (f) Is $J \subseteq R$?

3.7
Degree of a Polynomial; Prime Polynomials

In the factoring patterns of algebra we are primarily concerned with the factoring of *polynomials,* and for our purposes it will be convenient to employ the phrase "prime polynomial." In order to clarify our use of this terminology, we must first define what is meant by the *degree* of a polynomial. In a polynomial containing *only one literal symbol* the largest exponent of that symbol is called the *degree* of the polynomial. For example,

$$3x^4 - 5x^3 + 6x^2 - 3x + 2 \text{ is a polynomial of degree 4}$$

When a polynomial contains *more than one literal symbol,* the degree of the polynomial is the greatest value of the *sum* of the exponents of the literal symbols in any individual term of the polynomial. For example,

$$2x^3y^2 - y^4 + 6x^4 + 3x^2y - 5y^2 + 7xy \text{ is a polynomial of degree } \mathbf{5}$$

Polynomials are conventionally arranged in descending (less frequently in ascending) order of the degrees of their terms. Let us consider two polynomials and list the degree of each term.

Example 1:

$$5x^3 - 2x^2 + 3x + 5$$

Term	Degree of the term
$5x^3$	3
$-2x^2$	2
$3x$	1
5	0

1. Since the *greatest* degree of any term is 3, the degree of the polynomial is 3.
2. Note that the degree of the last term, 5, is defined to be *zero.* This is the degree of any nonzero term which consists solely of a specific number. For ex-

ample, an expression containing the single term 7 may be considered to be a polynomial of degree zero. By definition, 7 and $7x^0$ are the same number,* but the reason for this definition must be deferred to a study of the laws of exponents. No degree is defined for an expression consisting of the single term 0.

Example 2:

$$2a^3b^3 - 4a^5 + 5a^3b + 3ab - 6a + 4$$

Term	Degree of the term
$2a^3b^3$	6 (the *sum* of the exponents of the literal symbols)
$-4a^5$	5
$5a^3b$	4
$3ab$	2
$-6a$	1
4	0

Since the greatest degree of any term is 6, the degree of this polynomial is 6.

Exercise 35:

State the degree of each of the following polynomials:

(1) $5x^4 - 2x^3 + x^2 - 3x + 10$
(2) $3x^2y^4 - 2x^3y^2 + 4x^2y^2 - 6xy$
(3) $2b^7 - 5b^6 + 3b^4 - 2b^2 + 3$
(4) $6y^3$
(5) $x - 3$
(6) $xy - 3$
(7) $2x^2 - 3xy + 4y^2 - 7x + 4y + 9$
(8) $2x - 4y + 1$
(9) 8
(10) $x^4 - 2x^2y^2 + y^4 - 3x^2 + 4y - 10$

Prime Polynomials   In the section on factoring polynomials of the type $ax^2 + bx + c$ (Section 3.5), we found that there is no guarantee that such trinomials can always be factored. In the factoring process, these trinomials behave in the same way as the natural numbers. In seeking prime polynomial factors of a trinomial of this type or the

---

*When we state this, we are assuming that x is not equal to zero.

prime factors of a natural number, we find that one of the following three things can happen:

1. The number may have identical factors.
    The trinomial may have identical factors.
2. The number may have factors which are not identical.
    The trinomial may have factors which are not identical.
3. The number may have no factors other than itself and one, in which case we say the number is a prime number.
    The trinomial may have no factors other than itself and a constant, in which case we say the trinomial is a prime polynomial.

Examples:

(1) $9 = (3)(3)$ $\qquad\qquad\qquad$ $\begin{cases} \text{two identical} \\ \text{factors} \end{cases}$
$\quad\ \ x^2 - 4x + 4 = (x - 2)(x - 2)$

(2) $35 = (5)(7)$ $\qquad\qquad\quad$ $\begin{cases} \text{two factors,} \\ \text{not identical} \end{cases}$
$\quad\ \ x^2 - 4x - 12 = (x - 6)(x + 2)$

(3) $31 = 31$ $\qquad\qquad\qquad\quad$ $\begin{cases} \text{prime} \\ \text{prime polynomial} \end{cases}$
$\quad\ \ x^2 - 2x + 10 = x^2 - 2x + 10$

In the factoring patterns which we shall consider, it will always be our purpose to factor polynomials until we have arrived at a product of *prime polynomial* factors. The definition of a prime polynomial is given in the screened section below.

> **A prime polynomial is a polynomial of degree at least one which *cannot* be rewritten as the product of polynomials each of lesser degree and having rational numerical coefficients.** *

Observe that the prime integer 3 cannot be considered as a prime polynomial because, even though 3 can be considered as a polynomial, its degree would be 0. In general, an integer is not a prime polynomial.

On the other hand, *every polynomial of degree 1 is* a prime polynomial. For example, $x - 1$, x, $2x + 3$, $x + 3$ are prime polynomials. From what we have just said, $6x + 18$ is also a prime polynomial since it is of degree 1. But

$$6(x + 3) = 6x + 18 \quad \text{so} \quad 6x + 18 = 6(x + 3)$$

Does this mean that we are in trouble, that because $6x + 18 = 6(x + 3)$, then $6x + 18$ is not a prime polynomial, contrary to what we just said? No. It is true that,

---

*The phrase "rational numerical coefficients" means numerical coefficients that are integers or else quotients of integers with nonzero divisors such as $\frac{3}{4}$, $-\frac{7}{8}$, or $\frac{2}{5}$.

in the product $6(x + 3)$, the factor 6 can be considered as a polynomial of degree 0 which is less than the degree of $6x + 18$, but $x + 3$ is of degree 1 which is *not* less than the degree of $6x + 18$.

Examples:

(1) $x + 10$ is a prime polynomial. So are

$$2(x + 10) = 2x + 20$$

$$\frac{1}{5}(x + 10) = \frac{1}{5}x + 2$$

$$-3(x + 10) = -3x - 30$$

(2) $x^2 - 10x + 25$ is *not* a prime polynomial because
$x^2 - 10x + 25 = (x - 5)(x - 5)$
which is the product of two polynomials *each* of lesser degree and having rational coefficients.

Note: The factoring patterns considered in this chapter are concerned only with polynomials *which contain more than one term.* We do not as a rule factor monomials or monomial factors. It is true that the polynomial $5x^2y$ may be rewritten as $(5x)(x)(y)$ and that the product $6(x + 3)$ may be rewritten as $(2)(3)(x + 3)$, but there is generally no advantage to be gained by doing so.

Exercise 36:

Show that the following polynomials are not prime by writing each one as the product of polynomials of lesser degree:

(1) $x^2 + 2x - 35$ (2) $y^4 - 7y^2 - 8$
(3) $15x^6 + 7x^3 - 2$ (4) $7a^4 + 31a^2 + 12$
(5) $10x^8 + 3x^4 - 4$ (6) $8y^{10} - 14y^5 + 3$
(7) $x^2y^4 - xy^2 - 2$ (8) $6a^2 - ab - 2b^2$

3.8
Common Factors

The factoring pattern derived from the third special product involves a reexamination of the distributive axiom, which was discussed in Section 2.6.

$$a(b + c) = ab + ac$$
$$(a + b)(c + d) = a(c + d) + b(c + d)$$

We shall read the two equations above literally forwards and backwards in the following discussion of common factors.

Common Monomial Factors   Let us begin by taking another look at a product that has already been encountered, namely that of a monomial times a multinomial.

Examples:

(1) $a(b + c + d) = ab + ac + ad$
(2) $2a(x + y) = 2ax + 2ay$
(3) $5w^2(x + y - z) = 5w^2x + 5w^2y - 5w^2z$

When a monomial is multiplied by a multinomial, as in the above examples, the product has only one distinguishing feature: the monomial factor shows up clearly in every term of the final result.

$$5w^2(x + y - z) = \underline{5w^2}x + \underline{5w^2}y - \underline{5w^2}z$$

While this is an obvious result of the distributive axiom, the *reversal* of such a product becomes a most important process in the finding of prime polynomial factors.

> Since $a(b + c + d) = ab + ac + ad$
> then $ab + ac + ad = a(b + c + d)$

When we encounter a multinomial such as $ab + ac + ad$ it is essential that we recognize two things:

1. The presence of the common factor a in *every* term of the number.
2. The fact that such a multinomial can always be factored as the product of a monomial and a multinomial.

The technique for reversing such a product is illustrated below.

Consider the following number:

$$2x + 6$$

If we *divide* this number by 2, we have

$$\frac{2x + 6}{2} = \frac{2x}{2} + \frac{6}{2} = \underline{x + 3}$$

Then if we multiply this result by the *same* number, 2, we have

$$2(x + 3)$$

Therefore

$$2x + 6 = 2(x + 3)$$

The process of factoring a common factor from a multinomial amounts to doing just that: dividing and multiplying the multinomial by the *same number*. This cannot possibly change the value of the multinomial, but it does give us a factored form of it. Let us take another look at the above example. Since $2x + 6$ is itself a prime polynomial, is there any advantage to be gained by writing it as $2(x + 3)$? From the prime polynomial point of view, the answer is, "Not much." However, for technical reasons, as you will see in what follows, the form $2(x + 3)$ is much easier to deal with than $2x + 6$. For this reason a good rule to follow when factoring polynomials is the following: *look first for common **numerical** factors*. For our purposes it will be sufficient to look first for common numerical factors which are *integers*, such as the common integral factor, 2, in the above example.

Example:

(1) In the number $2ax - 4ay$, each term is clearly divisible by 2 and by a; so $2a$ is a common factor.

(2) If we divide $2ax - 4ay$ by $2a$ we have

$$\frac{2ax - 4ay}{2a} = \frac{2ax}{2a} - \frac{4ay}{2a} = x - 2y$$

(3) Now if we multiply that quotient by the *same* number, $2a$, we have

$$2a(x - 2y)$$

(4) Therefore $2ax - 4ay = 2a(x - 2y)$.
[Note that on the right side of the equal mark we have three *factors* which are 2 and a and $(x - 2y)$.]

The foregoing is the procedure we follow in factoring common factors from a multinomial, and it is just as simple as it looks. However, it is just as important as it is simple, because, for many numbers, we cannot find the prime factors very easily unless we *first* recognize any common factors. Furthermore, we wish to find and factor from the multinomial *all* common factors present in every term of the number.

Suppose that we wish to find the prime factors of the number

$$6x^2y - 12xy^2$$

1. *Incomplete:* we could divide and multiply by 6.

$$6x^2y - 12xy^2 = 6(x^2y - 2xy^2)$$

2. *Incomplete:* we could divide and multiply by $6x$.

$$6x^2y - 12xy^2 = 6x(xy - 2y^2)$$

3. *Complete:* we could divide and multiply by $6xy$.

$$6x^2y - 12xy^2 = \underline{6xy(x - 2y)}$$

In Example 3 above the multinomial factor, $x - 2y$, is a *prime polynomial*. We achieved this only by finding and factoring from the multinomial *all* common factors present in every term.

Examples:

(1) $3x - 18 = 3(x - 6)$	(2) $2a + 2 = 2(a + 1)$
(3) $5b^5 - 10b^3 = 5b^3(b^2 - 5)$	(4) $a^2 + ab = a(a + b)$
(5) $a^3b^4 - a^4b^2 = a^3b^2(b^2 - a)$	(6) $2xy - x = x(2y - 1)$
(7) $x^2y + xy^2 = xy(x + y)$	(8) $8a^4 - 12a^2 = 4a^2(2a^2 - 3)$

(9) $26a^2b - 39ab = 13ab(2a - 3)$
(10) $6y^5 + 24y^4 + 3y^3 = 3y^3(2y^2 + 8y + 1)$
(11) $18ax^3 - 36ax^2 + 27ax = 9ax(2x^2 - 4x + 3)$
(12) $72m^4n^3 - 18m^3n^4 - 6m^2n^2 = 6m^2n^2(12m^2n - 3mn^2 - 1)$
(13) $33a^3 + 9a^2 + 6ab - 12ac = 3a(11a^2 + 3a + 2b - 4c)$
(14) $54a^4b + 42ab^5 = 6ab(9a^3 + 7b^4)$
(15) $28x^4 - 35x^3 - 56x^2 = 7x^2(4x^2 - 5x - 8)$
(16) $84x^2y^3z^2 - 48x^3y^2z^4 + 12x^2y^2z^2 = 12x^2y^2z^2(7y - 4xz^2 + 1)$

Note: In Example 1 above the number $3x - 18$ was factored by first dividing the number by 3 and then multiplying the quotient by 3; i.e.,

$$3x - 18 = 3(x - 6)$$

The student should observe that we could also have obtained factors for this number by dividing and multiplying by -3; i.e.,

$$3x - 18 = -3(-x + 6)$$

In like manner, a number such as $4a^2 - 8a$ may be factored by taking either $4a$ or $-4a$ as the common factor. Thus, we may write

$$4a^2 - 8a = 4a(a - 2)$$
or
$$4a^2 - 8a = -4a(-a + 2)$$

The choice of dividing and then multiplying by the *negative* of a common factor is always available whenever a common factor is present. The choice is open in factoring expressions having only common *monomial* factors, but it becomes an important decision in a subsequent discussion of common *multinomial* factors; so the choice should be noted and remembered. This point is illustrated in problems 41 through 50 below and will come up again in the later discussion.

Exercise 37:

Find the prime factors and all common numerical factors which are integers.

(1) $5a - 10$ (2) $5ax - 10ay$
(3) $3ax^2 - 6ay^2 + 12a$ (4) $6x^2y + 12xy^2 - 2xy$

(5) $4a - 8$ (6) $4ay - 8y$

(7) $2a + 6$ (8) $2a^2 + 6a$

(9) $x^2 + x$ (10) $4x^2 + 12x$

(11) $b^3 + b$ (12) $b^3 + b^2$

(13) $6b^3x + 3b^2y$ (14) $y^4 + y^2$

(15) $4a^2y^2 - 8ay$ (16) $3x^2y^3 - 12x^3y^2 + 6xy$

(17) $24a^3b^2c - 16ab^3c^2 + 32a^2bc^2$ (18) $3abc - 6ab + 15bc - 12ac$

(19) $x^3y + xy^3$ (20) $4a^5b + 20ab^5$

(21) $2ax^2 - 4ax + 20a$ (22) $3y^3 - 6y^2 + 18y$

(23) $18x^3 + 63x^2y$ (24) $21x^2y^3 - 7x^3y^2 + 35xy^2$

(25) $15a^3b^2c - 10ab^3c^2 + 25abc$ (26) $45r^3t - 18rt^3$

(27) $20x^3y^2 + 8x^2y^4 - 24x^2y^2$ (28) $22a^2 - 4ab + 10ac - 2a$

(29) $9r^4t^3 - 6r^3t^4 + 21r^2t^2$ (30) $56x^3y^2 - 16x^4y$

(31) $54a^4b - 18ab^4 - 27a^3b^2 + 9ab$ (32) $44m^4n - 121m^2n^4$

(33) $45a^3b^2c^3 - 108a^4b^3c^2 + 54a^2b^2c^2$

(34) $75x^2y^3z^4 + 125xy^2z^3 - 50x^4yz^2$

(35) $42a^2b - 35b^2c + 21abc$

(36) $34x^4y + 51x^3y^3 - 17x^2y$

(37) $6a^2bc^2 - 54a^3bc^2 + 48a^4b^2c^3 - 30ab^3c^2$

(38) $x^3y^3z^2 - 4x^5y^2z^3 + 36x^4y^3z^4$

(39) $98y^6 - 49y^5 + 14y^4 - 42y^3$

(40) $24a^5b^3 - 15a^4b^4 + 21a^3b^5 - 30a^2b^6$

Read the *Note* preceding this exercise and factor each of the following numbers in *two* ways:

(41) $4a - 20$ (42) $3b - 15$

(43) $x^2 - xy$ (44) $7a^2 - 14a$

(45) $2x^3 + 4x^2$ (46) $a^2b - a^2$

(47) $-6mn + 2m^2$ (48) $-2x^2 + 4x^3 - 8x^4$

(49) $-x^4y^2 - x^3y^3$ (50) $9a^4 + 18a^3 - 63a^2$

Common Multinomial Factors   All of the numbers in the preceding exercise involved factoring from each multinomial every common monomial factor. But the fact that a multinomial has a common factor simply means that the *same factor* appears in every term; there is certainly no necessity that a common factor be a monomial; a common factor may be a multinomial. Compare the following examples:

1. $ax + bx$ The common factor is $\underline{x}$

2. $a(x + y) + b(x + y)$ The common factor is $\underline{(x + y)}$

We can factor Example 2 above in exactly the same manner that we factor Example 1; i.e., *divide* each term of Example 2 by the common factor $(x + y)$ and then *multiply* the quotient by the same number:

1. $\dfrac{a\cancel{(x+y)}}{\cancel{(x+y)}} + \dfrac{b\cancel{(x+y)}}{\cancel{(x+y)}} = a + b$

2. Then, multiplying the quotient $(a + b)$ by the divisor $(x + y)$, we have

$$(x + y)(a + b)$$

3. Thus $a(x + y) + b(x + y) = (x + y)(a + b)$

$$\text{or } (a + b)(x + y)$$

> *Note:* In the resulting product, the common multinomial factor *must* be enclosed in parentheses.

In following the procedure described above, the student should observe that we are merely applying to the distributive axiom the symmetric property of the equality relation; e.g.,

since $\qquad (a + b)(x + y) = a(x + y) + b(x + y)$

then $\qquad a(x + y) + b(x + y) = (a + b)(x + y)$

In the following examples the common factor in each term is underlined:

(1) $a(\underline{2x - y}) + b(\underline{2x - y}) = (2x - y)(a + b)$
(2) $3x(\underline{a - b}) - 2y(\underline{a - b}) = (a - b)(3x - 2y)$
(3) $m(\underline{a + b}) + n(\underline{a + b}) - p(\underline{a + b}) = (a + b)(m + n - p)$
(4) $a(\underline{x + y + z}) + b(\underline{x + y + z}) = (x + y + z)(a + b)$

When the terms of an expression contain multinomial factors other than a *common* multinomial factor, it is wise to use *brackets* to enclose the quotient obtained by dividing each term by the common factor. The resulting product can then be simplified by removing the parentheses inside the brackets. This advice should be heeded faithfully because the application of it can help considerably in determining the *correct signs* of the terms in the simplified factors. Study the following examples of this situation carefully.

Example 1:

$$x(a + b) - (y + 4)(a + b)$$

The common factor in each term is $(\underline{a + b})$.

Write this factor and then use *brackets* to enclose the quotient obtained when each term is divided by $(a + b)$.

$$x(a + b) - (y + 4)(a + b) = (a + b)[x - (y + 4)]$$

Now simplify the last factor by removing the parentheses inside the brackets and replacing the brackets by parentheses.

$$(a + b)[x - (y + 4)] = (a + b)(x - y - 4)$$

Thus $\qquad x(a + b) - (y + 4)(a + b) = (a + b)(x - y - 4)$

Example 2:

$$(5a - 3)(b + 2) - (4a + 3)(b + 2)$$

The common factor in each term is $(b + 2)$.

Dividing and multiplying by $(b + 2)$ we obtain the following:

$$(b + 2)[(5a - 3) - (4a + 3)]$$

Remove the parentheses in the second factor and replace the brackets by parentheses.

$$(b + 2)(5a - 3 - 4a - 3)$$

Now combine like terms in the second factor.

$$(b + 2)(a - 6)$$

Thus $(5a - 3)(b + 2) - (4a + 3)(b + 2) = (b + 2)(a - 6)$

Example 3:

$$2x^2(a + b) + 3x(a + b)$$

The common factors in each term are x and $(a + b)$.

Divide and multiply the expression by $x(a + b)$.

$$x(a + b)[2x + 3] = x(a + b)(2x + 3)$$

Thus $2x^2(a + b) + 3x(a + b) = x(a + b)(2x + 3)$

Example 4:

$$(a + b) - (x + y)(a + b)$$

The common factor in each term is $(a + b)$.

$$(a + b)[1 - (x + y)] = (a + b)(1 - x - y)$$

Thus $(a + b) - (x + y)(a + b) = (a + b)(1 - x - y)$

Example 5:   *Note this example carefully.*

Common factors may sometimes be found by rewriting an expression in a different but equivalent form. For example, $x(b - a)$ and $-x(a - b)$ are the same number because both products equal $bx - ax$.

We may write

$$(a - b) + x(b - a)$$
$$\text{as}$$
$$(a - b) - x(a - b)$$

In the latter form the common factor in each term is $(a - b)$. Thus

$$(a - b) + x(b - a) = (a - b) - x(a - b)$$
$$= (a - b)(1 - x)$$

Exercise 38:

Find the prime factors.

(1) $x(a + 2) + y(a + 2)$

(2) $c(m + n) - d(m + n)$

(3) $a(x - y) + b(y - x)$ *Note:* See Example 5.

(4) $2x(a + b) - 5y(a + b)$

(5) $3a(2x + 5) + 2b(2x + 5)$

(6) $4x^2(3a - b) + 7y(b - 3a)$

(7) $x^2(2a + b) + y^2(2a + b) - z^2(2a + b)$

(8) $a(3x + y) - 2b(3x + y) + 4c(3x + y)$

(9) $a^2(x^2 + y - 4) + b^2(x^2 + y - 4)$

(10) $x(a + b - c) - y^2(a + b - c)$

(11) $2a(x + y) - (a + b)(x + y)$

(12) $3x(2a + b) + (x + 2y)(2a + b)$

(13) $2a^2(x - y) - (a^2 - b^2)(x - y)$

(14) $(a + b)(x - y) + (4a + b)(x - y)$

(15) $(4a - b)(3x + y) - (2a - 3b)(3x + y)$

(16) $3x^2(4a + b) - 5x(4a + b)$

(17) $6a^2(x - 2y) + 3a(2y - x)$

(18) $8x^2(3a + b) - 4(x + 3)(3a + b)$

(19) $4x^2(a + b + c) - 2x^3(a + b + c) + 6x^4(a + b + c)$

(20) $(x + y) + (x + y)(x - y)$

(21) $(2a - b)(2x - y) + (b - 2a)$

(22) $x(3a + b) - 2x^2(3a + b) + 5x^3(3a + b)$

In seeking prime factors, it is frequently necessary to factor a number *more than once.*

Example:

$$45 = \quad (9) \quad (5)$$

$$(3) \; (3) \; (5) \qquad \text{prime factors of 45}$$

Since the factor 9 was *not* prime, it was necessary to factor it.

Similar repetitions occur often in algebraic operations, particularly when there are common factors present. But the problem can be greatly simplified if you will make yourself remember the following: *always look for the common factor first.*

Examples:

(1) Factor the number $16ax^2 + 16ax + 4a$.
    The common factors are 4 and a.

$$16ax^2 + 16ax + 4a = 4a(4x^2 + 4x + 1)$$

We now have three factors: they are 4 and a and $4x^2 + 4x + 1$. We do not refactor monomial factors, such as 4, but the multinomial factor $4x^2 + 4x + 1$ is *not prime;* it is a trinomial perfect square and can be factored.

$$16ax^2 + 16ax + 4a = 4a(4x^2 + 4x + 1)$$
$$= 4a(2x + 1)(2x + 1)$$

(2) $6x^3 + 15x^2 - 36x = 3x(2x^2 + 5x - 12)$
    $= 3x(2x - 3)(x + 4)$

(3) $x^2(a + b) - 5x(a + b) + 6(a + b) = (a + b)(x^2 - 5x + 6)$
    $= (a + b)(x - 3)(x - 2)$

Exercise 39:

Find the prime factors of the following; check for common factors, including numerical factors that are integers:

(1) $2y^3 + 10y^2 - 48y$

(2) $3x^5 + 9x^3 - 162x$

(3) $10ax^2 + 5ax - 15a$

(4) $24xy^3 + 14xy^2 - 24xy$

(5) $3a^3x + 6a^2x + 3ax$

(6) $4x^3y + 24x^2y + 32xy$

(7) $2a^2 - 4a - 70$

(8) $a^3b - 14a^2b + 24ab$

(9) $5x^3 - 50x^2 + 125x$

(10) $x^3y - 3x^2y^2 - 10xy^3$

(11) $a^4 + 2a^3 + a^2$

(12) $18x^3 - 3x^2 - 3x$

(13) $a^2(x + y) + 3a(x + y) - 10(x + y)$

(14) $x^2(2a + b) + 2x(2a + b) + (2a + b)$

3.9
The Product of the Sum and the Difference of the Same Two Numbers

This special product is easier to write in symbolic language than it is to say in plain English. However, you should *learn* to say it in English (over and over) if only to put yourself in the admirable position of knowing what you are talking about.

Let a and b be any two numbers. Then $a + b =$ the sum of the two numbers; $a - b =$ the difference of the *same* two numbers; and $(a + b)(a - b) =$ the *product* of the sum and the difference of the same two numbers.

When we multiply these two numbers, we get a surprisingly simple result:

$$(a + b)(a - b) = a^2 - ab + ab - b^2$$
$$= a^2 - b^2$$

In like manner:

1. $(x + y)(x - y) = x^2 - xy + xy - y^2$
    $$= x^2 - y^2$$

2. $(r + s)(r - s) = r^2 - rs + rs - s^2$
    $$= r^2 - s^2$$

3. $(m + n)(m - n) = m^2 - mn + mn - n^2$
    $$= m^2 - n^2$$

Repeating the foregoing products in condensed form, we have the following:

$$(a + b)(a - b) = \mathbf{a^2 - b^2}$$
$$(x + y)(x - y) = \mathbf{x^2 - y^2}$$
$$(r + s)(r - s) = \mathbf{r^2 - s^2}$$
$$(m + n)(m - n) = \mathbf{m^2 - n^2}$$

In every boxed example we multiplied the sum and the difference of the same two numbers, and every result follows the same pattern — each answer is *the difference of two squares*. This pattern always occurs with this type of product, and, knowing this, we can get such products by inspection.

Examples:

(1) $(7 + 4)(7 - 4) = 7^2 - 4^2$
    $$= 49 - 16$$
    $$= 33$$
    or $(7 + 4)(7 - 4) = (11)(3) = 33$

(2) $(2a + b)(2a - b) = (2a)^2 - (b)^2$
    $$= 4a^2 - b^2$$

(3) $(x + 3y)(x - 3y) = (x)^2 - (3y)^2$
    $$= x^2 - 9y^2$$

(4) $(5a + 1)(5a - 1) = (5a)^2 - (1)^2$
    $$= 25a^2 - 1$$

(5) $(3x + 4y)(3x - 4y) = 9x^2 - 16y^2$

(6) $\left(\dfrac{x}{3} + 4\right)\left(\dfrac{x}{3} - 4\right) = \dfrac{x^2}{9} - 16$

(7) $\left(a + \dfrac{1}{3}\right)\left(a - \dfrac{1}{3}\right) = a^2 - \dfrac{1}{9}$

Looking over the above examples we can conclude that, to multiply the sum and the difference of the same two numbers, we simply *square each of the numbers and put a minus sign between them*. If $\triangle$ and $\square$ represented numbers, then

$$(\triangle + \square)(\triangle - \square) = \triangle^2 - \square^2$$

But remember that this pattern appears *only* if we multiply the *sum* and the *difference* of the *same* two numbers. Compare the following products carefully:

or	1. $(a + b)(a - b) = a^2 - b^2$ 2. $(a - b)(a + b) = a^2 - b^2$ (since multiplication is commutative)
but	3. $(a + b)(a + b) = a^2 + 2ab + b^2$ 4. $(a - b)(a - b) = a^2 - 2ab + b^2$

In products 1 and 2 we have the product of a sum and a difference, but in products 3 and 4 we are back to special product number one — the square of a binomial.

$$(a + b)(a + b) = (a + b)^2 = a^2 + 2ab + b^2$$
$$(a - b)(a - b) = (a - b)^2 = a^2 - 2ab + b^2$$

Exercise 40:

Multiply the following by inspection:

(1) $(5 + 2)(5 - 2)$ (2) $(10 - 3)(10 + 3)$

(3) $(c - d)(c + d)$ (4) $(c - d)(c - d)$

(5) $(x + 3y)(x - 3y)$ (6) $(2a + b)(2a - b)$

(7) $(3x - y)(3x + y)$ (8) $(3x - y)(3x - y)$

(9) $(4x + 3y)(4x - 3y)$ (10) $(4x + 3y)(4x + 3y)$

(11) $\left(a + \dfrac{1}{2}\right)\left(a - \dfrac{1}{2}\right)$ (12) $\left(2x + \dfrac{1}{3}\right)\left(2x - \dfrac{1}{3}\right)$

(13) $\left(\dfrac{4y}{5} - 5\right)\left(\dfrac{4y}{5} + 5\right)$ (14) $(3a + 5b)(3a - 5b)$

(15) $(7x - y)(7x + y)$ (16) $(9x - 1)(9x - 1)$

(17) $(a^2 + 3)(a^2 - 3)$ (18) $(x^3 + 6)(x^3 - 6)$

(19) $\left(\dfrac{2}{3}a + 2\right)\left(\dfrac{2}{3}a - 2\right)$ (20) $(y - 8)(y - 8)$

3.10
Factoring the Difference of Two Squares

Every statement of equality, by the symmetric property, is always true in reverse;

hence we have the following:

$$\text{Since} \qquad (a + b)(a - b) = a^2 - b^2$$

$$\text{then} \qquad a^2 - b^2 = (a + b)(a - b)$$

Looking at the last statement we can conclude that *the difference of two squares is always equal to the product of the sum and the difference of the same two numbers.* Now, what two numbers?

$$a^2 - b^2 = (a + b)(a - b)$$

Looking carefully at this product in reverse, we shall work backwards from the product to the factors.

1. $a^2 - b^2$ is the *difference* of two squares.
    a. The previous special product demonstrates that such a number has two factors which are the *sum* and the *difference* of the *same* two numbers.
    b. Let $\triangle$ and $\square$ represent these two numbers.

2. Then $a^2 - b^2 = (\triangle + \square)(\triangle - \square)$.
    a. The number represented by $\triangle$ in these factors must be the *same* in each factor, and the product of these two identical numbers must be a^2.
    b. Then the symbol $\triangle$ should be a since $(a)(a) = a^2$.

3. $a^2 - b^2 = (a + \square)(a - \square)$
    a. The number represented by $\square$ in these factors must be the *same* in each factor, and the product of these two identical numbers must be b^2.
    b. Then the number represented by $\square$ should equal b, since $(b)(b) = b^2$.

4. $a^2 - b^2 = (a + b)(a - b)$

Similarly:
$$4x^2 - 25y^2 = (? + ?)(? - ?)$$
    a. The first number in each factor should be $2x$, because $(2x)(2x) = 4x^2$.
    b. The second number in each factor should be $5y$, because $(5y)(5y) = 25y^2$.
Then $4x^2 - 25y^2 = (2x + 5y)(2x - 5y)$.

In like manner:
1. $16x^2 - 1 = (4x + 1)(4x - 1)$
2. $81a^2 - 16b^2 = (9a + 4b)(9a - 4b)$
3. $\dfrac{x^2}{4} - \dfrac{y^2}{9} = \left(\dfrac{x}{2} + \dfrac{y}{3}\right)\left(\dfrac{x}{2} - \dfrac{y}{3}\right)$
4. $25 - 16 = (5 + 4)(5 - 4)$
5. $a^4 - 9 = (a^2 + 3)(a^2 - 3)$
6. $x^6 - 49 = (x^3 + 7)(x^3 - 7)$
7. $144a^2 - 1 = (12a + 1)(12a - 1)$

Exercise 41:

Factor the following numbers:

(1) $r^2 - t^2$ (2) $m^2 - n^2$

(3) $4x^2 - y^2$ (4) $a^2 - 9b^2$

(5) $9x^2 - 49$ (6) $b^2 - 1$

(7) $b^4 - 25$ (8) $b^6 - 36$

(9) $b^8 - 100$ (10) $16a^2 - 9b^4$

(11) $\dfrac{x^2}{25} - 1$ (12) $\dfrac{4a^2}{25} - b^2$

(13) $25a^2 - 36b^2$ (14) $4x^2 - 49y^2$

(15) $169a^2 - 625b^2$ (16) $x^4 - 25y^2$

(17) $a^4 - b^2$ (18) $b^6 - 9c^2$

(19) $\dfrac{x^4}{9} - 1$ (20) $a^2 - \dfrac{1}{4}$

Any number which is the difference of two squares has factors which are the sum and the difference of the same two numbers. There is no requirement that the squares be the squares of *monomials*. For example, consider the following:

$$x^2 - (a + b)^2$$

This expression is the difference of two squares and can be factored accordingly. Note that the last square in this expression is the square of the *binomial a + b*. To factor the expression we must find the sum and difference of two numbers whose *squares* are x^2 and $(a + b)^2$; obviously the two numbers we need are x and $(a + b)$. *To avoid careless errors in signs* we use *brackets* to denote the product of the sum and the difference of x and $(a + b)$. The factors can then be simplified by removing the parentheses inside the brackets. To illustrate:

$$x^2 - (a + b)^2 = [x + (a + b)][x - (a + b)]$$
$$= (x + a + b)(x - a - b)$$

Following are more examples of factoring expressions of this type:

1. $a^2 - (r - s)^2 = [a + (r - s)][a - (r - s)]$
    $$= (a + r - s)(a - r + s)$$

2. $(x + y)^2 - 9 = [(x + y) + 3][(x + y) - 3]$
    $$= (x + y + 3)(x + y - 3)$$

3. $(a - b)^2 - (x + y)^2 = [(a - b) + (x + y)][(a - b) - (x + y)]$
    $$= (a - b + x + y)(a - b - x - y)$$

4. $(a - 4)^2 - (b + 2)^2 = [(a - 4) + (b + 2)][(a - 4) - (b + 2)]$
$$= (a - 4 + b + 2)(a - 4 - b - 2)$$
$$= (a + b - 2)(a - b - 6)$$

Exercise 42:

Factor the following numbers:

(1) $a^2 - (x - 2)^2$

(2) $36 - (a + b)^2$

(3) $(2x + y)^2 - 16$

(4) $1 - (2a - b)^2$

(5) $(3a + b)^2 - c^2$

(6) $49 - (a - b)^2$

(7) $(x - 3)^2 - 64$

(8) $(3x + 2)^2 - 1$

(9) $r^2 - (3x - y)^2$

(10) $36 - (x - y)^2$

(11) $(m + n)^2 - (r + s)^2$

(12) $(2a + b)^2 - (x - y)^2$

(13) $(x - 3)^2 - (y - 2)^2$

(14) $(2a + 3)^2 - (3b - 1)^2$

CAUTION!

In factoring a polynomial we are usually seeking its *prime* polynomial factors. For this reason the student should be on constant guard for two things:

1.   The presence of *common factors*.
2.   The possibility that a factor is not *prime* — i.e., that it can itself be factored.

Examples:

(1) $x^3y - xy^3 = xy(x^2 - y^2)$
$$= xy(x + y)(x - y)$$
(2) $2a^2 - 72 = 2(a^2 - 36)$
$$= 2(a + 6)(a - 6)$$
(3) $a^4 - 3a^2 - 4 = (a^2 + 1)(a^2 - 4)$
$$= (a^2 + 1)(a + 2)(a - 2)$$
(4) $x^4 - 16 = (x^2 + 4)(x^2 - 4)$
$$= (x^2 + 4)(x + 2)(x - 2)$$
(5) $x^4 - 10x^2 + 9 = (x^2 - 1)(x^2 - 9)$
$$= (x + 1)(x - 1)(x + 3)(x - 3)$$
(6) $8a^2 - 2(b - 1)^2 = 2[4a^2 - (b - 1)^2]$
$$= 2[2a + (b - 1)][2a - (b - 1)]$$
$$= 2(2a + b - 1)(2a - b + 1)$$

Exercise 43:

Find the *prime* factors of the following:

(1) $a^4 - 81$

(2) $3x^2 - 27$

$(3)\quad 2a^3 - 32ab^2$ $\qquad\qquad\qquad$ $(4)\quad 2x^9 - 32x$

$(5)\quad x^4 - 17x^2 + 16$ $\qquad\qquad$ $(6)\quad a^4 - 21a^2 - 100$

$(7)\quad 2x^5 - 6x^3 - 8x$ $\qquad\qquad$ $(8)\quad 3a^6 - 6a^4 + 3a^2$

$(9)\quad x^5y - xy^5$ $\qquad\qquad\qquad$ $(10)\quad 8x^5 + 18x^3 - 5x$

$(11)\quad x^4 - x^2$ $\qquad\qquad\qquad$ $(12)\quad a^5 - a^3$

$(13)\quad 18x^2 - 2(y + 3)^2$ $\qquad\quad$ $(14)\quad 3x(a + b)^2 - 27x^3$

$(15)\quad ab(x + y)^2 - a^3b^3$ $\qquad$ $(16)\quad x^3(2a - b)^2 - x^5$

3.11
The Sum or Difference of Two Cubes

Special product number five does not look particularly special at first glance — or even at second glance. It has to do with a *binomial* and a related *trinomial* whose product assumes an unexpectedly simple form.

The binomial factor may be either the sum or the difference of two numbers, such as $a + b$ or $a - b$. The relation of the trinomial factor to the binomial factor is always the same, but the relationship is not spectacularly obvious to the uninitiated. Look over the following examples of this product and see if you can discern a repeating relationship between each binomial and trinomial factor.

Examples:

$(1)\quad (a + b)(a^2 - ab + b^2)$

$(2)\quad (x + y)(x^2 - xy + y^2)$

$(3)\quad (a - b)(a^2 + ab + b^2)$

$(4)\quad (x - y)(x^2 + xy + y^2)$

A rather quick look should convince you that the first and last terms of each *trinomial* factor are the squares of the first and last terms of each *binomial* factor.

A longer look at the remaining term (the middle term) of the trinomial factor might be necessary to make it evident that in every example the middle term is actually the *product* of the two terms in the binomial factor with the sign of the product *reversed*.

When we multiply these numbers, we get the following results:

1. $(a + b)(a^2 - ab + b^2) = a^3 - a^2b + ab^2 + a^2b - ab^2 + b^3$
$$= \underline{a^3 + b^3}$$

2. $(x + y)(x^2 - xy + y^2) = x^3 - x^2y + xy^2 + x^2y - xy^2 + y^3$
$$= \underline{x^3 + y^3}$$

3. $(a - b)(a^2 + ab + b^2) = a^3 + a^2b + ab^2 - a^2b - ab^2 - b^3$
$$= \underline{a^3 - b^3}$$

4. $(x - y)(x^2 + xy + y^2) = x^3 + x^2y + xy^2 - x^2y - xy^2 - y^3$
$$= \underline{x^3 - y^3}$$

Rewriting these examples in condensed form, we have:

$$
\begin{array}{ll}
1. & (a + b)(a^2 - ab + b^2) = \mathbf{a^3 + b^3} \\
2. & (x + y)(x^2 - xy + y^2) = \mathbf{x^3 + y^3} \\
3. & (a - b)(a^2 + ab + b^2) = \mathbf{a^3 - b^3} \\
4. & (x - y)(x^2 + xy + y^2) = \mathbf{x^3 - y^3}
\end{array}
$$

A careful study of the repeating pattern in the above examples should enable you to supply the missing factors in the following exercise.

Exercise 44:

Supply the missing trinomial factor.

$$
\begin{array}{ll}
(1) \quad (m + n)(\qquad) = m^3 + n^3 \\
(2) \quad (r + s)(\qquad) = r^3 + s^3 \\
(3) \quad (a + 1)(\qquad) = a^3 + 1^3 \text{ or } a^3 + 1 \\
(4) \quad (b + 2)(\qquad) = b^3 + 2^3 \text{ or } b^3 + 8 \\
(5) \quad (y + 3)(\qquad) = y^3 + 27 \\
(6) \quad (m - n)(\qquad) = m^3 - n^3 \\
(7) \quad (r - s)(\qquad) = r^3 - s^3 \\
(8) \quad (a - 1)(\qquad) = a^3 - 1 \\
(9) \quad (b - 2)(\qquad) = b^3 - 8 \\
(10) \quad (y - 3)(\qquad) = y^3 - 27
\end{array}
$$

The special feature of the type of product illustrated in the preceding exercise is the result which it produces. It is the only kind of product that will equal either the *sum* or the *difference* of *two cubes*. Consequently, if we write this product in reverse, we have a pattern for factoring *any* number that is the sum or the difference of two cubes.

$$
\begin{array}{ll}
1. & a^3 + b^3 = (a + b)(a^2 - ab + b^2) \\
2. & x^3 + y^3 = (x + y)(x^2 - xy + y^2) \\
3. & a^3 - b^3 = (a - b)(a^2 + ab + b^2) \\
4. & x^3 - y^3 = (x - y)(x^2 + xy + y^2)
\end{array}
$$

In like manner:

$$
\begin{array}{ll}
5. & m^3 - n^3 = (m - n)(m^2 + mn + n^2) \\
6. & y^3 - 1 = (y - 1)(y^2 + y + 1) \\
7. & b^3 - 8 = (b - 2)(b^2 + 2b + 4) \\
8. & m^3 + n^3 = (m + n)(m^2 - mn + n^2) \\
9. & x^3 + 27 = (x + 3)(x^2 - 3x + 9) \\
10. & 8a^3 + 1 = (2a + 1)(4a^2 - 2a + 1)
\end{array}
$$

Therefore, the sum or difference of two cubes is always the product of a binomial and a trinomial factor in which the binomial and trinomial have the same relationship illustrated in each of the foregoing examples.

It is possible for some numbers to fit two different factoring patterns. For example, consider the following:

$$x^6 - y^6 \qquad \begin{aligned} x^6 &= (x^2)^3 \quad or \quad (x^3)^2 \\ y^6 &= (y^2)^3 \quad or \quad (y^3)^2 \end{aligned}$$

The number $x^6 - y^6$ *can* be written $(x^2)^3 - (y^2)^3$, the difference of two *cubes*. But the number $x^6 - y^6$ can also be written $(x^3)^2 - (y^3)^2$, the difference of two *squares*.

If we factor $x^6 - y^6$ *both* ways, we obtain

1. $\begin{aligned} x^6 - y^6 = (x^2)^3 - (y^2)^3 &= (x^2 - y^2)(x^4 + x^2y^2 + y^4) \\ &= \underline{(x + y)(x - y)(x^4 + x^2y^2 + y^4)} \end{aligned}$

2. $\begin{aligned} x^6 - y^6 = (x^3)^2 - (y^3)^2 &= (x^3 + y^3)(x^3 - y^3) \\ &= (x + y)(x^2 - xy + y^2)(x - y)(x^2 + xy + y^2) \\ &= \underline{(x + y)(x - y)(x^2 - xy + y^2)(x^2 + xy + y^2)} \end{aligned}$

In the last factoring process we have actually come up with the *prime factors* of $x^6 - y^6$. In number 1, the factors are correct, but the trinomial factor $(x^4 + x^2y^2 + y^4)$ is *not* prime.

$$x^4 + x^2y^2 + y^4 = (x^2 + xy + y^2)(x^2 - xy + y^2)$$

But rather than work out this factoring process, we shall simply agree, when such a choice arises, to treat *all* such numbers as the difference of two *squares* instead of the difference of two *cubes*, since the former provides the easier way to find the prime factors.

Study the following examples carefully, remembering the words of warning given at the end of Section 3.10. *Always* look out for the following:

1. the presence of a *common* factor.
2. the possibility that a factor can itself be *factored*.

Examples:

(1) $\begin{aligned} b^6 - 1 = (b^3)^2 - 1 &= (b^3 + 1)(b^3 - 1) \\ &= \underline{(b + 1)(b^2 - b + 1)(b - 1)(b^2 + b + 1)} \end{aligned}$

(2) $\begin{aligned} b^6 + 1 = (b^2)^3 + 1 \\ = \underline{(b^2 + 1)(b^4 - b^2 + 1)} \end{aligned}$

(3) $\begin{aligned} 2x^4 + 16x = 2x(x^3 + 8) \\ = \underline{2x(x + 2)(x^2 - 2x + 4)} \end{aligned}$

$$(4) \quad 4ay^3 - 4a^4 = 4a(y^3 - a^3)$$
$$= \underline{4a(y - a)(y^2 + ay + a^2)}$$

Exercise 45:

Find the prime factors of the following:

(1) $a^3 + 1$
(2) $a^3 - 1$

(3) $x^3 - 64$
(4) $x^3 + 64$

(5) $r^3 + s^3$
(6) $x^3 - y^6$

(7) $8y^3 + 27$
(8) $x^6 + 1$

(9) $b^6 + 64$
(10) $a^3 - 125$

(11) $a^{12} + b^{12}$
(12) $x^3 - y^9$

Study the four examples preceding this exercise; then find the prime factors of the following:

(13) $3a^5 - 81a^2$
(14) $2b^3 + 16$

(15) $x^6 - 1$
(16) $a^7 + 8a$

(17) $16x - 2x^4$
(18) $x^4y - xy^4$

(19) $24a^4 - 3a$
(20) $81x^3y + 3y$

(21) $a^6 - 64$
(22) $x^9 + 1$

(23) $2x^4y - 16xy^4$
(24) $a^5 - a^2$

(25) $x^6 + x^3$
(26) $2b^4 - 250b$

(27) $2a^4 - 54a$
(28) $x^5 - x^2y^3$

3.12
Factoring by Grouping

In retrospect, all preceding sections of this chapter have relentlessly demonstrated the fact that your ability to factor a number depends ultimately upon your ability to *recognize*, of the five patterns considered, *which* pattern (or patterns) the number fits. One would know, for instance, that the factors of $a^2 - 144$ are $(a + 12)(a - 12)$ only if he recognized $a^2 - 144$ as the difference of two squares.

After we have become familiar with the five patterns, we can consider a slightly more sophisticated procedure which is called factoring by grouping. This process involves *grouping the terms* of a multinomial in such a way as to make it fit one of the five patterns, even though the prospects may not look very promising at first glance. This is done by literally undoing the procedure followed in removing signs of grouping (see Section 2.7). Instead of removing parentheses and/or brackets, we *insert* such signs in a series of terms in order to make the resulting expression fit one of the factoring patterns. For an illustration of this technique, study the following example.

Example 1:

Consider the multinomial

$$x^2 - 6x + 9 - y^2$$

If we place *parentheses* around the first three terms of this expression we have

$$(x^2 - 6x + 9) - y^2$$

and the resulting expression is the *same as the original one;* i.e.,

$$(x^2 - 6x + 9) - y^2 = x^2 - 6x + 9 - y^2$$

However, in the form

$$(x^2 - 6x + 9) - y^2$$

the three terms enclosed in parentheses should be easily recognizable as a trinomial perfect square; therefore this number can be written as follows:

$$(x^2 - 6x + 9) - y^2 = (x - 3)^2 - y^2$$

The latter expression, $(x - 3)^2 - y^2$, is the difference of two squares and can be factored by the method discussed in Section 3.10.

$$\begin{aligned}(x^2 - 6x + 9) - y^2 &= (x - 3)^2 - y^2 \\ &= [(x - 3) + y][(x - 3) - y] \\ &= (x - 3 + y)(x - 3 - y)\end{aligned}$$

Repeating these steps in condensed form we have the following:

$$\begin{aligned}x^2 - 6x + 9 - y^2 &= \\ (x^2 - 6x + 9) - y^2 &= \\ (x - 3)^2 - y^2 &= \\ [(x - 3) + y][(x - 3) - y] &= \\ (x - 3 + y)(x - 3 - y)\end{aligned}$$

CAUTION!

In the above example the *first term* placed inside the parentheses had a *positive* coefficient, and the insertion of the parentheses did not affect the *signs* of the terms enclosed. This is *not true* when the first term enclosed has a *negative* coefficient. Look over the next example carefully.

Example 2:

Consider the multinomial

$$16 - a^2 + 2ab - b^2$$

We *cannot* rewrite this number as $16 - (a^2 + 2ab - b^2)$ because the insertion of the parentheses behind the minus sign *reverses* the sign of each term placed inside. In other words, $16 - (a^2 + 2ab - b^2)$ is equal to $16 - a^2 - 2ab + b^2$, which is clearly *not* the same number we started with. To avoid this kind of error in grouping terms, proceed as follows: in grouping terms behind a *minus sign*, you must first *reverse the sign of every term placed inside the parentheses*. Thus, to place the last three terms of $16 - a^2 + 2ab - b^2$ in parentheses, we reverse the sign of every term placed inside.

$$16 - a^2 + 2ab - b^2 = 16 - (a^2 - 2ab + b^2)$$

(Note that if the parentheses on the right are removed, the two expressions are exactly the same.)

Since the expression inside the parentheses, $a^2 - 2ab + b^2$, equals $(a - b)^2$, we proceed as follows:

$$\begin{aligned}
16 - a^2 + 2ab - b^2 &= 16 - (a^2 - 2ab + b^2) \\
&= 16 - (a - b)^2 \\
&= [4 + (a - b)][4 - (a - b)] \\
&= (4 + a - b)(4 - a + b)
\end{aligned}$$

Following are more examples of factoring by grouping:

(1) $\begin{aligned}[t]
a^2 - b^2 - 4b - 4 &= a^2 - (b^2 + 4b + 4) \\
&= a^2 - (b + 2)^2 \\
&= [a + (b + 2)][a - (b + 2)] \\
&= (a + b + 2)(a - b - 2)
\end{aligned}$

(2) $\begin{aligned}[t]
9x^2 - 12x + 4 - y^2 &= (9x^2 - 12x + 4) - y^2 \\
&= (3x - 2)^2 - y^2 \\
&= [(3x - 2) + y][(3x - 2) - y] \\
&= (3x - 2 + y)(3x - 2 - y)
\end{aligned}$

(3) $\begin{aligned}[t]
2b + a^2 - b^2 - 1 &= a^2 - b^2 + 2b - 1 \\
&= a^2 - (b^2 - 2b + 1) \\
&= a^2 - (b - 1)^2 \\
&= [a + (b - 1)][a - (b - 1)] \\
&= (a + b - 1)(a - b + 1)
\end{aligned}$

(4) $\begin{aligned}[t]
a^2 + 2a + 1 - x^2 - 4xy - 4y^2 &= (a^2 + 2a + 1) - (x^2 + 4xy + 4y^2) \\
&= (a + 1)^2 - (x + 2y)^2 \\
&= [(a + 1) + (x + 2y)][(a + 1) - (x + 2y)] \\
&= (a + 1 + x + 2y)(a + 1 - x - 2y)
\end{aligned}$

Numbers which can be factored by grouping often do not contain three terms which form a trinomial perfect square. Frequently expressions can be factored by grouping the terms in such a way as to reveal a common factor. Again it should be stressed that when parentheses are imposed on terms following a minus sign, the sign of every term placed inside the parentheses must be reversed. Consider the following example:

$$2a + 2b - ax - bx$$

Grouping the terms by two's, we have

$$(2a + 2b) - (ax + bx)$$

Then, using the distributive axiom, remove the common factor present in *each grouping*.

$$(2a + 2b) - (ax + bx) = 2(a + b) - x(a + b)$$

The two terms separated by the minus sign contain the *common factor* $(a + b)$, and the expression can be factored by the method discussed in Section 3.8.

$$2(a + b) - x(a + b) = (a + b)(2 - x)$$

Repeating the steps of this example in condensed form we have the following:

$$
\begin{aligned}
2a + 2b - ax - bx &= \\
(2a + 2b) - (ax + bx) &= \\
2(a + b) - x(a + b) &= \\
(a + b)(2 - x)
\end{aligned}
$$

Following are further examples of grouping the terms of an expression in order to reveal a common factor:

(1) $x^2 - xy + 3x - 3y = (x^2 - xy) + (3x - 3y)$
$$= x(x - y) + 3(x - y)$$
The common factor in each term is $\underline{(x - y)}$
$$= (x - y)(x + 3)$$

(2) $3x + 12 - xy - 4y = (3x + 12) - (xy + 4y)$
$$= 3(x + 4) - y(x + 4)$$
The common factor is $\underline{(x + 4)}$
$$= (x + 4)(3 - y)$$

(3) $a + c - ab - bc = (a + c) - (ab + bc)$
$$= (a + c) - b(a + c)$$
The common factor is $\underline{(a + c)}$
$$= (a + c)(1 - b)$$

(4) $a - b - a^2 + b^2 = (a - b) - (a^2 - b^2)$
$$= (a - b) - (a + b)(a - b)$$
The common factor is $\underline{(a - b)}$
$$= (a - b)[1 - (a + b)]$$
$$= (a - b)(1 - a - b)$$

The different examples given of factoring by grouping illustrate the fact that terms in an expression may be grouped in a variety of ways (such as by two's or three's) to produce a recognizable factoring pattern. This may require, at first, some trial runs on the part of the student because the first grouping might not produce a recognized pattern. Look over the following example in which two different groupings are tried. Factor

$$a^2 - b^2 - 4a + 4$$

Grouping the terms by two's we obtain

$$(a^2 - b^2) - (4a - 4)$$

Factoring each group we obtain

$$(a + b)(a - b) - 4(a - 1)$$

This gets us nowhere because the two terms in the last expression, $(a + b)(a - b)$ and $-4(a - 1)$, do *not* contain a common factor. While there is nothing incorrect in the above procedure, there is nothing productive in it either because we have wound up with two *terms* separated by a minus sign instead of the *product* we were seeking. Grade on this effort: "A" for technique but "F" for strategy.

Now let us take the *same* number

$$a^2 - b^2 - 4a + 4$$

and rearrange the terms in this manner:

$$a^2 - 4a + 4 - b^2$$

Then group the first *three* terms to obtain

$$(a^2 - 4a + 4) - b^2$$

Note that $a^2 - 4a + 4 = (a - 2)^2$. Thus

$$
\begin{aligned}
(a^2 - 4a + 4) - b^2 &= (a - 2)^2 - b^2 \\
&= [(a - 2) + b][(a - 2) - b] \\
&= (a - 2 + b)(a - 2 - b)
\end{aligned}
$$

To summarize, factoring by grouping requires a careful examination of the terms in an expression and some preliminary consideration of possible patterns which the terms might fit. If the first grouping fails to produce a recognizable pattern, then a different grouping should be tried. Above all, the student should note carefully the *difference* in the two results obtained by the different groupings in the foregoing example:

(1) $a^2 - b^2 - 4a + 4 = (a + b)(a - b) - 4(a - 1)$
(2) $a^2 - b^2 - 4a + 4 = (a - 2 + b)(a - 2 - b)$

Both of the statements above are true, but *only the second one is correct* if our goal is to *factor $a^2 - b^2 - 4a + 4$.*

At this point you may be assailed by dismal visions of grouping the same terms

endlessly in a possibly futile attempt to find factors. Not so. There are only five factoring patterns to be considered, and the multinomials to be factored contain a
finite number of terms. In this limited situation, the chances are excellent that, for
any specific problem on grouping, all possibilities will be exhausted long before you
are. To the question "When is it permissible to give up?" the answer for the exercises
in this chapter is simple — "Never." And once you have worked through the
exercises, it is to be hoped that you will have acquired enough rudimentary expertise
in the art to make the question unnecessary.

Following are additional examples of factoring by grouping:

(1) $ac + bc - ay - by =$
     $(ac + bc) - (ay + by) =$
     $c(a + b) - y(a + b) =$ [The common factor is $(a + b)$]
     $(a + b)(c - y)$

(2) $x^2 - y^2 - 6y - 9 =$
     $x^2 - (y^2 + 6y + 9) =$
     $x^2 - (y + 3)^2 =$ (This expression is the difference of two squares)
     $[x + (y + 3)][x - (y + 3)] =$
     $(x + y + 3)(x - y - 3)$

(3) $a - b + a^2 - 2ab + b^2 =$
     $(a - b) + (a^2 - 2ab + b^2) =$
     $(a - b) + (a - b)(a - b) =$ [The common factor is $(a - b)$]
     $(a - b)[1 + (a - b)] =$
     $(a - b)(1 + a - b)$

(4) $4a + 2b - 8a^2 - 8ab - 2b^2$
     The common factor in each term is 2.

     $2[2a + b - 4a^2 - 4ab - b^2]$

     Grouping the terms *inside* the brackets, we have

     $2[(2a + b) - (4a^2 + 4ab + b^2)]$
              or
     $2[(2a + b) - (2a + b)(2a + b)]$

     The common factor inside the brackets is $\underline{(2a + b)}$.

     $2\{(2a + b)[1 - (2a + b)]\}$
     $2\{(2a + b)(1 - 2a - b)\} = 2(2a + b)(1 - 2a - b)$

Exercise 46:

Factor by grouping.

(1) $ar + at + br + bt$ (2) $x^2 - 2x + 1 - y^2$
(3) $ax + bx - ay - by$ (4) $2a + 4b - ac - 2bc$

(5) $a^2 - b^2 + 2b - 1$

(6) $a^2 - 6a + 9 - b^2$

(7) $a - b - a^2 + b^2$

(8) $x^2 - y^2 - x - y$

(9) $x^2 - y^2 - 8x + 16$

(10) $3a + 3b + 6c + ax + bx + 2cx$

(11) $a + b + a^2 + 3ab + 2b^2$

(12) $x - y + x^2 - 2xy + y^2$

(13) $r^2 - s^2 + 4s - 4$

(14) $m + n - m^2 - 2mn - n^2$

(15) $a^2 + 2ab + b^2 - a - b$

(16) $2ax + 4bx - 2ay - 4by$

(17) $3a + 3b - 3a^2 - 6ab - 3b^2$

(18) $2x^2 - 2y^2 - 3x - 3y$

(19) $a^2 - b^2 + 10a + 25$

(20) $x^2 - 4y^2 - x + 2y$

Exercise 47:

Review Problems
Determine the products by inspection.

(1) $(2x - 4)(3x + 2)$

(2) $(3a - 5)(3a + 5)$

(3) $(a - 5b)^2$

(4) $3x^2(3x^3 - 2x^2 + 5x)$

(5) $(x + 2)(x^2 - 2x + 4)$

(6) $(5a - 7b)(2a - 3b)$

(7) $(4x - 3y)(4x + 3y)$

(8) $(4x - 3y)(4x - 3y)$

(9) $(4x + 3y)^2$

(10) $\left(x - \dfrac{1}{2}\right)^2$

(11) $\left(2a + \dfrac{1}{3}\right)\left(2a - \dfrac{1}{3}\right)$

(12) $(x^2 + 1)(x^2 - 1)$

(13) $(x^2 + 1)^2$

(14) $(x^2 + 1)(x^4 - x^2 + 1)$

Find the prime factors; check for common factors including numerical factors that are integers.

(15) $3x^2y^2 - 6x^3y + 12xy^3 - 3xy$

(16) $10x^2 - 13x - 3$

(17) $9x^2 + 30x + 25$

(18) $16x^2 - 1$

(19) $x^3 + 8$

(20) $6a^2 - 23a - 4$

(21) $81b^2 - 25c^4$

(22) $a^2 + a + \dfrac{1}{4}$

(23) $12b^2 - 29b + 14$

(24) $64a^2 - 1$

(25) $64a^3 - 1$

(26) $25x^2 - 40x + 16$

(27) $3x^2 + 6x$

(28) $\dfrac{x^2}{9} - 25$

(29) $2ax^2 - 4ax + 2a$

(30) $a^5b - ab^5$

(31) $2x^5 - 16x^3 - 18x$

(32) $8a^3b - 50ab^3$

(33) $16x^4 - 1$

(34) $16x^4 - 2x$

(35) $8a^2 + 34ab + 21b^2$

(36) $a^8 - 1$

(37) $a(b + c) + 2(b + c)$

(38) $2a(x - y) - 3b(y - x)$

(39) $(x + y)^2 - 25$

(40) $3a^2(2x + y) - 6ab(2x + y)$

(41) $(x - 2y) -$
         $(x + 3y)(x - 2y)$

(42) $16 - (2a + b)^2$

(43) $x^2(3a - b) - y^2(3a - b)$

(44) $2a(3a + b)^2 - 4a^2(3a + b)$

(45) $(2x - 5y)^2 - 9$

(46) $(a + b)^2 - (c + d)^2$

(47) $9a(4x + y) - 3a^2(4x + y)$

(48) $a^2 - (b - c)^2$

(49) $3x + 3y + bx + by$

(50) $ab + ac - bx - cx$

(51) $ab - 2ac - 3b + 6c$

(52) $x + y - ax - ay$

(53) $ac + bc - a - b$

(54) $a + b + a^2 - b^2$

(55) $x - y - x^2 + y^2$

(56) $a^2 - 10a + 25 - b^2$

(57) $16 - x^2 + 2xy - y^2$

(58) $x^2 - y^2 - 6x + 9$

(59) $x^2 - y^2 - x - y$

(60) $a^2 - b^2 + 2b - 1$

Chapter Test 3

I. Define a prime number.

II. If S is the set of the first twelve primes, then, listed by elements,

$$S = \underline{\hspace{5in}}$$

III. The prime factorization of 660 is $\underline{\hspace{5in}}$.

IV. In the polynomial $2x^4 - 3x^3y^3 + 4x^2y^2 - 5xy^4 + 7$
1. The degree of the third term is $\underline{\hspace{1in}}$.
2. The degree of the polynomial is $\underline{\hspace{1in}}$.

V. Give the products by inspection.
1. $(3x + 2)(3x - 2)$
2. $(a - 2)(a^2 + 2a + 4)$
3. $(3b - 1)(2b + 5)$
4. $(2x + 7)^2$
5. $\left(a + \dfrac{1}{2}\right)\left(a - \dfrac{1}{2}\right)$
6. $(3a + 4b)(3a + 4b)$
7. $(3a + 4b)(3a - 4b)$
8. $(2x + 1)(4x^2 - 2x + 1)$
9. $(5x - 2)(3x - 4)$
10. $\left(x - \dfrac{1}{2}\right)^2$

VI. Find the prime factors of the following:
1. $a^4 - 4a^2 + 4$
2. $x^2 - 9$
3. $a^2 - 7a + 10$
4. $6x^2 - 13x - 5$
5. $a^3 + 8$
6. $x^4 - 6x^2 - 16$
7. $a^3b - ab^3$
8. $x^6 - 9$
9. $x^6 - 27$
10. $2a^3 + 5a^2 - 3a$
11. $a(x - y) + 3(x - y)$
12. $x(3y + 1) - 4(3y + 1)$
13. $x^4 - 3x^2 - 4$
14. $x^4 - 64$
15. $x^4 - 64x$
16. $(a + b)^2 - 1$
17. $x^2 - 2x + 1 - y^2$
18. $ax + bx - 3a - 3b$

19. $a^2 - b^2 - a - b$
20. $a^2 - b^2 + 8a + 16$
21. $(x + 2)(x + 5) - y(x + 5)$
22. $16 - a^2 + 2ab - b^2$
23. $2x + 2y - x^2 + y^2$
24. $16 - (x + y)^2$
25. $ab - 3ac - 3b + 9c$
26. $2a^4b - 16a^2b - 18b$

VII. $T =$ the set of positive integral divisors of 14. The set T listed by elements is

$$T = \underline{\hspace{5cm}}$$

Fill in the following blanks with the correct symbol $\in, \notin, \subseteq, \nsubseteq$:

1. $3 \underline{\hspace{1cm}} T$
2. $\{1, 2, 4\} \underline{\hspace{1cm}} T$
3. $\{2, 5\} \underline{\hspace{1cm}} T$
4. $\{2\} \underline{\hspace{1cm}} T$
5. $-2 \underline{\hspace{1cm}} T$
6. $7 \underline{\hspace{1cm}} T$

4
OPERATIONS WITH FRACTIONS

4.1
The Meaning of Fractions

As counting procedures became more sophisticated, man found it necessary to invent fractions. The necessity arose some time before recorded history; in fact, it might well have presented itself along with man's first primeval notion to acquire property and bequeath it to his descendents. The idea that 1/2 means *one of two equal parts* is quite readily grasped by two people who have just inherited one measure of land between them, not to mention two women before Solomon disputing possession of one infant.

The ancient Babylonians divided the face of the full moon into 240 equal sections and then used the numbers 5, 10, 20, 40, and 80 to tell how many of those sections were visible on the first five successive days of the new moon.* This was going on before 1000 B.C., and fractions have continued to be of practical usefulness ever since. On the fifth night after the new moon the Babylonians could see, according to their computations, 80/240 or 1/3 of the moon's disc. The fraction 1/3 means the same thing today that it meant three thousand years ago: that is, one of three equal parts of something, and that something could be any finite measurable entity, such as the disc of the moon, the kinetic energy of a particle, the price of potatoes, or the brightness of a star.

Every fraction is the quotient of two numbers. If a and b are any two numbers (except that b cannot equal zero), a/b is a fraction in which a is the *numerator* (or dividend) of the fraction, and b is the *denominator* (or divisor).

4.2
The Signs of Fractions

Every fraction has three signs. When we write 4/2 we mean $+4/+2$, and the sign in front of the fraction is also understood to be positive.

*Florian Cajori, *A History of Mathematics* (New York: The Macmillan Company, 1893), p. 5.

$$\text{Therefore } \frac{4}{2} \text{ means } +\frac{+4}{+2}$$

The *three signs* of every fraction are

1. the sign of the numerator;
2. the sign of the denominator;
3. the sign in front of the fraction.

By using the fraction 4/2, we can demonstrate that if any *two* of these three signs are *reversed*, the fraction keeps the *same value*.

$$+\frac{+4}{+2} = +\left(\frac{+4}{+2}\right) = +(+2) = \underline{2}$$

Reverse the sign of the *numerator* and the *denominator*.

$$+\frac{-4}{-2} = +\left(\frac{-4}{-2}\right) = +(+2) = \underline{2}$$

Therefore, $+\dfrac{+4}{+2} = +\dfrac{-4}{-2} = \underline{2}$

$$+\frac{+4}{+2} = +\left(\frac{+4}{+2}\right) = +(+2) = \underline{2}$$

Reverse the sign of the *numerator* and the sign *in front of* the fraction.

$$-\frac{-4}{+2} = -\left(\frac{-4}{+2}\right) = -(-2) = +2 = \underline{2}$$

Therefore, $+\dfrac{+4}{+2} = -\dfrac{-4}{+2} = \underline{2}$

$$+\frac{+4}{+2} = +\left(\frac{+4}{+2}\right) = +(+2) = \underline{2}$$

Reverse the sign of the *denominator* and the sign *in front of* the fraction.

$$-\frac{+4}{-2} = -\left(\frac{+4}{-2}\right) = -(-2) = +2 = \underline{2}$$

Therefore, $+\dfrac{+4}{+2} = -\dfrac{+4}{-2} = \underline{2}$

In each of the three examples above, we changed *two* of the three signs and the value of the fraction remained the same.

$$+\frac{+4}{+2} = +\frac{-4}{-2} = -\frac{-4}{+2} = -\frac{+4}{-2} = 2$$

Be sure to remember that, in order to change the sign of any *multinomial*, the sign of *every term* in the multinomial must be changed (see Section 2.5). When the numerator or denominator of a fraction is a multinomial, we can change the sign of the numerator or the denominator only by changing the sign of *every term* of the multinomial.

Examples:

(1) $+\dfrac{+4+2}{+2} = +\left(\dfrac{+4+2}{+2}\right) = +\left(\dfrac{+6}{+2}\right) = +(+3) = \underline{3}$

   $+\dfrac{-4-2}{-2} = +\left(\dfrac{-4-2}{-2}\right) = +\left(\dfrac{-6}{-2}\right) = +(+3) = \underline{3}$

   $-\dfrac{-4-2}{+2} = -\left(\dfrac{-4-2}{+2}\right) = -\left(\dfrac{-6}{+2}\right) = -(-3) = \underline{3}$

   $-\dfrac{+4+2}{-2} = -\left(\dfrac{+4+2}{-2}\right) = -\left(\dfrac{+6}{-2}\right) = -(-3) = \underline{3}$

(2) $+\dfrac{a-b}{a} = -\dfrac{-a+b}{a} = -\dfrac{b-a}{a}$

   $+\dfrac{a-b}{a} = -\dfrac{a-b}{-a}$

   $+\dfrac{a-b}{a} = +\dfrac{-a+b}{-a} = +\dfrac{b-a}{-a}$

   Therefore, $\dfrac{a-b}{a} = -\dfrac{b-a}{a} = -\dfrac{a-b}{-a} = \dfrac{b-a}{-a}$

Exercise 48:

Change each of the following to an equal fraction having a positive sign in front of the fraction:

(1) $-\dfrac{7}{8}$

(2) $-\dfrac{-2a}{a+b}$

(3) $-\dfrac{3a}{-5b}$

(4) $-\dfrac{-x}{x+y}$

(5) $-\dfrac{-2}{3}$

(6) $-\dfrac{a-b}{b-a}$

(7) $-\dfrac{-5x}{7y}$

(8) $-\dfrac{-2x-5y}{2x+5y}$

Exercise 49:

Change the following to equal fractions having only positive signs in the numerator:

(1) $\dfrac{-a}{-b}$

(2) $\dfrac{-x}{2x-y}$

(3) $\dfrac{-3a}{5a+2}$

(4) $\dfrac{-x-2y}{-y}$

(5) $\dfrac{-5a-b}{2a-3b}$

(6) $-\dfrac{-x-y}{x+y}$

(7) $\dfrac{-b-3c}{b+3c}$

(8) $-\dfrac{-4a-b}{2a-3b}$

4.3
Rational Numbers; Irrational Numbers; One-to-One Correspondence

Rational Numbers   As noted in Section 1.3, fractions such as $\dfrac{3}{4}$ or $-\dfrac{5}{8}$ are real numbers and, like the integers, they label by number points on the number line. The integers, although infinite in extent, do not identify by number all the points on the number line. Since the set of real numbers can be matched one to one with *every point* on such a line, the set would have to include numbers that represent points on the number line which lie between the points designated by the integers. For example, if we divide the distance between 0 and 1 on the number line into *two* equal parts, then the point that marks that division lies exactly halfway between 0 and 1 and is represented by the number $\dfrac{1}{2}$. The point that lies exactly halfway between 0 and -1 would then be represented by the number $-\dfrac{1}{2}$ (see Figure 4.1). Thus $\dfrac{1}{2}$ and $-\dfrac{1}{2}$ are real numbers because each one represents the exact location of a point on the number line.

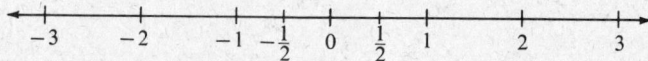

Figure 4.1

The uniform distances between any two integers on the number line may be divided into any desired number of equal parts, and for each point of such sub-divisions, one and only one real number will designate its location. For example, suppose that we chop the distance between 2 and 3 into five equal parts and consider what number would designate the location of the *second point* of the five equal sub-divisions. This point would lie 2 units to the right of zero *plus* $\dfrac{2}{5}$ of the distance between 2 and 3; thus the real number describing the location of this point would be $2+\dfrac{2}{5}$, which we write (rather sloppily) in arithmetic notation as $2\dfrac{2}{5}$. Since $2\dfrac{2}{5}$ really means $2+\dfrac{2}{5}$, it also means $\dfrac{10}{5}+\dfrac{2}{5}$ or $\dfrac{12}{5}$. This latter form of the number, $\dfrac{12}{5}$, is the most convenient way to write it in performing arithmetic computations. In similar fashion, we could chop the distance between -2 and -3 into five equal parts, and the real number designating the second of the five equal subdivisions would be $-2\frac{2}{5}$ or $-\frac{12}{5}$. (See Figure 4.2)

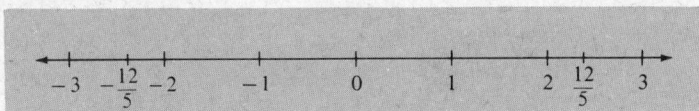

Figure 4.2

These numbers, $\frac{1}{2}$, $-\frac{1}{2}$, $\frac{12}{5}$, and $-\frac{12}{5}$, are all special kinds of fractions or ratios; they belong to a set of numbers which can all be expressed as *the quotient of two integers*, with the provision that the divisor cannot be zero. In symbolic language, all of these particular fractions can be written in the following form:

$$\frac{a}{b} \qquad \left(\begin{array}{l} a \text{ is an } integer \text{ and } b \\ \text{is a } nonzero\ integer \end{array}\right)$$

When we consider all possibilities for numbers that can be written in this form, it eventually becomes evident that *every integer* can also be written in this same form.

Examples:

$$6 = \frac{6}{1} = \frac{12}{2} = \frac{18}{3}, \text{ etc.}$$

$$-24 = -\frac{24}{1} = -\frac{48}{2} = -\frac{72}{3}, \text{ etc.}$$

$$105 = \frac{105}{1} = \frac{210}{2} = \frac{315}{3}, \text{ etc.}$$

Consequently, we can classify all of the real numbers which we have encountered thus far by saying that all of them can be written as the quotient of an integer and a nonzero integer. All such real numbers are called **rational numbers**, the meaning of the word rational in this sense being derived from the word *ratio*.

Definition: Rational numbers are numbers which can be written in the form $\frac{a}{b}$, where a and b are integers and b cannot equal zero.

Exercise 50:

Locate the points on the number line represented by the following numbers:

(1) 3, -7, 5, 4, -10, -12

(2) $\frac{3}{4}$, $-5\frac{2}{3}$, $4\frac{3}{5}$, $-2\frac{7}{10}$, 2.7

From this simple beginning, the real number system grew through a series of enlargements into the set of numbers which represent all the points on a straight line continuing indefinitely. As we have seen, this would obviously include all the points to the right or left of zero whose distance and direction from zero could be designated by a rational number. The question then arises, does this set of rational numbers account for *all* of the points on the number line? The answer is No. It can be shown that there are points on that line whose distance and direction from zero cannot be represented by any rational number. We shall return to this question at the end of this section.

It is a good idea at this point to consider again a peculiar characteristic of the rational numbers, a characteristic that can be illustrated by expressing them as decimals. If we take a series of quotients of integers and do what the symbol indicates (that is, divide the numerator by the denominator), we get the following results:

1. $\frac{1}{2}$ means $1 \div 2$ or $2\overline{)1.0000000}\ldots = .500000\overline{0}\ldots$

2. $\frac{2}{3}$ means $2 \div 3$ or $3\overline{)2.000000}\ldots = .66666\overline{6}\ldots$

3. $\frac{5}{18}$ means $5 \div 18$ or $18\overline{)5.00000000}\ldots = .277777\overline{7}\ldots$

4. $\frac{5}{33}$ means $5 \div 33$ or $33\overline{)5.0000000000}\ldots = .15151515\overline{15}\ldots$

5. $\frac{4}{5}$ means $4 \div 5$ or $5\overline{)4.000000}\ldots = 8.00000\overline{0}\ldots$

6. $\frac{47}{111}$ means $47 \div 111$ or $111\overline{)47.000000000}\ldots = .423423\overline{423}\ldots$

7. $\frac{1}{7}$ means $1 \div 7$ or $7\overline{)1.000000000000}\ldots = .142857142857\overline{142857}\ldots$

The bars placed over the figures in the above examples indicate in each case a repeating sequence which continues indefinitely. Looking over the above results carefully, we see that in numbers 2, 3, 4, 6, and 7 we have a *repeating decimal which never ends.* As a matter of fact, if we consider zero as a repeating figure, then *every* result above may be correctly called a repeating decimal which never ends. Accepting zero as a repeating figure, we find that every integer can also be written as a repeating decimal.

Examples:

$$-6 = -\frac{6}{1} = -6.00000\overline{0}\ldots$$

$$125 = \frac{125}{1} = 125.0000\overline{0}\ldots$$

Exercise 51:

Show that each of the following is equal to a repeating decimal which never ends:

(1) -3 (2) $\dfrac{1}{3}$ (3) $\dfrac{5}{6}$

(4) $\dfrac{2}{7}$ (5) $\dfrac{2}{11}$ (6) 8

(7) $\dfrac{3}{5}$ (8) $\dfrac{17}{33}$ (9) $\dfrac{3}{4}$

If you did the preceding exercise correctly you might be tempted to hazard a guess (and you would be right) that *every rational number* can be written as a repeating decimal. While it is never a part of intelligence to rely on guesswork, lacking the necessary skills at this point to prove the matter, we shall be forced for the present to take it on faith that *every rational number is equal to some repeating decimal.* Furthermore this statement has the marvelous property of being true in reverse (as noted earlier many statements are not!), and it can also be proved that *every repeating decimal is equal to a rational number.*

With this information we can answer conclusively the following question: If the rational numbers are numbers that can be written as the quotient of an integer and a nonzero integer, which of the real numbers are rational numbers?

The *rational numbers* are the following:

1. Any fraction that can be written as the quotient of an integer and a nonzero integer.
2. Every integer, because each can be written as the quotient of an integer and a nonzero integer.
3. Every non-ending, *repeating* decimal, because each can be written as the quotient of an integer and a nonzero integer.

For convenient reference we shall henceforth use the letter Q to represent the set of all rational numbers. As noted in the above statements, this set, Q, contains all of the integers as well as all ratios of integers (excluding zero in the denominator), and all nonending repeating decimals. We can construct a diagram of the set, Q using the region enclosed by any plane figure. For example, let the region enclosed by a semi-circle represent the set of rational numbers (see Figure 4.3).

Figure 4.3

The set Q and its relationship to the sets N and J are shown in Figure 4.4.

Figure 4.4

Note that the diagram previously constructed for the sets N and J is completely contained within the semi-circle representing Q. This arrangement shows that all natural numbers and integers are included in the set of rational numbers. Furthermore, the "Q semi-circle," by enclosing points outside of N and J, shows that there are rational numbers which are neither natural numbers nor integers; the set Q also includes ratios of integers (such as $\frac{2}{3}$, $\frac{5}{33}$, $\frac{19}{37}$, etc.) each of which, as indicated earlier, can be represented by the non-ending repeating decimals: $.66666\overline{6} \ldots$, $.151515\overline{15} \ldots$, $.513513\overline{513} \ldots$.

Exercise 52:

The statement "$6 \in N, J, Q$, and R" is read "the number 6 belongs to the set of natural numbers, the set of integers, the set of rational numbers, and the set of real numbers." Refer to the diagram shown in Figure 4.4 and list all of the sets (N, J, Q, and R) to which each of the following numbers belongs:

(1) -5 (2) $\frac{5}{8}$ (3) $.6$ (4) 0

(5) 24 (6) $3\frac{4}{5}$ (7) $.181818\overline{18} \ldots$ (8) $4.7321321\overline{321}$

(9) $-\frac{4}{7}$ (10) $.999999\overline{9} \ldots$

Irrational Numbers All of the real numbers that we have thus far encountered are contained in the set of rational numbers. However, in later work you will encounter other real numbers which are not rational, that is, numbers that *cannot* be expressed as the quotient of two integers, and therefore numbers that are never equal to *repeating* decimals. A good example of such a number, and probably a familiar one, is $\sqrt{2}$.

Remember that $\sqrt{36} = 6$ because $(6)(6) = 36$; hence $\sqrt{36}$ is a rational number, and so is $\sqrt{25}$ or $\sqrt{16}$. But real numbers such as $\sqrt{2}$, $\sqrt{15}$, $\sqrt[3]{6}$, or $\sqrt[4]{7}$ are not rational; their decimal equivalents are non-ending decimals that never repeat, and they are called *irrational numbers*. As an example, computed to ten decimal places, we have

$$\sqrt{2} = 1.4142135623\ldots$$

and no matter how far the computation is continued, the decimal will never end, nor will it ever repeat. Although such numbers may seem a bit strange at first, it is stranger still that the irrational numbers were known and used by the human race centuries before the negative numbers.

The irrational numbers, a set that we shall henceforth designate by the letter H, will merit further study in later work. For the present we shall merely note their inclusion in the set of real numbers; thus the set of real numbers is the set of all infinite (non-ending) decimals, *both repeating and non-repeating*.

If we employ a semi-circle as a diagram of the set H (Figure 4.5), we may combine

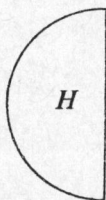

Figure 4.5

it with the earlier "Q semi-circle" to represent a diagram of the set of real numbers, R.

When the elements of two sets are combined to make a third set, the resulting set is called the "union" of the two sets, and the operation is designated by the symbol $\cup$. Hence the union of H and Q is written

$$H \cup Q$$

Since union of these two sets is equivalent to R, we write

$$H \cup Q = R$$

A diagram can be very useful in presenting a composite view of the real number system. As a beginning, let the region enclosed by a circle represent the elements of R (Figure 4.6). But R is made up of two mutually exclusive (disjoint) subsets, Q and H. We can diagram this situation by dividing the "R-circle" into two separate regions so that the two regions enclosed by the following figure, but *not* the boundaries, represent all of the elements of R (Figure 4.7).

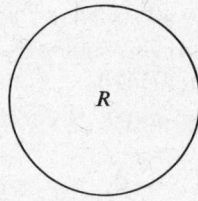

Figure 4.6

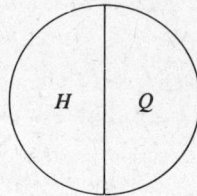

$$H \cup Q = R$$

Figure 4.7

From this diagram we can see at a glance that H and Q are disjoint and that both H and Q are subsets of R. The diagram can also be used to show the relationships of N and J to R. See Figure 4.8.

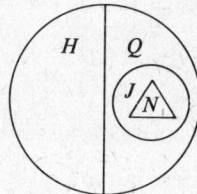

$$H \cup Q = R$$

Figure 4.8

One-to-One Correspondence   The foregoing sections have described the make-up of the real number system in its entirety. This system is actually the evolutionary development of an ancient counting device which preceded all numbers — that of one-to-one correspondence. If a classroom, for example, is equipped with chairs for students, and if each chair is occupied by a student and all students are seated, we can conclude, by matching chairs with students and without counting either, that there are the same number of chairs in the room as there are students. A child who counts off objects on his fingers is making a one-to-one correspondence between his fingers and the objects counted. In this way, counting began before numbers. A prehistoric shepherd who had no notion of numbers but who wished to account for the safe return of his flock from grazing might well have done so by setting aside one stone for each animal that he turned out to pasture in the morning, and then, in the evening, removing from the stack one stone for each animal that returned. Thus any stone or stones remaining in the original stack would represent that many missing animals. In like manner, the real numbers can be paired in one-to-one correspondence with the points on a straight line. Each number of the system by means of symbols (3, 7/8, $\sqrt{5}$, .464646) and signs (+ or −) gives the *distance and direction* of the point from zero. Every real number is either rational or irrational and every one, except zero, must be either positive or negative.

Exercise 53:

(1) Study the diagram in Figure 4.8 and answer the following questions yes or no:

(a) Is $N \subseteq J$? (b) Is $H \subseteq R$? (c) Is $J \subseteq Q$?
(d) Is $N \subseteq H$? (e) Is $H \subseteq Q$? (f) Is $Q \subseteq R$?

(2) Define a rational number.

(3) Can a real number be both rational and irrational?

4.4
Multiplication and Reduction of Fractions

We can investigate any operation with fractions, such as multiplication, by literally drawing a picture of the situation. Suppose, for example, that we wanted to get 1/3 of 1/2 of something. For the sake of simplicity, we will let the "something" be an apple pie (Figure 4.9).

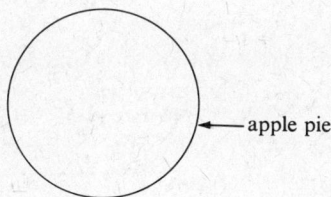
←— apple pie

Figure 4.9

The statement, 1/3 *of* 1/2 means, in plain arithmetic, 1/3 *times* 1/2; so if we take 1/3 of 1/2 of an apple pie, we are actually multiplying 1/3 by 1/2. In order to get 1/2 of the pie, we shall divide the pie into two equal parts, so that each part represents 1/2 of the pie (Figure 4.10).

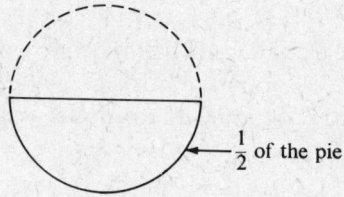

$\frac{1}{2}$ of the pie

Figure 4.10

To get 1/3 of 1/2, we must now divide the 1/2 of the pie into *three equal parts*, as shown in Figure 4.11.

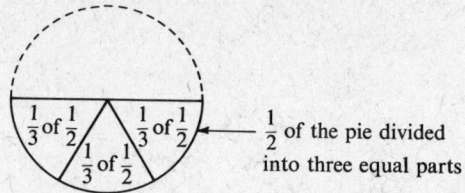

$\frac{1}{2}$ of the pie divided into three equal parts

Figure 4.11

Each wedge of the three equal parts above represents exactly 1/3 of 1/2 of the pie. Now that we have a picture of 1/3 of 1/2 of this pie, how much of the whole pie have we actually got?

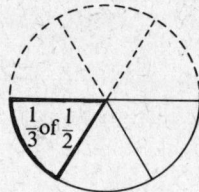

Figure 4.12

As Figure 4.12 shows, when we take 1/3 of 1/2 of the pie, we actually come out with *one* of *six equal parts* of the whole pie, which is exactly 1/6 of the pie. Therefore, when we take 1/3 of 1/2 of anything, we will wind up every time with 1/6 of the whole thing.

$$\frac{1}{3} \text{ of } \frac{1}{2} = \frac{1}{6}$$

$$or \quad \frac{1}{3} \cdot \frac{1}{2} = \frac{1}{6}$$

This example shows what happens when fractions are multiplied together. Because (1/3)(1/2) has to equal 1/6, we need to ask this question: since 1/6 is the *right* answer, how do we multiply (1/3)(1/2) so that we will be sure to come out with 1/6?

$$\frac{1}{3} \cdot \frac{1}{2} \text{ means } \frac{1 \cdot 1}{3 \cdot 2} = \frac{1}{6}$$

The above example tells the whole story of multiplying fractions. We *multiply all of the numerators together and then multiply all of the denominators together*. Since this is guaranteed to give the correct result, it is clearly a sensible way to handle the matter.

For future reference we shall pause here and remind ourselves of the symmetric axiom of equalities. Note this *carefully*.

$$\text{If } \quad \frac{1}{3} \cdot \frac{1}{2} = \frac{1 \cdot 1}{3 \cdot 2}$$

$$\text{then } \frac{1 \cdot 1}{3 \cdot 2} = \frac{1}{3} \cdot \frac{1}{2}$$

Examples:

(1) $\dfrac{1}{2} \cdot \dfrac{1}{2} = \dfrac{1 \cdot 1}{2 \cdot 2} = \dfrac{1}{4}$

(2) $\dfrac{2}{5} \cdot \dfrac{3}{7} = \dfrac{2 \cdot 3}{5 \cdot 7} = \dfrac{6}{35}$

(3) $2 \cdot \dfrac{2}{5} = \dfrac{2}{1} \cdot \dfrac{2}{5} = \dfrac{2 \cdot 2}{1 \cdot 5} = \dfrac{4}{5}$

Note: In multiplying integers by fractions, you can avoid errors by writing each integer in the form of a fraction; e.g., $2 = \dfrac{2}{1}$, $3 = \dfrac{3}{1}$, $4 = \dfrac{4}{1}$, etc. In like manner, $a = \dfrac{a}{1}$, $x^2 = \dfrac{x^2}{1}$, $2x = \dfrac{2x}{1}$, etc.

(4) $\dfrac{5}{7} \cdot 3 = \dfrac{5}{7} \cdot \dfrac{3}{1} = \dfrac{5 \cdot 3}{7 \cdot 1} = \dfrac{15}{7}$

(5) $\dfrac{2}{3} \cdot \dfrac{1}{5} \cdot \dfrac{2}{7} = \dfrac{(2)(1)(2)}{(3)(5)(7)} = \dfrac{4}{105}$

(6) $\dfrac{a}{b} \cdot \dfrac{c}{d} = \dfrac{a \cdot c}{b \cdot d} = \dfrac{ac}{bd}$

(7) $a \cdot \dfrac{a}{b} = \dfrac{a}{1} \cdot \dfrac{a}{b} = \dfrac{a \cdot a}{1 \cdot b} = \dfrac{a^2}{b}$

(8) $\dfrac{x}{y} \cdot \dfrac{a}{b} \cdot \dfrac{c}{d} = \dfrac{x \cdot a \cdot c}{y \cdot b \cdot d} = \dfrac{xac}{ybd} = \dfrac{acx}{bdy}$

The fraction bar (called the vinculum) always has the effect of putting *parentheses* around both the numerator and the denominator of any fraction.

$$\frac{a+b}{x+y} \text{ means } \frac{(a+b)}{(x+y)}$$

Examples:

(1) $\dfrac{a+b}{x+y} \cdot \dfrac{a-b}{x-y} = \dfrac{(a+b)}{(x+y)} \cdot \dfrac{(a-b)}{(x-y)}$

$\qquad\qquad = \dfrac{(a+b)(a-b)}{(x+y)(x-y)} = \dfrac{a^2-b^2}{x^2-y^2}$

(2) $\dfrac{x-2}{x+4} \cdot \dfrac{x+1}{x-3} = \dfrac{(x-2)}{(x+4)} \cdot \dfrac{(x+1)}{(x-3)}$

$\qquad\qquad = \dfrac{(x-2)(x+1)}{(x+4)(x-3)} = \dfrac{x^2-x-2}{x^2+x-12}$

(3) $(x-3) \cdot \dfrac{x+2}{x-5} = \dfrac{(x-3)}{1} \cdot \dfrac{(x+2)}{(x-5)}$

$\qquad\qquad = \dfrac{(x-3)(x+2)}{1(x-5)}$

$\qquad\qquad = \dfrac{x^2-x-6}{x-5}$

Exercise 54:

Multiply the following:

(1) $\dfrac{1}{3} \cdot \dfrac{2}{5}$

(2) $5 \cdot \dfrac{3}{4}$

(3) $\dfrac{2}{3} \cdot \dfrac{4}{7} \cdot \dfrac{-4}{5}$

(4) $\dfrac{3}{5} \cdot 7$

(5) $-\dfrac{1}{2} \cdot -\dfrac{1}{3}$

(6) $3 \cdot \dfrac{-3}{4}$

(7) $\dfrac{1}{a} \cdot b$

(8) $x \cdot \dfrac{x}{y}$

(9) $\dfrac{a}{b} \cdot \dfrac{x}{y}$

(10) $\dfrac{a}{b} \cdot \dfrac{c}{d} \cdot \dfrac{-x}{y}$

(11) $2x \cdot \dfrac{x^2}{y}$

(12) $a \cdot \dfrac{-a}{b} \cdot \dfrac{-c}{d}$

(13) $\dfrac{3a}{b} \cdot \dfrac{-5a^2}{b} \cdot \dfrac{-a}{4b}$

(14) $\dfrac{5x}{y} \cdot \dfrac{2x}{3y^2} \cdot \dfrac{-4}{z}$

(15) $\dfrac{x-2}{x+5} \cdot \dfrac{x+3}{x-1}$

(16) $\dfrac{a+b}{a-b} \cdot \dfrac{a+2b}{a-3b}$

(17) $\dfrac{x-4}{x+7} \cdot (2x+3)$ (See Example 3)

(18) $\dfrac{x-y}{x+y} \cdot (x-y)$

(19) $(a-4) \cdot \dfrac{a-3}{a+7}$

(20) $\dfrac{2x+3}{x-2} \cdot \dfrac{1}{x+3}$

When we begin to perform operations with fractions, we suddenly have reason to appreciate the marvelous convenience of the number *one* in the multiplication process. It has a property in multiplication that no other real number has: when any number is multiplied by *one*, the number remains identically the same.

$$(7)(1) = 7 \qquad\qquad \left(\dfrac{2}{3}\right)(1) = \dfrac{2}{3}$$
$$(-3)(1) = -3 \qquad (a+b)(1) = a+b$$

For this reason the number one is called the *multiplicative identity*, and the property of this number in multiplication can be stated symbolically in the following manner:

There is an element $1 \in R$, $1 \neq 0$, such that for every $a \in R$, $(a)(1) = a$.

The statement $1 \neq 0$ above is read "1 is not equal to 0"; in other words, the *multiplicative* identity is required to be different from the *additive* identity (see Section 1.9).

Furthermore, this number *one* can always be written as a number divided by itself.

$$\dfrac{2}{2} = 1 \qquad \dfrac{3}{3} = 1 \qquad \dfrac{x}{x} = 1 \qquad \dfrac{a+b}{a+b} = 1$$
$$\qquad\qquad\qquad\qquad (x \text{ cannot} = 0) \quad (a+b \text{ cannot} = 0)$$

By using this fact and the substitution axiom, which states that a quantity may be substituted for *its equal* in any expression, we can show that every fraction has infinitely many equivalent forms.

Substitution Axiom

If $a = b$, then b may be substituted for a in any mathematical expression.

Example:

(1) $\dfrac{2}{3} \cdot 1 = \dfrac{2}{3}$

But $1 = \dfrac{2}{2}$

Then $\dfrac{2}{3} \cdot 1 = \dfrac{2}{3} \cdot \dfrac{2}{2} = \dfrac{2 \cdot 2}{3 \cdot 2} = \dfrac{4}{6}$

Therefore $\dfrac{2}{3} = \dfrac{4}{6}$

In the same problem, replace 1 by 3/3.

Then $\dfrac{2}{3} \cdot 1 = \dfrac{2}{3} \cdot \dfrac{3}{3} = \dfrac{2 \cdot 3}{3 \cdot 3} = \dfrac{6}{9}$

Therefore $\dfrac{2}{3} = \dfrac{6}{9}$

In the same problem, replace 1 by 4/4.

Then $\dfrac{2}{3} \cdot 1 = \dfrac{2}{3} \cdot \dfrac{4}{4} = \dfrac{2 \cdot 4}{3 \cdot 4} = \dfrac{8}{12}$

Therefore $\dfrac{2}{3} = \dfrac{8}{12}$

We can repeat the above process endlessly, by letting $1 = 5/5, 6/6, 7/7, 8/8$, etc., and show that

$$\frac{2}{3} = \frac{4}{6} = \frac{6}{9} = \frac{8}{12} = \frac{10}{15} = \frac{12}{18} = \frac{14}{21}, \text{ etc.}$$

Every fraction in the box above has the same value. But one of them is simpler than all the others because it involves the smallest numbers, and that, of course, is the fraction 2/3. The fraction 2/3 is expressed in the lowest integral terms which can equal that particular ratio. In operations with fractions we make sure that each one is expressed in its lowest terms. The process of getting all fractions to their lowest terms is called the reduction of fractions.

To reduce a fraction to its lowest terms, we *factor* both the numerator and the denominator and look for the multiplicative identity.

Examples:

(1) $\dfrac{4}{6} = \dfrac{(2)(2)}{(3)(2)} = \dfrac{2}{3} \cdot \dfrac{2}{2} = \dfrac{2}{3} \cdot 1 = \dfrac{2}{3}$

so $\dfrac{4}{6} = \dfrac{2}{3}$

(2) $\dfrac{21}{35} = \dfrac{(3)(7)}{(7)(5)} = \dfrac{(3)(7)}{(5)(7)} = \dfrac{3}{5} \cdot \dfrac{7}{7} = \dfrac{3}{5} \cdot 1 = \dfrac{3}{5}$

so $\dfrac{21}{35} = \dfrac{3}{5}$

(3) $\dfrac{x^2}{xy} = \dfrac{x \cdot x}{x \cdot y} = \dfrac{x}{x} \cdot \dfrac{x}{y} = 1 \cdot \dfrac{x}{y} = \dfrac{x}{y}$

so $\dfrac{x^2}{xy} = \dfrac{x}{y}$

(4) $\dfrac{a^2 - b^2}{a^2 - 2ab + b^2} = \dfrac{(a + b)(a - b)}{(a - b)(a - b)} = \dfrac{a + b}{a - b} \cdot \dfrac{a - b}{a - b} = \dfrac{a + b}{a - b} \cdot 1 = \dfrac{a + b}{a - b}$

so $\dfrac{a^2 - b^2}{a^2 - 2ab + b^2} = \dfrac{a + b}{a - b}$

In the foregoing examples, we have taken the long way around to show what goes on when a fraction is reduced to a simpler form. In such operations, we usually speed things up a bit by leaving out the last two steps. When the *same factor* appears in both the numerator and denominator of a fraction, it always means that the rest of the fraction has been multiplied by *one*, and we may therefore take what is left as a simpler form of the same fraction. This is why we can cancel common *factors* (but never common *terms*). Repeating two of the above examples we have

1. $\dfrac{4}{6} = \dfrac{(2)(2)}{(3)(2)} = \dfrac{2}{3} \cdot \dfrac{2}{2} = \dfrac{2}{3} \cdot 1 = \dfrac{2}{3}$

or $\dfrac{4}{6} = \dfrac{(2)(\cancel{2})^{1}}{(3)(\cancel{2})_{1}} = \dfrac{2}{3}$

2. $\dfrac{x^2}{xy} = \dfrac{(x)(x)}{(x)(y)} = \dfrac{x}{x} \cdot \dfrac{x}{y} = 1 \cdot \dfrac{x}{y} = \dfrac{x}{y}$

or $\dfrac{x^2}{xy} = \dfrac{(\cancel{x})^{1}(x)}{(\cancel{x})_{1}(y)} = \dfrac{x}{y}$

Examples:

Study the following examples of reduction of fractions carefully:

(1) $\dfrac{9}{36} = \dfrac{(\cancel{3})^{1}(\cancel{3})^{1}}{(\cancel{3})_{1}(\cancel{3})_{1}(4)} = \dfrac{1}{4}$

(2) $\dfrac{a^2b}{ab^2} = \dfrac{(\cancel{a})^{1}(a)(\cancel{b})^{1}}{(\cancel{a})_{1}(b)(\cancel{b})_{1}} = \dfrac{a}{b}$

(3) $\dfrac{8x^3y}{18xy^2} = \dfrac{(4)(\cancel{2})(\cancel{x})(x)(x)(\cancel{y})}{(9)(\cancel{2})(\cancel{x})(y)(\cancel{y})} = \dfrac{4x^2}{9y}$

(4) $\dfrac{3a + 6b}{a + 2b} = \dfrac{3\cancel{(a + 2b)}}{1\cancel{(a + 2b)}} = \dfrac{3}{1} = 3$

(5) $\dfrac{x^2 + 4x}{2x + 8} = \dfrac{x\cancel{(x + 4)}}{2\cancel{(x + 4)}} = \dfrac{x}{2}$

(6) $\dfrac{x^2 - 9}{x^2 - x - 6} = \dfrac{(x + 3)\cancel{(x - 3)}}{(x + 2)\cancel{(x - 3)}} = \dfrac{x + 3}{x + 2}$

(7) $\dfrac{4a - 20}{a^2 - 7a + 10} = \dfrac{4\cancel{(a - 5)}}{(a - 2)\cancel{(a - 5)}} = \dfrac{4}{a - 2}$

Exercise 55:

Reduce the following fractions to their lowest terms:

(1) $\dfrac{18}{39}$

(2) $\dfrac{24}{36}$

(3) $\dfrac{-15}{75}$

(4) $\dfrac{-5}{-35}$

(5) $\dfrac{35}{5}$

(6) $\dfrac{-28}{21}$

(7) $\dfrac{-15}{3}$

(8) $\dfrac{-3x}{-6y}$

(9) $\dfrac{4a}{5a}$

(10) $\dfrac{-2x}{4x}$

(11) $\dfrac{-2x^2}{4x}$

(12) $\dfrac{-2x}{4x^2}$

(13) $\dfrac{-3b}{-7b}$

(14) $\dfrac{-8b^2}{2b}$

(15) $\dfrac{6xy}{3x}$

(16) $\dfrac{-3x}{-6xy}$

(17) $\dfrac{x^3y}{xy^3}$

(18) $\dfrac{xy}{x^2 + xy}$

(19) $\dfrac{2a}{a^2 + ab}$

(20) $\dfrac{3x^2 - 6xy}{3x}$

(21) $\dfrac{2x + 4}{5x + 10}$

(22) $\dfrac{x}{x^2 - x}$

(23) $\dfrac{2a}{3a^2 + 6a}$

(24) $\dfrac{4b^2 - 8b}{2b}$

(25) $\dfrac{2x - 6}{x - 3}$ (26) $\dfrac{x^2 + 5x + 6}{2x + 6}$ (27) $\dfrac{a^2 - b^2}{ab - b^2}$

(28) $\dfrac{b^2 - a^2}{a - b}$ (29) $\dfrac{x^2 - xy + y^2}{x^3 + y^3}$ (30) $\dfrac{a^2b - b^3}{a^2 - 2ab + b^2}$

(31) $\dfrac{m - n}{n - m}$ (32) $\dfrac{6x^2 - 5x - 6}{9x^2 - 4}$ (33) $\dfrac{a^3 - 8}{5a^2 - 11a + 2}$

(34) $\dfrac{28x^2 - x - 2}{35x^2 - 38x + 8}$ (35) $\dfrac{6x^3 - 28x^2 + 16x}{3x^3 - 48x}$ (36) $\dfrac{x^2 - y^2}{x^3 - y^3}$

When fractions are multiplied, all common factors in the numerator and denominator of the product must be divided out in order to get the simplest possible result.

Consider the example

$$\frac{1}{3} \cdot \frac{2}{5} \cdot \frac{3}{4} \cdot \frac{5}{7}$$

When we multiply all the numerators and all the denominators we have the following:

$$\frac{1}{3} \cdot \frac{2}{5} \cdot \frac{3}{4} \cdot \frac{5}{7} = \frac{(1)(2)(3)(5)}{(3)(5)(4)(7)}$$

We now have four *factors* in the numerator and also four *factors* in the denominator; all of the common factors must be divided out.

$$\frac{1}{3} \cdot \frac{2}{5} \cdot \frac{3}{4} \cdot \frac{5}{7} = \frac{(1)(2)(3)(5)}{(3)(5)(4)(7)} = \frac{\overset{1 \cdot 1 \cdot 1}{(1)(\cancel{2})(\cancel{3})(\cancel{5})}}{\underset{1 \cdot 1 \cdot 2}{(\cancel{3})(\cancel{5})(\cancel{4})(7)}} = \frac{1}{14}$$

Since multiplying fractions means that all of the numerators are factors and all of the denominators are factors, it would be considerably more intelligent to divide out the common factors *before* we multiply rather than afterwards.

$$\frac{1}{3} \cdot \frac{2}{5} \cdot \frac{3}{4} \cdot \frac{5}{7} = \frac{(1)(2)(3)(5)}{(3)(5)(4)(7)} = \frac{\overset{1 \cdot 1 \cdot 1}{(1)(\cancel{2})(\cancel{3})(\cancel{5})}}{\underset{1 \cdot 1 \cdot 2}{(\cancel{3})(\cancel{5})(\cancel{4})(7)}} = \frac{1}{14}$$

$$\text{or} \quad \frac{1}{\cancel{3}} \cdot \frac{\overset{1}{\cancel{2}}}{\cancel{5}} \cdot \frac{\overset{1}{\cancel{3}}}{\cancel{4}} \cdot \frac{\overset{1}{\cancel{5}}}{7} = \frac{1}{14}$$

Consequently, when we multiply fractions, we divide out the common factors *first*. With algebraic fractions, the easiest way to find all of the common factors is to factor each numerator and denominator into its prime factors, and also to factor common numerical (integral) factors from any multinomial, before anything else is done.

Examples:

(1) $\dfrac{3x^2}{y^2}\cdot\dfrac{2x^3}{6y^3}\cdot\dfrac{5y^4}{3x} = \dfrac{\overset{1}{\cancel{3}}xx}{\underset{1\cdot1}{\cancel{y}\cancel{y}}}\cdot\dfrac{\overset{1\cdot1}{\cancel{2}xx\cancel{x}}}{\underset{1\cdot1\cdot1}{\cancel{3}\cdot\cancel{2}\cancel{y}\cancel{y}y}}\cdot\dfrac{\overset{1\cdot1\cdot1\cdot1}{5\cancel{y}\cancel{y}\cancel{y}\cancel{y}}}{\underset{1\cdot1}{3\cancel{x}}} = \dfrac{5x^4}{3y}$

(2) $\dfrac{x^2-4}{x^2-4x+4}\cdot\dfrac{x^2-5x+6}{2x-6} = \dfrac{(x+2)\overset{1}{\cancel{(x-2)}}}{\underset{1}{\cancel{(x-2)}}\underset{1}{\cancel{(x-2)}}}\cdot\dfrac{\overset{1}{\cancel{(x-2)}}\overset{1}{\cancel{(x-3)}}}{2\underset{1}{\cancel{(x-3)}}} = \dfrac{x+2}{2}$

(3) $\dfrac{a^3-8}{a^2-6a+8}\cdot\dfrac{a^2-8a+16}{a^2-2a-8}\cdot\dfrac{3a+9}{3a^2+6a+12} =$

$\dfrac{\overset{1}{\cancel{(a-2)}}\overset{1}{\cancel{(a^2+2a+4)}}}{\underset{1}{\cancel{(a-4)}}\underset{1}{\cancel{(a-2)}}}\cdot\dfrac{\overset{1}{\cancel{(a-4)}}\overset{1}{\cancel{(a-4)}}}{\underset{1}{\cancel{(a-4)}}(a+2)}\cdot\dfrac{\overset{1}{\cancel{3}}(a+3)}{\underset{1}{\cancel{3}}\underset{1}{\cancel{(a^2+2a+4)}}} = \dfrac{a+3}{a+2}$

(4) $5\cdot\dfrac{3x+6}{10x^2+30x+20} = \dfrac{5}{1}\cdot\dfrac{3(x+2)}{10(x^2+3x+2)} = \dfrac{5}{1}\cdot\dfrac{3(x+2)}{10(x+2)(x+1)}$

$= \dfrac{\overset{1}{\cancel{5}}(3)\overset{1}{\cancel{(x+2)}}}{\underset{2}{\cancel{(10)}}\underset{1}{\cancel{(x+2)}}(x+1)} = \dfrac{3}{2(x+1)}$

(5) $x\left(1+\dfrac{1}{x}\right) = \dfrac{x}{1}\left(1+\dfrac{1}{x}\right)$

$= \dfrac{x}{1}\cdot1+\dfrac{\overset{1}{\cancel{x}}}{1}\cdot\dfrac{1}{\underset{1}{\cancel{x}}}$

$= x+1$

(6) $(a+b)\left(\dfrac{1}{a^2-b^2}\right) = \dfrac{a+b}{1}\cdot\dfrac{1}{a^2-b^2}$

$= \dfrac{\overset{1}{\cancel{(a+b)}}}{1}\cdot\dfrac{1}{\underset{1}{\cancel{(a+b)}}(a-b)}$

$= \dfrac{1}{a-b}$

Exercise 56:

Multiply the following:

(1) $\left(\dfrac{2}{3}\right)^3$

(2) $\dfrac{3x^2}{4y}\cdot\dfrac{8y^3}{2x^4}\cdot\dfrac{6x^2}{y^2}$

(3) $\dfrac{3}{x} \cdot \dfrac{x^2 - x}{3x - 3}$

(4) $\dfrac{a}{2} \cdot \dfrac{2x^2 + 6}{a^2 + a}$

(5) $(x - 1)\left(\dfrac{2}{x^2 - 1}\right)$

(6) $(a + b)\left(\dfrac{4}{a^2 - b^2}\right)$

(7) $\left(\dfrac{3a}{a^2 + 6a + 8}\right)(a + 2)$

(8) $\left(\dfrac{2y}{x^3 - y^3}\right)(x - y)$

(9) $b\left(a + \dfrac{1}{b}\right)$

(10) $xy\left(\dfrac{1}{x} - \dfrac{1}{y}\right)$

(11) $2a\left(1 - \dfrac{1}{2a}\right)$

(12) $ab\left(\dfrac{2}{a} + \dfrac{3}{b}\right)$

(13) $\dfrac{a}{2x - 4} \cdot \dfrac{x^2 - 4}{ax - 2a}$

(14) $\dfrac{a + 3}{a - 2} \cdot \dfrac{a^2 + 3a - 10}{a^2 - 9}$

(15) $\dfrac{2x^2 + 5x + 2}{6x^2 - 4x} \cdot \dfrac{x^3 - 25x}{6x^2 + 33x + 15} \cdot \dfrac{3x - 2}{x^2 + 7x + 10}$

(16) $\dfrac{a^3 - 1}{a^2 - 1} \cdot \dfrac{a + 3}{a^3 + a^2 + a} \cdot \dfrac{a^2 - 3a}{a^2 - 9}$

(17) $\dfrac{2b^2 - 11b - 21}{3b^2 + 11b - 4} \cdot \dfrac{9b^2 - 1}{3b^2 - 20b - 7}$

(18) $\dfrac{6x^2 - 7x - 3}{15x^2 - x - 2} \cdot \dfrac{5x^2 + 3x - 2}{2x^2 - 3x}$

(19) $\left(\dfrac{2x + 4}{x^2 + 5x + 6}\right)(x^2 - 9)\left(\dfrac{x}{x - 3}\right)$

(20) $\left(\dfrac{a^2 - 5a + 4}{a^2 - 4}\right)\left(\dfrac{a^2 - 4a + 4}{a^2 - a}\right)(a^2 + 3a + 2)$

(21) $\dfrac{x^2 + 2x + 4}{3x^2 + 5x - 2} \cdot \dfrac{x^2 - 4}{3x^2 - 9x} \cdot \dfrac{6x^2 - 2x}{x^3 - 8}$

(22) $\dfrac{2a^2 + 3a - 20}{4a^2 - 20a + 25} \cdot \dfrac{a^2 - 4a}{a^2 - 16} \cdot \dfrac{2a - 5}{3a^2 + 12a}$

4.5
Division of Fractions

For every element $a \in R$ *except* zero, there is another element of R which is called the *reciprocal* or the *multiplicative inverse* of the given element. Excluding zero, every real number has one multiplicative inverse just as every human being has one head. The relationship between a number and its multiplicative inverse is that the *product* of the two is always equal to the number *one*. The following table lists numbers in the left column and the multiplicative inverse (or reciprocal) of each number in the center column.

Number	Multiplicative inverse or reciprocal	Relationship between the two numbers
7	$\dfrac{1}{7}$	$7 \cdot \dfrac{1}{7} = \dfrac{7}{7} = 1$
3	$\dfrac{1}{3}$	$3 \cdot \dfrac{1}{3} = \dfrac{3}{3} = 1$
-2	$-\dfrac{1}{2}$	$-2 \cdot -\dfrac{1}{2} = -\dfrac{-2}{-2} = 1$
$\dfrac{3}{4}$	$\dfrac{4}{3}$	$\dfrac{3}{4} \cdot \dfrac{4}{3} = \dfrac{12}{12} = 1$
$-\dfrac{2}{5}$	$-\dfrac{5}{2}$	$-\dfrac{2}{5} \cdot -\dfrac{5}{2} = \dfrac{10}{10} = 1$
a	$\dfrac{1}{a}$	$a \cdot \dfrac{1}{a} = \dfrac{a}{a} = 1$
$\dfrac{1}{b}$	b	$\dfrac{1}{b} \cdot b = \dfrac{b}{b} = 1$
$x + y$	$\dfrac{1}{x + y}$	$(x + y) \cdot \dfrac{1}{(x + y)} = \dfrac{x + y}{x + y} = 1$
$\dfrac{a + b}{ab}$	$\dfrac{ab}{a + b}$	$\dfrac{\frac{1}{(a+b)}}{\frac{(ab)}{1}} \cdot \dfrac{\frac{1}{(ab)}}{\frac{(a+b)}{1}} = 1$

To state this fact symbolically we would write

For every $a \in R$ except $a = 0$, there is a number $1/a \in R$ such that $a \cdot 1/a = 1$

The number 0 cannot have a multiplicative inverse because $1/0$ is undefined (see Section 1.9).

The importance of all this springs from the fact that division is the inverse of multiplication. When we divide one number by another, we actually *multiply* the dividend by the multiplicative inverse of the *divisor*.

Consider the following:

$$12 \div 3$$

dividend   divisor

The *divisor* is the number 3.

The multiplicative inverse of 3 is $\frac{1}{3}$.

$12 \div 3$ *means* $12 \cdot \frac{1}{3}$

$$12 \div 3 = 12 \cdot \frac{1}{3} = \frac{12}{1} \cdot \frac{1}{3} = \frac{12}{3} = 4$$

The example above illustrates the process of division as it applies to all numbers, and that process is to *multiply the dividend by the multiplicative inverse of the divisor*.

Examples:

(1) $27 \div 9 = 27 \cdot \frac{1}{9} = \frac{\overset{3}{\cancel{27}}}{1} \cdot \frac{1}{\underset{1}{\cancel{9}}} = 3$

(2) $20 \div 4 = 20 \cdot \frac{1}{4} = \frac{\overset{5}{\cancel{20}}}{1} \cdot \frac{1}{\underset{1}{\cancel{4}}} = 5$

(3) $20 \div \frac{1}{4} = 20 \cdot 4 = 80$

(4) $a \div b = a \cdot \frac{1}{b} = \frac{a}{1} \cdot \frac{1}{b} = \frac{a}{b}$

(5) $\frac{1}{5} \div \frac{3}{5} = \frac{1}{\underset{1}{\cancel{5}}} \cdot \frac{\overset{1}{\cancel{5}}}{3} = \frac{1}{3}$

(6) $\frac{2}{7} \div \frac{4}{5} = \frac{\overset{1}{\cancel{2}}}{7} \cdot \frac{5}{\underset{2}{\cancel{4}}} = \frac{5}{14}$

(7) $\frac{3}{4} \div 4 = \frac{3}{4} \cdot \frac{1}{4} = \frac{3}{16}$

(8) $\frac{a^2 + ab}{a - b} \div (a + b) = \frac{a\overset{1}{\cancel{(a+b)}}}{(a-b)} \cdot \frac{1}{\underset{1}{\cancel{(a+b)}}} = \frac{a}{a - b}$

(9) $\dfrac{a^2 - 9}{a^2 + 4a + 4} \div \dfrac{a - 3}{a + 2} = \dfrac{a^2 - 9}{a^2 + 4a + 4} \cdot \dfrac{a + 2}{a - 3}$

$$= \dfrac{(a + 3)\overset{1}{\cancel{(a - 3)}}}{\cancel{(a + 2)}(a + 2)} \cdot \dfrac{\overset{1}{\cancel{(a + 2)}}}{\cancel{(a - 3)}} = \dfrac{a + 3}{a + 2}$$

In dividing fractions, therefore, we simply take the multiplicative inverse of the divisor and then proceed exactly as we did in multiplying fractions.

Examples:

(1) $\dfrac{x^2 - 4}{2x + 4} \cdot \dfrac{x^2 - 5x + 6}{x^2 - 4x + 4} \div \dfrac{3x - 9}{x - 2}$

$\dfrac{x^2 - 4}{2x + 4} \cdot \dfrac{x^2 - 5x + 6}{x^2 - 4x + 4} \cdot \dfrac{x - 2}{3x - 9}$

$\dfrac{\overset{1}{\cancel{(x + 2)}}\overset{1}{\cancel{(x - 2)}}}{2\cancel{(x + 2)}} \cdot \dfrac{\overset{1}{\cancel{(x - 3)}}\overset{1}{\cancel{(x - 2)}}}{\cancel{(x - 2)}\cancel{(x - 2)}} \cdot \dfrac{(x - 2)}{3\cancel{(x - 3)}} = \dfrac{x - 2}{6}$

(2) $\dfrac{6a^2 - a - 2}{3a^2 - 5a + 2} \div (2a + 1)$

$\dfrac{(3a - 2)(2a + 1)}{(a - 1)(3a - 2)} \cdot \dfrac{1}{(2a + 1)}$

$\dfrac{\overset{1}{\cancel{(3a - 2)}}\overset{1}{\cancel{(2a + 1)}}}{(a - 1)\cancel{(3a - 2)}} \cdot \dfrac{1}{\cancel{(2a + 1)}} = \dfrac{1}{a - 1}$

Exercise 57:

Perform the indicated operations.

(1) $\dfrac{3}{7} \div \dfrac{4}{7}$

(2) $\dfrac{3}{8} \div 4$

(3) $\dfrac{3}{5} \div \dfrac{5}{6}$

(4) $18 \cdot \dfrac{2}{3} \div \dfrac{1}{3}$

(5) $\dfrac{3}{8} \div 8$

(6) $9 \div \dfrac{3}{7}$

(7) $8 \cdot \dfrac{3}{4} \div 4$

(8) $\dfrac{2}{3} \cdot 12 \div \dfrac{1}{2}$

(9) $\dfrac{2x}{y^2} \div \dfrac{x}{y}$

(10) $\dfrac{a^2}{b} \div a$

(11) $\dfrac{1}{2} \cdot \dfrac{3}{5} \div \dfrac{2}{5}$

(12) $\dfrac{4x^2 y}{5} \cdot \dfrac{3x}{4y^2} \div \dfrac{x^2}{2y}$

(13) $\dfrac{x^2 - 4}{x^2 - 4x + 3} \div \dfrac{3x + 6}{x^2 - 9}$ (14) $\dfrac{x^2 + 4x + 4}{x + 3} \div (x + 2)$

(15) $(2a^2 + 6a)\left(\dfrac{a^2 - 4}{6a^2}\right) \div \dfrac{2 - a}{3a}$ (16) $\dfrac{2a^2 b}{5} \cdot \dfrac{15b}{a^3} \div \dfrac{6}{b^2}$

(17) $\dfrac{a^2 - 3a - 4}{a + 2} \div (a - 4)$ (18) $\dfrac{y^2 - 7y + 10}{y + 3} \div (y - 2)$

(19) $\left(\dfrac{x^2 - 16}{x^2 + 4x + 4}\right)(x + 2) \div (x + 4)$ (20) $\dfrac{a - 1}{2a + 6} \cdot \dfrac{a^2 + a + 1}{a^2 + 6a + 9} \div \dfrac{a^3 - 1}{a^2 - 9}$

(21) $\dfrac{x^3 + 27}{2x + 1} \cdot \dfrac{2x^2 - 5x - 3}{2x^2 - 6x + 18} \div (x^2 - 9)$

(22) $\dfrac{x^2 - 7x + 12}{x^2 - 4} \cdot \dfrac{x^2 + x - 2}{3x - 2} \div \dfrac{2x - 6}{x - 2}$

(23) $\dfrac{2x^2 - 5x + 2}{2x^2 + 11x + 15} \cdot \dfrac{2x^2 + 9x + 10}{x^3 - 8} \div \dfrac{2x^2 + 3x - 2}{x^2 + 2x + 4}$

(24) $\dfrac{a^2 - 4a}{3a^2 - 2a - 1} \cdot \dfrac{a^2 + 7a + 12}{a^2 - 16} \div \dfrac{a^2 + 3a}{3a + 1}$

(25) $\dfrac{x^2 - 5x - 6}{x^2 - 3x + 2} \cdot \dfrac{4x^2}{x^2 + 2x + 1} \div \dfrac{2x^2 - 12x}{x^2 - 1}$

(26) $\dfrac{5a^2 + 9a - 2}{3a + 15} \cdot \dfrac{2a^2 - 50}{3a^2 + 7a + 2} \div \dfrac{5a - 1}{6a^2 + 2a}$

4.6
Addition and Subtraction of Fractions

There is a small phenomenon connected with the addition (or subtraction) of fractions, and that is that some students persist in trying to make a mystery of it. Or, at least, they view the addition of fractions as something quite different from all other addition. This is pure, unthinking nonsense. Addition means the same thing all the time. And since subtraction means to reverse the sign of the subtrahend and add, we shall limit our discussion to addition.

The addition of fractions in algebra, trigonometry, calculus, or anywhere is exactly the same process that you followed in adding fractions in Grades Four and Five. If you remember fifth-grade arithmetic, then you already know all there is to know about adding *any* fractions; but you may not know that you know it.

The only problem in adding fractions is that the student must be acutely conscious of *what addition actually means*. As pointed out repeatedly in Section 2.4, in addition we can combine only those *terms which are exactly alike*. Now, why is this true? The answer is simple: it is true because we accept the truth of the distributive axiom.

If you will look back over the procedure for factoring common factors (Section 3.8), it may become evident to you that the distributive axiom is an infallible guide in the matter of addition: e.g.,

$$ab + ac = a(b + c)$$

Thus

(1) $4x + 2x = 6x$ *because*

$$4x + 2x = x(4 + 2) = x(6) = 6x$$

In like manner

(2) $\dfrac{4}{7} + \dfrac{2}{7} = \dfrac{6}{7}$ *because*

$$\frac{4}{7} + \frac{2}{7} = \frac{1}{7}(4 + 2) = \frac{1}{7}(6) = \frac{6}{7}$$

or

$$\frac{4}{7} + \frac{2}{7} = \frac{4 + 2}{7} = \frac{6}{7}$$

And

(3) $\dfrac{4}{9} + \dfrac{5}{9} - \dfrac{1}{9} = \dfrac{8}{9}$ *because*

$$\frac{4}{9} + \frac{5}{9} - \frac{1}{9} = \frac{1}{9}(4 + 5 - 1) = \frac{1}{9}(8) = \frac{8}{9}$$

or

$$\frac{4}{9} + \frac{5}{9} - \frac{1}{9} = \frac{4 + 5 - 1}{9} = \frac{8}{9}$$

In the above examples we were adding quantities that were *all* x's, or *all* sevenths, or *all* ninths. When the terms to be added are *not* exactly the same kinds of things, we cannot combine them. Note the following:

(4) $4x + 2y = 4x + 2y$

(5) $\dfrac{4}{7} + \dfrac{2}{5} = \dfrac{4}{7} + \dfrac{2}{5}$

In Example 5 above we cannot combine sevenths and fifths any more than we can combine x's and y's in Example 4. In any fraction, the *denominator* serves as a label which identifies the kind of thing we have. Fractions, then, are the same kinds of things *only when their denominators are exactly alike*. Therefore, the addition of fractions boils down to one simple common-sense routine:

Never combine fractions by addition or subtraction unless their denominators are exactly alike.

Please stop right here and read (and reread) the screened statement above until you are confident that it is lodged in your permanent memory. Then go back and study Examples 2 and 3 above carefully. These examples illustrate that fractions having the same (or a common) denominator may be added by adding the terms of their *numerators* and putting the result over the *common denominator*. Thus

$$\frac{5}{11} + \frac{4}{11} - \frac{2}{11} = \frac{5+4-2}{11} = \frac{7}{11}$$

so

$$\frac{5}{x} + \frac{4}{x} - \frac{2}{x} = \frac{5+4-2}{x} = \frac{7}{x}$$

In like manner,

$$\frac{5}{a+b} + \frac{4}{a+b} - \frac{2}{a+b} = \frac{5+4-2}{a+b} = \frac{7}{a+b}$$

But remember that in addition only similar terms can be combined; hence

$$\frac{x}{a+b} + \frac{4}{a+b} - \frac{2}{a+b} = \frac{x+4-2}{a+b} = \frac{x+2}{a+b}$$

where the result, $\dfrac{x+2}{a+b}$, is the simplest result obtainable.

When the numerators of fractions to be added contain *more than one term*, extreme care must be taken to avoid errors in signs. Compare the following two examples very carefully, and remember that every fraction bar implies a parentheses around numerator and denominator.

Examples:

(1) $$\frac{2x}{a+b} - \frac{5}{a+b} = \frac{(2x) - (5)}{a+b} = \frac{2x-5}{a+b}$$

(2) $$\frac{2x}{a+b} - \frac{x-5}{a+b} = \frac{(2x) - (x-5)}{a+b} = \frac{2x-x+5}{a+b} = \frac{x+5}{a+b}$$

In Example 2 above note that the minus sign in front of the fraction $\dfrac{x-5}{a+b}$ affects the sign of *every term in the numerator* when the fractions are combined.

Following are further examples of adding fractions that have the same denominators:

(3) $$\frac{2}{13} + \frac{5}{13} - \frac{3}{13} = \frac{2+5-3}{13} = \frac{4}{13}$$

(4) $\dfrac{a}{13} + \dfrac{b}{13} - \dfrac{c}{13} = \dfrac{a + b - c}{13}$

(5) $\dfrac{2}{13} + \dfrac{5x}{13} - \dfrac{3}{13} = \dfrac{2 + 5x - 3}{13} = \dfrac{5x - 1}{13}$

(6) $\dfrac{2}{a} + \dfrac{5x}{a} - \dfrac{3}{a} = \dfrac{2 + 5x - 3}{a} = \dfrac{5x - 1}{a}$

(7) $\dfrac{2}{3a + b} - \dfrac{5x}{3a + b} - \dfrac{3}{3a + b}$

$\qquad = \dfrac{2 + 5x - 3}{3a + b}$

$\qquad = \dfrac{5x - 1}{3a + b}$

(8) $\dfrac{2}{3a + b} + \dfrac{5x}{3a + b} - \dfrac{2x - 3}{3a + b}$

$\qquad = \dfrac{2 + 5x - (2x - 3)}{3a + b}$

$\qquad = \dfrac{2 + 5x - 2x + 3}{3a + b}$

$\qquad = \dfrac{3x + 5}{3a + b}$

(9) $\dfrac{2x - 4}{3a + b} - \dfrac{5x - 1}{3a + b} + \dfrac{6x + 2}{3a + b}$

$\qquad = \dfrac{(2x - 4) - (5x - 1) + (6x + 2)}{3a + b}$

$\qquad = \dfrac{2x - 4 - 5x + 1 + 6x + 2}{3a + b}$

$\qquad = \dfrac{3x - 1}{3a + b}$

Exercise 58:

Add the following:

(1) $\dfrac{1}{5} + \dfrac{2}{5}$

(2) $\dfrac{5}{10} + \dfrac{4}{10} - \dfrac{2}{10}$

(3) $\dfrac{1}{x} + \dfrac{2}{x}$

(4) $\dfrac{5}{a} + \dfrac{4}{a} - \dfrac{2}{a}$

(5) $\dfrac{1}{x + 3} + \dfrac{2}{x + 3}$

(6) $\dfrac{5}{7} + \dfrac{4}{7} - \dfrac{2}{7}$

(7) $\dfrac{x}{5} + \dfrac{2}{5}$

(8) $\dfrac{a}{7} + \dfrac{b}{7} - \dfrac{c}{7}$

(9) $\dfrac{3}{5} - \dfrac{2}{5}$

(10) $\dfrac{3}{x} + \dfrac{4}{x} - \dfrac{2}{x}$

(11) $\dfrac{x}{a} - \dfrac{2}{a}$

(12) $\dfrac{a}{x} + \dfrac{b}{x} - \dfrac{c}{x}$

(13) $\dfrac{x}{x+3} - \dfrac{2}{x+3}$

(14) $\dfrac{x}{x+1} + \dfrac{3x}{x+1} - \dfrac{5}{x+1}$

(15) $\dfrac{x}{(x+2)(x-1)} - \dfrac{2}{(x+2)(x-1)}$

(16) $\dfrac{x}{(x+1)(x-4)} + \dfrac{3x}{(x+1)(x-4)} - \dfrac{5}{(x+1)(x-4)}$

(17) $\dfrac{2b+1}{b-3} + \dfrac{b-1}{b-3}$

(18) $\dfrac{x}{7} + \dfrac{3x}{7} - \dfrac{5}{7}$

(19) $\dfrac{2b+1}{b-3} - \dfrac{b-1}{b-3}$ (See Example 2 above)

(20) $\dfrac{a+4}{a-1} + \dfrac{2a-3}{a-1}$

(21) $\dfrac{4a+3}{a+1} + \dfrac{2a-4}{a+1}$

(22) $\dfrac{a+4}{a-1} - \dfrac{2a-3}{a-1}$

(23) $\dfrac{4a+3}{a+1} - \dfrac{2a-4}{a+1}$

(24) $\dfrac{4x+1}{x+1} + \dfrac{3x-1}{x+1}$

(25) $\dfrac{2a}{a+1} - \dfrac{a-3}{a+1} + \dfrac{2a-4}{a+1}$

(26) $\dfrac{4x+1}{x+1} - \dfrac{3x-1}{x+1}$

(27) $\dfrac{6}{(x+3)(x+2)} - \dfrac{3}{(x+3)(x+2)} + \dfrac{2}{(x+3)(x+2)}$

(28) $\dfrac{2x+3}{(x+2)(x-1)} - \dfrac{x-2}{(x+2)(x-1)}$

(29) $\dfrac{3x+4}{(x+3)(x+2)} - \dfrac{2x-1}{(x+3)(x+2)} + \dfrac{5x+1}{(x+3)(x+2)}$

(30) $\dfrac{3x-1}{(x+2)(x-1)} - \dfrac{2x+3}{(x+2)(x-1)} + \dfrac{5x}{(x+2)(x-1)}$

(31) $\dfrac{4x^2+1}{(x+3)(x+2)} - \dfrac{x^2-2}{(x+3)(x+2)} - \dfrac{x^2+3}{(x+3)(x+2)}$

(32) $\dfrac{5a+3}{(a+1)(a-2)} - \dfrac{a-4}{(a+1)(a-2)} - \dfrac{2a-1}{(a+1)(a-2)}$

Reducing Answers to the Simplest Form   In the foregoing examples of adding fractions, all the results of the addition led to fractions that were in their simplest forms,
i.e., fractions that could not be reduced. It frequently happens, however, that when
fractions are added or subtracted, the initial result is *not* in its simplest form, and
when that occurs, the initial answer *must be reduced*. Note the following illustrations
carefully:

$$\frac{2}{9} + \frac{1}{9} + \frac{2+1}{9} = \frac{3}{9}$$

The initial result, $\frac{3}{9}$, has the common factor 3 in both numerator and denominator,
so we write

$$\frac{2}{9} + \frac{1}{9} = \frac{2+1}{9} = \frac{3}{9} = \frac{\cancel{(3)}^{1}(1)}{\cancel{(3)}(3)_{1}} = \frac{1}{3}$$

Similarly,

$$\frac{x^2}{x^2 - y^2} + \frac{xy}{x^2 - y^2} = \frac{x^2 + xy}{x^2 - y^2} = \frac{x\cancel{(x+y)}^{1}}{\cancel{(x+y)}(x-y)_{1}} = \frac{x}{x-y}$$

Study the following examples of adding fractions in which the initial result must be
reduced to a simpler form:

(1) $\dfrac{7}{10} + \dfrac{9}{10} - \dfrac{4}{10} = \dfrac{7+9-4}{10} = \dfrac{12}{10} = \dfrac{(6)\cancel{(2)}^{1}}{(5)\cancel{(2)}_{1}} = \dfrac{6}{5}$

(2) $\dfrac{a}{a+b} + \dfrac{b}{a+b} = \dfrac{a+b}{a+b} = \dfrac{1\cancel{(a+b)}^{1}}{1\cancel{(a+b)}_{1}} = \dfrac{1}{1} = 1$

(3) $\dfrac{a^2}{a+b} - \dfrac{b^2}{a+b} = \dfrac{a^2 - b^2}{a+b} = \dfrac{\cancel{(a+b)}^{1}(a-b)}{\cancel{(a+b)}_{1}} = a - b$

(4) $\dfrac{2(a-2)}{2a^2+a-3} + \dfrac{7}{2a^2+a-3} = \dfrac{2(a-2)+7}{2a^2+a-3} = \dfrac{2a-4+7}{2a^2+a-3}$

$= \dfrac{\overset{1}{\cancel{(2a+3)}}}{\underset{1}{\cancel{(2a+3)}}(a-1)}$

$= \dfrac{1}{a-1}$

(5) $\dfrac{8(x+3)}{(x+2)(x-6)} - \dfrac{6(x+6)}{(x+2)(x-6)} = \dfrac{8(x+3)-6(x+6)}{(x+2)(x-6)}$

$= \dfrac{8x+24-6x-36}{(x+2)(x-6)}$

$= \dfrac{2x-12}{(x+2)(x-6)}$

$= \dfrac{2\overset{1}{\cancel{(x-6)}}}{(x+2)\underset{1}{\cancel{(x-6)}}}$

$= \dfrac{2}{x+2}$

Exercise 59:

Add the following, and when necessary reduce the result to simplest form:

(1) $\dfrac{5}{8} - \dfrac{3}{8}$ (2) $\dfrac{4}{5} + \dfrac{3}{5}$

(3) $\dfrac{5}{9} + \dfrac{7}{9}$ (4) $\dfrac{5}{7} + \dfrac{9}{7}$

(5) $\dfrac{3}{10} + \dfrac{7}{10} - \dfrac{2}{10}$ (6) $\dfrac{3}{5x} - \dfrac{2}{5x}$

(7) $\dfrac{x}{2} + \dfrac{3x+6}{2}$ (8) $\dfrac{3}{5x} - \dfrac{3-2x}{5x}$

(9) $\dfrac{4y+5}{6y} - \dfrac{2y+3}{6y}$ (10) $\dfrac{5x-1}{x+4} - \dfrac{x-5}{x+4}$

(11) $\dfrac{a^2}{(a+b)(a+3b)} - \dfrac{b^2}{(a+b)(a+3b)}$

(12) $\dfrac{ab}{(a+b)(a-b)} - \dfrac{b^2}{(a+b)(a-b)}$

(13) $\dfrac{2x}{5x+10} + \dfrac{3x}{5x+10}$ (14) $\dfrac{2(x+3)}{5x+10} + \dfrac{3(x-2)}{5x+10}$

(15) $\dfrac{(x + 1)(x + 2)}{x^2 - 1} + \dfrac{2(x - 4)}{x^2 - 1}$ (16) $\dfrac{(x + 2)(x + 3)}{5x^2 + 12x + 7} - \dfrac{(x + 1)(x - 1)}{5x^2 + 12x + 7}$

Adding Fractions That Have Different Denominators   The preceding discussion stressed that fractions cannot be added or subtracted unless their denominators are exactly alike and illustrated the technique of combining fractions having a common (the same) denominator.

But we frequently need to add fractions whose denominators are *not* the same. For example, if I own a lot that is 4/7 of an acre and acquire some adjoining land that is 2/5 of an acre, how much land do I have? In short,

$$\frac{4}{7} + \frac{2}{5} = ?$$

As these fractions are expressed here, they cannot be combined because their denominators are not alike. But as we observed in Section 4.4, every fraction can be written in infinitely many *equivalent* forms. Using the multiplicative identity, $1 = 2/2,\ 3/3,\ 4/4,\ 5/5,\ 6/6,\ 7/7,\ 8/8$, etc., we can show the following:

1. $\dfrac{4}{7} = \dfrac{8}{14} = \dfrac{12}{21} = \dfrac{16}{28} = \boxed{\dfrac{20}{35}} = \dfrac{24}{42} = \dfrac{28}{49} = \dfrac{32}{56} \cdots$

2. $\dfrac{2}{5} = \dfrac{4}{10} = \dfrac{6}{15} = \dfrac{8}{20} = \dfrac{10}{25} = \dfrac{12}{30} = \boxed{\dfrac{14}{35}} = \dfrac{16}{40} \cdots$

As the circled forms indicate

$$\frac{4}{7} = \frac{20}{35} \text{ and } \frac{2}{5} = \frac{14}{35}$$

in which the denominators of the equivalent forms are exactly alike. Consequently we can combine 4/7 and 2/5 if we use equivalent forms having the *same denominator*.

$$\frac{4}{7} + \frac{2}{5} = \frac{20}{35} + \frac{14}{35} = \frac{1}{35}(20 + 14) = \frac{1}{35}(34) = \frac{34}{35}$$

or

$$\frac{4}{7} + \frac{2}{5} = \frac{20}{35} + \frac{14}{35} = \frac{20 + 14}{35} = \frac{34}{35}$$

The above example is an accurate summary of what goes on any time that any fractions are added or subtracted.

The process of making the denominators exactly alike is called finding the *least common denominator* (L.C.D.), because, for every batch of fractions that we could wish to add, there is always a *smallest number* that will serve as a common denominator for all of the fractions. The following process is a foolproof method for finding the least common denominator for any bunch of fractions whatsoever; consequently, if you intend to add fractions of all kinds correctly, now is the time to learn this process.

Let us consider the following problem:

$$\frac{1}{2} + \frac{2}{3} - \frac{5}{6} + \frac{4}{9} - \frac{3}{8}$$

Step One: Recopy the problem just as it is, *except* that each *denominator* is written in its prime factors.

$$\frac{1}{(2)} + \frac{2}{(3)} - \frac{5}{(2)(3)} + \frac{4}{(3)(3)} - \frac{3}{(2)(2)(2)}$$

Step Two: Look over the prime factors in all of the denominators and ask this question: What are the *different* factors that actually appear in all of the denominators? The answer here is that the only *different* factors are (2) and (3).

Step Three: List the different factors —

2 and 3

and ask this question about each one of them: What is the *greatest* number of times that this factor occurs as a factor in any *single one* of the denominators? Answer: The factor (2) occurs *three times* in the last denominator, and the factor (3) occurs *twice* in the next-to-last denominator.

Step Four: Then the factor (2) must be used *three times* as a factor in the L.C.D.; and the factor (3) must be used *twice* as a factor in the L.C.D.

$$\begin{aligned} \text{L.C.D.} &= (2)(2)(2)(3)(3) \\ &= 72 \end{aligned}$$

This means that each of the original fractions can be changed into an *equivalent fraction* having 72 as a denominator.

Step Five: Using the *multiplicative identity*, change each fraction into an equivalent fraction having the same denominator, 72.

$$\frac{1}{(2)} + \frac{2}{(3)} - \frac{5}{(2)(3)} + \frac{4}{(3)(3)} - \frac{3}{(2)(2)(2)}$$

$$\frac{1}{2} \cdot \frac{36}{36} + \frac{2}{3} \cdot \frac{24}{24} - \frac{5}{(2)(3)} \cdot \frac{12}{12} + \frac{4}{(3)(3)} \cdot \frac{8}{8} - \frac{3}{(2)(2)(2)} \cdot \frac{9}{9}$$

$$\frac{36}{72} + \frac{48}{72} - \frac{60}{72} + \frac{32}{72} - \frac{27}{72}$$

Step Six: The above fractions all have the same denominator; therefore, we can combine them by adding the numerators.

$$\frac{36}{72} + \frac{48}{72} - \frac{60}{72} + \frac{32}{72} - \frac{27}{72} = \frac{36 + 48 - 60 + 32 - 27}{72}$$

$$= \frac{29}{72}$$

The six steps above, which you may shorten at will when you are thoroughly familiar with the process, will enable you to add correctly any fractions on the face of the earth. Let us take some algebraic fractions to be added and *repeat these same six steps.* Add the following:

$$\frac{2}{x^2 - 4} + \frac{3}{x + 2} - \frac{1}{2x - 4}$$

Step One: Recopy the problem just as it is except that we write each *denominator* in its prime factors and include also the factoring of common numerical factors that are integers from any multinomial.

$$\frac{2}{(x + 2)(x - 2)} + \frac{3}{(x + 2)} - \frac{1}{2(x - 2)}$$

Step Two: Look over the factors in all of the denominators and ask this question: What are the *different factors* which appear in all of the denominators? The answer here is that the *three different factors* are

$$(x + 2) \text{ and } (x - 2) \text{ and } (2)$$

Step Three: List all of the different factors —

$$(2) \text{ and } (x + 2) \text{ and } (x - 2)$$

and ask this question about each one of them: What is the *greatest* number of times that each factor occurs as a factor in any *single one* of the denominators? Answer: The factor (2) occurs *once* in the last denominator. The factor $(x + 2)$ occurs *once* in the first denominator and *once* in the second denominator (the *greatest* number of times it occurs in any *single* denominator is *once*). The factor $(x - 2)$ occurs *once* in the first denominator and *once* in the last denominator.

Step Four: Then the factors (2), $(x + 2)$, and $(x - 2)$ must each occur as factors *once* in the L.C.D.

$$\text{L.C.D.} = 2(x + 2)(x - 2)$$

This means that each of the original fractions can be changed into an *equivalent* fraction having $2(x + 2)(x - 2)$ as a denominator.

Step Five: Using the multiplicative identity, change each fraction into an equivalent fraction having the *same denominator,* $2(x + 2)(x - 2)$.

$$\frac{2}{(x+2)(x-2)} + \frac{3}{(x+2)} - \frac{1}{2(x-2)}$$

$$\frac{2}{(x+2)(x-2)}\cdot\frac{2}{2} + \frac{3}{(x+2)}\cdot\frac{2(x-2)}{2(x-2)} - \frac{1}{2(x-2)}\cdot\frac{(x+2)}{(x+2)}$$

$$\frac{(2)(2)}{2(x+2)(x-2)} + \frac{(3)(2)(x-2)}{2(x+2)(x-2)} - \frac{1(x+2)}{2(x+2)(x-2)}$$

Step Six: The above fractions all have the same denominator; therefore we can combine them by adding their numerators.

$$\frac{(2)(2)}{2(x+2)(x-2)} + \frac{(3)(2)(x-2)}{2(x+2)(x-2)} - \frac{1(x+2)}{2(x+2)(x-2)} = \frac{4+6(x-2)-1(x+2)}{2(x+2)(x-2)}$$

$$= \frac{4+6x-12-x-2}{2(x+2)(x-2)}$$

$$= \frac{5x-10}{2(x+2)(x-2)}$$

The final answer in the foregoing example is not in its simplest form. Hence we reduce this fraction by factoring the numerator and dividing out the common factor.

$$\frac{5x-10}{2(x+2)(x-2)} = \frac{5\cancel{(x-2)}^{1}}{2(x+2)\cancel{(x-2)}_{1}} = \frac{5}{2(x+2)} \text{ or } \frac{5}{2x+4}$$

Summarizing the previous example in condensed form, we have the following:

$$\frac{2}{x^2-4} + \frac{3}{x+2} - \frac{1}{2x-4} = \frac{2}{(x+2)(x-2)} + \frac{3}{(x+2)} - \frac{1}{2(x-2)}$$

$$= \frac{2}{(x+2)(x-2)}\cdot\frac{2}{2} + \frac{3}{(x+2)}\cdot\frac{2(x-2)}{2(x-2)} - \frac{1}{2(x-2)}\cdot\frac{(x+2)}{(x+2)}$$

$$= \frac{(2)(2)}{2(x+2)(x-2)} + \frac{(3)(2)(x-2)}{2(x+2)(x-2)} - \frac{1(x+2)}{2(x+2)(x-2)}$$

$$= \frac{4+6(x-2)-1(x+2)}{2(x+2)(x-2)}$$

$$= \frac{4+6x-12-x-2}{2(x+2)(x-2)}$$

$$= \frac{5x-10}{2(x+2)(x-2)}$$

$$= \frac{5\cancel{(x-2)}^{1}}{2(x+2)\cancel{(x-2)}_{1}}$$

$$= \frac{5}{2(x+2)} \text{ or } \frac{5}{2x+4}$$

In the following problem, trace the *same six steps:*

$$\frac{x}{x^2 - 9} + \frac{3}{x^2 + 6x + 9} - \frac{1}{2x - 6}$$

1. $\dfrac{x}{(x + 3)(x - 3)} + \dfrac{3}{(x + 3)(x + 3)} - \dfrac{1}{2(x - 3)}$

2. $(x + 3)$ and $(x - 3)$ and (2)
   These are the only *different* factors in all of the denominators.

3. $(x + 3)$ *twice*
   Greatest number of times this factor occurs in a single denominator.
   $(x - 3)$ *once*
   Greatest number of times this factor occurs in a single denominator.
   (2) *once*
   Greatest number of times this factor occurs in a single denominator.

4. L.C.D. $= 2(x + 3)(x + 3)(x - 3)$

5. $\dfrac{x}{(x + 3)(x - 3)} \qquad + \qquad \dfrac{3}{(x + 3)(x + 3)} \qquad - \qquad \dfrac{1}{2(x - 3)}$

   $\dfrac{x}{(x + 3)(x - 3)} \cdot \dfrac{2(x + 3)}{2(x + 3)} + \dfrac{3}{(x + 3)(x + 3)} \cdot \dfrac{2(x - 3)}{2(x - 3)} - \dfrac{1}{2(x + 3)} \cdot \dfrac{(x + 3)(x - 3)}{(x + 3)(x - 3)}$

6. $\dfrac{(2)(x)(x + 3)}{2(x + 3)(x + 3)(x - 3)} + \dfrac{(2)(3)(x - 3)}{2(x + 3)(x + 3)(x - 3)} - \dfrac{1(x + 3)(x + 3)}{2(x + 3)(x + 3)(x - 3)}$

   $$= \frac{2x(x + 3) + 6(x - 3) - 1(x + 3)(x + 3)}{2(x + 3)(x + 3)(x - 3)}$$

   $$= \frac{2x^2 + 6x + 6x - 18 - 1(x^2 + 6x + 9)}{2(x + 3)(x + 3)(x - 3)}$$

   $$= \frac{2x^2 + 6x + 6x - 18 - x^2 - 6x - 9}{2(x + 3)(x + 3)(x - 3)}$$

   $$= \frac{x^2 + 6x - 27}{2(x + 3)(x + 3)(x - 3)}$$

   $$= \frac{(x + 9)\overset{1}{\cancel{(x - 3)}}}{2(x + 3)(x + 3)\underset{1}{\cancel{(x - 3)}}}$$

   $$= \frac{x + 9}{2(x + 3)(x + 3)} \text{ or } \frac{x + 9}{2x^2 + 12x + 18}$$

In adding fractions and integers, it is generally simpler to put all terms in fractional form. Consider the following:

Example 1:

$$\frac{2}{x - 1} + \frac{3}{x^2 - 1} + 4$$

1. $\dfrac{2}{(x-1)} + \dfrac{3}{(x+1)(x-1)} + \dfrac{4}{1}$

2. $(x-1)$ and $(x+1)$ and (1)

3. $(x-1)$ *once*
   $(x+1)$ *once*
   (1) *once*

4. L.C.D. $= (1)(x+1)(x-1) = (x+1)(x-1)$

5. $\dfrac{2}{(x-1)} + \dfrac{3}{(x+1)(x-1)} + \dfrac{4}{1}$

$$= \dfrac{2}{(x-1)} \cdot \dfrac{(x+1)}{(x+1)} + \dfrac{3}{(x+1)(x-1)} + \dfrac{4}{1} \cdot \dfrac{(x+1)(x-1)}{(x+1)(x-1)}$$

$$= \dfrac{2(x+1)}{(x+1)(x-1)} + \dfrac{3}{(x+1)(x-1)} + \dfrac{4(x+1)(x-1)}{(x+1)(x-1)}$$

6. $\dfrac{2(x+1)}{(x+1)(x-1)} + \dfrac{3}{(x+1)(x-1)} + \dfrac{4(x+1)(x-1)}{(x+1)(x-1)}$

$$= \dfrac{2(x+1) + 3 + 4(x+1)(x-1)}{(x+1)(x-1)}$$

$$= \dfrac{2x + 2 + 3 + 4(x^2 - 1)}{(x+1)(x-1)}$$

$$= \dfrac{2x + 2 + 3 + 4x^2 - 4}{(x+1)(x-1)}$$

$$= \dfrac{4x^2 + 2x + 1}{(x+1)(x-1)}$$

Example 2: $a + \dfrac{3}{a}$

$a + \dfrac{3}{a} = \dfrac{a}{1} + \dfrac{3}{a}$ L.C.D $= (1)(a) = a$

$$= \dfrac{a}{1} \cdot \dfrac{a}{a} + \dfrac{3}{a}$$

$$= \dfrac{a^2}{a} + \dfrac{3}{a}$$

$$= \dfrac{a^2 + 3}{a}$$

Example 3: $\dfrac{x^2 + y^2}{x + y} + x - y$

$\dfrac{x^2 + y^2}{x + y} + x - y = \dfrac{x^2 + y^2}{x + y} + \dfrac{x}{1} - \dfrac{y}{1}$

$$= \dfrac{x^2 + y^2}{x + y} + \dfrac{x - y}{1} \qquad \text{L.C.D.} = 1(x + y) = (x + y)$$

$$= \dfrac{x^2 + y^2}{x + y} + \dfrac{(x - y)}{1} \cdot \dfrac{(x + y)}{(x + y)}$$

$$= \dfrac{x^2 + y^2}{x + y} + \dfrac{(x - y)(x + y)}{x + y}$$

$$= \frac{x^2 + y^2 + (x - y)(x + y)}{x + y}$$

$$= \frac{x^2 + y^2 + x^2 - y^2}{x + y}$$

Exercise 60:

Add the following:

(1) $\dfrac{3}{5} - \dfrac{3}{4} + \dfrac{1}{2} - \dfrac{3}{10}$

(2) $\dfrac{a}{b} + \dfrac{c}{d}$

(3) $\dfrac{3}{5} + \dfrac{1}{x}$

(4) $\dfrac{2}{7} - \dfrac{1}{5} + 3$

(5) $\dfrac{1}{a} + \dfrac{1}{b} + \dfrac{1}{c}$

(6) $\dfrac{3}{x^2} - \dfrac{2}{xy} + \dfrac{3}{y^2}$

(7) $\dfrac{2}{a^2} - \dfrac{3}{ab} + 2$

(8) $x + \dfrac{1}{x}$

(9) $3 + \dfrac{x}{y}$

(10) $\dfrac{1}{a} - \dfrac{1}{b}$

(11) $1 - \dfrac{a}{b}$

(12) $y + \dfrac{2}{y}$

(13) $\dfrac{3}{a} + \dfrac{5}{a^2 + 3a}$

(14) $\dfrac{2}{x^2 - 4x} - \dfrac{3}{x}$

(15) $\dfrac{2}{x} - \dfrac{3}{x^2 + 2x}$

(16) $\dfrac{a}{a^2 - 9} + \dfrac{2}{a^2 + 6a + 9} - \dfrac{1}{2a - 6}$

(17) $\dfrac{3}{x^2 - 16} - \dfrac{2}{x + 4} - \dfrac{1}{x - 4}$

(18) $\dfrac{x}{x^2 - 5x + 6} + \dfrac{2}{3x - 9} - \dfrac{3}{x - 2}$

(19) $\dfrac{3x}{x^3 - 1} + \dfrac{2}{x^2 + x + 1}$

(20) $\dfrac{x + 4}{x^2 - 4x - 12} + \dfrac{2}{x - 6} - \dfrac{3}{x^2 + 4x + 4}$

(21) $\dfrac{a^2 + b^2}{a + b} + a - b$

(22) $2 + \dfrac{1}{x + y}$

(23) $3 - \dfrac{1}{x - y}$

(24) $\dfrac{2}{x + y} + x - y$

(25) $\dfrac{2}{a^2 - 4} + \dfrac{3}{a^2 - 2a}$

(26) $\dfrac{4}{3a + 3} + \dfrac{2}{a^2 - 1} + a$

(27) $\dfrac{2}{x^2 + 3x} - \dfrac{1}{x^2 + 6x + 9} + \dfrac{2}{3x + 9}$

(28) $\dfrac{3x}{2x + 1} + \dfrac{5}{6x^2 - 4x} - \dfrac{x + 2}{6x^2 - x - 2}$

(29) $\dfrac{a}{a^3 - 8} + \dfrac{2}{a - 2} - \dfrac{a + 3}{a^3 + 2a + 4}$

(30) $\dfrac{2x - 1}{x^2 - 2x - 15} + \dfrac{3}{x - 5} + 2x$

(31) $\dfrac{1}{2x^2 - x - 3} - \dfrac{2}{3x^2 + 2x - 1} + \dfrac{3x}{6x^2 - 11x + 3}$

(32) $\dfrac{x - 1}{x^2 - 4} - \dfrac{x + 1}{x^2 + 4x + 4} + \dfrac{2}{x + 2}$

Exercise 61:

Perform the indicated operations.

(1) $\dfrac{2}{3} + \dfrac{1}{2}$ (2) $\dfrac{2}{3} - \dfrac{1}{2}$

(3) $\dfrac{2}{3} \cdot \dfrac{1}{2}$ (4) $\dfrac{2}{3} \div \dfrac{1}{2}$

(5) $\dfrac{a}{b} + \dfrac{c}{d}$ (6) $\dfrac{a}{b} - \dfrac{c}{d}$

(7) $\dfrac{a}{b} \cdot \dfrac{c}{d}$ (8) $\dfrac{a}{b} \div \dfrac{c}{d}$

(9) $\dfrac{2}{x} + y$ (10) $\dfrac{2}{x} - y$

(11) $\dfrac{2}{x} \cdot y$ (12) $\dfrac{2}{x} \div y$

(13) $\dfrac{3y}{x^2} + \dfrac{x}{2y} - \dfrac{y}{3x}$ (14) $\dfrac{3y}{x^2} \cdot \dfrac{x}{2y} \div \dfrac{y}{3x}$

(15) $\dfrac{x}{3x^2 + 5x - 2} + \dfrac{2}{3x - 1} - \dfrac{1}{x + 2}$

(16) $\dfrac{x}{3x^2 + 5x - 2} \cdot \dfrac{2}{3x - 1} \div \dfrac{1}{x + 2}$

(17) $\dfrac{x - 3}{x^2 - 1} - \dfrac{2x}{x - 1} + \dfrac{x - 1}{x + 1}$

(18) $\dfrac{x - 3}{x^2 - 1} \cdot \dfrac{2x}{x - 1} \div \dfrac{x - 1}{x + 1}$

(19) $\dfrac{a - 1}{a^2 + a - 2} + \dfrac{2a + 4}{a - 1} - \dfrac{2a}{a + 2}$

(20) $\dfrac{a-1}{a^2+a-2} \cdot \dfrac{2a+4}{a-1} \div \dfrac{2a}{a+2}$

(21) $\left(\dfrac{3a}{b} + \dfrac{b^2}{2a}\right)\dfrac{a^2}{b}$ (22) $\left(\dfrac{x}{y^2} + \dfrac{y}{x}\right) \div \dfrac{2y}{x^2}$

(23) $\dfrac{3a}{b} + \dfrac{b^2}{2a} \cdot \dfrac{a^2}{b}$ (24) $\dfrac{x}{y^2} + \dfrac{y}{x} \div \dfrac{2y}{x^2}$

4.7
Complex Fractions

Since the vinculum (or fraction bar) in a fraction always indicates *division*, the division of any two quantities, including two fractions, can be written with a fraction bar. For example:

$$\frac{1}{x} \div \frac{1}{y} \text{ can be written} \frac{\dfrac{1}{x}}{\dfrac{1}{y}}$$

The form on the right in the example above is called a *complex fraction*, and it is simply another way to write the quotient of two quantities involving fractions.

$$\text{Since } \frac{1}{x} \div \frac{1}{y} = \frac{\dfrac{1}{x}}{\dfrac{1}{y}}$$

$$\text{then } \frac{\dfrac{1}{x}}{\dfrac{1}{y}} = \frac{1}{x} \div \frac{1}{y}$$

Many students exhibit an adverse reaction to complex fractions. My advice to them would be that, if they don't like fractions stacked up in this way, then unstack them.

$$\frac{\dfrac{1}{x}}{\dfrac{1}{y}} = \frac{1}{x} \div \frac{1}{y} = \frac{1}{x} \cdot \frac{y}{1} = \frac{y}{x}$$

We could arrive at the same result faster by making intelligent use of the multiplicative identity. In the complex fraction $\dfrac{\dfrac{1}{x}}{\dfrac{1}{y}}$, both of the *denominators*, x and y, will divide into the number xy. Actually xy is the least common denominator for the two fractions, $1/x$ and $1/y$. We can set $1 = xy/xy$ and then multiply the original complex fraction by *one* in this form.

$$\frac{\dfrac{1}{x} \cdot \dfrac{xy}{xy}}{\dfrac{1}{y} \cdot \dfrac{xy}{xy}} = \frac{\dfrac{1}{x}(xy)}{\dfrac{1}{y}(xy)} = \frac{\dfrac{1}{\cancel{x}} \cdot \dfrac{\cancel{x}y}{1}}{\dfrac{1}{\cancel{y}} \cdot \dfrac{x\cancel{y}}{1}} = \frac{y}{x}$$

Either of these methods may be used to simplify a complex fraction.

Caution: Remember that the fraction bar has the effect of putting parentheses about the numerator and denominator of any fraction, and in the order of operations, the parentheses must be simplified *first*.

Example 1:

$$\frac{1 - \dfrac{2}{3}}{1 + \dfrac{1}{2}}$$

A. First Method (unstack the fractions)

$$\frac{1 - \dfrac{2}{3}}{1 + \dfrac{1}{2}} = \frac{\left(1 - \dfrac{2}{3}\right)}{\left(1 + \dfrac{1}{2}\right)} = \left(1 - \frac{2}{3}\right) \div \left(1 + \frac{1}{2}\right)$$

Follow the order of operations and work inside each parentheses *first*.

$$\left(1 - \frac{2}{3}\right) \div \left(1 + \frac{1}{2}\right) = \left(\frac{3}{3} - \frac{2}{3}\right) \div \left(\frac{2}{2} + \frac{1}{2}\right)$$

$$= \left(\frac{1}{3}\right) \div \left(\frac{3}{2}\right)$$

$$= \frac{1}{3} \cdot \frac{2}{3}$$

$$= \frac{2}{9}$$

B. Second Method (multiply the complex fraction by a convenient form of the number one)

$$\frac{1 - \frac{2}{3}}{1 + \frac{1}{2}}$$

For the two fractions, 2/3 and 1/2, the L.C.D. = 6. Set 1 = 6/6 and multiply the complex fraction by 1 in this form.

$$\frac{1 - \frac{2}{3}}{1 + \frac{1}{2}} = \frac{\left(1 - \frac{2}{3}\right)}{\left(1 + \frac{1}{2}\right)} \cdot \frac{6}{6} = \frac{\left(1 - \frac{2}{3}\right)6}{\left(1 + \frac{1}{2}\right)6} = \frac{6 - 4}{6 + 3} = \frac{2}{9}$$

Example 2:

$$\frac{\dfrac{1}{x^2} + \dfrac{1}{y}}{1 - \dfrac{1}{xy}}$$

A. First Method: Perform the indicated arithmetic, working *inside* each set of parentheses first.

$$\frac{\dfrac{1}{x^2} + \dfrac{1}{y}}{1 - \dfrac{1}{xy}} = \frac{\left(\dfrac{1}{x^2} + \dfrac{1}{y}\right)}{\left(1 - \dfrac{1}{xy}\right)} = \left(\frac{1}{x^2} + \frac{1}{y}\right) \div \left(1 - \frac{1}{xy}\right)$$

$$= \left(\frac{y + x^2}{x^2 y}\right) \div \left(\frac{xy - 1}{xy}\right)$$

$$= \frac{(y + x^2)}{(x^2 y)} \cdot \frac{(xy)}{(xy - 1)}$$

$$= \frac{\overset{1}{\cancel{xy}}(y + x^2)}{\underset{x}{\cancel{x^2 y}}(xy - 1)}$$

$$= \frac{y + x^2}{x(xy - 1)}$$

$$= \frac{y + x^2}{x^2 y - x}$$

B. Second Method: Multiply the complex fraction by the number 1 written in the form $\dfrac{x^2 y}{x^2 y}$.

$$\frac{\dfrac{1}{x^2} + \dfrac{1}{y}}{1 - \dfrac{1}{xy}} = \frac{\left(\dfrac{1}{x^2} + \dfrac{1}{y}\right)}{\left(1 - \dfrac{1}{xy}\right)} \cdot \frac{x^2 y}{x^2 y} = \frac{x^2 y\left(\dfrac{1}{x^2} + \dfrac{1}{y}\right)}{x^2 y\left(1 - \dfrac{1}{xy}\right)}$$

$$= \frac{y + x^2}{x^2 y - x}$$

Exercise 62:

Simplify the complex fractions.

(1) $\dfrac{\dfrac{1}{5}}{\dfrac{3}{5}}$

(2) $\dfrac{\dfrac{3}{7}}{\dfrac{3}{8}}$

(3) $\dfrac{1 - \dfrac{1}{3}}{1 + \dfrac{2}{3}}$

(4) $\dfrac{\dfrac{1}{a}}{\dfrac{1}{b}}$

(5) $\dfrac{\dfrac{x}{y^2}}{\dfrac{2x}{y^2}}$

(6) $\dfrac{2 + \dfrac{1}{x}}{1 - \dfrac{1}{y}}$

(7) $\dfrac{\dfrac{1}{a} + \dfrac{1}{b}}{\dfrac{1}{c}}$

(8) $\dfrac{\dfrac{1}{x} - \dfrac{1}{y}}{\dfrac{1}{x}}$

(9) $\dfrac{\dfrac{2}{a} - \dfrac{3}{b}}{\dfrac{1}{ab}}$

(10) $\dfrac{\dfrac{3}{x + y}}{\dfrac{2}{x - y}}$

(11) $\dfrac{\dfrac{x + y}{x}}{\dfrac{x - y}{x}}$

(12) $\dfrac{1 + \dfrac{a}{b}}{1 - \dfrac{a}{b}}$

(13) $\dfrac{\dfrac{1}{x} - \dfrac{1}{y}}{2 + \dfrac{1}{x^2 y^2}}$

(14) $\dfrac{2 - \dfrac{1}{x^2}}{2 + \dfrac{1}{y^2}}$

(15) $\dfrac{\dfrac{1}{a + b}}{\dfrac{1}{a - b}}$

(16) $\dfrac{2 + \dfrac{1}{x + y}}{3 - \dfrac{1}{x - y}}$

(17) $\dfrac{\dfrac{x}{x + y}}{\dfrac{y}{x^2 - y^2}}$

(18) $\dfrac{1 + \dfrac{2}{a + 3}}{3 - \dfrac{2}{a - 3}}$

(19) $\dfrac{\dfrac{1}{x + 1}}{2 - \dfrac{1}{x - 1}}$

(20) $\dfrac{2 - \dfrac{4}{a^2 - b^2}}{1 + \dfrac{2}{a + b}}$

Chapter Test 4

I. Change each fraction to an equal fraction having only positive signs in the numerator and denominator, and reduce each equivalent fraction to its simplest form.

1. $\dfrac{-6}{8}$

2. $\dfrac{-x - 2y}{x + 2y}$

II. Reduce each fraction to its simplest form.

1. $\dfrac{18}{27}$

2. $\dfrac{ab}{a}$

3. $\dfrac{x}{xy}$

4. $\dfrac{4a^2b}{6ab^3}$

5. $\dfrac{2x - 6}{x^2 - 5x + 6}$

6. $\dfrac{ax + ay + bx + by}{x^2 - y^2}$

III. Perform the indicated operations and simplify.

1. $\dfrac{4}{9} + \dfrac{1}{6}$

2. $\dfrac{4}{9} - \dfrac{1}{6}$

3. $\dfrac{4}{9} \cdot \dfrac{1}{6}$

4. $\dfrac{4}{9} \div \dfrac{1}{6}$

5. $\dfrac{3}{5} + 2$

6. $\dfrac{3}{5} \cdot 2$

7. $x + \dfrac{1}{x}$

8. $x \div \dfrac{1}{x}$

9. $\dfrac{1}{x} - \dfrac{1}{y}$

10. $\dfrac{1}{x} \cdot \dfrac{1}{y}$

11. $\dfrac{1}{x} \div \dfrac{1}{y}$

12. $\dfrac{1}{ab} + \dfrac{1}{a}$

13. $\dfrac{1}{ab} + a$

14. $\dfrac{1}{ab} \cdot a$

15. $\dfrac{3x}{y^2} + \dfrac{y}{2x} - \dfrac{y}{3y}$

16. $\dfrac{3x}{y^2} \cdot \dfrac{y}{2x} \div \dfrac{x}{3y}$

17. $\dfrac{1}{x + y} + \dfrac{1}{x}$

18. $\dfrac{1}{x + y} - \dfrac{1}{x^2 - y^2}$

19. $\dfrac{a}{a + 1} + \dfrac{2}{a - 3} - \dfrac{3a}{a^2 - 2a - 3}$

20. $\dfrac{2}{a + 1} \cdot \dfrac{2}{a - 3} \div \dfrac{3a}{a^2 - 2a - 3}$

21. $ab\left(\dfrac{1}{a} + \dfrac{1}{b}\right)$

22. $(x + y) + \dfrac{1}{x^2y^2}$

23. $\dfrac{3}{3a + 3b - ac - bc} + \dfrac{a}{a + b}$

24. $\dfrac{x^2 - 9}{2x - 4} \cdot \dfrac{x^2 - 7x + 10}{x^2 + 6x + 9} \div \dfrac{x - 3}{x - 5}$

25. $\dfrac{x^2}{y}\left(\dfrac{3x}{y} + \dfrac{y}{3x}\right)$

IV. Simplify the following:

1. $\dfrac{\dfrac{2}{5}}{\dfrac{3}{10}}$ 2. $\dfrac{\dfrac{3}{a^2}}{\dfrac{2}{ab}}$ 3. $\dfrac{1 + \dfrac{1}{x}}{1 - \dfrac{1}{x}}$ 4. $\dfrac{\dfrac{1}{x-y}}{\dfrac{3}{x^2-y^2}}$

V. List all the sets N, J, Q, and R to which each of the following numbers belongs:

1. $.66666\ldots$ 2. 7.9 3. $5\dfrac{3}{4}$

4. 13 5. $\dfrac{3}{8}$ 6. -14

7. 2684 8. $-\dfrac{6}{11}$ 9. $3.2121212\overline{1}\ldots$

5
FIRST-DEGREE EQUATIONS IN ONE UNKNOWN

The concept of the equation provided the science of mathematics with an admirable workhorse. Perhaps no single concept yet uncovered by man has been of greater usefulness nor more far-reaching in its applications. The intelligent use of equations in mathematics and the physical sciences has helped significantly to create for *Homo sapiens* an expanding technology that constantly pushes at the frontiers of our knowledge about our planet, its physical laws, the solar system, and the galaxies beyond. In its general utility to mankind, it is possible that the equation really is the greatest thing since the wheel, and you will find that the wheel itself can be expressed as an equation.

Yet the basic idea of an equation is exceedingly simple. An equation is a *statement of equality between two quantities*, which means that an equation must have an *equal mark:*

$$\frac{x}{2} + \frac{x}{3} = 1 \quad \text{This is an equation.}$$

$$\frac{x}{2} + \frac{x}{3} - 1 \quad \text{This is } not \text{ an equation.}$$

Examples of equations:

(1) $4 + 2 = 6$

(2) $5x^3 - x^2 = 2x - 7$

(3) $2x - 3y + 4 = 0$

(4) $C = 2\pi r$

(5) $3 = 3$

(6) $y = \sin 2x$

(7) $2^y = 8$

(8) $\log x + \log x = \log 6$

When we wish to state that two quantities are not equal, we use the symbol $\neq$.

Examples:

(1) $4 + 2 \neq 10$

(2) $|-3| \neq -3$

(3) $\dfrac{x + 1}{x + 1} \neq 5 \quad (x \neq -1)$

5.1
Conditional Equations and Identities

Equations involving variables fall into three classes: *identities*, *conditional equations*, and *impossible equations*.

1.   An equation such as the following:

$$(x + 1)^2 = x^2 + 2x + 1$$

is always true for *any* value of x.

Examples:

(a)   Let $x = 1$
$$(1 + 1)^2 = (1)^2 + 2(1) + 1$$
$$(2)^2 = 1 + 2 + 1$$
$$4 = 4$$

(b)   Let $x = 2$
$$(2 + 1)^2 = (2)^2 + 2(2) + 1$$
$$(3)^2 = 4 + 4 + 1$$
$$9 = 9$$

(c)   Let $x = -5$
$$(-5 + 1)^2 = (-5)^2 + 2(-5) + 1$$
$$(-4)^2 = 25 - 10 + 1$$
$$16 = 16$$

No matter how long we continue substituting values for x, this equation, $(x + 1)^2 = x^2 + 2x + 1$, will be true for any value. This equation, as well as any other equation that is true for *all* admissible values of the variable, is called an *identity*. Other examples of identities are $x + x = 2x$ and $x/2 + x/3 = 5x/6$. These two identities, as well as the example above, are true for all $x \in R$. The phrase "all admissible values" of the variable simply means all values for which the terms in the identity have meaning. The identity

$$\frac{1}{x} + 1 = \frac{1 + x}{x}$$

is meaningless when $x = 0$ because the terms $\dfrac{1}{x}$ and $\dfrac{1 + x}{x}$ are not defined when $x = 0$ (see Section 1.9). Thus this last identity is true for all $x \in R$ except $x = 0$.

2. Now consider the equation
$$x + 2 = 5$$

This equation is a true statement only if $x = 3$. Or we might have an equation such as the following:
$$x^2 = 36$$

This equation is true only if $x = 6$ or $x = -6$. Equations which are true only for a certain value (or values) of the variable are called *conditional equations*.

3. Finally, we might have an equation such as the following:
$$x + 1 = x$$

This equation is never going to be a true statement for *any* value of x. In other words, this equation is a lie, or, as such equations are sometimes called, an *impossible equation*. Other examples of impossible equations are
$$|x| = -10 \quad \text{and} \quad \frac{x + 5}{x + 5} = 3$$

5.2
The Equality Axioms

An equal mark asserts that two quantities mean exactly the same thing. Since every equation must have an equal mark, every equation must have two sides: the quantity on the left side of the equal mark and the quantity on the right side of the equal mark. In the equation
$$3x + 5 = 7x$$

the left side is the multinomial $3x + 5$ and the right side is the monomial $7x$.

It is important that you cultivate an instantaneous reflex reaction to equations *as equations* because the mere existence of an equation endows it with properties that belong to *all* equations but *only* to equations. These properties are stated in equality axioms.

An axiom has been called a "self-evident" truth. Actually, an axiom is a statement whose truth is accepted without proof, and it generally defines some basic and unchanging relationship drawn from our observation and experience. It is *axiomatic*, for instance, that any number is equal to itself. We accept this as a true statement, but we make no attempt to prove it. We state this axiom symbolically in the following way:

> **For any number $a \in R$, $a = a$.**

This is called the *reflexive axiom*. We shall label it E_r and henceforth refer to it by that label.

The equality axioms therefore are basic *assumptions* we make about the particular characteristics with which all equations are endowed. From that point on we agree that all other facts (or theorems) which we can deduce about equations are predicated on the truth of these assumptions. But please remember that the *equality* axioms apply to *equations*. The most wrong-headed algebra on earth results when some novice blithely applies an equality axiom to an expression that is not an equation in the first place.

Since the equality axioms apply to *all* equations, the meaning of each one can be illustrated with very simple equations. In the following discussion we shall do just that and then summarize the axioms in symbolic language.

1. Consider the equation

$$4 + 2 = 6$$

If we reverse the sides of this equation we obtain

$$6 = 4 + 2$$

and the resulting statement is true. This is an example of the *symmetric axiom*, which has already been encountered and which we shall henceforth label E_s.

2. Consider the equations

$$5 + 7 = 12$$

and

$$12 = (3)(4)$$

By the reflexive axiom, $12 = 12$, and since $5 + 7$ and $(3)(4)$ both mean the same thing as 12, we conclude that $5 + 7 = (3)(4)$: e.g.,

$$\text{since } 5 + 7 = 12$$
$$\text{and} \quad 12 = (3)(4)$$
$$\text{then } 5 + 7 = (3)(4)$$

This is an example of the *transitive axiom*, which is based on the assumption that if two quantities are both equal to the *same* quantity, then they are equal to each other. For convenient reference we shall label this equality axiom E_t.

3. Consider the equation

$$4 + 2 = 6$$

If the *same* number (*any* number) is added to both sides of the equation the resulting statement is true. To illustrate this, let us select a number at random, such as -2, and add it to both sides.

$$4 + 2 = 6$$
$$4 + 2 + (-2) = 6 + (-2)$$
$$4 + 2 - 2 = 6 - 2$$
$$4 = 4$$

This is an example of the *addition axiom* and will be labeled E_a.

4.   Consider the equation

$$6 + 4 = 10$$

If we multiply both sides of this equation by the *same* number, and again *any* number will do, the resulting statement is true. Multiplying both sides first by 3 and then by $\frac{1}{2}$, we have the following:

$$6 + 4 = 10$$
$$3(6 + 4) = 3(10)$$
$$3(6) + 3(4) = 3(10)$$
$$18 + 12 = 30$$
$$30 = 30$$

$$6 + 4 = 10$$
$$\frac{1}{2}(6 + 4) = \frac{1}{2}(10)$$
$$\frac{1}{2}(6) + \frac{1}{2}(4) = \frac{1}{2}(10)$$
$$3 + 2 = 5$$
$$5 = 5$$

This is an example of the *multiplication axiom,* which we shall henceforth refer to as E_m.

Restating these equality axioms in symbolic language, we have the following:

Equality Axioms $a, b,$ and $c \in R$	
E_r The *reflexive axiom:* $a = a$	
E_s The *symmetric axiom:* If $a = b$, then $b = a$	
E_t The *transitive axiom:* If $a = b$ and $b = c$, then $a = c$	
E_a The *addition axiom:** If $a = b$, then $a + c = b + c$	
E_m The *multiplication axiom:** If $a = b$, then $ac = bc$	

*These two axioms are postulated for simplicity. In many books, using the reflexive and substitution axioms, they are proved as theorems.

Throughout the remainder of this text the equality axioms will be referred to as E_r, E_s, E_t, E_a, and E_m. They will be constant companions in our work with equations, and they should be memorized.*

5.3
First-Degree Equations in One Unknown

In this chapter we shall consider equations which have only one unknown (or variable) and which can be reduced to a form having only nonnegative integral exponents, the greatest of which is one.

Examples:

(1) $x + 2 = 5$

(2) $y - 6 = 2y + 3$

(3) $5z - 8 = 3z$

(4) $3(a - 4) - 2(a + 1) - 7 = 0$

(5) $\dfrac{5m}{3} + \dfrac{m}{2} = 5$

In every example above, there is only one variable in each equation, and the greatest exponent of any variable is *one*. You will find that in all work with equations, you should always be acutely aware of two things:

1. the number of different variables in the equation, and

2. the greatest degree of any term involving the variables. (In an equation whose terms form a polynomial, the *degree* of any term is the sum of the exponents of the variables in the term.)

Our purpose with these first-degree equations in one unknown is to find the value of the unknown that will make the equation true or that will "satisfy the equation." Such a value is also called a *root* of the equation. All first-degree equations in one variable that are not identities have at most one root, which is why we refer to the variable as *an* unknown.

At this point the total sum of our knowledge about the behavior of equations is contained in the equality axioms previously stated. But happily, *that is all we need to know* to solve these first-degree equations.

According to E_m, we may multiply both sides of an equation by the same number, c, without altering the truth of the equation. However, in division we actually *multiply* by the multiplicative inverse of the divisor. Therefore we can also *divide* both

---

*With the introduction of the equality axioms, the student will have acquired a background sufficient to identify the field postulates and begin to study proofs of the laws of algebra considered thus far in the text. For courses designed to include algebraic proofs, the Appendix in this text should follow this chapter; a detailed and elementary discussion is given there of closure, binary operations, the field postulates, and proofs of elementary theorems.

sides of any equation by any number, c (except that $c \neq 0$), since we will actually be *multiplying* by $1/c$. We can restate E_m in the following manner: *both sides of an equation may be multiplied or divided by the same number provided that the divisor is not zero.* The following examples illustrate how we can use E_m to solve equations, i.e., find the value of the unknown that satisfies the equation.

Examples:

(1) $2x = 6$

Divide both sides by 2 (using E_m).

$$\frac{\overset{1}{\cancel{2}x}}{\underset{1}{\cancel{2}}} = \frac{6}{2}$$

$$x = 3$$

We can check this result by substituting 3 for x in the original equation.

$$2x = 6$$
$$2(3) = 6$$
$$6 = 6$$

(2) $3y = -15$

Divide both sides by 3 (using E_m).

$$\frac{\overset{1}{\cancel{3}y}}{\underset{1}{\cancel{3}}} = \frac{-15}{3}$$

$$y = -5$$

Check:

$$3y = -15$$
$$3(-5) = -15$$
$$-15 = -15$$

(3) $-7a = -28$

$$\frac{\overset{1}{\cancel{-7}a}}{\underset{1}{\cancel{-7}}} = \frac{-28}{-7} \quad \text{(using } E_m\text{)}$$

$$a = 4$$

Check:

$$-7a = -28$$
$$-7(4) = -28$$
$$-28 = -28$$

(4) $\quad -4b = 5$

$$\frac{\overset{1}{\cancel{-4}}b}{\underset{1}{\cancel{-4}}} = \frac{5}{-4} \quad \text{(using } E_m\text{)}$$

$$b = -\frac{5}{4}$$

Check:

$$-4b = 5$$

$$-4\left(-\frac{5}{4}\right) = 5$$

$$\frac{\overset{-1}{\cancel{-4}}}{1} \cdot -\frac{5}{\underset{1}{\cancel{4}}} = 5$$

$$(-1)(-5) = 5$$

$$5 = 5$$

(5) $\quad \dfrac{3}{5}x = -10$

Multiply both sides by the multiplicative inverse of $\dfrac{3}{5}$, which is $\dfrac{5}{3}$.

$$\frac{\overset{1}{\cancel{5}}}{\underset{1}{\cancel{3}}} \cdot \frac{\overset{1}{\cancel{3}}}{\underset{1}{\cancel{5}}}x = \frac{5}{3} \cdot \frac{-10}{1} \quad \text{(using } E_m\text{)}$$

$$x = \frac{-50}{3}$$

Check:

$$\frac{3}{5}x = -10$$

$$\frac{\cancel{3}}{5} \cdot \frac{-50}{\cancel{3}} = -10$$

$$\frac{-50}{5} = -10$$

$$-10 = -10$$

(6) $\quad .3x = 1.2$

Note: $\quad .3x = \dfrac{3}{10}x$ and $1.2 = 1 + \dfrac{2}{10} = \dfrac{10}{10} + \dfrac{2}{10} = \dfrac{12}{10}$

$$\frac{3}{10}x = \frac{12}{10}$$

Multiply both sides by the multiplicative inverse of $\dfrac{3}{10}$, which is $\dfrac{10}{3}$.

$$\frac{\overset{1}{\cancel{10}}}{\underset{1}{\cancel{3}}} \cdot \frac{\overset{1}{\cancel{3}}}{\underset{1}{\cancel{10}}}x = \frac{\overset{1}{\cancel{10}}}{\underset{1}{\cancel{3}}} \cdot \frac{\overset{4}{\cancel{12}}}{\underset{1}{\cancel{10}}}$$

$$x = 4$$

Check:

.3x = 1.2
.3(4) = 1.2
1.2 = 1.2

Repeating the first four examples in condensed form we have the following:

(1) $2x = 6$	$\dfrac{2x}{2} = \dfrac{6}{2}$	$x = 3$
(2) $3y = -15$	$\dfrac{3y}{3} = \dfrac{-15}{3}$	$y = -5$
(3) $-7a = -28$	$\dfrac{-7a}{-7} = \dfrac{-28}{-7}$	$a = 4$
(4) $-4b = 5$	$\dfrac{-4b}{-4} = \dfrac{5}{-4}$	$b = -\dfrac{5}{4}$

All of the above equations follow the same pattern, or *form*. If we let x be the unknown and then let a and b represent any two fixed numbers, or *constants* (except that $a \neq 0$), we can represent the form of each of the above equations in the following manner:

$$ax = b \qquad (a \neq 0)$$

This form is the pattern *to which every conditional first-degree equation in one unknown can be reduced*, and therefore it is the symbolic statement, or *standard form*, of all such equations. After we have written these first-degree equations in this form, we can solve the equation by dividing both sides of it by the *coefficient of the unknown*.

$$ax = b \qquad (a \neq 0)$$
$$\frac{\cancel{a}x}{\cancel{a}} = \frac{b}{a}$$
$$x = \frac{b}{a}$$

Exercise 63:

Solve the following equations:

(1) $3x = 27$ (2) $-5x = 30$ (3) $7a = -42$

(4) $-4b = -48$ (5) $24c = 12$ (6) $-10x = 30$

(7) $3y = 2$ (8) $4a = -7$ (9) $-2b = -1$

(10) $-5m = -3$ (11) $\frac{1}{2}x = \frac{3}{4}$ (12) $\frac{1}{3}x = \frac{2}{3}$

(13) $-\frac{3}{5}x = \frac{1}{10}$ (14) $-14a = 14$ (15) $25b = -25$

(16) $-x = 4$ (17) $\frac{2}{3}y = \frac{8}{9}$ (18) $-\frac{3}{4}x = 16$

5.4
Equations and Set Notation; Solution Set

In some types of equations there will be more than one value of the unknown that will satisfy the equation. For example:

$$x + 2 = 5 \text{ is true only if } x = 3$$

but

$$x^2 = 25 \text{ is true if } x = 5 \text{ or if } x = -5$$

Some equations have an infinite number of solutions. For this reason we consider the solution of an equation to be a *set* of numbers, and we refer to the unknown more properly as a *variable* whose values are the elements of a finite or an infinite set of numbers. The *solution set* for any equation is the set of *all* numbers which satisfy the equation.

Examples:

(1) $x + 2 = 5$ The *solution set* is $\{3\}$
(2) $x^2 = 25$ The *solution set* is $\{5, -5\}$

A type of set notation different from those illustrated in Section 3.6 is frequently used to denote the solution set of an equation. The type of notation used provides a means of describing a set by a *rule* or procedure rather than by listing the elements of the set. This is done by using a dummy symbol, such as x, y, a, b, etc. to represent *any* element of the set; the symbol is followed by a vertical bar, |, which is translated "such that." The bar is followed by the rule that generates the elements of the set. Finally, the entire statement is enclosed in *braces*. For example, the statement

$$\{x \mid x + 2 = 5\}$$

is read "the set of all elements x such that every x in the set is a solution of the equation $x + 2 = 5$." This particular set contains only one element, the number 3; so we could also write

$$\{x \mid x + 2 = 5\} = \{3\}$$

In the above statement the set in braces to the left of the equal mark is defined by rule; the same set is listed by elements on the right of the equal mark.

The rule notation for sets, in which the rule is often stated as an equation, is useful in many ways and should be learned promptly and thoroughly. Following are other examples of the same notation:

Examples:

(1) $\{a \mid a \in N\}$

"The set of all elements a such that a is a natural number." The same set listed by elements is $\{1, 2, 3, 4, 5, \ldots\}$. Hence

$$\{a \mid a \in N\} = \{1, 2, 3, 4, 5, \ldots\}$$

(2) $\{x \mid x^2 = 36\} = \{6, -6\}$

In the remaining examples the sets are described first by a statement and then by a rule:

(3) *Statement:* The set of integers, J
Rule: $\{b \mid b \in J\}$

(4) *Statement:* The solution set of $2x + 3 = 5$
Rule: $\{x \mid 2x + 3 = 5\}$

(5) *Statement:* The solution set of $6y - 5 = 1$
Rule: $\{y \mid 6y - 5 = 1\}$

One advantage of using set notation with equations is that the solution set represents all possible solutions. For example, in the equation $x^2 = 16$ it is correct to say that $x = 4$, since 4 is a solution; but the solution *set* for this equation is $\{4, -4\}$. Thus, if we write $\{x \mid x + 4 = 9\}$ the statement means "the set of *all x*'s such that $x + 4 = 9$." We can then show that the number 5 is the *only* solution by writing

$$\{x \mid x + 4 = 9\} = \{5\}$$

The two sets above contain exactly the same elements (in this case the element 5) and are therefore equal (see Section 3.6). The solutions of many different types of equations can be stated with admirable precision in this manner; so this is an excellent time to get acquainted with the proper notation.

Examples:

(1) Solve $2x = 6$

Solution: $x = 3$ *or*

$\{x \mid 2x = 6\} = \{3\}$

$\downarrow$ $\downarrow$

the set of all   *solution set*

elements x

such that

$2x = 6$

(2) Solve $3y = -15$

Solution: $y = -5$ *or*

$\{y \mid 3y = -15\} = \{-5\}$

$\downarrow$ $\downarrow$

the set of all   *solution set*

elements y such

that $3y = -15$

Exercise 64:

State the following equations as sets of numbers and give the solution set for each one:

(1) $7x = 63$ (2) $-4y = 20$

(3) $6a = -48$ (4) $-5m = -45$

(5) $ax = b$ (6) $10y = -40$

(a and b are constants, $a \neq 0$)

(7) $\frac{1}{2}a = 4$ (8) $\frac{1}{3}b = 9$

5.5
Reducing First-Degree Equations to the Form $ax = b$

Equality axiom E_a states that we may add the same number to both sides of an equation without altering the truth of the equality.

$$E_a \qquad a,b, \text{ and } c \in R$$
$$\text{If } a = b, \text{ then } a + c = b + c$$

We can use this axiom to rearrange the terms of an equation.

Examples:

(1) Find the solution set for $\{x \mid 4x = 2x + 12\}$.

In the equation $4x = 2x + 12$ add $-2x$ to both sides.

$$4x + (-2x) = 2x + 12 + (-2x) \qquad \text{(using } E_a)$$
$$4x - 2x = 2x + 12 - 2x$$

Combine like terms.

$$2x = 12$$

Divide both sides of the equation by 2 (using E_m).

$$\frac{2x}{2} = \frac{12}{2}$$
$$x = 6$$

Check:
$$4x = 2x + 12$$
$$4(6) = 2(6) + 12$$
$$24 = 12 + 12$$
$$24 = 24$$

Therefore, $\{x \mid 4x = 2x + 12\} = \{6\}$.

(2) Find the solution set for $\{x \mid 8 + 3x = 7x\}$.

Add $-3x$ to both sides of the equation (using E_a).

$$8 + 3x - 3x = 7x - 3x$$

Combine the like terms.

$$8 = 4x$$

Divide both sides of the equation by 4 (using E_m).

$$\frac{8}{4} = \frac{4x}{4}$$
$$2 = x$$
$$x = 2 \qquad \text{(using } E_s)$$

so $\{x \mid 8 + 3x = 7x\} = \{2\}$.

(3) Find the solution set for $\{y \mid 3y - 14 = 8y + 1\}$.

$$3y - 14 = 8y + 1$$
$$3y - 8y = 1 + 14 \qquad \text{(using } E_a)$$
$$-5y = 15$$
$$\frac{-5y}{-5} = \frac{15}{-5} \qquad \text{(using } E_m)$$
$$y = -3$$

so $\{y \mid 3y - 14 = 8y + 1\} = \{-3\}$.

When an equation contains parentheses, we perform the indicated operations before we apply the equality axioms.

Example:

$$3(x + 2) - 5(x - 4) = -2$$
$$3x + 6 - 5x + 20 = -2$$

Combine like terms.

$$-2x + 26 = -2$$
$$-2x = -2 - 26 \quad \text{(using } E_a)$$
$$-2x = -28$$
$$\frac{-2x}{-2} = \frac{-28}{-2} \quad \text{(using } E_m)$$
$$x = 14$$

so $\{x \mid 3(x + 2) - 5(x - 4) = -2\} = \{14\}$.

Summary: To solve a first-degree equation in one unknown:

1. **Perform all operations indicated by parentheses.**
2. **Using E_a put all terms involving the unknown on one side of the equal mark and all other terms on the other side of the equal mark.**
3. **Combine the like terms on each side.**
4. **Using E_m divide both sides of the equation by the coefficient of the unknown.**

Examples:

(1) $$2(y + 8) - 3(y - 4) = 3$$
$$2y + 16 - 3y + 12 = 3$$
$$-y + 28 = 3$$
$$-y = 3 - 28 \quad \text{(using } E_a)$$
$$-y = -25$$
$$\frac{-y}{-1} = \frac{-25}{-1} \quad \text{(using } E_m)$$
$$y = 25$$
$$\{x \mid 2(y + 8) - 3(y - 4) = 3\} = \{25\}$$

(2) $$(x - 2)(x + 3) = x^2 - 7x + 10$$
$$x^2 + x - 6 = x^2 - 7x + 10$$
$$x^2 + x - x^2 + 7x = 10 + 6 \quad \text{(using } E_a)$$
$$x^2 - x^2 + x + 7x = 16$$
$$8x = 16$$
$$\frac{8x}{8} = \frac{16}{8} \quad \text{(using } E_m)$$
$$x = 2$$
$$\{x \mid (x - 2)(x + 3) = x^2 - 7x + 10\} = \{2\}$$

Exercise 65:

Find the solution set of each equation.

(1) $-5a = -75$ (2) $3x = 4$

(3) $5x - 6 = 14$ (4) $2b = 8b + 30$

(5) $3y + 4 = -8$ (6) $2w = 5w - 9$

(7) $4x - 12 = -2x$ (8) $11y + 4 = -18$

(9) $x + 1 = 2x - 6$ (10) $a - 3 = 2a + 7$

(11) $3b + 2 = 6 - 2b$ (12) $6x - 5 = 2x - 5$

(13) $5y - 6 = 2y + 6$ (14) $2b + 8 = 7b - 2$

(15) $7x + 13 = 2x + 4$ (16) $3(x - 1) = 12$

(17) $2(x - 3) = 7$ (18) $3(2y + 4) = 12$

(19) $-2(3b - 1) = -7$ (20) $\frac{1}{3}(6x + 12) = -4$

(21) $2(a - 3) - 3(2a - 2) = 24$ (22) $4(2x - 3) - 2 = 5(x + 1)$

(23) $3(b + 2) - (b - 8) = 3(b + 5)$

(24) $\frac{1}{4}(4x + 8) = 3$

(25) $\frac{1}{2}(2x + 4) - \frac{1}{3}(3x - 9) = 2x + 1$

(26) $2(x^2 - 3) - 4(x + 1) = 2x^2 + 9$

(27) $\frac{3}{4}(8a - 4) + 2(a + 2) = -7$

(28) $(x - 3)(x + 4) = x^2 - 5x + 6$

(29) $2(3 + 2x) - 5(x - 1) = 12$

(30) $(x + 3)(x + 5) = (x + 8)(x - 1)$

(31) $(2x - 1)(x - 3) = (x - 8)(2x - 1)$

(32) $x^2 - 9 = (x - 4)(x - 3)$

Give the solution sets of the following:

(33) $\{x \mid 3x = -18\} = \{\ \ \}$

(34) $\{y \mid 4y - 8 = 8\} = \{\ \ \}$

(35) $\{a \mid 2a - 5 = 4a + 7\} = \{\ \ \}$

(36) $\{b \mid 3(b - 1) = 5(b + 2)\} = \{\ \ \}$

(37) $\{m \mid m^2 + 4 = (m - 1)(m + 5)\} = \{\ \ \}$

(38) $\{x \mid qx = p\} = \{\ \ \}$ $(q \neq 0)$

(39) $\{y \mid 2y = k\} = \{\ \ \}$

(40) $\{a \mid ma - r = 0\} = \{\ \ \}$ $(m \neq 0)$

5.6
Equations Containing Fractions

If an equation contains fractions, equality axiom E_m may be used to simplify the situation. It is anathema in some quarters to say "clear the fractions from the equation," but if E_m is put to intelligent use, that is exactly what happens. However, remember that E_m is an *equality* axiom and it can be used to clear fractions only from *equations.*

The routine is quite simple. As we observed in Section 4.6, for any given bunch of fractions there is always a least common denominator (L.C.D.). *In equations containing fractions, we find the L.C.D. for all of the fractions and then use it as a convenient multiplier for both sides of the equation.*

Examples:

(1) $\dfrac{x}{3} + \dfrac{x}{6} = 2$

For the fractions in this equation, the number 6 is the L.C.D. Using E_m, multiply both *sides* of the equation by 6.

$$\frac{x}{3} + \frac{x}{6} = 2$$

$$6\left(\frac{x}{3} + \frac{x}{6}\right) = 6(2) \qquad \text{(using } E_m\text{)}$$

$$\frac{6}{1}\cdot\frac{x}{3} + \frac{6}{1}\cdot\frac{x}{6} = 12$$

$$\overset{2}{\underset{1}{\frac{\cancel{6}}{1}}}\cdot\frac{x}{\underset{1}{\cancel{3}}} + \overset{1}{\underset{1}{\frac{\cancel{6}}{1}}}\cdot\frac{x}{\underset{1}{\cancel{6}}} = 12$$

$$2x + x = 12$$

$$3x = 12$$

$$\frac{\cancel{3}x}{\cancel{3}} = \frac{12}{3}$$

$$x = 4$$

Check:

$$\frac{x}{3} + \frac{x}{6} = 2 \qquad \frac{12}{6} = 2$$

$$\frac{4}{3} + \frac{4}{6} = 2 \qquad 2 = 2$$

$$\frac{8}{6} + \frac{4}{6} = 2$$

Therefore, the solution set is $\{4\}$.

In solving the equation given in Example 1, we multiplied both sides of the equation by the number, or *constant*, 6. After the multiplication was done, the original equation, $x/3 + x/6 = 2$, was changed to $2x + x = 12$. These two equations are called *equivalent equations* because they have the same solution set, namely $\{4\}$. Any time both sides of an equation are multiplied by a nonzero constant, the resulting equation will always have the same solution set and will be an equivalent equation for the original equation.

But this is not necessarily true when we multiply both sides of an equation by the *unknown* (or variable). The resulting equation is called a derived equation, but it may *not* be an equivalent equation because multiplication by the variable may introduce a solution (also called a root) that does not satisfy the original equation. It sometimes happens that the original equation has no solution at all. For this reason every solution or root of an equation that is found after multiplying by any expression containing the variable must be checked in the *original* equation.

(2) $\dfrac{2}{x-2} + 2 = \dfrac{x}{x-2}$

The L.C.D. for the fractions in this equation is the number $x - 2$. Using E_m, multiply both sides of the equation by $x - 2$.

$$\frac{2}{x-2} + 2 = \frac{x}{x-2}$$

$$\frac{2}{\cancel{(x-2)}} \cdot \frac{\cancel{(x-2)}}{1} + 2(x-2) = \frac{x}{\cancel{(x-2)}} \cdot \frac{\cancel{(x-2)}}{1}$$

$$2 + 2x - 4 = x$$

$$2x - x = 4 - 2 \qquad \text{(using } E_a\text{)}$$

$$x = 2$$

When we check the *original equation* for $x = 2$, we get a meaningless result since division by zero is undefined.

$$\frac{2}{x-2} + 2 = \frac{x}{x-2}$$

$$\frac{2}{2-2} + 2 = \frac{2}{2-2}$$

$$\frac{2}{0} + 2 = \frac{2}{0} \qquad \textbf{Nonsense}$$

Therefore the original equation has no solution; i.e., the solution set has no elements.

The solution set is $\varnothing$.

(3) $3 - \dfrac{3}{x-1} = \dfrac{2x}{x-1}$

The L.C.D. is $x - 1$. Multiplying both sides by this number we have

$$3(x - 1) - \frac{3}{\cancel{(x-1)}} \cdot \frac{\cancel{(x-1)}}{1} = \frac{2x}{\cancel{(x-1)}} \cdot \frac{\cancel{(x-1)}}{1}$$
$$3x - 3 - 3 = 2x$$
$$3x - 2x = 3 + 3 \qquad \text{(using } E_a\text{)}$$
$$x = 6$$

Check:

$$3 - \frac{3}{x - 1} = \frac{2x}{x - 1}$$
$$3 - \frac{3}{6 - 1} = \frac{2(6)}{6 - 1}$$
$$3 - \frac{3}{5} = \frac{12}{5}$$
$$\frac{15}{5} - \frac{3}{5} = \frac{12}{5}$$
$$\frac{12}{5} = \frac{12}{5}$$

Therefore, the solution set is $\{6\}$.

Summary

When the same quantity is added to both sides of an equation, the result is an equivalent equation having the same solution set.

When both sides of an equation are multiplied or divided by any nonzero con-stant, the result is an equivalent equation having the same solution set.

When both sides of an equation are multiplied by an unknown (or variable), the result is a derived equation and may or may not have the same solution set. There-fore all roots obtained after multiplying an equation by a variable must be checked in the original equation.

(4) $\quad \dfrac{1}{2} + \dfrac{3}{x} = \dfrac{1}{4}$

For the fractions in this equation the L.C.D. is the number $4x$. Using E_m, multiply both sides of the equation by $4x$.

$$\frac{1}{2} + \frac{3}{x} = \frac{1}{4}$$
$$4x\left(\frac{1}{2} + \frac{3}{x}\right) = 4x\left(\frac{1}{4}\right)$$

$$\frac{\overset{2}{\cancel{4}x}}{1}\cdot\frac{1}{\cancel{2}} + \frac{\overset{1}{4\cancel{x}}}{1}\cdot\frac{3}{\cancel{x}} = \frac{\overset{1}{\cancel{4}x}}{1}\cdot\frac{1}{\cancel{4}}$$

$$2x + 12 = x$$
$$2x - x = -12$$
$$x = -12$$

Check:

$$\frac{1}{2} + \frac{3}{x} = \frac{1}{4}$$

$$\frac{1}{2} + \frac{3}{-12} = \frac{1}{4}$$

$$\frac{1}{2} - \frac{3}{12} = \frac{1}{4}$$

$$\frac{1}{2} - \frac{1}{4} = \frac{1}{4}$$

$$\frac{2}{4} - \frac{1}{4} = \frac{1}{4}$$

$$\frac{1}{4} = \frac{1}{4}$$

Therefore, the solution set is $\{-12\}$.

(5) $\frac{2}{3}x - \frac{1}{5}x = 14$

Note: $\frac{2}{3}x = \frac{2}{3}\cdot x = \frac{2}{3}\cdot\frac{x}{1} = \frac{2x}{3}$

$\frac{1}{5}x = \frac{1}{5}\cdot x = \frac{1}{5}\cdot\frac{x}{1} = \frac{x}{5}$

$\frac{2}{3}x - \frac{1}{5}x = 14$ can be written

$\frac{2x}{3} - \frac{x}{5} = 14$

L.C.D. = 15. Multiply both sides by 15.

$$\frac{\overset{5}{\cancel{15}}}{1}\cdot\frac{2x}{\cancel{3}} - \frac{\overset{3}{\cancel{15}}}{1}\cdot\frac{x}{\cancel{5}} = 15(14)$$

$$5(2x) - 3(x) = 15(14)$$
$$10x - 3x = 210$$
$$7x = 210$$
$$x = \frac{210}{7}$$
$$x = 30$$

The solution set is $\{30\}$.

(6) $\dfrac{2}{x-3} - \dfrac{1}{x+2} = \dfrac{3}{x^2 - x - 6}$

$\dfrac{2}{x-3} - \dfrac{1}{x+2} = \dfrac{3}{(x-3)(x+2)}$

L.C.D. $= (x-3)(x+2)$.

Multiply both sides of the equation by $(x-3)(x+2)$.

$$\dfrac{(x-3)(x+2)}{1} \cdot \dfrac{2}{(x-3)} - \dfrac{(x-3)(x+2)}{1} \cdot \dfrac{1}{(x+2)} =$$

$$\dfrac{(x-3)(x+2)}{1} \cdot \dfrac{3}{(x-3)(x+2)}$$

$$2(x+2) - 1(x-3) = 3$$

$$2x + 4 - x + 3 = 3$$

$$2x - x = 3 - 3 - 4$$

$$x = -4$$

The solution set is possibly $\{-4\}$, but we must check first since we multiplied both sides of the equation by an expression involving the variable.

Check:

$$\dfrac{2}{x-3} - \dfrac{1}{x+2} = \dfrac{2}{x^2 - x - 6}$$

$$\dfrac{2}{-4-3} - \dfrac{1}{-4+2} = \dfrac{3}{(-4)^2 - (-4) - 6}$$

$$\dfrac{2}{-7} - \dfrac{1}{-2} = \dfrac{3}{16 + 4 - 6}$$

$$-\dfrac{2}{7} + \dfrac{1}{2} = \dfrac{3}{14}$$

$$-\dfrac{4}{14} + \dfrac{7}{14} = \dfrac{3}{14}$$

$$\dfrac{3}{14} = \dfrac{3}{14}$$

Therefore, the solution set is $\{-4\}$.

   The foregoing examples all illustrate the procedure which should be followed in solving any fractional equations. By using E_m, with the L.C.D. as a convenient multiplier, clear all of the fractions from the equation *first*. Remember that when the variable is used as a multiplier, all roots must be checked in the original equations.

Exercise 66:

Find the solution set of each equation.

(1) $\dfrac{x}{3} - \dfrac{1}{2} = 1$ (2) $\dfrac{y}{3} + \dfrac{y}{2} = 5$

(3) $\dfrac{5}{x} - \dfrac{2}{3} = \dfrac{1}{6}$ (4) $\dfrac{1}{2} + \dfrac{1}{4} = \dfrac{3}{x}$

(5) $\dfrac{3}{4}a - \dfrac{1}{2}a = -3$ (6) $\dfrac{b}{5} - 3 = \dfrac{b}{2}$

(7) $\dfrac{y}{7} + y = \dfrac{1}{2}$ (8) $\dfrac{3}{x-1} = \dfrac{2}{3}$

(9) $\dfrac{3}{2y-1} = \dfrac{-3}{4}$ (10) $\dfrac{2}{3x+2} = \dfrac{3}{5}$

(11) $\dfrac{5x}{x+4} = 2 - \dfrac{2}{x+4}$ (12) $\dfrac{3}{y+3} - 2 = \dfrac{-y}{y+3}$

(13) $\dfrac{2}{x+5} + \dfrac{1}{x-5} = \dfrac{3}{x^2-25}$ (14) $3 - \dfrac{1}{x-3} = \dfrac{5x}{x-3}$

(15) $\dfrac{2}{y+1} = 4 - \dfrac{2y}{y+1}$ (16) $\dfrac{2}{x+2} - \dfrac{3}{x-2} = \dfrac{-10}{x^2-4}$

(17) $\dfrac{1}{2a+1} + \dfrac{2}{2a-1} = \dfrac{-2}{4a^2-1}$

(18) $\dfrac{4}{x-4} - \dfrac{1}{x+5} = \dfrac{-3}{x^2+x-20}$

(19) $\dfrac{1}{2a+3} - \dfrac{2}{a-1} = \dfrac{2}{2a^2+a-3}$

(20) $\dfrac{2}{x-1} + \dfrac{2x}{(x+4)(x-1)} = \dfrac{2}{x+4} + \dfrac{2}{(x+4)(x-1)}$

(21) $\dfrac{3}{x} - \dfrac{1}{x-3} = \dfrac{5}{x^2-3x}$

(22) $\dfrac{3}{2y-1} - \dfrac{1}{3y+1} = \dfrac{11}{6y^2-y-1}$

5.7
Absolute-Value Equations; Literal Equations

In this section we shall investigate two types of equations which will occur in later mathematics courses and which are useful in related courses in the sciences and the mathematics of business.

Absolute-Value Equations In mathematics we frequently encounter equations involving absolute values. At this point you should review carefully Section 1.4, which explains in detail that the absolute value of a number means its distance from zero (on the number line), with no regard for direction. Hence $|3| = 3$ and $|-3| = 3$.

Now let us consider algebraic equations involving the concept of absolute value. Keep in mind that the object is to find the corresponding solution set (that is, the set of *all* numbers that satisfy the equation). One such equation is $|x| = 6$. Even though 6 is *a* solution of this equation, $\{6\}$ is NOT the solution set because there is another number whose absolute value is 6. We must remember that there are two numbers whose distance from zero is 6 units (Figure 5.1).

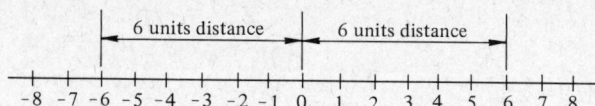

Figure 5.1

Therefore $\{6, -6\}$ is the correct solution set.

In general, for any positive number p, the equation $|x| = p$ will have two solutions, namely p and $-p$.

Examples:

    (1) $|x| = 5$ has the solution set $\{5, -5\}$.
    (2) $|x| = 127$ has the solution set $\{127, -127\}$.

However, the equation $|x| = 0$ has a single solution. That is, its solution set is $\{0\}$. Only zero has an absolute value equal to zero. Note that the equation $|x| = -3$ has *no* solution, for the absolute value of a number is never negative. Thus its solution set is the null set, or the empty set, denoted by the symbol $\emptyset$.

Now consider a slightly more complicated equation.

$$|x + 1| = 6$$

As was discussed earlier, there are two numbers whose absolute value is 6 (namely 6 and -6). So if $x + 1 = 6$ or if $x + 1 = -6$, the equation is satisfied. We know that $x + 1 = 6$ if $x = 5$, and that $x + 1 = -6$ if $x = -7$. So it appears that $\{5, -7\}$ is the solution set. Indeed, both numbers satisfy the equation.

$$|5 + 1| = |6| = 6 \quad \text{and} \quad |-7 + 1| = |-6| = 6$$

In general, no matter what first-degree algebraic expression is within the vertical bars, if the equation states that its absolute value is equal to a *positive* number, there are *two* solutions to be considered. This is emphasized by the fact that the absolute value disregards the direction of a number from zero. We must consider distance to the left as well as to the right. Consider the equation

$$|2x - 5| = 7$$

To solve this equation, our knowledge of absolute value reminds us that $|7| = 7$ and $|-7| = 7$. If $2x - 5 = 7$, the equation is satisfied, but also if $2x - 5 = -7$, the equation is satisfied. Thus, to solve the absolute-value equation, we set up **two** corresponding first-degree equations.

$$2x - 5 = 7 \qquad \text{or} \qquad 2x - 5 = -7$$
$$2x = 12 \qquad\qquad\qquad 2x = -2$$
$$x = 6 \qquad\qquad\qquad x = -1$$

Note that 6 and -1 are both solutions. If $x = 6$, we have $|2x - 5| = |2(6) - 5| = |12 - 5| = |7| = 7$. Likewise, if $x = -1$, we have $|2x - 5| = |2(-1) - 5| = |-2 - 5| = |-7| = 7$. Therefore $|2x - 5| = 7$ has the solution set $\{6, -1\}$.

Examples:

(1) $|x - 3| = 2$

$$x - 3 = 2 \qquad \text{or} \qquad x - 3 = -2$$
$$x = 5 \qquad\qquad\qquad x = 1$$

Check:

If $x = 5$, then $|x - 3| = |5 - 3| = |2| = 2$, and
if $x = 1$, then $|x = 1| = |1 - 3| = |-2| = 2$.

Therefore $\{5, 1\}$ is the solution set.

(2) $|2 - 5x| = 1$

$$2 - 5x = 1 \qquad \text{or} \qquad 2 - 5x = -1$$
$$-5x = -1 \qquad\qquad\qquad -5x = -3$$

$$x = \frac{-1}{-5} \qquad\qquad\qquad x = \frac{-3}{-5}$$

$$x = \frac{1}{5} \qquad\qquad\qquad x = \frac{3}{5}$$

Check:

If $x = \dfrac{1}{5}$, then $\left|2 - 5\left(\dfrac{1}{5}\right)\right| = |2 - 1| = |1| = 1$, and

if $x = \dfrac{3}{5}$, then $\left|2 - 5\left(\dfrac{3}{5}\right)\right| = |2 - 3| = |-1| = 1$

Therefore $\left\{\dfrac{1}{5}, \dfrac{3}{5}\right\}$ is the solution set.

(3) $\left|\dfrac{3x + 2}{4}\right| = 1$

$\dfrac{3x + 2}{4} = 1$ or $\dfrac{3x + 2}{4} = -1$

$3x + 2 = 4$ $3x + 2 = -4$

$\qquad 3x = 2$ $3x = -6$

$\qquad\quad x = \dfrac{2}{3}$ $x = -2$

Check:

If $x = \dfrac{2}{3}$, then $\left|\dfrac{3\left(\dfrac{2}{3}\right) + 2}{4}\right| = \left|\dfrac{2 + 2}{4}\right| = \left|\dfrac{4}{4}\right| = |1| = 1$, and

if $x = -2$, then $\left|\dfrac{3(-2) + 2}{4}\right| = \left|\dfrac{-6 + 2}{4}\right| = \left|\dfrac{-4}{4}\right| = |-1| = 1$

Therefore $\left\{\dfrac{2}{3}, -2\right\}$ is the solution set.

Note carefully that *in each of the preceding examples the solution set had two elements.* This will be true for all such equations that have solutions except one, and that single exception is the case in which the absolute value of an expression is equal to *zero.* The reason for this exception is that zero is the only real number that is neither positive nor negative; as noted earlier, if $|x| = 0$, $x = 0$ is the only solution. Study carefully the following examples of this.

(4) $|x - 5| = 0$

This equation is satisfied only if $x - 5 = 0$; that is, $x = 5$.
If $x = 5$, then $|5 - 5| = |0| = 0$.

Therefore $\{5\}$ is the solution set.

(5) $|3x + 2| = 0$

$3x + 2 = 0$

$\qquad 3x = -2$

$\qquad\quad x = -\dfrac{2}{3}$

If $x = -\dfrac{2}{3}$, then $\left|3\left(-\dfrac{2}{3}\right) + 2\right| = |-2 + 2| = |0| = 0.$

Therefore $\left\{-\dfrac{2}{3}\right\}$ is the solution set.

The foregoing examples illustrate the procedure that should be used in solving absolute-value equations. For algebraic expressions of the first degree, if the absolute value is equal to a positive number, we must remember to consider two solutions. If the absolute value is zero, we expect only one solution. And if the absolute value is equal to a negative number, the equation is obviously impossible because the definition of absolute value is contradicted, and the solution set is the empty set.

Literal Equations   If l and w represent the length and width of a rectangle, and A stands for the area of the rectangle, then $A = lw$. This equation (or formula) has three quantities represented by letters. Such an equation, in which the quantities are represented by letters, is called a *literal equation*.

Since a literal equation *is* an equation, we can use the equality axioms to solve it for any *one* of the variables present, or specifically, to isolate any one of the variables on one side of the equal mark.

When the variable appears to the first degree only, we solve for that variable in exactly the same way that we solve *any* first-degree equation in one unknown. No matter how many variables are present in the equation, if we wish to isolate just *one* of them, we consider it to be the only unknown and treat everything else as *constants* or fixed numbers.

Examples:

Study the following examples carefully:

(1) $A = lw$ (solve for w)

$lw = A$ (using E_s)

$\dfrac{\cancel{l}w}{\cancel{l}} = \dfrac{A}{l}$ (using E_m)   ($l \neq 0$)

$w = \dfrac{A}{l}$ (solution for w)

or $\{w \mid A = lw\} = \left\{\dfrac{A}{l}\right\}$

$A = lw$ (solve for l)

$lw = A$ (using E_s)

$\dfrac{l\cancel{w}}{\cancel{w}} = \dfrac{A}{w}$ (using E_m)   ($w \neq 0$)

$l = \dfrac{A}{w}$ (solution for l)

or $\{l \mid A = lw\} = \left\{\dfrac{A}{w}\right\}$

(2) $al - w = t$ (solve for l)

$al = t + w$ (using E_a)

$$\frac{\cancel{a}l}{\cancel{a}} = \frac{t+w}{a} \qquad \text{(using E}_m) \qquad (a \neq 0)$$

$$l = \frac{t+w}{a} \qquad \text{(solution for } l)$$

or $\qquad \{l \mid al - w = t\} = \left\{\dfrac{t+w}{a}\right\}$

(3) $\dfrac{w}{b} - 3t = \dfrac{l}{a} \qquad$ (solve for a) $\qquad (a \neq 0,\ b \neq 0)$

Using E$_m$ and the L.C.D., clear the fractions from the equation *first*.

$$\frac{a\cancel{b}}{1} \cdot \frac{w}{\cancel{b}} - \frac{ab}{1} \cdot 3t = \frac{ab}{1} \cdot \frac{l}{a}$$

$$aw - 3abt = bl$$

$$a(w - 3bt) = bl \qquad \qquad \text{(factor the left side}$$
$$\text{to isolate } a)$$

$$\frac{a(\cancel{w - 3bt})}{(\cancel{w - 3bt})} = \frac{bl}{(w - 3bt)} \qquad \text{(using E}_m)$$

$$a = \frac{bl}{w - 3bt}$$

or $\qquad \left\{a \;\middle|\; \dfrac{w}{b} - 3t = \dfrac{l}{a}\right\} = \left\{\dfrac{bl}{w - 3bt}\right\}$

The obvious advantage of using set notation in literal equations is that it designates quite clearly which variable is the one to be isolated.

Exercise 67:

(1) Find the solution sets of the following equations:

(a) $|x| = 3$ (b) $|2x| = 10$ (c) $|3x| = 2$

(d) $|x - 1| = 0$ (e) $|x - 2| = 5$ (f) $|x + 3| = 1$

(g) $|2x - 3| = 6$ (h) $|3x - 1| = 4$ (i) $|5x + 4| = 1$

(j) $|3 + 2x| = 3$ (k) $|1 - 2x| = 5$ (l) $|2 - 3x| = 5$

(m) $\left|\dfrac{x}{3}\right| = 2$ (n) $\left|\dfrac{x}{5}\right| = 4$ (o) $\left|\dfrac{x + 1}{2}\right| = 3$

(p) $\left|\dfrac{x - 4}{3}\right| = 2$ (q) $\left|\dfrac{3x + 2}{7}\right| = 0$ (r) $\left|\dfrac{4x - 5}{2}\right| = 3$

(2) Find the solution set for each of the following. No literal numbers used as divisors can be zero.

(a) $\{x \mid ax + b = c\}$ (b) $\{r \mid rt = d\}$

(c) $\{w \mid lwh = V\}$ (d) $\left\{h \;\middle|\; V = \dfrac{\pi r^2 h}{3}\right\}$

(e) $\left\{a \mid \dfrac{1}{a} + b = \dfrac{c}{a}\right\}$ (f) $\left\{b \mid \dfrac{1}{a} + b = \dfrac{c}{a}\right\}$

(g) $\{t \mid at - bw = s\}$ (h) $\{w \mid at - bw = s\}$

(i) $\{x \mid ax + bx = c\}$ (j) $\{y \mid cy + dy = e\}$

(k) $\{a \mid ab - d = ac\}$ (l) $\{m \mid mp - t = m\}$

(m) $\{x \mid x + 5y = 3\}$ (n) $\{y \mid x + 5y = 3\}$

(o) $\{x \mid 3x + 2y = 7\}$ (p) $\{y \mid 3x + 2y = 7\}$

(q) $\{a \mid 5a - 7b = 4\}$ (r) $\{b \mid 5a - 7b = 4\}$

(s) $\{x \mid 4x - 5y = 8\}$ (t) $\{y \mid 4x - 5y = 8\}$

(u) $\left\{a \mid \dfrac{1}{a} + \dfrac{1}{b} = \dfrac{1}{c}\right\}$ (v) $\left\{b \mid \dfrac{1}{a} + \dfrac{1}{b} = c\right\}$

(w) $\left\{r \mid P = \dfrac{A}{1 + r}\right\}$ (x) $\left\{F \mid C = \dfrac{5}{9}(F - 32)\right\}$

5.8
Stated Problems

One practical use of algebraic symbolism and operations is that they provide us with a convenient way of solving problems. Representing an unknown number with a symbol, such as x, enables us to represent *operations involving the number* as well as the number itself. For example, if x *years* is a man's age, then five years ago he was $x - 5$ years old, and two years hence he will be $x + 2$ years old.

In order to solve problems based on different facts stated about numbers, you must *first* learn the technique of translating such statements into symbolic language. This technique is illustrated in the following statements:

A. x is a number

(1)   One-half of the number $\dfrac{1}{2}x$ or $\dfrac{x}{2}$

(2)   Twice the number $2x$

(3)   Five less than half the number $\dfrac{x}{2} - 5$

(4)   Twice the number increased by 7 $2x + 7$

(5)   The additive inverse of the number $-x$

(6)   Thirty percent of x $.30x$ or $.3x$

(7)   Forty-five percent of x $.45x$

(8)   The grade average for test grades of 80
      and x $\dfrac{80 + x}{2}$

(9)   The amount of acid in 35 cubic centimeters
      (cm³) of solution that is $x\%$ acid $35\left(\dfrac{x}{100}\right)$ or $.35x$ cm³

(10) A price of x dollars reduced by 25% $x - .25x$ dollars

(11) Two consecutive integers if the first one is x x and $x + 1$

(12) The sum of two consecutive integers if the
first one is x $x + x + 1$ or $2x + 1$

(13) The square of x x^2

(14) The sum of the squares of x and 3 $x^2 + 3^2$ or $x^2 + 9$

(15) The square of the sum of x and 3 $(x + 3)^2$ or $x^2 + 6x + 9$

(16) x decreased by 5% of x $x - .05x$

(17) x increased by 60% of x $x + .60x$ or $x + .6x$

(18) Thirty-five percent of 15 more than x $.35(x + 15)$

(19) The price of 5 kilograms (kg) of candy at
x cents/kg * $5x$ cents

(20) The price per kilogram of candy if
x kilograms cost $4.00 $\dfrac{400}{x}$ cents/kg

(21) One year's interest on x dollars invested at
6% per year $.06x$

If an object moves at a constant rate of speed for a specific time, the distance traveled is equal to the rate multiplied by the time. If r is the rate, t is the time, and d is the distance, then

   (1) $rt = d$

Equation (1) may be written in the alternate forms

   (2) $t = \dfrac{d}{r}$ [Dividing both sides of (1) by r]

   (3) $r = \dfrac{d}{t}$ [Dividing both sides of (1) by t]

These equations enable us to express symbolically the *distance*, *rate*, or *time* of a moving object. Note the following examples:

B. A car travels at the rate of x kilometers per hour (km/h) *
   (1) The rate of a car traveling 8 km/h faster $x + 8$ km/h
   (2) The rate of a car traveling 10 km/h slower $x - 10$ km/h
   (3) The distance the car travels in 6 hours $6x$ kilometers
   (4) The time the car travels if it goes 600 kilometers $\dfrac{600}{x}$ hours

C. x km/h = the rate a man rows in still water, and 3 km/h = the rate of the current in which he rows
   (1) The rate at which the man rows *upstream* (against the $x - 3$ km/h
        current)
   (2) The rate at which the man rows *downstream* (with the $x + 3$ km/h
        current)

*An explanation of metric units and their relations to English units are shown in Table 8.1 on pages 300-301.

(3) Twice the man's rate downstream $2(x + 3)$ km/h

(4) The distance the man rows upstream in 4 hours $4(x - 3)$ kilometers

(5) The distance the man rows downstream in 5 hours $5(x + 3)$ kilometers

(6) The time the man travels if he rows 10 kilometers
     upstream $\dfrac{10}{x - 3}$ hours

(7) The time the man travels the same 10 kilometers
     back downstream $\dfrac{10}{x + 3}$ hours

(8) The distance the man rows upstream in 30 minutes
     Note: 30 minutes $= \dfrac{1}{2}$ hour $\dfrac{1}{2}(x - 3)$ kilometers

(9) The distance the man rows downstream in 40 minutes
     Note: 40 minutes $= \dfrac{40}{60}$ hour $= \dfrac{2}{3}$ hour $\dfrac{2}{3}(x + 3)$ kilometers

Exercise 68:

x is a nonzero number; express each of the following in terms of x.

(1) A number three times as large as x

(2) A number three more than x

(3) The number x decreased by 3

(4) 7 less than x

(5) 5 more than twice x

(6) 4 less than half of x

(7) 8 less than $\dfrac{1}{3}$ of x

(8) The cube of x

(9) The sum of the squares of x and 4

(10) The square of the sum of x and 4

(11) The sum of three consecutive integers if the first is x

(12) Sixty-five percent of x

(13) Forty percent of x

(14) A price of x dollars reduced by 20%

(15) A year's interest on x dollars invested at $5\dfrac{1}{2}\%$

(16) A selling price based on a cost of x dollars and a mark-up of 15%

(17) The amount of alcohol remaining if from 32 liters of a solution that is
      16% alcohol, x liters are drained off

(18) The amount of acid present in x liters of a solution that is 8% acid

(19) The price of 10 kilograms of beef at x cents/kg

(20) The price per kilogram of beef that is $20.00 for x kilograms

(21) The length of a rectangle that is x meters wide and 6 meters longer than it is wide

(22) The area of a rectangle that is x meters wide and 6 meters longer than it is wide

(23) The distance a car travels in 8 hours at x km/h

(24) The time a car travels if it covers 200 kilometers at x km/h

(25) The rate of a car that travels 300 kilometers in x hours

(26) The rate a plane travels at x kilometers flying speed against a 30-km/h headwind.

(27) The rate a plane travels at a flying speed of 200 km/h with a tailwind of x km/h

(28) The time a plane travels if it flies 800 kilometers at a flying speed of x km/h against a headwind of 40 km/h

(29) The distance a plane travels in x hours at a flying speed of 300 km/h against a headwind of 35 km/h

(30) The part of a job that can be done in one day if x days are required to complete the job

(31) The number of cents in x dimes

(32) The multiplicative inverse of $x - 2$

(33) The product of x and its additive inverse

(34) The grade average for test grades of 75, 86, and x

(35) A two-digit number in which the unit's digit is 3 and the ten's digit is x

   Note: In the two-digit number 46, the unit's digit is 6 and the ten's digit is 4; the *number* is $4(10) + 6$.

(36) The perimeter of a rectangle whose width is x meters and whose length is twice the width.

When we seek to evaluate an unknown number on the basis of certain facts known about the number, we *first* symbolize the number itself and *then* its relationships to the known facts.

If we have enough facts to set up an equation involving the unknown number, then we have only to solve the equation to find the number. In all problems taken from statements of facts about an unknown number, such as x, the *ultimate* problem is to find x; but the *immediate* problem is to *find an equation involving x*. We can't very well "solve for x" unless we have something to solve! Outlined below is a recommended procedure which should lead you to the desired equation.

1. Select a symbol for the unknown number, such as x.
2. If x represents a distance or length, *draw a figure* representing the situation. Label all known parts, and then, where possible, label unknown parts of the figure in terms of x.

3. Represent in symbolic language all known facts about the number *in terms of* x.
4. Tabulate all of the symbolized facts and look for *two equal quantities*.
5. Connect the two equal quantities with an equal mark and solve the resulting equation.

Examples of stated problems:

I. If a certain number is doubled and then increased by 9, the result is 17. What is the number?
   (1) Let x = the number.
   (2) $2x$ = the number doubled.
   (3) $2x + 9$ = the number doubled and increased by 9.
   (4) 17 = the number doubled and increased by 9.
       Statements (3) and (4) represent the *same quantity:*
       $$2x + 9 = 17 \qquad \text{(using } E_t)$$
       $$2x = 17 - 9$$
       $$2x = 8$$
       $$x = 4, \text{ the number.}$$

II. John made 67 on the first math quiz. What grade must he make on the second quiz to bring his grade average up to 75?
   (1) x = the grade on the second quiz.
   (2) 67 = the grade on the first quiz.
   (3) $\dfrac{x + 67}{2}$ = the average of the two grades.
   (4) 75 = the average of the two grades.
       $$\frac{x + 67}{2} = 75 \qquad \text{(using } E_t)$$
       $$x + 67 = 150$$
       $$x = 150 - 67$$
       $$x = 83, \text{ the grade required for a 75 average.}$$

III. In a two-digit number the unit's digit is twice as large as the ten's digit. The number is equal to 12 more than four times the unit's digit. What is the number?
   (1) Let x = the ten's digit.
   (2) Then $2x$ = the unit's digit.
   (3) And $10x + 2x$ = the number.
   (4) $4(2x)$ or $8x$ = four times the unit's digit.
   (5) $8x + 12$ = 12 more than four times the unit's digit = the number.
       Statements (3) and (5) represent the same quantity.
       $$10x + 2x = 8x + 12$$
       $$12x = 8x + 12$$

$$12x - 8x = 12$$
$$4x = 12$$
$$x = 3 \quad \text{(the ten's digit)}$$
$$2x = 6 \quad \text{(the unit's digit)}$$
$$36 = \text{the number.}$$

IV. The length of a rectangle is 5 meters longer than its width (Figure 5.2). If the perimeter of the rectangle is 54 meters, what are the dimensions (length and width) of the rectangle?

Figure 5.2

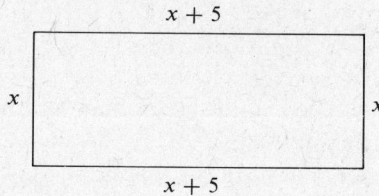

(1) Let x meters = the width of the rectangle.
(2) Then $x + 5$ meters = the length of the rectangle.
(3) $x + x + 5 + x + x + 5$ meters = the perimeter of the rectangle.
     $4x + 10$ meters = the perimeter.
(4) 54 meters = the perimeter.
$$4x + 10 = 54 \quad \text{(using } E_t)$$
$$4x = 54 - 10$$
$$4x = 44$$
$$x = 11 \text{ meters, width of the rectangle.}$$
$$x + 5 = 16 \text{ meters, length of the rectangle.}$$

Note that the preceding problem is much easier to represent in terms of x when a figure is drawn to represent the rectangle. *Whenever possible*, a figure or diagram should be drawn because even the simplest drawing can often clarify the conditions of a problem enormously. For another illustration of this, see Example VI.

V. Bill can take the office inventory in 5 hours. Joe can take the same inventory in 4 hours. How many hours would they need to take the inventory if they worked together?
(1) Since Bill can do the inventory in 5 hours, in *one* hour he can complete 1/5 of it.
(2) Since Joe can do the inventory in 4 hours, in *one* hour he can complete 1/4 of it.
(3) In *one* hour the two working together can complete $1/5 + 1/4$ of the inventory.
(4) Let x hours = the time required to take the inventory working together.
(5) Then in *one* hour the two working together can complete $1/x$ of the inventory.

Statements (3) and (5) both mean the *same thing*:

$$\frac{1}{5} + \frac{1}{4} = \frac{1}{x} \qquad \text{(using } E_t\text{)}$$

$$\overset{4}{\frac{\cancel{20}x}{1}} \cdot \frac{1}{\cancel{5}} + \overset{5}{\frac{\cancel{20}x}{1}} \cdot \frac{1}{\cancel{4}} = \frac{20\cancel{x}}{1} \cdot \frac{1}{\cancel{x}}$$

$$4x + 5x = 20$$
$$9x = 20$$
$$x = \frac{20}{9} \qquad \text{or } 2\frac{2}{9} \text{ hours to take the inventory together.}$$

VI. A man drives for 5 hours at a certain rate of speed, then increases his speed by ten km/h and drives 3 hours longer. If he travels 430 kilometers on the trip, at what two speed rates does he drive?

In straight-line motion (rate of speed)(time) = distance.

$$(r)(t) = (d)$$

Let x km/h equal the first rate of speed, and sketch the following diagram:

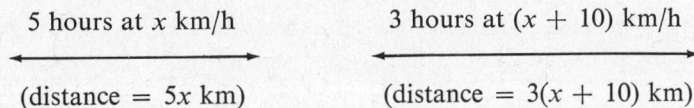

5 hours at x km/h   3 hours at $(x + 10)$ km/h

←──────────────────→   ←──────────────────→

(distance = $5x$ km)   (distance = $3(x + 10)$ km)

(1) Let x km/h = his first rate of speed.
(2) 5 hours = time he drove at that rate.
(3) Then $5x$ kilometers = *distance* he traveled at x km/h.
(4) $x + 10$ km/h = his second rate of speed.
(5) 3 hours = time he drove at that rate.
(6) Then $3(x + 10)$ kilometers = *distance* he traveled at $x + 10$ km/h.
(7) From statements (3) and (6),
     $5x + 3(x + 10)$ = total distance traveled.
(8) 430 kilometers = total distance traveled.
     $5x + 3(x + 10) = 430 \qquad$ (using E_t)
     $5x + 3x + 30 = 430$
     $8x + 30 = 430$
     $8x = 430 - 30$
     $8x = 400$
     $x = 50$ km/h, first speed.
     $x + 10 = 60$ km/h, second speed.

VII. A 100-cm³ solution of acid and water is 40% acid. How many cm³ of pure water should be added to make a solution that is 16% acid?

(1) 100 cm³ = amount of solution that is 40% acid.
(2) Then .40(100) = 40 cm³ = the total amount of acid in the solution.

(3) Let x cm^3 = the amount of pure water to be added.

(4) Then $(100 + x)$ cm^3 = the amount of new solution that is 16% acid.

(5) And $.16(100 + x)$ cm^3 = the total amount of acid in the solution.

Statements (2) and (5) represent the same quantity; therefore

$$.16(100 + x) = .40(100) \qquad \text{(using } E_t)$$
$$16 + .16x = 40$$
$$16(100) + .16x(100) = 40(100) \qquad (E_m)$$
$$1600 + 16x = 4000$$
$$16x = 4000 - 1600 \qquad (E_a)$$
$$16x = 2400$$
$$x = \frac{2400}{16} \qquad (E_m)$$
$$x = 150 \text{ cm}^3 \text{ of pure water to be added.}$$

Success in solving stated problems amounts to selecting a symbol for the unknown number and then carefully tabulating its relations to known quantities in terms of the selected symbol. To do this you must be willing to *read* the problem until you know clearly what the problem is (one or two rapid scannings will *not* accomplish this).

A thorough understanding of the problem, a careful tabulation of the known facts in symbolic language, and frequent illustrative drawings will effectively remove guesswork and confusion from the solving of stated problems. *Nothing else will.*

Summary

1. Read the problem carefully and repeatedly.
2. Decide what quantity you are looking for and give it a name, such as x.
3. Draw a figure representing the conditions given in the problem whenever possible.
4. Tabulate all facts known about x in terms of x.
5. Look over the tabulated facts and find two equal quantities.
6. Write the equation and solve it.

CAUTION! Read the following *before* attempting the next exercise:

The first forty of the stated problems which follow are arranged by type into four groups so that you may encounter, in stepwise fashion, the various techniques that are helpful in following the advice given in the Summary above. For example, in the two groups involving geometric figures and motion, *a pictorial diagram should always be drawn*.

Because it is easier to learn new techniques in the simplest possible context, the grouped problems have more instructional than practical value. In other words, for these beginning problems, the problem itself is not significant, but *the manner in which you reach a solution to the problem is tremendously significant* if you ever intend to apply mathematical skills to a wide variety of practical problems. You must be keenly aware at the outset that "The Answer" to any of these initial problems is *not important*. What *is* important is the realization on your part that a careful and logical analysis of the problem, translated into symbolic language and precisely written down, is the only way to acquire the skills necessary to attack and solve the more practical problems which follow the grouped sections.

Those of you who have the good sense to resist the rather childish tendency to rush headlong toward The Answer and concentrate instead on making a careful analysis of the problem will make two enormous gains: (1) You will learn that there is no substitute for clear thinking; and (2) You will acquire the initial techniques necessary to work all different kinds of stated problems.

Exercise 69:

Solve the following problems:

Number Problems

(1) If a number is doubled and then increased by 6, the result is 24. Find the number.

(2) The sum of three consecutive integers is 30. Find the numbers.

(3) The difference between two numbers is 6. The difference between the squares of the numbers is 84. Find the numbers.

(4) A number is divided by 5 and the quotient decreased by 10. The result is 20. What is the number?

(5) If a number is increased by $\frac{3}{4}$ of itself, the result is 21. What is the number?

(6) If the square of a number is increased by 24, the result is equal to the square of the sum of the number and 2. What is the number?

(7) The sum of two numbers is 15. The larger number is equal to the smaller number doubled and decreased by 3. What are the numbers?

(8) The sum of two numbers is 10. If the smaller number is squared and increased by 20, the result is equal to the square of the larger number. What are the numbers?

(9) In a two-digit number the unit's digit is one more than the ten's digit. If the number is increased by 4, the result is equal to eight times the unit's digit. What is the number?

(10) In a two-digit number the unit's digit is twice as large as the ten's digit. If the number is decreased by 36, the result is equal to three times the ten's digit. What is the number?

Problems About Plane Figures

(11) A rectangle is twice as long as it is wide. The perimeter of the rectangle is 54 centimeters. Find the length and width of the rectangle.

(12) The length of a rectangle is 8 meters more than twice its width. If the perimeter of the rectangle is 52 meters, what are its dimensions?

(13) A rectangle is 4 meters longer than it is wide. The perimeter is eight times the width. What are the dimensions of the rectangle?

(14) The lengths of two sides of a triangle are equal. The remaining side is 4 centimeters shorter, and the perimeter of the triangle is 14 centimeters. Find the lengths of the sides.

(15) A rectangle is three times as long as it is wide. The perimeter of the rectangle is 64 meters. Find the length and width of the rectangle.

(16) A rectangular field is twice as long as it is wide and is fenced on all four sides. The field is divided into two equal parts by another fence running lengthwise across it. If the total amount of fencing is 320 meters find the dimensions of the field.

(17) A rectangular playground is to be built against the side of a school building, and the remaining three sides are to be fenced in. The playground is designed to be 20 meters longer than it is wide, and its perimeter is 160 meters. If the longer side of the playground is placed against the building, how much fencing will be required?

(18) The diagonal of a rectangle is 8 centimeters longer than its width. The length of the rectangle is 12 centimeters. Find the width. Hint: The sum of the squares of the two sides of a right triangle equals the square of the hypotenuse.

(19) In a triangle one side is twice as long as the shortest side. The remaining side is 6 meters longer than the shortest side, and the perimeter is 34 meters. What are the lengths of the sides of the triangle?

(20) A lot is in the shape of a right triangle with a front of 150 meters and a depth of 100 meters. A rectangular building 75 meters long is to be built on the lot with the front of the building on the 150-meter side of the lot. What is the greatest width this building could have? Hint: Look for a pair of similar triangles and find an equation in the ratios of their corresponding sides.

(21) The shortest side of a triangle is 50 percent the length of the longest side and the remaining side is 70 percent the length of the longest side. If the perimeter of the triangle is 22 meters, find the lengths of the sides.

(22) The lengths of the sides of a triangle are three consecutive integers. If the perimeter is 2 meters longer than twice the longest side, find the lengths of the sides.

Motion Problems

(23) A car is driven 3 hours at a certain speed and then its speed is increased by 10 km/h for the next 2 hours. If the total distance traveled was 300 kilometers, at what two rates was the car driven?

(24) A man walked to town at 4 km/h and rode the bus back at 40 km/h. The combined trip required $1\frac{1}{4}$ hours. How long and how far did he walk?

(25) Two cars, which are 200 kilometers apart, start at the same time and travel toward each other. One car travels 12 km/h faster than the other. If they meet in 2 hours, how fast was each car traveling?

(26) The distance between two towns, A and B, is 330 kilometers. A train leaves town A and travels toward town B at 50 km/h. An hour later a second train leaves town B and travels toward town A at 60 km/h. At what distance from A will the two trains meet?

(27) In a stream whose current flows at a rate of 2 km/h, a man rowed upstream and returned to his starting point in 4 hours. If the man rows at a rate of 4 km/h in still water, how long did he row upstream? How far did he travel upstream?

(28) A man rowed upstream for 2 hours and returned to his starting point in 1 hour. If he rows at a rate of 6 km/h in still water, what was the rate of the stream?

(29) A riverboat travels at a rate that is 80 percent faster than the rate of the current of the river. If the boat travels 56 kilometers downstream in 5 hours, what is the rate of the boat and the rate of the current?

(30) A pilot flew north against a headwind of 40 km/h and made the trip in 5 hours. Returning south at the same speed flying with a tailwind of 40 km/h, he made the return trip in 3 hours. What was the flying speed of the plane?

(31) A pilot flew east against a headwind of 60 km/h and then made the return trip with a tailwind of 40 km/h. The flying speed of the plane was 360 km/h, and the round trip required 7 hours. How long and how far did he fly east?

(32) An agressive house cat named Woochow is lounging against the side of a pole to which he is attached with an 18-meter chain. He watches a squirrel approach to within 2 meters of the pole and promptly gives chase. Woochow runs at a rate of 9 meters per second (m/s) and the squirrel at 8 meters per second. Which will Woochow reach first, the squirrel or the end of the chain? At the start of the chase what is the squirrel's life expectancy?

Solution and Mixture Problems

(33) A 100-cm³ solution of alcohol and water is 35% alcohol. How many cm³ of pure water must be added to make a solution that is 7% alcohol?

(34) How many cm³ of water should be added to 32 cm³ of alcohol to make a solution that is 64% alcohol?

(35) A solution of salt and water is 30% salt. How many cm³ of this solution should be mixed with pure water to make 18 cm³ of a solution that is 10% salt?

(36) A dairy manager plans to produce 100 liters of low-fat milk. If 100 liters of whole milk are 5% butterfat, how many liters of the whole milk must be removed and replaced with skimmed milk in order to make 100 liters of milk that is 2% butterfat?

(37) Two kinds of candy sell for 45 cents and 60 cents/kg. How many kilograms of each should be used to make 45 kilograms of a mixture to sell for 50 cents/kg?

(38) One collection of miscellaneous Christmas cards is priced at 30 cents a card and another type is priced at 50 cents a card. How many of each should be placed in assortment boxes of 50 cards each to sell for $18.00 per box?

(39) A florist sells roses for $1.50 each and lilies for $1.00 each. How many of each should be included in an arrangement of two dozen flowers to sell for $34.00 if the price of greenery in the arrangement is $5.00?

(40) A dairy sells pure cream for 75 cents per 160-cm^3 container and skimmed milk for 25 cents. How many cm^3 of each should be mixed to make coffee cream that will sell for 50 cents/160 cm^3?

Miscellaneous Problems

(41) The Brown's household budget requires their food costs to average $180.00 a month for each quarter of the year. With Thanksgiving and Christmas in mind, Mrs. Brown economized during October and spent $143.00 for groceries. If she expects the December grocery costs to exceed the November costs by $25.00, how much should she spend for food in November?

(42) A nurse is instructed to prepare 100 cm^3 of a 9% saline (salt and water) solution. In the supply room she finds solutions that are 3% salt and 15% salt respectively. How many cm^3 of each solution should she mix to get a solution of the volume and percentage required?

(43) The relationship between degrees Fahrenheit (F) and degrees Celsius (°C) is °F = 9/5°C + 32. Solve this equation for °C. Change 149 degrees Fahrenheit to degrees Celsius.

(44) Tom's grades in mathematics are 75, 67 and 71. The final exam counts 1/3 of his course grade. What should he make on the final exam to average 75 in the course?

(45) In a front-yard enterprise, Johnny is selling lemonade for 5 cents a glass, and Karl is selling bottled drinks for 10 cents each. If they make 50 sales and take in $3.65, how many of each did they sell?

(46) Mr. Smith lives 15 kilometers from his office, where he has free parking. He figures that it costs him 10 cents/km to drive his car. He can ride the bus (both ways) for 80 cents a day. Figuring 20 work days to a month, how many days a month can he drive his car and hold his transportation costs to $27.00 a month?

(47) Bob can do a job in 3 hours. Jack can do the same job in 2 hours. If they both work together, how long will it take them to get the job done?

(48) An intake pipe can fill a swimming pool in 8 hours. A second pipe can fill it in 6 hours. If both pipes are open, how many hours will be required to fill the pool?

(49) A merchant wishes to sell an assortment of 10-cent and 15-cent Christmas cards in boxes of 100. How many of each price should be included in a box to sell for $12.00?

(50) A solution of acid and water is 16% acid. How many cm³ of this solution should be mixed with pure water to make 32 cm³ of a solution that is 10% acid?

(51) A cylindrical can has a surface area of 32π cm². If the radius of the top is 2 cm, what is the height of the can?

(52) A rectangular box with a top is twice as long as it is wide and is 10 cm high. The surface area of the box is 180 cm² more than four times the square of the width. Find the length and width of the box.

(53) Mr. Drake, a retired draftsman, receives $200. a month from his retirement fund and $180. a month from Social Security. He has $40,000 from savings and the sale of property which he wants to invest in bonds that will yield enough interest to give him a total average monthly income of $600. What amount of interest per year must he get from his investment? At what rate should the $40,000 be invested to pay the desired interest?

(54) On his income Mr. Black pays 2% state income tax, 3% state and city sales tax, 5% retirement fund, $470. a year social security tax, and $960. a year for property taxes. He also pays 30% federal income tax after deductions amounting to $2400. Assuming no change in taxes, what income does Mr. Black need in order to have $20,000 a year after paying taxes and retirement?

(55) The Carter family plans to build a circular swimming pool with a surrounding tile walk whose width is $\frac{1}{4}$ the radius of the pool. They want to use a 10-meter-square section of their backyard to build the pool and walk. What should the radius of the pool be?

5.9
Expressing Repeating Decimals as Fractions

Any number which can be written as the quotient of an integer and a nonzero integer — i.e., any *rational* number — can also be expressed in an equivalent decimal form if the indicated division is done. Illustrations of this procedure given in Section 4.3 led us to conclude that all decimals produced by such division were non-ending (or infinite) *repeating* decimals where, in some cases, the repeating digit was zero.

Examples:

(1) $\frac{1}{3} = .333\overline{3} \ldots$

(2) $\frac{8}{27} = .296296\overline{296} \ldots$

(3) $\frac{3}{8} = .37500\overline{0} \ldots$

(4) $\frac{124}{55} = 2.25454\overline{54} \ldots$

(5) $\frac{6}{1} = 6.0000\overline{0} \ldots$

The notation used above is slightly redundant. In previous examples we have used the bar above the numbers to indicate the pattern in which the digits repeat and the three dots at the end to indicate that the pattern repeats indefinitely. Actually, the bar above the repeating sequence is sufficient to indicate both the pattern and its non-ending repetition; so henceforth, in writing infinite repeating decimals, we shall delete the three dots and use the bar only. However, in order to continue emphasizing that the indicated pattern *is* a repeating one, we shall depart from convention to the point of repeating each sequence at least once (if not oftener) before placing the bar above the digits. Thus, the decimal equal to $\frac{1}{3}$ will be written $.333\overline{3}$ and that equal to $\frac{8}{27}$ will be written $.296\overline{296}$.

If it is possible to express ratios of integers as infinite repeating decimals, then the question arises, how can the procedure be reversed? For example, how would one show that the infinite decimal $.333\overline{3}$ is the *same number* as $\frac{1}{3}$? There is a surprisingly simple way to do this. The techniques employed in solving stated problems can be used to find an equivalent fraction for any infinite repeating decimal.

We shall also make use of Euclid's statement that "if equals are added to equals, the results are equal." For example, suppose we know that $a = b$ and $c = d$;

$$\text{then} \qquad a = b$$
$$\text{and} \qquad a + c = b + c \qquad \text{E}_a$$

In the second equation, using the substitution axiom, replace c *by its equal, d*, on the right side.

$$\text{Thus} \qquad a + c = b + d$$

This amounts to saying that if two equations are both true equations, then they may be added, left side to left side and right side to right side, and the resulting equation is also true.

$$\begin{aligned} \text{If} \quad & a = b \\ \text{and} \quad & c = d \\ \text{then} \quad & a + c = b + d \end{aligned}$$

Since subtracting one number from another means that we *add* the negative of the subtrahend to the minuend, we may also subtract equals from equals and obtain results that are equal. Consider the following examples:

(1) $4 + 2 = 6$
(2) $3 = 1 + 2$

If we *add* Equation (2) to Equation (1), left side to left side and right side to right side, we are adding equals to equals, and the result will be a true equation:

(1) $4 + 2 = 6$
(2) $\underline{ 3 = 1 + 2}$
         $4 + 2 + 3 = 6 + 1 + 2$
         $9 = 9$

In like manner we may *subtract* Equation (2) from Equation (1) and obtain a true equation:

(1) $4 + 2 = 6$
         $\underline{-3 = -1 - 2}$
         $4 + 2 - 3 = 6 - 1 - 2$
         $3 = 3$

Now, suppose that we wish to find the fraction equal to the infinite decimal $.333\overline{3}$. We make the plausible assumption that *any* decimal expansion, whether terminating in zeros or not, may be multiplied by a real number, such as 10, or a multiple of 10 such as 100, 1000, 10,000, etc. Then we proceed as follows:

(1) Let x = the fraction that is equal to $.333\overline{3}$
(2) Then $x = .333\overline{3}$
(3) And $10x = 3.333\overline{3}$ (Using E_m)

Subtract Equation (2) from Equation (3)

$$\begin{aligned} 10x &= 3.333\overline{3} \\ \underline{x} &= \underline{.333\overline{3}} \\ 9x &= 3 \\ x &= \frac{3}{9} = \frac{(1)(\not{3})}{(3)(\not{3})} = \frac{1}{3} \end{aligned}$$

Therefore $.333\overline{3} = \dfrac{1}{3}$

Thus, to express any infinite repeating decimal as a fraction, we first set up an equation in which a symbol such as x represents the equivalent fraction. Then a second equation is obtained by multiplying both sides of the first one by some multiple of 10 (10, 100, 1000, etc.) so that the *repeating digits* following the decimal point are *identical* in both equations. (To make the corresponding digits which follow the decimal point identical, it may sometimes be necessary to multiply the original equation by some multiple of 10 more than once.) If we subtract one equation from another one which has identical repeating digits on the right side, the repeating part of each decimal will vanish. The fraction represented by x can then be found by dividing both sides of the resulting equation by the coefficient of x. To illustrate this, we shall take Examples (2) and (4) from the list shown at the beginning of this discussion and reverse the procedure.

A. Find the fraction equal to $.296296\overline{296}$
    (1) Let $x = .296296\overline{296}$
    (2) Then $1000x = 296.296\overline{296}$
    Subtract Equation (1) from Equation (2)

$$1000x = 296.296\overline{296}$$
$$x = .296\overline{296}$$
$$999x = 296$$
$$x = \frac{296}{999} = \frac{(8)(37)}{(27)(37)} = \frac{8}{27}$$

Therefore $.296296\overline{296} = \frac{8}{27}$

B. Find the fraction equal to $2.2545\overline{454}$
    (1) Let $x = 2.2545\overline{454}$
    (2) Then $10x = 22.545\overline{454}$
    (3) And $1000x = 2254.545\overline{454}$
    Subtract Equation (2) from Equation (3)

$$1000x = 2254.545\overline{454}$$
$$10x = 22.545\overline{454}$$
$$990x = 2232$$
$$x = \frac{2232}{990} = \frac{(124)(18)}{(55)(18)} = \frac{124}{55}$$

Therefore $2.2545\overline{454} = \frac{124}{55}$

Following are more examples:

C. Find the fraction equal to $.2727\overline{27}$
    (1) Let $x = .2727\overline{27}$
    (2) Then $100x = 27.2727\overline{27}$
    Subtract Equation (1) from Equation (2)

$$100x = 27.2727\overline{27}$$
$$x = .2727\overline{27}$$
$$\overline{99x = 27}$$
$$x = \frac{27}{99} = \frac{(3)(\cancel{9})}{(11)(\cancel{9})} = \frac{3}{11}$$

Therefore $.2727\overline{27} = \frac{3}{11}$

D. Find the fraction equal to $3.162162\overline{162}$
   (1) Let $x = 3.162162\overline{162}$
   (2) Then $1000x = 3162.162\overline{162}$

Subtract Equation (1) from Equation (2)

$$1000x = 3162.162\overline{162}$$
$$x = 3.162\overline{162}$$
$$\overline{999x = 3159}$$
$$x = \frac{3159}{999} = \frac{(117)(\cancel{27})}{(37)(\cancel{27})} = \frac{117}{37}$$

Therefore $3.162162\overline{162} = \frac{117}{37}$

The fact that adding two true equations will produce a true equation can lead to an interesting result with regard to those infinite decimals which have *zero* as a repeating digit. Consider the following example:

(1) $\frac{1}{3} = .3333\overline{3}$

(2) $\frac{2}{3} = .6666\overline{6}$

Add Equation (1) to Equation (2)

$$\frac{1}{3} = .3333\overline{3}$$

$$\frac{2}{3} = .6666\overline{6}$$

$$\overline{\frac{3}{3} = .9999\overline{9}}$$

or $1 = .9999\overline{9}$

Thus $1.0000\overline{0} = .9999\overline{9}$

If this result looks unbelievable at first sight, look again. In fact, take a long look. There are some mathematical subtleties in the above addition that we are blithely ignoring. For example, we add two numbers by beginning with the rightmost digit of each one, but the numbers we "added" on the right sides of the equations above have *no* rightmost digit! Such addition is questionable, to say the least, but the result it produces, i.e., $1 = .9999\overline{9}$, is worth considering, if only to stir the imagination or perhaps prod an intuitive suspicion. The final result might not look quite so wild if the student reflects that the statement does *not* say that $1 = .999999$ or $1 = .999999999999$, but that $1 = .9999\overline{9}$, where the number on the right has an *infinite* succession of nines following the decimal point.

Another approach to the same situation could be taken by finding the fraction equal to $.9999\overline{9}$, using the method previously discussed. We proceed as follows:

(1) Let $x = .9999\overline{9}$
(2) Then $10x = 9.9999\overline{9}$
  Subtract Equation (1) from Equation (2)

$$10x = 9.999\overline{9}$$
$$x = .999\overline{9}$$
$$\overline{9x = 9}$$
$$x = \frac{9}{9} \text{ or } \frac{1}{1} \text{ or } 1 \text{ or } 1.000\overline{0}$$

Therefore $1.000\overline{0} = .9999\overline{9}$

By using more advanced notions and techniques than can be employed here, we can demonstrate (rigorously, we think) that 1 and $.9999\overline{9}$ actually are the same number. But for the present we shall proceed on the assumption that such is the case and then consider the consequences. There are some interesting ones. Consider the following:

(1) $2 = 1 + 1$
  Then $2 = 1 + .999\overline{9}$ (Using the substitution axiom)
  $2 = 1.999\overline{9}$
  or $2.0000\overline{0} = 1.9999\overline{9}$

(2) $3 = 2 + 1$
  Then $3 = 2 + .999\overline{9}$
  $3 = 2.999\overline{9}$
  or $3.000\overline{0} = 2.999\overline{9}$

(3) $.87500\overline{0} = .87400\overline{0} + .00100\overline{0}$

$$\left[\begin{array}{l} \text{Note: } 1.000\overline{0} = .999\overline{9} \\ \text{Multiply both sides of the equation by } .001 \\ 1.000\overline{0}(.001) = .999\overline{9}(.001) \\ .00100\overline{0} = .000999\overline{9} \end{array}\right]$$

Substitute $.000999\overline{9}$ for $.00100\overline{0}$ in the original equation to obtain
  $.87500\overline{0} = .87400\overline{0} + .000999\overline{9}$ (Using the substitution axiom).
  $.87500\overline{0} = .874999\overline{9}$

The foregoing examples suggest that any infinite decimal which has *zero* as a repeating digit can also be expressed in an equivalent decimal form which has *nine* as the repeating digit. Thus, if we find the fraction equal to $.25000\overline{0}$ and the fraction equal to $.24999\overline{9}$, we should wind up with the same fraction. To illustrate:

(1) $.25000\overline{0} = .25 = \dfrac{25}{100} = \dfrac{(1)\cancel{(25)}}{(4)\cancel{(25)}} = \dfrac{1}{4}$ or

Let $x = .25000\overline{0}$
Then $100x = 25.000\overline{0}$
and $1000x = 250.000\overline{0}$
Subtract the second equation from the last one

$$1000x = 250.000\overline{0}$$
$$100x = 25.000\overline{0}$$
$$\overline{900x = 225}$$

$$x = \frac{225}{900} = \frac{1\cancel{(225)}}{4\cancel{(225)}} = \frac{1}{4}$$

(2) Now try the same method with $.24999\overline{9}$
Let $x = .24999\overline{9}$
Then $100x = 24.999\overline{9}$
and $1000x = 249.999\overline{9}$
Subtract the second equation from the last one

$$1000x = 249.999\overline{9}$$
$$100x = 24.999\overline{9}$$
$$\overline{900x = 225}$$

$$x = \frac{225}{900} = \frac{(1)\cancel{(225)}}{(4)\cancel{(225)}} = \frac{1}{4}$$

Thus $.25000\overline{0} = .24999\overline{9} = \dfrac{1}{4}$

Exercise 70:

Express each of the infinite repeating decimals as an equivalent fraction.

(1) $.444\overline{4}$ (2) $.555\overline{5}$

(3) $.777\overline{7}$ (4) $.888\overline{8}$

(5) $.7000\overline{0}$ (6) $.6999\overline{9}$

(7) $.6363\overline{63}$ (8) $.81818\overline{1}$

(9) $2.51515\overline{1}$ (10) $4.36363\overline{6}$

(11) $.372727\overline{72}$ (12) $1.181818\overline{18}$

(13) $.261261\overline{261}$ (14) $.711711\overline{711}$

(15) $.2315315\overline{315}$ (16) $1.414414\overline{414}$

(17) $4.999\overline{9}$ (18) $7.999\overline{9}$

(19) $.75000\overline{0}$ (20) $.74999\overline{9}$

(21) $.128000\overline{0}$ (22) $.127999\overline{9}$

Chapter Test 5

I. Find the solution set of each of the following equations:

1. $2b + 3 = 7$

2. $\frac{1}{3}y = \frac{2}{9}$

3. $|3 - 2x| = 7$

4. $3x + 7 = 5x - 9$

5. $3(y - 1) - 2(y + 3) = 0$

6. $\left|\frac{x}{3} - \frac{x}{5}\right| = 2$

7. $\frac{3}{a + 1} - 1 = \frac{a}{a + 1}$

8. $\frac{2}{x + 2} + 5 = \frac{5x}{x - 2}$

II. Find the solution set of each of the following. All literal symbols represent nonzero numbers.

1. $\{x \mid ax + b = c\}$

2. $\{x \mid 2x + 3y = 6\}$

3. $\left\{y \left| \frac{a}{y} + c = \frac{b}{y}\right.\right\}$

4. $\{a \mid ab - 2c = ad\}$

III. Express each of the following quantities in terms of the symbol x:
1. The yearly income of a man who makes x dollars a month.
2. The number of hours required to travel 500 kilometers at x km/h.
3. Thirty-two percent of x.
4. The grade average for test grades of 86, 92, and x.
5. The price per pound of beef that costs $15.00 for x pounds.
6. The amount of alcohol in a 50-cm³ solution that is $x\%$ alcohol.

IV. In each of the following problems a symbol is assigned to represent a specific quantity. Using that symbol as suggested, show how to set up an equation that will solve the problem. Then solve the resulting equation.

1. A hospital laboratory technician is asked to prepare 300 cm³ of a solution of water and glucose that is 5% glucose. The lab supply is available in a concentrated solution that is 25% glucose. How many cc of this solution and how many cm³ of pure water must be mixed to obtain the required solution?

    Let x cm³ = the amount of 25% glucose solution used

2. The Browns are planning a new home which will cost $48 a square meter to build. The lot they want costs $9500, and they want to reserve an additional $900. for landscaping. If they can spend $38,000 for house, lot, and landscaping, how large a house can they build?

    Let x square meters = the number of square meters in the house

V. Express each decimal as an equivalent fraction.
1. $.54\overline{54}$ 2. $.60\overline{00}$ 3. $.59\overline{99}$

6
LINEAR EQUATIONS, FUNCTIONS AND GRAPHS

The phrase "linear equations" sprang from part of a significant mathematical romance which took place in the early seventeenth century and which established a relationship between straight lines and certain kinds of equations. The alliance began with René Descartes in the dual role of a creative mathematician and an exceedingly efficient matchmaker. It led to the union of *equations* with *sets of points* which in turn describe geometric patterns (straight lines and curves) that are visual manifestations of the equations themselves. It is our purpose in this chapter to investigate equations which represent straight lines and which are therefore called *linear equations*.

Descartes began by devising a scheme for locating the precise position of a particular point out of the infinity of points which make up a plane surface. The scheme was named for him and, for all of its simplicity, has the imposing name of the rectangular Cartesian coordinate system.

6.1
The Rectangular Cartesian Coordinate System

We may think of space as being composed of infinitely many slices stacked together, much like a loaf of sliced bread, except that our slices of space have no depth. If we figuratively take out a slice of this space, we have a flat, smooth surface that extends forever in two dimensions, length and width, but that has no depth at all. Such a slice is what we refer to when we speak of a *plane* or a *plane surface*.

The surface of a table, a floor, a ceiling, a blackboard, and a sheet of paper can all be regarded as concrete representations of planes. Each one is, of course, a partial representation only, because the surfaces of floors, ceilings, blackboards, and sheets of paper are all finite in extent while a plane goes on forever. But any one of them can represent a portion of a plane, and that portion can be used to study the properties of a plane.

A plane (two dimensional: length and width) can be considered as an aggregation of infinitely many straight lines; and each line (one dimensional: length only) we may regard as being made up of infinitely many points (a point has no dimensions). Thus when we set ourselves to mentally dissecting a plane, we arrive at its elemental component, a *point*, and every plane becomes a set of infinitely many points.

We can chart our way across a plane in the same manner that we chart our way across the unmarked expanses of the oceans: by imposing arbitrary *direction lines* to guide us. On the plane surface shown in Figure 6.1, we have imposed two direction lines perpendicular to each other. The arbitrary imposition of these lines in no way distorts the physical properties of the plane nor the position of any point in it, just as lines of latitude and longitude, while tremendously useful to us, do not in any way rearrange the oceans. These lines, called *axes*, divide the plane into four parts, called *quadrants*, which are numbered counterclockwise in the manner shown in Figure 6.1.

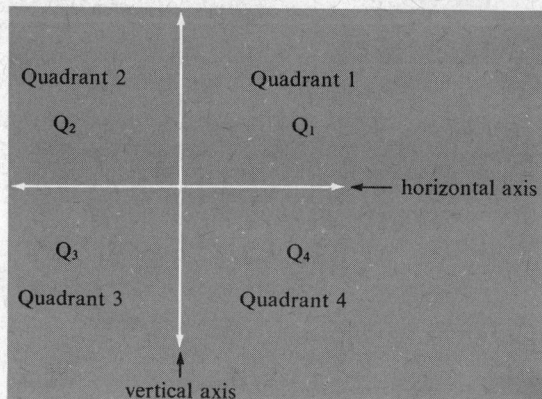

Quadrant 2

Q_2

Quadrant 1

Q_1

horizontal axis

Q_3

Q_4

Quadrant 3

Quadrant 4

vertical axis

Figure 6.1

Since the real numbers can be paired in a one-to-one correspondence with the points on a straight line (Section 4.3), we can mark off the two axes in uniform graduations of length, and use the real numbers to locate any point on either axis. We take the intersection point of the two axes (called the origin) as a starting point and begin here with the number zero. On the horizontal axis we reproduce the number line: every point on this axis to the *right* of zero (the origin) is identified by a positive number, and every point on this axis to the left of the origin is identified by a negative number. On the vertical axis the same scheme is repeated, the positive numbers identifying all points on this axis *above* the origin, and the negative numbers all points *below* the origin (see Figure 6.2).

From Figure 6.2 we see that the numbers on the horizontal axis mark the number of units that a point lies to the *right* or *left* of the vertical axis, while the numbers on the vertical axis mark the location of the point *above* or *below* the horizontal axis. We shall distinguish between the two axes by giving them two different names: the horizontal axis will be the *x-axis* or the *x-direction* line, and the vertical axis will be the *y-axis* or the *y-direction* line.

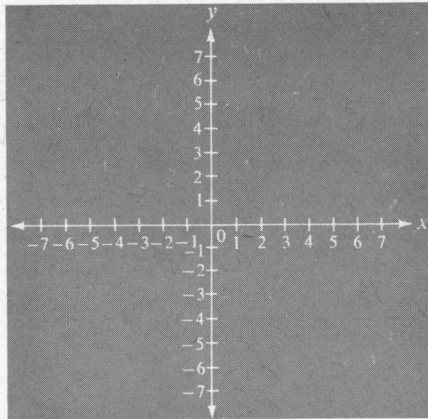

Figure 6.2

Now let us consider some points, *A*, *B*, and *C*, in a plane, all of which are *three* units to the right of the vertical axis (see Figure 6.3). The *x*-direction number of all of these points is 3, and they all lie in a straight line.

The *x*-direction number of a point is called the *abscissa* of the point. The straight line through points *A*, *B*, and *C* marks the location of *all* the points in this plane which have an abscissa of 3, or all of the points where *x* = 3. Consequently, when we have only the *abscissa* of a point, it could be any one of infinitely many points lying in a straight line parallel to the *y*-axis.

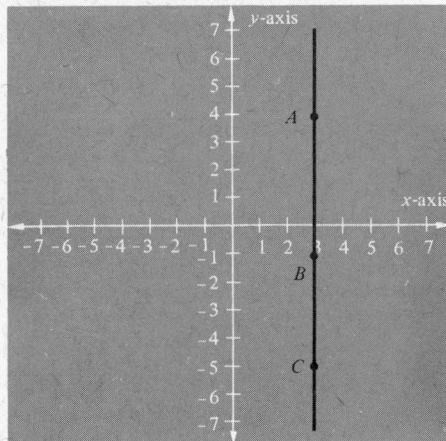

Figure 6.3

The *y*-direction number of a point is called the *ordinate* of the point. On the same graph, we shall locate some points, *M*, *N*, and *P*, all of which lie two units above the *x*-axis; that is, each point has an ordinate equal to 2 (see Figure 6.4).

All of the points M, N, and P lie in a straight line parallel to the x-axis and two units above it. The line through these points marks the position of every point in the plane whose ordinate is 2, or all the points where $y = 2$.

The two straight lines in Figure 6.4 represent respectively all the points in the plane where $x = 3$ and all the points where $y = 2$. Since two straight lines which are parallel respectively to two intersecting lines can cross only once, the intersection point, T, of these lines is unique: *it is the only point in the plane where $x = 3$ and $y = 2$ at the same time.* Consequently, if we have both the abscissa and ordinate of a point, the two numbers single out one point only in the plane. These two numbers (abscissa and ordinate) are called the *coordinates* of the point, and, taken together, they locate the exact position of the point.

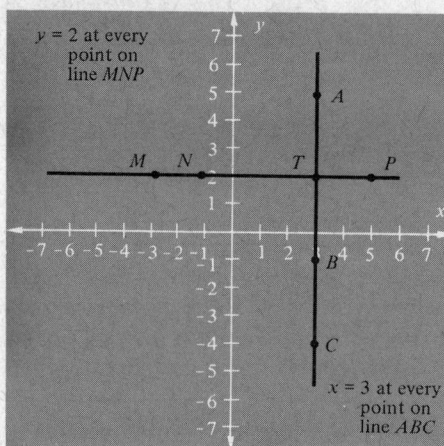

Figure 6.4

This was the scheme devised by Descartes to locate points in a plane, a coordinate system made up of *ordered pairs* of numbers. The word *ordered* is essential because we must know which of the two coordinates is the abscissa (the x-direction number) and which is the ordinate (the y-direction number). Thus two numbers, *given in a specific order*, are required to locate a point. For this reason we agree to write coordinates of points in the following way: the abscissa is written *first* and then the ordinate; the two numbers are separated by a comma and enclosed in parentheses. For example, the point whose abscissa is 3 and whose ordinate is 2 is written (3,2).

Examples:

Locate the following points:

(1) (2,5)

(2) (5,2)

(3) (−3,4)

(4) (4,−3)

(5) (−5,0)

(6) (0,−5)

(7) (2,0)

(8) (0,2)

(9) (−1,−6)

(10) (−6,−1)

To locate these points, we start at the origin, whose coordinates are (0,0), and count off the abscissa in the indicated direction (right or left) on the x-axis; then, in a perpendicular path from the abscissa, we count off the ordinate in the indicated direction (up or down) (see Figure 6.5).

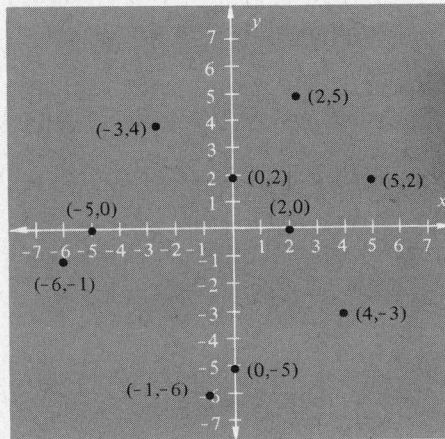

Figure 6.5

Because the real numbers can be paired in one-to-one correspondence with the points on a line, we have seen that by using two lines, one horizontal and one vertical, we can designate every point in the plane by *an ordered pair of numbers* called respectively the abscissa and the ordinate of the point. We can refer symbolically to any point in a plane by its coordinates (x,y).

Exercise 71:

Locate the following points:

(1) (3,5) (3,−5) (−3,5) (−3,−5)

(2) (4,0) (0,4) (−4,0) (0,−4)

(3) (−4,−5) $\left(-\frac{1}{2},\frac{3}{7}\right)$ $\left(\frac{4}{5},-1\right)$ (3,−4)

(4) (1,3) (3,1) $\left(-\frac{1}{2},2\right)$ (−.5,−3)

(5) $\left(\frac{4}{7},-5\right)$ $\left(\frac{2}{3},-3\right)$ (1,0) (0,−1)

(6) In what quadrant must a point lie if both of its coordinates are positive numbers?

(7) In which quadrants do the coordinates of a point have the following signs: (−,+)? (+,−)? (−,−)?

(8) What is the ordinate of every point on the x-axis?

(9) What is the abscissa of every point on the y-axis?

(10)   On a graph, plot the location of all the points whose abscissas are -4. Plot the location of all the points whose ordinates are -3. At what point do these two lines intersect?

6.2
First-Degree Equations in Two Variables

The rectangular coordinate system provides us with a framework with which to interpret the situation we encounter when we consider the behavior of a first-degree equation in two variables. We may of course use any symbols we please for the two different variables; if we elect to use x and y, then such an equation would have these two variables with any numerical coefficients (except that they cannot both be zero), and the exponents of the variables would be nonnegative integers of which the greatest is one.

Examples:

First-degree equations in two variables.

(1) $2x - 3y = 7$

(2) $5x - 2y - 8 = 0$

(3) $x = 2y - 3$

(4) $y = 4x + 1$

(5) $3x - 8y = 2$

(6) $y = -5x + 6$

(7) $4y - 10x - 13 = 0$

(8) $12 = 5x - 4y$

Now let us consider one such equation:

$$2x - y = 6$$

A *solution* of this equation must be a value for x and a value for y that satisfy the equation.

For instance, $x = 3$ and $y = 0$ is a *solution* of the equation because $2x - y = 6$ is true when $x = 3$ and $y = 0$.

$$2x - y = 6$$
$$2(3) - 0 = 6$$
$$6 - 0 = 6$$
$$6 = 6$$

Here we should reflect at length on two things about this equation, $2x - y = 6$, and the solution, $x = 3$ and $y = 0$.

1.   The pair of numbers, $x = 3$ and $y = 0$, satisfies the equation $2x - y = 6$; and when $x = 3$, y *must be* 0 in order for the equation to be true. These two numbers, 3 and 0, are an *ordered pair of numbers*; that is, they cannot be interchanged because the equation is *not* true when $x = 0$ and $y = 3$.

2.   This equation, $2x - y = 6$, does not restrict the values of x and y to any *specific* pair of real numbers (such as 3 and 0). It states, in effect, that x and y can be *any pair* of numbers which fits the situation: that is, *2 times the x-number minus the y-number must always equal 6*. What we have here therefore is a relationship between two variables that is valid for *any value of x* provided that the proper value of y is matched with it.

In view of this last observation, we can select values for x at random from the real number system and then find the value of y that satisfies the relationship described by the equation, $2x - y = 6$.

Examples:

(1) Let $\quad x = 0$

$2x - y = 6$

$2(0) - y = 6$

$0 - y = 6$

$-y = 6$

$y = -6$

[When $x = 0$, y must $= -6$]

Then the *ordered pair of numbers* $(0, -6)$ is a solution of $2x - y = 6$.

(2) Let $\quad x = 1$

$2x - y = 6$

$2(1) - y = 6$

$2 - y = 6$

$-y = 6 - 2$

$-y = 4$

$y = -4$

[When $x = 1$, y must $= -4$]

The ordered pair of numbers $(1, -4)$ is a solution of $2x - y = 6$.

(3) Let $\quad x = -3$

$2x - y = 6$

$2(-3) - y = 6$

$-6 - y = 6$

$-y = 6 + 6$

$-y = 12$

$y = -12$

[When $x = -3$, y must $= -12$]

The ordered pair of numbers $(-3, -12)$ is a solution of $2x - y = 6$.

This can obviously go on forever; we have only to select values for x from the real number system, and there is an infinite supply. For each value that we assign to x, there will be *one and only one* value of y that will match with it to satisfy the equation. Consequently, *the solutions of this equation form an infinite set of ordered pairs of numbers*. In the table below we have listed fifteen ordered pairs of numbers, all of which are solutions of the equation $2x - y = 6$.

x	-4	$-\dfrac{3}{2}$	$-\dfrac{1}{2}$	-1	0	1	2	3	4	5	6	$\dfrac{22}{3}$	8	9	10
y	-14	-9	-7	-8	-6	-4	-2	0	2	4	6	$\dfrac{26}{3}$	10	12	14

There are infinitely many more ordered pairs of real numbers which satisfy this equation, as many, in fact, as there are numbers in the real number system.

We know from our study of the rectangular coordinate system that *an ordered pair of numbers identifies one point in the plane*. Then the ordered pairs of numbers which satisfy the equation, $2x - y = 6$, identify a *set* of points in the plane. We shall select from the given table of solutions a convenient subset of these ordered pairs, rewrite them in coordinate notation (x,y), and then locate on a graph the points identified by the coordinates. From the table of solutions for $2x - y = 6$ consider the following subset of solutions:

$$\{(0,-6),\quad (1,-4),\quad (2,-2),\quad (3,0),\quad (4,2),\quad (5,4),\quad (6,6)\}$$

These pairs of numbers identify the points plotted in Figure 6.6.

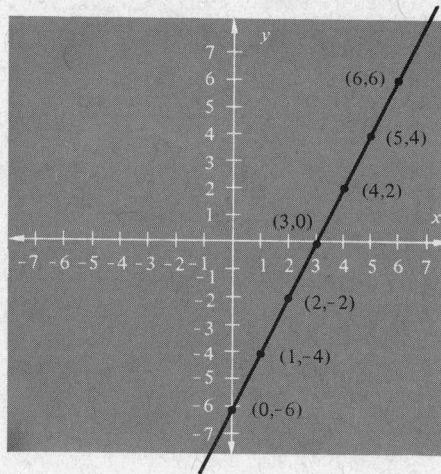

Figure 6.6

The subset of solutions for $2x - y = 6$, which we selected to locate the points on the graph in Figure 6.6, identifies a set of points which lie in a *straight line*. In fact, these points all lie on the *particular* straight line which crosses the x-axis at the point $(3,0)$ and the y-axis at the point $(0,-6)$. Although the proof is beyond the scope of this text, it can be proved that *every ordered pair of numbers which satisfies the equation $2x - y = 6$ identifies a point which lies on this same straight line*. Therefore the straight line shown in Figure 6.6 and the equation $2x - y = 6$ both represent the same ordered pairs of numbers or the same set of points. We can summarize this situation in the following manner:

The solution set of the equation $2x - y = 6$ is the infinite set of ordered pairs which identifies all points (x,y) in a plane such that $2x - y = 6$, or

$$\{(x,y) \mid 2x - y = 6\}$$

It can be proved that every first-degree equation in two variables represents an infinite set of ordered pairs of numbers (or points in a plane) which identifies a particular *straight line*. The *solution set* for such an equation has as its elements the coordinates of all the points on the line, and therefore the solution set is an infinite set. We can never list them all by coordinates, but we can show infinitely many of them by graphing the line which they represent.

Examples:

(1) $\{(x,y) \mid x - 2y = 5\}$

This is the set of all points (x,y) in a plane such that $x - 2y$ is always equal to 5. We can use the equality axioms to rewrite the equation $x - 2y = 5$ in a more convenient form for computing a subset of the infinitely many solutions.

$$
\begin{aligned}
x - 2y &= 5 \\
x - 5 &= 2y && \text{E}_a \\
2y &= x - 5 && \text{E}_s \\
y &= \frac{x - 5}{2} && \text{E}_m
\end{aligned}
$$

x	-3	-1	1	3	5
y	-4	-3	-2	-1	0

The graph of this equation is shown in Figure 6.7.

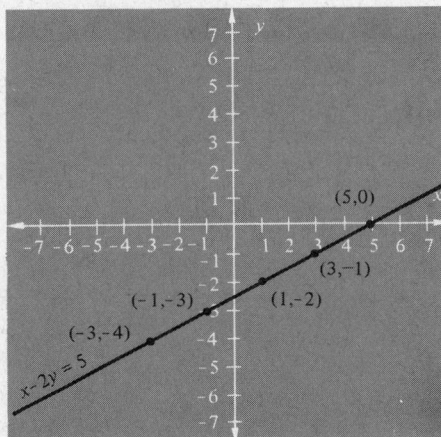

Figure 6.7

(2) $\{(x,y) \mid y = 3\}$

This is the set of all points (x,y) in a plane such that the ordinate, y, of every point is always equal to 3.

Note: Since the variable x does not appear in the equation $y = 3$, we may consider that the coefficient of x is zero:

$0x + y = 3$ means the same thing as $y = 3$

In this equation, no matter what value x has, y is *always* 3. Figure 6.8 is the graph of this equation.

x	-1	0	2	4	7
y	3	3	3	3	3

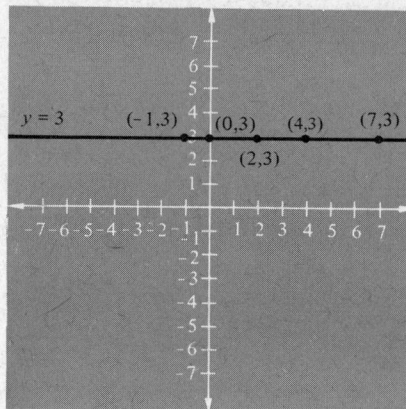

Figure 6.8

(3) $\{(x,y) \mid 2x + 3y = 12\}$

$$2x + 3y = 12$$
$$3y = 12 - 2x \qquad E_a$$
$$y = \frac{12 - 2x}{3} \qquad E_m$$

x	-3	0	3	4	6
y	6	4	2	$\dfrac{4}{3}$	0

The graph of this equation is Figure 6.9.

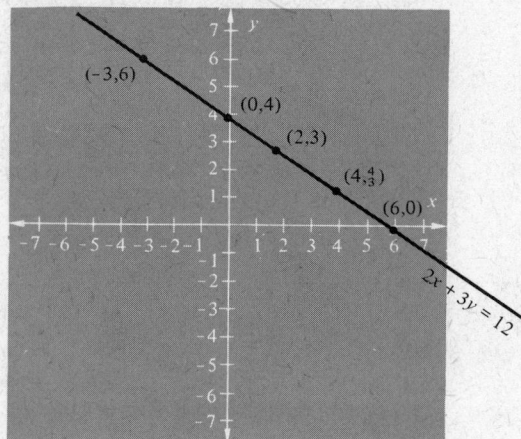

Figure 6.9

Exercise 72:

Compute the coordinates of five points for each of the following sets of points; plot the points and graph the line:

(1) $\{(x,y) \mid x + y = 6\}$ (2) $\{(x,y) \mid x - y = 2\}$

(3) $\{(x,y) \mid 2x + y = 4\}$ (4) $\{(x,y) \mid x - 3y = 6\}$

(5) $\{(x,y) \mid 3x + 2y = 12\}$ (6) $\{(x,y) \mid 5x - y = 10\}$

(7) $\{(x,y) \mid 4x + 3y = 8\}$ (8) $\{(x,y) \mid 3x - 5y = 5\}$

(9) $\{(x,y) \mid x = 4\}$ (10) $\{(x,y) \mid y = -5\}$

(11) $\{(x,y) \mid 3x - 4y = 0\}$ (12) $\{(x,y) \mid 2x - 3y - 12 = 0\}$

6.3
Operations With Sets

In order to investigate further the sets of points defined by first-degree equations in one or two variables, we need to consider two operations which can be performed with sets.

One operation, called *union*, was introduced in Section 4.3. To review briefly, from two sets A and B, another set can be made of all elements which belong either to A or to B or to both. This operation is designated by the symbol $\cup$, and $A \cup B$ is read "the union of A and B." In symbolic language

$$A \cup B = \{x \mid x \in A \quad or \quad x \in B\}$$

Examples:

(1) $A = \{\text{Jack, Mary}\}$
         $B = \{\text{James, Jane, Joe}\}$
$A \cup B = \{\text{Jack, Mary, James, Jane, Joe}\}$

(2) $T = \{-2, -1, 2, 3, 5\}$
        $W = \{-1, 0, 1, 2, 3\}$
$T \cup W = \{-2, -1, 0, 1, 2, 3, 5\}$

From any two (or more) sets another set can also be formed which includes only those elements which belong to *both* (or all) of the sets. This operation is called the *intersection* of the sets and is indicated by the symbol $\cap$. If A and B are two sets, then

$$A \cap B$$

means the intersection of A and B. In symbolic language

$$A \cap B = \{x \mid x \in A \text{ and } x \in B\}$$

Examples:

(1) $A = \{\text{Jack, James, Mary}\}$
         $B = \{\text{Joe, James, Mary, Bob}\}$
$A \cap B = \{\text{James, Mary}\}$

(2) $D = \{2, 4, 6, 8\}$
            $E = \{1, 2, 3, 4, 5, 6, 7\}$
    $D \cap E = \{2, 4, 6\}$

(3) $W = \{(1,2), (4,3), (-3,7), (5,8)\}$
          $V = \{(-1,3), (5,8), (2,4), (7,-3)\}$
   $W \cap V = \{(5,8)\}$

The student should recall from Section 4.3 that two sets are *disjoint* if they have no elements in common and that the set which has no elements is called the null set, designated by the symbol $\varnothing$. Consequently, we conclude the following:

1.   If A and B are disjoint sets, then

      $A \cap B = \varnothing$

2.   If A and B are any sets, then

      $A \cap \varnothing = \varnothing$
      $B \cap \varnothing = \varnothing$

Exercise 73:

(1) $A = \{3, 4, 5, 7, 10\}$
    $B = \{1, 2, 3, 4\}$
    List by elements:
    (a) $A \cup B$
    (b) $A \cap B$

(2) $A = \{(2,-1), (-5,0), (8,2), (4,7)\}$
    $B = \{(2,8), (-3,5), (4,7), (0,-5), (2,-1)\}$
    List by elements:
    (a) $A \cup B$
    (b) $A \cap B$

(3) $A = \{(-1,2), (3,0), (5,7)\}$
    $B = \{(1,-3), (2,-1), (3,5)\}$
    List by elements:
    (a) $A \cup B$
    (b) $A \cap B$

(4) $A = \{-5, -1, 0, 3, 7\}$
    $B = \{-1, 0, 2, 5, 7\}$
    $C = \{-5, -3, 0, 2, 7\}$
    List by elements:
    (a) $A \cup B$
    (b) $A \cup C$
    (c) $B \cup C$
    (d) $A \cap B$
    (e) $A \cap C$
    (f) $B \cap C$

6.4
Simultaneous Solutions of Systems of Linear Equations

The preceding sections have demonstrated the fact that a first-degree equation in one or two variables defines a set of points which lie in a straight line. Two such equations would represent two straight lines which, if placed on the same graph, would present one of three possible situations.

1. The two lines could be parallel.
2. The two lines could be the same line and coincide at every point.
3. The two lines could intersect at one and only one point.

If the two lines are *parallel*, they would have no point in common, and the intersection of the two sets of points would be the null set. In this case the two equations are said to be *inconsistent*.

Case I: *An Inconsistent System*

Let us consider the lines defined by the equations $x + y = 4$ and $x + y = 7$. (See Figure 6.10.)

Let $A = \{(x,y) \mid x + y = 4\}$ and $B = \{(x,y) \mid x + y = 7\}$. By computing the values for x and y indicated below, we find that $\{(0,4), (1,3), (2,2), (4,0) \ldots\}$ is a subset of A, and $\{(0,7), (1,6), (2,5), (3,4), (7,0) \ldots\}$ is a subset of B.

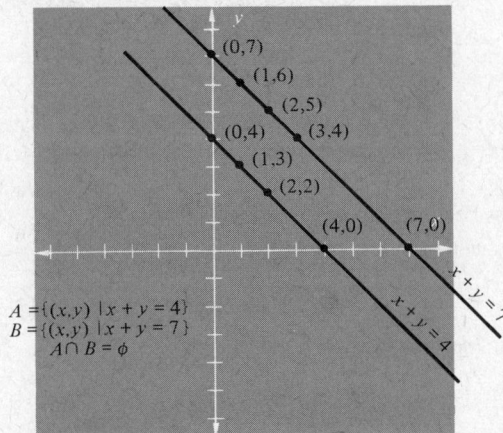

Figure 6.10

When the points represented by these ordered pairs are plotted on a graph, the lines produced are parallel. Therefore they have no point in common and the two sets A and B will never contain the same ordered pair. Consequently $A \cap B = \varnothing$.

If the two equations both define the same line, then their graphs will coincide at every point, and the intersection of the two sets of points will simply be every point on the line. Such a system is called a *dependent* system (see Figure 6.11).

Case II: *A Dependent System*

Let $A = \{(x,y) \mid x - 2y = 4\}$ and $B = \{(x,y) \mid 3x - 6y = 12\}$. Then $\{(-2,-3), (0,-2), (2,-1), (4,0), \ldots\}$ is a subset of A, and $\{(-2,-3), (0,-2), (2,-1), (4,0), \ldots\}$ is a subset of B.

The ordered pairs are the *same* for both sets; therefore the two equations define the same line and their intersection is every point on the line. $A \cap B = A$ or $A \cap B = B$.

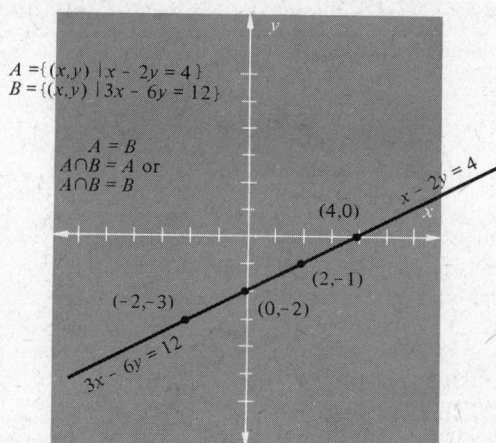

Figure 6.11

If the two lines defined by the two equations are not parallel and are not the same line, then they can intersect at one and only one point. The ordered pair of numbers which identify the intersection point will appear in both solution sets, and that pair will be the *only* element which belongs to both sets. Such a system of equations is called an *independent* system (see Figure 6.12).

Case III: *An Independent System*

Let $A = \{(x,y) \mid x + y = 3\}$ and $B = \{(x,y) \mid 3x - y = 1\}$. Then $\{(0,3), \underline{(1,2)}, (2,1), (3,0), \ldots\}$ is a subset of A, and $\{(-1,-4), (0,-1), \underline{(1,2)}, (2,5) \ldots\}$ is a subset of B. $A \cap B = \{(1,2)\}$

When two first-degree equations form an independent system, there will always be one and only one ordered pair of numbers which belongs to both sets of points and identifies the intersection point. The coordinates of this point represent the *simultaneous solution* of the system since they are the only pair of numbers that satisfy both equations.

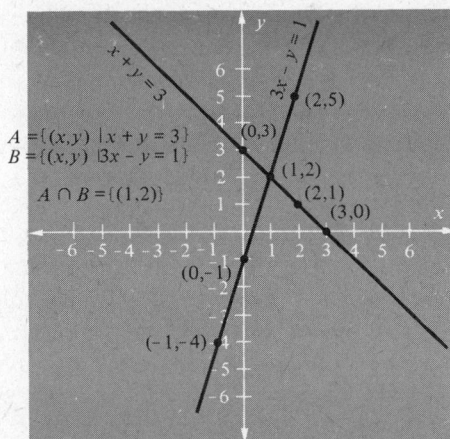

Figure 6.12

Exercise 74:

Graph the following systems and classify them as (1) inconsistent, (2) dependent, or (3) independent. Let A and B represent the two sets of points defined by each system, and find $A \cap B$ for each one.

(1) $\{(x,y) \mid x + 2y = 7\}$ and $\{(x,y) \mid 3x - y = 7\}$
(2) $\{(x,y) \mid x - 2y = -2\}$ and $\{(x,y) \mid 3x - 6y = -6\}$
(3) $\{(x,y) \mid 3x - 2y = 9\}$ and $\{(x,y) \mid 2x + 4y = -10\}$
(4) $\{(x,y) \mid y = -2\}$ and $\{(x,y) \mid 2x + y = 4\}$
(5) $\{(x,y) \mid 2x - y = 6\}$ and $\{(x,y) \mid 6x - 3y = -8\}$
(6) $\{(x,y) \mid x + 2y = -1\}$ and $\{(x,y) \mid 3x + 5y = -4\}$
(7) $\{(x,y) \mid 2x - y = 12\}$ and $\{(x,y) \mid 6x - 3y = 10\}$
(8) $\{(x,y) \mid 4x - 3y = -4\}$ and $\{(x,y) \mid 2x + y = 3\}$
(9) $\{(x,y) \mid x - 3y = 6\}$ and $\{(x,y) \mid 4x - 12y = 24\}$
(10) $\{(x,y) \mid 3x - 4y = 3\}$ and $\{(x,y) \mid 4x + 2y = 15\}$
(11) $\{(x,y) \mid x - 5y = 10\}$ and $\{(x,y) \mid 2x - 10y = -6\}$
(12) $\{(x,y) \mid 3x - 4y = 0\}$ and $\{(x,y) \mid 6x + 2y = 5\}$

6.5
Algebraic Solutions of Simultaneous Linear Equations

In Exercise 74, problems 1, 3, 4, 6, 8, 10, and 12 are illustrations of the fact that when two first-degree equations in one or two variables form an independent system, the two straight lines which the equations define will always have one and only one point in common. Therefore the ordered pair of numbers which identify that one point is the only pair of numbers that satisfies both equations. This is the reason, of course, that the coordinates of this intersection point are called *the* simultaneous solution of the system, since it is clear that there is but *one* such solution.

Knowing this in advance, we can find the coordinates of that one intersection point for any independent system by making practical use of the equality axioms (Section 5.2) and the substitution axiom (Section 4.4). In fact, the simultaneous solution can be found more conveniently by employing algebraic methods rather than by the graphing method of the preceding section.

The algebraic methods for finding the simultaneous solution amount fundamentally to making use of the axioms and common sense at the same time. In devising such a method, one simple fact must be kept firmly in mind: *to get a unique value for a variable one must arrange somehow to get an equation which contains that variable only.* Granted this ultimate objective, one may employ the axioms in any feasible way that will accomplish it, and there are several ways that it can be done. Following are two such methods that will isolate the *one* ordered pair of numbers which will satisfy both equations of an independent linear system.

I. First Method: *Addition or Subtraction of the Two Equations*

A. Consider the system of lines from problem 1, Exercise 74, whose equations are

(1) $x + 2y = 7$
(2) $3x - y = 7$

Since we have two variables in each equation and we need an equation having only *one* variable, our purpose is to find some logical way to get rid of one of these variables, and it doesn't matter in the least which one.

B. In this method, by using E_m, we first make the numerical coefficient of the *same* variable in each equation *alike* in absolute value but *opposite* in sign. In Equation (1) of this system the coefficient of y is 2. In Equation (2) the coefficient of y is -1. We shall therefore multiply both sides of Equation (2) by the number 2.

(2) $3x - y = 7$
       $2(3x - y) = 2(7)$ (by using E_m)
         $6x - 2y = 14$

The above equation forms a dependent system with $3x - y = 7$. It identifies the same set of points and represents the same line. For this reason it is called an *equivalent equation* for $3x - y = 7$.

C. Recopy the system, replacing $3x - y = 7$ by the equivalent equation $6x - 2y = 14$.

(1) $x + 2y = 7$
(2) $6x - 2y = 14$

Since Equation (2) guarantees that $6x - 2y$ is the same as 14, if we add $6x - 2y$ to the *left* side of Equation (1) and 14 to the *right* side of Equation (1), we are actually adding the *same* number to both sides of Equation (1). In short, using the fact that when equals are added to equals the results are equal, we add Equations (1) and (2) above.

(1) $x + 2y = 7$
      $6x - 2y = 14$
(2) $7x + 0 = 21$ (using E_a)
        $7x = 21$

D.   The resulting equation

$$7x = 21$$

has only *one* variable, and its solution will be the abscissa (the x-value) of the intersection point of the two lines. The proof of this statement will be given at the conclusion of this problem.

    $7x = 21$
    $x = 3$ the abscissa of the intersection point

E.   To find the ordinate of the intersection point, we use the substitution axiom and replace x by 3 in *either* of the original equations.

(1) $x + 2y = 7$
     $3 + 2y = 7$
        $2y = 7 - 3$
        $2y = 4$
         $y = 2$ the ordinate of the intersection point

(2) $3x - y = 7$
   $3(3) - y = 7$
    $9 - y = 7$
      $-y = 7 - 9$
      $-y = -2$
        $y = 2$ the ordinate of the intersection point

F.   In Step E we have shown that *both* equations are true when $x = 3$ and $y = 2$; therefore the simultaneous solution of the system is the ordered pair (3,2).

Repeating the steps outlined above in condensed form, we have the following:

A. (1) $x + 2y = 7$
    (2) $3x - y = 7$

B. $3x - y = 7$ (Multiply both sides of $3x - y = 7$ by 2)
        $2(3x - y) = 2(7)$
        $6x - 2y = 14$ → equivalent equation for $3x - y = 7$

C. (1) $x + 2y = 7$
    (2) $6x - 2y = 14$ (Add the two equations)
         ―――――――――
         $7x + 0 = 21$
         $7x = 21$

D. $7x = 21$ (Solve for x)
    $x = 3$ → the abscissa of the intersection point

E. (1) $x + 2y = 7$ (Substitute 3 for x in one of the original
         $3 + 2y = 7$ equations)
         $2y = 4$
         $y = 2$ → the ordinate of the intersection point

    (2) $3x - y = 7$ (Check by substituting 3 for x in the other
         $3(3) - y = 7$ equation)
         $9 - y = 7$
         $y = 2$ → the ordinate of the intersection point

F. The simultaneous solution = (3,2).

Look over the results of the foregoing method and note carefully that by eliminating the variable y in Step C, we obtained a *unique* value for $x [x = 3]$ which, when substituted in the original equations, gave the *same* value for y in each equation $[y = 2]$. We have unquestionably found a pair of numbers, 3 for x and 2 for y, that satisfies *both* equations, and we know from solution by graphing that *only one* pair of numbers will satisfy both of them simultaneously. Consequently, the ordered pair (3,2) must be the simultaneous solution of the system. The method obviously works; now, the question is *why?*

To answer this question, we need to represent *any* independent system of linear equations symbolically, and then show why this method always produces the co-ordinates of the intersection point. To do this, we proceed in the following manner:

Proof of the Method of Addition and Subtraction

1. Let the following equations represent any independent system of linear equations:

    (1) $ax + by + c = 0$
    (2) $dx + ey + f = 0$

    and let the ordered pair (x_1, y_1) be the coordinates of the intersection point. Then (x_1, y_1) is a solution of both equations; i.e., it makes the left side of (1) and (2) both equal to zero.

2. If h and k are any constants, the equation

(3) $h(ax + by + c) + k(dx + ey + f) = 0$
is also true for $x = x_1$ and $y = y_1$ since

$$h(0) + k(0) = 0$$

for *all* values of h and k.

3. Consequently, if h and k are chosen so that in Equation (3) the coefficient of y (or x) vanishes, Equation (3) would reduce to the form

$$x = p \qquad (\text{or } y = q)$$

and since either equation must be true for $x = x_1$ and $y = y_1$, $p = x_1$ (or $q = y_1$).

In the worked-out example we found the simultaneous solution of the system by eliminating the variable y. You should verify that with this system we could just as easily have eliminated the variable x and arrived at the same solution.

Exercise 75:

Use addition or subtraction to eliminate one of the variables and find the simultaneous solution of the following systems:

(1) Exercise 74, problems
3, 4, 6, 8 and 10

(2) $x + 2y = 6$
$2x - 3y = 5$

(3) $3x - 2y = 7$
$2x + 5y = -11$

(4) $3x + 2y = -2$
$5x - y = 14$

(5) $2x + 3y = 8$
$3x - 2y = 12$

(6) $9x - y = 1$
$3x + 5y = 11$

(7) $\frac{1}{2}x + y = \frac{1}{2}$

$9x - 6y = 1$

(8) $3x + 4y = 4$

$2x - \frac{2}{3}y = 1$

(9) $\frac{1}{2}x + \frac{1}{3}y = 4$

$\frac{3}{2}x - \frac{1}{2}y = 3$

(10) $5x - 4y = 1$
$10x + 6y = 9$

II. Second Method: *Substitution*

Another way of eliminating one of the two variables in a system of linear equations is to use the substitution axiom, which states that a quantity may be substituted for its equal in any mathematical expression. Using the same system of equations that we solved by the first method, we have the following:

A. (1) $x + 2y = 7$
(2) $3x - y = 7$

Select *either one* of the two equations and solve it for *either one* of the two variables. As a matter of plain common sense, select an equation which has a variable that is easy to isolate.

(1) $x + 2y = 7$
$x = 7 - 2y$ (using E_a)

The equation $x = 7 - 2y$ states explicitly that *at every point* on the line $x + 2y = 7$, the abscissa (x-value) must be equal to $7 - 2y$.

B. Then, if there is *any point* on the second line, $3x - y = 7$, which has the *same coordinates* as a point on the line $x + 2y = 7$, *at that point* x must be equal to $7 - 2y$. Consequently, at any point that lies on both lines (the intersection point)

$$x = 7 - 2y$$

C. To find this point, substitute $7 - 2y$ for x in the equation of the *second* line.

(2) $3x - y = 7$
$3(7 - 2y) - y = 7$
$21 - 6y - y = 7$
$-6y - y = 7 - 21$
$-7y = -14$
$y = 2$ the ordinate of the intersection point

D. To find the x-value of the intersection point use the fact (from Step B) that $x = 7 - 2y$ at the intersection point. Knowing that $y = 2$ at that point, we replace y by 2.

$x = 7 - 2y$
$x = 7 - 2(2)$
$x = 7 - 4$
$x = 3$ abscissa of the intersection point

E. The simultaneous solution = (3,2).

Condensing the preceding steps in the method of substitution, we have the following:

A. (1) $x + 2y = 7$
(2) $3x - y = 7$

B. (1) $x + 2y = 7$
$x = 7 - 2y$ at the intersection point

C. (2) $3x - y = 7$
$3(7 - 2y) - y = 7$
$21 - 6y - y = 7$
$-7y = -14$
$y = 2$ at the intersection point

D. $x = 7 - 2y$ (from Step B)

   $x = 7 - 2(2)$

   $x = 7 - 4$

   $\underline{x = 3}$ abscissa of the intersection point

E.   Simultaneous solution = (3,2)

Instead of selecting Equation (1) and solving for x in the above example, we could just as easily have selected Equation (2) and solved it for y. When using the method of substitution, always begin by selecting an equation of the system which has a variable that is easy to isolate.

Exercise 76:

Using the method of substitution find the simultaneous solution of the problems in Exercise 75.

Simultaneous Linear Equations in Three or More Variables   Up to now, we have been considering linear equations in two variables, such as

$$3x + 2y = 7$$

There is no need to stop at two variables. Linear equations in three variables, such as

$$4x + y - 3z = 2$$

occur often in mathematics and applications. The geometric interpretation of the solution of a linear equation in two variables is simple because the equations are those of straight lines, but the geometric interpretation of the solutions of a linear equation in three variables is a bit more complicated and will be deferred to a later course. Actually, we will restrict our attention to the algebraic solution of systems of linear equations in three variables, and even more precisely, to the simultaneous solution of systems consisting of three linear equations in three variables.

For example, consider

$$(1) \quad 2x + y - z = 1$$
$$(2) \quad 3x - y + 2z = 7$$
$$(3) \quad x + y + z = 6$$

Before proceeding, let us outline our plan of attack. The ultimate purpose of any algebraic attack on this system is precisely the same as that stated for systems of two linear equations; namely, to get a unique value for a variable we must arrange somehow to get an equation that contains that variable only.

First, we will choose two different pairs of equations from the three in the system, and from each pair we shall eliminate the *same variable*, using the same method of addition and subtraction employed in solving systems of two linear equations. The result will be a system of two linear equations in two variables. We will solve that system, then substitute the resulting values of the two variables into any of the original equations to obtain the value of the third variable. Let us see how it works.

A. Eliminate z. Add Equation (3) to Equation (1).

$$
\begin{array}{rl}
(1) & 2x + y - z = 1 \\
(3) & \underline{x + y + z = 6} \\
 & 3x + 2y = 7
\end{array}
$$

Then, using a *different* pair of equations from the system | Equations (2) and (3)], subtract Equation (3) multiplied by 2 from Equation (2) to eliminate the same variable, z

$$
\begin{array}{rl}
(2) & 3x - y + 2z = 7 \\
(3) & \underline{-2x - 2y - 2z = -12} \\
 & x - 3y = -5
\end{array}
$$

B. Solve the system of two equations in two variables resulting from the operations in step A.

$$
\begin{array}{r}
3x + 2y = 7 \\
x - 3y = -5
\end{array}
$$

Multiply the second equation by -3 and add.

$$
\begin{array}{r}
3x + 2y = 7 \\
\underline{-3x + 9y = 15} \\
11y = 22 \\
y = 2
\end{array}
$$

Substitute $y = 2$ into the second equation

$$
\begin{array}{r}
x - 3(2) = -5 \\
x - 6 = -5 \\
x = 1
\end{array}
$$

Hence the solution to the system of two linear equations in x and y is $x = 1$, $y = 2$.

C. Solve for z. Substitute $x = 1$, $y = 2$ into Equation (1) of the original system. [You could substitute them into Equation (2) or Equation (3) just as well.]

$$
\begin{aligned}
\text{(1)} \quad 2x + y - z &= 1 \\
2(1) + 2 - z &= 1 \\
2 - z &= 1 \\
z &= 3
\end{aligned}
$$

D. Simultaneous solution: $x = 1$, $y = 2$, $z = 3$. (You should check to see if those values of x, y, and z actually satisfy the original system.)

   Of course, there was nothing magic about eliminating the z first. Let us solve the system again by eliminating the y first. This time we will leave out most of the directions — you should be able to fill them in.

$$
\begin{aligned}
\text{(1)} \quad 2x + y - z &= 1 \\
\text{(2)} \quad 3x - y + 2z &= 7 \\
\text{(3)} \quad x + y + z &= 6
\end{aligned}
$$

A. Eliminate y.

$$
\begin{array}{ll}
\text{(1)} \quad 2x + y - z = 1 & \text{(2)} \quad 3x - y + 2z = 7 \\
\text{(3)} \quad \dfrac{-x - y - z = -6}{x - 2z = -5} & \text{(3)} \quad \dfrac{x + y + z = 6}{4x + 3z = 13}
\end{array}
$$

B. Solve the system.

$$
\begin{aligned}
x - 2z &= -5 \\
4x + 3z &= 13 \\[4pt]
4x - 8z &= -20 \\
\dfrac{-4x - 3z}{ - 11z} &= \dfrac{-13}{-33} \\
z &= 3 \\[6pt]
x - 2(3) &= -5 \\
x - 6 &= -5 \\
x &= 1
\end{aligned}
$$

Solution: $x = 1$, $z = 3$

C. Solve for y.

$$
\begin{aligned}
\text{(3)} \quad x + y + z &= 6 \\
1 + y + 3 &= 6 \\
y + 4 &= 6 \\
y &= 2
\end{aligned}
$$

D. Simultaneous solution: $x = 1$, $y = 2$, $z = 3$.

We will conclude our discussion of systems of linear equations at this point. There is much more of interest and importance to be said about the subject, particularly about applications, but that will have to wait for a future course. Even so, if you master the techniques presented in this chapter, you will find that you will be able to deal with a surprisingly wide range of applications.

Exercise 77:

Find the simultaneous solutions of the following systems:

(1) $\begin{aligned} x + y + z &= 4 \\ y + 2z &= 5 \\ 3z &= 6 \end{aligned}$ 　　　　 (2) $\begin{aligned} x \quad\quad + z &= 0 \\ y + z &= -1 \\ x + y \quad\quad &= 1 \end{aligned}$

(3) $\begin{aligned} 2x + y - z &= 6 \\ x - y + 2z &= -1 \\ 3x + y + z &= 6 \end{aligned}$ 　　　　 (4) $\begin{aligned} -x + y - z &= 1 \\ 2x + y + 7z &= -2 \\ x - 3y + 2z &= -1 \end{aligned}$

(5) $\begin{aligned} 2x + 3y - z &= 1 \\ 4x - 3y + z &= 2 \\ 6x + 6y + 2z &= 7 \end{aligned}$ 　　　　 (6) $\begin{aligned} 2x + y + 2z &= -3 \\ x - 2y + z &= 6 \\ 3x - y + 5z &= 4 \end{aligned}$

6.6
Solution of Stated Problems with Two Variables

In Chapter 5, Section 5.8, we investigated the process of solving stated problems by translating the quantities in the problems into mathematical symbols, and then setting up an equation relating these quantities. At that time, in all of the equations which were derived from the problems, only one variable was employed. But it is quite permissible to use two variables (or more) in solving stated problems. In fact, it is not only permissible, it is often highly convenient and sometimes downright necessary to employ two variables instead of one. The process of symbolizing the quantities and deriving the equation is exactly the same. The only difference is that, when one uses *two* variables instead of one, then *two* independent equations relating the variables must be found in order to get a unique solution that fits the conditions of the problem simultaneously.

The first of the following examples is Example D from Section 5.8 worked with two variables instead of one.

A. The length of a rectangle is 5 meters longer than its width. If the perimeter of the rectangle is 54 meters, what are the dimensions of the rectangle?
   (1) Let x meters = the length of the rectangle.
   (2) Let y meters = the width of the rectangle.
   (3) $x - y = 5$ since x is 5 meters longer than y.
   (4) $x + y + x + y =$ the perimeter of the rectangle
   or $2x + 2y =$ the perimeter.

(5) 54 meters = the perimeter.
(6) from (4) and (5)

$$2x + 2y = 54$$

From (3) and (6)

$$x - y = 5$$
$$2x + 2y = 54$$

Solving this system by addition we have the following:

(1) $x - y = 5$
(2) $2x + 2y = 54$

(1) $2x - 2y = 10$ (equivalent equation for $x - y = 5$)
(2) $\underline{2x + 2y = 54}$
     $4x + 0 = 64$
     $ 4x = 64$
     $ x = 16$ meters, length of the rectangle

(1) $ x - y = 5$
     $16 - y = 5$
     $ -y = 5 - 16$
     $ -y = -11$
     $ y = 11$ meters, width of the rectangle

B. The sum of two numbers is 43 and their difference is 9. Find the numbers.
    (1) Let x = the larger number.
    (2) Let y = the smaller number.
    (3) $x + y$ = the sum of the numbers.
    (4) 43 = the sum of the numbers.
    (5) Then $x + y = 43$.
    (6) $x - y$ = the difference of the numbers.
    (7) 9 = the difference of the numbers.
    (8) Then $x - y = 9$

From (5) and (8)

$$x + y = 43$$
$$x - y = 9$$

Solving by addition we have

$$x + y = 43$$
$$\underline{x - y = 9}$$
$$2x + 0 = 52$$
$$2x = 52$$
$$x = 26 \quad \text{the larger number}$$

$$x + y = 43$$
$$26 + y = 43$$
$$y = 43 - 26$$
$$y = 17 \qquad \text{the smaller number}$$

C. At a constant airspeed a pilot flew from town A to town B against a head-wind in 3 hours and returned in 2 hours. If the distance between A and B is 600 kilometers, what was the rate of the wind and the airspeed of the plane?
    (1) Let x km/h = the airspeed of the plane.
    (2) Let y km/h = the rate of the wind.
    (3) Then $(x - y)$ km/h = the rate of the plane from A to B.
    (4) 3 hours = the time required to fly from A to B.
    (5) Then $3(x - y)$ kilometers = the distance from A to B.
    (6) 600 kilometers = the distance from A to B.
    (7) From statements (5) and (6)

$$3(x - y) = 600$$

or

$$3x - 3y = 600$$

    (8) $(x + y)$ km/h = the rate of the plane from B to A.
    (9) 2 hours = the time required to fly from B to A.
    (10) Then $2(x + y)$ kilometers = the distance from B to A.
    (11) From statements (6) and (10), using E_t,

$$2(x + y) = 600$$

or

$$2x + 2y = 600$$

From (7) and (11)

$$3x - 3y = 600 \quad \text{or} \quad x - y = 200$$
$$2x + 2y = 600 \quad \text{or} \quad x + y = 300$$

Solving by addition we have

$$x - y = 200$$
$$\underline{x + y = 300}$$
$$2x = 500$$
$$x = 250 \text{ km/h} = \text{airspeed of plane}$$
$$x + y = 300$$
$$250 + y = 300$$
$$y = 50 \text{ km/h} = \text{rate of the wind}$$

D. Two solutions of salt and water are respectively 30% salt and 14% salt. How many cm³ of each solution should be mixed to make 32 cm³ of a solution that is 20% salt?

    (1) Let x cm³ = the amount of 30% solution used.
    (2) Let y cm³ = the amount of 14% solution used.

(3) Then $(x + y)$ cm³ = the amount of 20% solution to be mixed.

(5) From (3) and (4)
$$x + y = 32$$

(6) .30x cm³ = the amount of salt present in x cm³ of 30% solution.

(7) .14y cm³ = the amount of salt present in y cm³ of 14% solution.

(8) $(.30 + .14y)$ cm³ = the total amount of salt in the 20% solution.

(9) .20(32) cm³ = the total amount of salt in the 20% solution.

(10) From (8) and (9)
$$.30x + .14y = .20(32)$$

or

$$30x + 14y = 640$$

From statements (5) and (10)
$$x + y = 32$$
$$30x + 14y = 640$$

Multiplying the first equation by -7 and dividing the second one by 2 we obtain

$$-7x - 7y = -224$$
$$\underline{15x + 7y = 320}$$
$$8x = 96$$
$$x = 12 \text{ cm}^3 \text{ of 30\% solution to be used in the mixture}$$
$$x + y = 32$$
$$12 + y = 32$$
$$y = 32 - 12$$
$$y = 20 \text{ cm}^3 \text{ of 14\% solution to be used in the mixture}$$

Exercise 78:

Solve the following using two variables:

Number Problems

(1) The sum of two numbers is 23 and their difference is 9. Find the numbers.

(2) The sum of two numbers is 11. If twice the smaller number is subtracted from the larger, the result is 2. What are the numbers?

(3) The sum of two numbers is 21. If the smaller number is divided by the larger number, the quotient is $\frac{3}{4}$. What are the numbers?

(4) In a two-digit number, the sum of the digits is 12. If the ten's digit is two times as large as the unit's digit, what is the number?

(5) A two-digit number is 18 more than the number with its digits reversed. If the ten's digit is twice the unit's digit, what is the original number?

(6) The sum of the digits of a two-digit number is 17. The number with the digits reversed is 9 less than the original number. What is the original number?

Problems About Plane Figures

(7)   The perimeter of a triangle is 19 cm. The sum of two sides of the triangle is 11 cm, and the remaining side is twice as long as the shortest side. What are the lengths of the sides?

(8)   The perimeter of a triangle is 20 meters. One side is 3 meters longer than another side, and the sum of these two sides is 2 meters greater than the remaining side. What are the lengths of the sides?

(9)   The perimeter of a rectangle is 56 cm. If the length is decreased by 6 cm and the width decreased by 4 cm, the new length will be twice the new width. What are the dimensions of the original rectangle?

(10) A fenced rectangular playground is 14 meters longer than it is wide. If a concrete walk 3 meters wide is built around the four sides inside the fence, the reduced perimeter of the playground will be 148 meters. What are the dimensions of the original playground?

(11) A rectangular area is to be fenced in against the side of a clubhouse. The longer side of the fence is opposite the clubhouse, and the area is divided into two equal parts by a cross fence parallel to the ends. If 600 meters of fencing are required, and the length of the enclosed area is 120 meters longer than the width, find the length and width of the enclosed area.

(12) A triangular-shaped lot has 80 front meters and is 60 meters deep. The largest rectangular building that can be placed on the lot, with the front of of the building on the 80-meter side, has a perimeter of 140 meters. What are the dimensions of the building? (Hint: Use similar triangles to obtain a second equation.)

Motion Problems

(13) Two motorists leave from opposite points and travel toward each other at different rates of speed. When they meet, Car A has traveled 300 kilometers, and Car B has traveled 275 kilometers. If the rate of Car B were increased by 10 km/h, the two would meet after Car A had gone 276 kilometers and Car B 299 kilometers. How fast was each car traveling?

(14) A man drove from town A to town B and returned in 12 hours. He drove to town B at an average speed of 55 km/h and returned at 65 km/h. How many hours did he drive each way? What is the distance between the two towns?

(15) Two cars left town A at the same time and traveled in opposite directions at different rates of speed. In 5 hours they were 600 kilometers apart. If one car traveled 10 km/h faster than the other, how fast was each car traveling?

(16) A pilot flew north against a headwind for 6 hours. On the return trip the wind had increased by 10 km/h, and he returned to his starting point in 4 hours. If the airspeed of the plane was 220 km/h, what were the rates of the headwind and the tailwind?

(17) A boat traveled upstream for 3 hours and returned to its starting point in 1 hour. If the rate of the boat was 6 km/h faster than the rate of the stream, what was the rate of each?

(18) A man rowed a boat 2 kilometers upstream, at which point a gust of wind blew his hat off. He took no notice and rowed on for 15 minutes more. Then, missing his hat, he reversed direction in order to retrieve it. He eventually caught up with his hat at the same point at which he had originally started rowing. What was the rate of the current? (Hint: Let x km/h equal the rate of the man (in still water) and let y km/h equal the rate of the current. Although only one equation is available, the rate of the current is independent of the rate of the man, and the variable x will divide out of the final equation.)

Solution and Mixture Problems

(19) A solution of acid and water is 20% acid. A second solution is 40% acid. How many cm^3 of each should be mixed to make 32 cm^3 of solution that is 35% acid?

(20) Skimmed milk which is free of butterfat is to be combined with cream that is 40% butterfat to make 100 liters of milk that is 6% butterfat. How many liters of each should be used in the mixture?

(21) A solution of alcohol and water is 30% alcohol. A second solution is 12% alcohol. How many cm^3 of each should be mixed to make 75 cm^3 of a solution that is 18% alcohol?

(22) Pure water and brine that is 30% salt are to be combined to make 100 liters of a solution that is 24% salt. How many liters of each should be used?

(23) A grocer mixes 45 kilograms of two kinds of candy to sell for 25 cents a kilogram. If one kind of candy sells for 20 cents a kilogram and the other kind sells for 35 cents a kilogram, how many kilograms of each kind did he put in the mixture?

(24) A florist sells mixed flowers in bunches of 15 for $4.80 per bunch. The bunches contain three different kinds of flowers which sell individually for 20 cents, 35 cents, and 50 cents each. If each bunch contains twice as many 20-cent flowers as 50-cent flowers, how many of each type should be included in a bunch?

Miscellaneous Problems

(25) A metropolitan bank wants to build a one-story suburban branch. A 20,000-square-meter rectangular building site for a building and adjacent parking lot is available for $20,000. The type of building they plan to build costs $23.00 per square meter, and the parking lot will cost $3.00 per square meter. If their overall budget for the site, building, and parking lot is $200,000, how large a building can they plan?

(26) Mr. Davis wishes to make a total gift of $700 to his private high school and college. The company that Mr. Davis works for will match his contribution to the college and will match 50% of his contribution to the private high school. Mr. Davis wants the total amount received by each

school to be the same. What part of his $700 should Mr. Davis give to each school?

(27) A dairy manager plans to market a light table cream that is a mixture of fat-free skimmed milk and pure cream. If skimmed milk sells for 16 cents per 16-cm^3 bottle and pure cream for 65 cents per 16-cm^3 bottle, how many cm^3 of each should be put into 16 cm^3 of table cream that is to sell for 37 cents?

(28) Jones and Smith invested a total of $12,000 in two different stocks. Jones received 6% interest on his investment the first year and Smith received 5 1/2% interest. If they received a total of $680 interest, how much money did each man invest?

(29) An apartment house contains 16 apartments having either one or two bedrooms. The two-bedroom apartments rent for $20.00 a month more than the one-bedroom apartments. All apartments are rented and the total rent income is $2,200 per month. If the rent of the two-bedroom apartments is increased $10 a month, the total rent income dollars per month would equal 226 times the number of one-bedroom apartments. How many one-bedroom apartments are in the building? What is the monthly rent of each type of apartment?

(30) Mr. Delaney wants to make an investment of $20,000. He plans to buy stock in Company A and Company B. The stock of Company A is a growth stock with small dividends. The stock of Company B is an income stock with higher paying dividends. Stock in Company A costs $50 per share and pays $.30 per share in dividends. The stock in Company B costs $80 per share and pays $4.00 per share in dividends. Mr. Delaney wants an income of $560. from the stock in the first year. How much of each stock should he buy?

(31) At a school fund-raising project tickets were sold to a spaghetti supper at the rate of $1.50 for adults and $.75 for students. $630 was collected the first week. The second and last week of the sale twice as many adults and half as many students bought tickets, and $855 was collected. How many adults and how many students bought tickets?

(32) At a sale of radios and record players a merchant sold record players for $30.00 more than radios. He made 50 sales and collected $1,350. If the number of each item sold had been reversed, he would have collected $300 more. How many of each did he sell?

(33) Tom's average for the first two quizzes in biology is 75. On the third and final quiz he can reach an average of 80 if he can improve his second quiz grade by 4 points. What were his first two quiz grades?

(34) A test pilot plans to test two new planes on flights of 3 hours each on the same day. Each plane will be flown at an airspeed of 600 km/h, and the pilot plans to fly due north of the airport and return by the same route. On the day the tests are scheduled there is a 40-km/h wind from the north. Assuming no change in the wind, how far north should the pilot fly in order to complete each test in 3 hours?

6.7
The Functional Relationship

A linear equation exemplifies a relationship between two variables which operates so that, given a value for one of the variables, the value of the other is determined.

When such a relationship does exist, we may designate one of the variables as the *independent* variable; the other becomes the *dependent* variable. While these variables may be arbitrary symbols in a mathematical exercise, such as $y = 3x + 2$, they may also represent physical quantities and their relationships, the study of which is of manifest interest and usefulness to humankind.

For example, suppose that the owner of an ornate gift shop in a high-rent district has determined that he can make a reasonable profit and cover all of his overhead expenses (costs of merchandise, building rent, utilities, shipping costs, employees' salaries, accounting costs, taxes, insurance, advertising, office supplies, etc.) if he makes the selling price of each item equal to twice its cost. Thus if x represents the cost of an item and y represents its selling price, then the equation $y = 2x$ represents the relation between the two quantities.

A table showing specific values for x and y could be constructed as follows:

x dollars (cost of item)	2	2.50	3	3.75	4	5	6.40	9	15	etc.
y dollars (selling price of item)	4	5	6	7.50	8	10	12.80	18	30	etc.

In the above table, x is a variable which represents the cost of any gift in the shop, and for each value of x there is one and only one value of y. Obviously any items which have the same cost will have the same selling price, but note that it is *not* possible for a single item to have two different prices. Thus we say that to each value of x, there corresponds a *unique* value of y. Since the selling price, y, depends upon the cost, x, the variable x is the independent variable and y is the dependent variable. Mathematically speaking, the procedure for determining the selling price (e.g., multiplying the cost by 2) is called a "function," and we say that "the selling price is a function of the cost." Any two variables may be represented by symbols, such as x and y, and when they are so represented, the idea of a function is contained in the following statement:

When two variables x and y are related so that, for *each admissible value of x*, there is *one and only one value of y*, then y is said to be a *function* of x.

A handy word for saying that there is only one of something is to say that it is *unique*. A thing cannot be "more" unique or "less" unique than another thing, any more than one dead fish can be more dead or less dead than another dead fish: both states, uniqueness and deadness, exist absolutely or not at all.

The *admissible* values of the independent variable, x, depend upon the nature of the function. In the gift shop illustration, for instance, both cost and selling price would be positive numbers. If x is the length of the edge of a cube, then x must be a positive number; if x represents the number of one of the days of the year, then x must be a positive integer whose values range from 1 to 365.

The word *function* is commonly abbreviated by a letter of the alphabet such as f, g, h, etc., or F, G, H, etc. For the gift shop items, instead of using the notation $y = 2x$ to express the relationship between the selling price, y, and the cost, x, we could write the following:

$$f(x) = 2x$$

where f symbolizes the procedure for computing the selling price and $f(x)$ represents the *value of f* at x.

In the notation $f(x) = 2x$ we have actually replaced the symbol for the selling price, y, by a more sophisticated symbol, $f(x)$. This latter notation has the advantage of specifying that the relation between the cost and the selling price is a *function* in which the independent variable is called x; e.g., for each value of x there is one and only one value for $f(x)$ or y.

The two notations for the value of the function at x are used interchangeably. The equations

$$f(x) = 2x$$

and

$$y = 2x$$

designate exactly the same function; i.e.,

$$y = f(x)$$

Novices with the notation $f(x)$ are notorious for confounding the function, f, with the value of the function, $f(x)$, or for confusing $f(x)$ with f *times* x. Hence it may be a good idea to belabor a bit the notion of a function. The fact that $f(x)$ represents *the value of the function at x* means that for a specific value of x, such as 2, *the value of the function at 2 is denoted by $f(2)$*. Consider the following examples:

(1) $f(x) = 3x$
Here f is the symbol for a procedure which tells us to "multiply x by 3" in order to compute $f(x)$.
Thus when x is 0, $f(0)$ is 3(0) or 0
when x is 1, $f(1)$ is 3(1) or 3
when x is 2, $f(2)$ is 3(2) or 6, etc.

(2) $f(x) = 2x + 4$

Here f is the symbol for a procedure which tells us to "multiply x by 2 and then add 4" to compute $f(x)$.

Thus when x is 0, $f(0)$ is $2(0) + 4$ or 4

when x is 1, $f(1)$ is $2(1) + 4$ or 6

when x is 2, $f(2)$ is $2(2) + 4$ or 8, etc.

(3) $f(x) = 6$

Here f is the symbol for a procedure which tells us "no matter what x is, $f(x)$ is 6."

Thus when x is 0, $f(0)$ is 6

when x is 1, $f(1)$ is 6

when x is 2, $f(2)$ is 6, etc.

While the small letter f is the most usual symbol for a function, *any* letter followed by another in parentheses may be used. Study the following examples carefully.

(1) $y = g(x)$

g is a function of x, and y is the value of g at x.

(2) $z = j(t)$

j is a function of t, and z is the value of j at t.

(3) $w = h(v)$

h is a function of v, and w is the value of h at v.

Exercise 79:

Following are ten examples of functions; the independent variable is given first and then a function whose value depends on that variable. The statement of each function, for the first three, is also symbolized in functional notation. Restate the last seven in functional notation.

The Independent Variable	*A Function of the Variable*
1. The weight (w) of a letter	The amount of postage (P) required for the letter.
$P = f(w)$	
2. The depth (d) of a submarine	The pressure (p) on the hull of the submarine
$p = f(d)$	

3. The variable x $2x + 5$

$$2x + 5 = f(x)$$

4. The number (n) of the day of the   The maximum temperature (T) on
    year   that day in a particular locality

$$? = ?$$

5. The variable z $6z^2 - 2z + 3$

$$? = ?$$

6. The distance (d) in miles traveled   The taxi fare (F) based on a fixed
    by taxi   cost per kilometer

$$? = ?$$

7. The temperature (T) of a gas in a   The pressure (p) of the gas against
    sealed container   the sides of the container

$$? = ?$$

8. The length (e) of the edge of a cube   The volume (V) of the cube

$$? = ?$$

9. The variable (t) $\sqrt{2t + 1}$

$$? = ?$$

10. The radius (r) of a circle   The area (A) of the circle

$$? = ?$$

Note carefully that each example given above satisfies our description of a function: for each admissible value of the independent variable there is one and only one value for the function. The set of admissible values of the independent variable is called the **domain** of the function, and the set of corresponding values for the function is called the **range** of the function. If we let D represent the domain of the function and F represent the range of the function, then D and F are two sets of numbers, and the function itself defines a specific type of relationship between the elements of the two sets; that is, each element of D is matched with one and only one element of F. Thus a function is a matching mechanism which can be defined in the following way:

Definition of a Function

A function is a set of ordered pairs in which each element of a set **A**, called the domain of the function, is matched with a unique element of another set, **B**. The set of all elements of **B** which are matched with elements of **A** is called the range of the function.

It is traditional to reserve the first element in each ordered pair for elements of A and to reserve the second element of the ordered pair for the matching element of B. The word *unique* in the above definition means that a function is a set of ordered pairs in which no two pairs can have the same *first* element.

The Graph of a Function   The expression $3x - 5$ defines a function of x since the value of $3x - 5$ depends solely on the value of x, and for each value assigned to x, there is one and only one value for $3x - 5$. We may give this function any symbolic name we please, and then list some corresponding values from the domain and the range. The set of ordered pairs computed in this way represents a set of points in the plane. By plotting these points, we determine the *graph* of the function.

If we choose to refer to $3x - 5$ as $f(x)$, we write $f(x) = 3x - 5$. Then $f(0)$ means "the value of $3x - 5$ when x is replaced by zero." We can then tabulate a subset of the corresponding elements from the domain and the range of the function in the following manner:

$$f(x) = 3x - 5 \qquad \text{Domain of } f: \{x \mid x \in R\}$$

$$
\begin{aligned}
&\text{1.} \quad f(0) = 3(0) - 5 & f(0) &= -5 \\
&\qquad\quad\; = 0 - 5 \\
&\qquad\quad\; = -5 \\[4pt]
&\text{2.} \quad f(1) = 3(1) - 5 & f(1) &= -2 \\
&\qquad\quad\; = 3 - 5 \\
&\qquad\quad\; = -2 \\[4pt]
&\text{3.} \quad f(2) = 3(2) - 5 & f(2) &= 1 \\
&\qquad\quad\; = 6 - 5 \\
&\qquad\quad\; = 1 \\[4pt]
&\text{4.} \quad f(3) = 3(3) - 5 & f(3) &= 4 \\
&\qquad\quad\; = 9 - 5 \\
&\qquad\quad\; = 4 \\[4pt]
&\text{5.} \quad f(-1) = 3(-1) - 5 & f(-1) &= -8 \\
&\qquad\quad\;\; = -3 - 5 \\
&\qquad\quad\;\; = -8 \\[4pt]
&\text{6.} \quad f(-2) = 3(-2) - 5 & f(-2) &= -11 \\
&\qquad\quad\;\; = -6 - 5 \\
&\qquad\quad\;\; = -11
\end{aligned}
$$

x	0	1	2	3	-1	-2	$\rightarrow$ elements from the domain of f
$f(x)$	-5	-2	1	4	-8	-11	$\rightarrow$ elements from the range of f

For the functions whose graphs will be discussed here, the domain of each is the set of real numbers. Thus the ordered pairs tabulated above are a *subset* of the infinite set of ordered pairs defined by the function.

For this same function we could just as easily replace $f(x)$ by the symbol y and write

$$y = 3x - 5$$

1. $x = 0$
$y = 3(0) - 5$
$y = 0 - 5$
$y = -5$

2. $x = 1$
$y = 3(1) - 5$
$y = 3 - 5$
$y = -2$

3. $x = 2$
$y = 3(2) - 5$
$y = 6 - 5$
$y = 1$

4. $x = 3$
$y = 3(3) - 5$
$y = 9 - 5$
$y = 4$

5. $x = -1$
$y = 3(-1) - 5$
$y = -3 - 5$
$y = -8$

6. $x = -2$
$y = 3(-2) - 5$
$y = -6 - 5$
$y = -11$

x	0	1	2	3	-1	-2	$\rightarrow$ elements from the domain
y	-5	-2	1	4	-8	-11	$\rightarrow$ elements from the range

In either case, regardless of whether we call the value of the function y or $f(x)$, for the same function we shall obviously get the *same ordered pairs of numbers*. These ordered pairs may be written as Cartesian coordinates in which an element of the domain is paired with its corresponding element from the range of the function. These coordinates may be expressed as a set of ordered pairs in either of the following ways, depending upon which of the two types of notation is used:

A subset of $\{(x, f(x)) \mid f(x) = 3x - 5\}$

$$= \{(0, -5), (1, -2), (2, 1), (3, 4), (-1, -8), (-2, -11), \ldots\}$$

or

A subset of $\{(x, y) \mid y = 3x - 5\}$

$$= \{(0, -5), (1, -2), (2, 1), (3, 4), (-1, -8), (-2, -11), \ldots\}$$

Plotting the points identified by these coordinates gives us the graph of the function. The identical graphs of the function $3x - 5$ in Figure 6.13 illustrate how the two types of notation are shown in the graph.

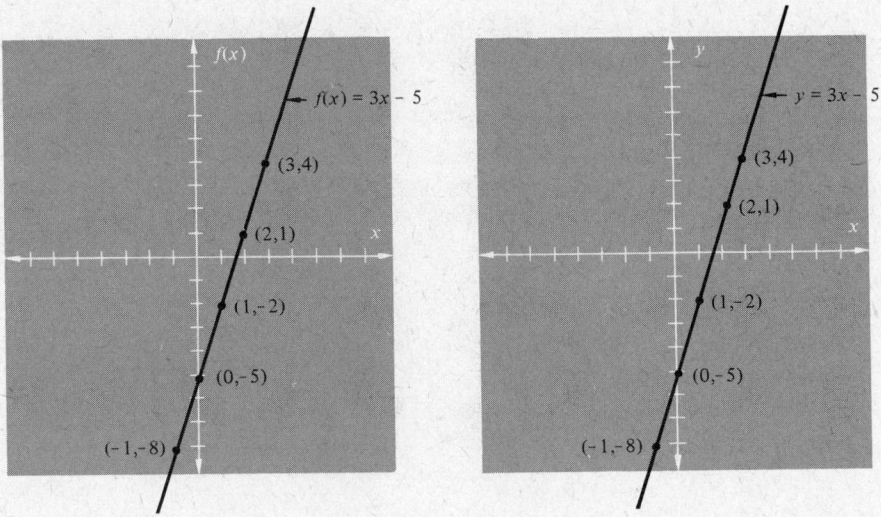

Figure 6.13

Example 1:

$$f(x) = x^2 + 2x - 8 \qquad \text{Domain of } f: \{x \mid x \in R\}$$

Find $f(-5), f(-4), f(-3), f(-2), f(-1), f(0), f(1), f(2), f(3)$.

$$f(x) = x^2 + 2x - 8$$
$$f(-5) = (-5)^2 + 2(-5) - 8$$
$$= 25 - 10 - 8$$
$$f(-5) = 7$$

$f(-5) = 7$	$f(-2) = -8$	$f(1) = -5$
$f(-4) = 0$	$f(-1) = -9$	$f(2) = 0$
$f(-3) = -5$	$f(0) = -8$	$f(3) = 7$

Express the ordered pairs as a set and graph the function (see Figure 6.14).

A subset of $\{(x, f(x)) \mid f(x) = x^2 + 2x - 8\}$
$$= \{(-5,7), (-4,0), (-3,-5), (-2,-8), (-1,-9),$$
$$(0,-8), (1,-5), (2,0), (3,7), \ldots\}$$

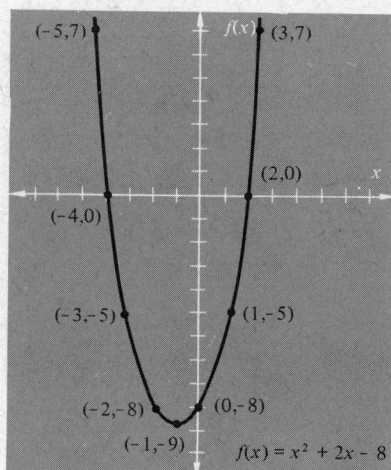

Figure 6.14

Our discussion of functions is restricted to *real-valued functions of real variables;* e.g., the domain and the range must each be a set of real numbers. In the two preceding examples of graphs of functions, both functions ($y = 3x - 5$ and $f(x) = x^2 + 2x - 8$) define an infinite set of ordered pairs since the *domain* of each is the set of all real numbers. The *range* of $y = 3x - 5$ is also the set of all real numbers while the range of $f(x) = x^2 + 2x - 8$ is an infinite *subset* of the real numbers; the graph of the latter function indicates that $f(x)$ can be any real number greater than or equal to -9, but $f(x)$ can never be less than -9.

We should note, however, that the definition of a function does *not* require that the set of ordered pairs be infinite. In the maximum-temperature function, $T = f(n)$ (problem 4 of Exercise 75), the domain of the function is clearly the set of positive integers from 1 to 365 inclusive. Thus the *graph* of this function would be 365 distinct points, the coordinates of each point indicating respectively the day of the year and the maximum temperature on that day. Compare the graphs in Examples 2 and 3.

Example 2:

$$f(x) = 2x \qquad \text{Domain of } f: \{x \mid x \text{ is a natural number less than 6}\}$$
$$f(1) = 2, \quad f(2) = 4, \quad f(3) = 6, \quad f(4) = 8, \quad f(5) = 10$$
$$\{(x, f(x) \mid f(x) = 2x\} = \{(1, 2), (2, 4), (3, 6), (4, 8), (5, 10)\}$$

Note that the domain of x is a set of *only five numbers,* $\{1, 2, 3, 4, 5\}$, and therefore the graph of the function is represented by the five points shown in Figure 6.15.

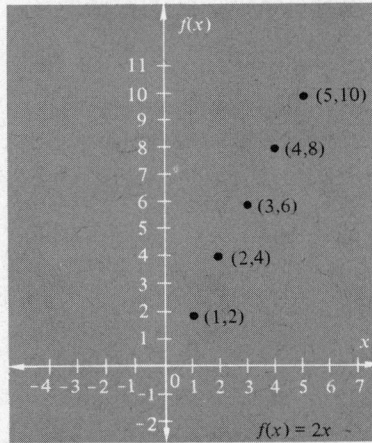

Figure 6.15

Example 3:

$$f(x) = 2x \qquad \text{Domain of } f: \{x \mid x \in R\}$$

$$f(-1) = -2, f\left(-\frac{1}{2}\right) = -1, f(0) = 0, f\left(\frac{1}{2}\right) = 1, f(1) = 2$$

A subset of $\{(x, f(x)) \mid f(x) = 2x\}$

$$= \left\{(-1, -2), \left(-\frac{1}{2}, -1\right), (0,0), \left(\frac{1}{2}, 1\right), (1,2) \ldots\right\}$$

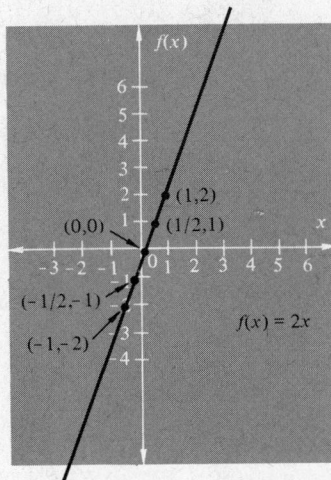

Figure 6.16

In the example above the domain of x is the set of real numbers. Thus the graph is the straight line shown in Figure 6.16 on which the coordinates of *every point* represent an ordered pair defined by the function $f(x) = 2x$.

Example 4:

$$g(x) = x^3 \qquad \text{Domain of } g: \{x \mid x \in R\}$$

$$g(-2) = -8, g(-1) = -1, g(0) = 0, g(1) = 1, g(2) = 8$$

A subset of $\{(x,g(x)) \mid g(x) = x^3\} = \{(-2,-8), (-1,-1), (0,0), (1,1), (2,8), \ldots\}$

The graph of this function is shown in Figure 6.17.

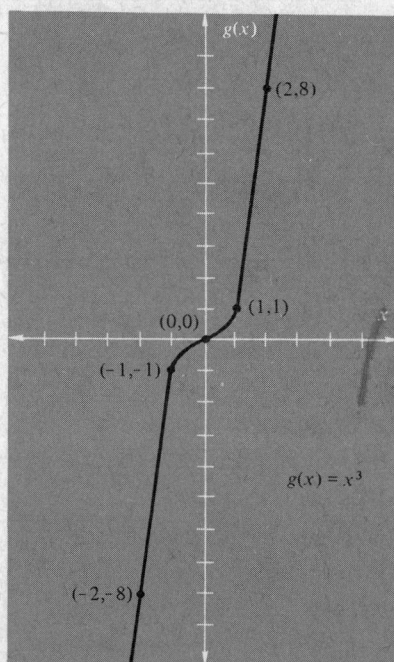

Figure 6.17

Example 5:

$$j(x) = \sqrt{25 - x^2} \qquad \text{Domain of } j: \{x \mid x \text{ is a nonnegative number less than or equal to 5}\}$$

Note that the domain of j is an infinite set because it includes every real number that is greater than or equal to 0 *and* less than or equal to 5.

$$j(0) = 5 \qquad j(3) = 4 \qquad j(4) = 3 \qquad j(5) = 0$$

We can show that the set of ordered pairs defined by j in the above computations is a subset of an infinite set of ordered pairs with fixed end points by using the following notation:

$$\{(0,5), \ldots , (x,j(x)), \ldots , (3,4), (4,3), (5,0)\}$$

The graph of j begins at $x = 0$, ends at $x = 5$, and is defined for every value of x between 0 and 5 (see Figure 6.18).

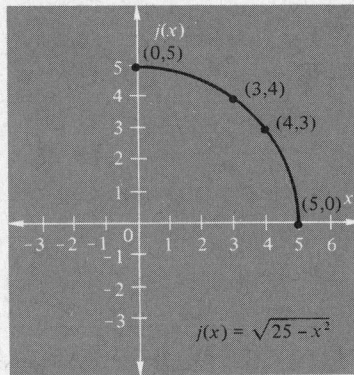

Figure 6.18

Exercise 80:

(1) $f(x) = x^2 - 4x + 3$ Domain: $x \in R$
Find $f(-1)$, $f(0)$, $f(1)$, $f(2)$, $f(3)$, $f(4)$, $f(5)$.
Express the ordered pairs as a set and graph the function.

(2) $y = 4 - x^2$ Domain: $x \in R$
Complete the table.

x	-3	-2	-1	0	1	2	3
y							

Express the ordered pairs as a set and graph the function.

(3) $f(x) = 2x + 3$ Domain: $x \in R$
Find $f(-3)$, $f(-2)$, $f(-1)$, $f(0)$, $f(1)$, $f(2)$, $f(3)$.
Express the ordered pairs as a set and graph the function.

(4) $y = \frac{1}{2}x - 6$ Domain: $x \in \{-6, -4, -2, 0, 2, 4, 6\}$
Complete the table.

x	-6	-4	-2	0	2	4	6
y							

Express the ordered pairs as a set and graph the function.

(5) $f(x) = x^2 + 2x - 3$ Domain: $x \in R$
Find $f(-4)$, $f(-3)$, $f(-2)$, $f(-1)$, $f(0)$, $f(1)$, $f(2)$.
Express the ordered pairs as a set and graph the function.

(6) $g(x) = 4x - x^2$ Domain: $\{x \mid x$ is a nonnegative real number less than
or equal to 4$\}$

Find $g(0)$, $g\left(\dfrac{1}{2}\right)$, $g(1)$, $g(2)$, $g\left(\dfrac{5}{2}\right)$, $g(3)$, $g(4)$. Express the ordered pairs as
a set and graph the function. (See Example 5.)

(7) $h(t) = t + 3$ Domain: $t \in \{0, 1, 2, 3, 4, 5, 6\}$
Express the ordered pairs as a set and graph the function.

(8) $j(x) = 25 - x^2$ Domain: $\{x \mid x$ is a nonnegative real number less than
or equal to 5$\}$

Find $j(0)$, $j(3)$, $j(4)$, $j(5)$. Express the ordered pairs as a set and graph the
function. (See Example 5.)

(9) $f(x) = x^3 - 2$ Domain: $x \in R$
Find $f(-2)$, $f(-1)$, $f(0)$, $f(1)$, $f(2)$.
Express the ordered pairs as a set and graph the function.

(10) $g(x) = \sqrt{x}$ Domain: $\{x \mid x$ is a nonnegative number less than or equal
to 9$\}$

Find $g(0)$, $g(1)$, $g(4)$, $g(9)$.
Express the ordered pairs as a set and graph the function. (See Example 5.)

Since we are considering only real-valued functions, such symbols as $f(x)$, $g(x)$,
$h(x)$, etc. represent real numbers. Thus we can perform the same operations with
specific values of functions that we perform with any real numbers. Consider the
following examples:

(1) $f(x) = 3x$ and $g(x) = x + 1$ Then $f(1) = 3$ and $g(4) = 5$
Thus $f(1) + g(4) = 3 + 5 = 8$
$f(1) - g(4) = 3 - 5 = -2$
$f(1)g(4) = (3)(5) = 15$
$\dfrac{f(1)}{g(4)} = \dfrac{3}{5}$

For these same functions, if a and b are real numbers we have the following:

$f(a) = 3a \qquad g(b) = b + 1$

Then $f(a) + g(b) = 3a + b + 1$

$f(a) - g(b) = 3a - (b + 1) \text{ or } 3a - b - 1$

$f(a)g(b) = 3a(b + 1) = 3ab + 3a$

$\dfrac{f(a)}{g(b)} = \dfrac{3a}{b + 1} \qquad \begin{array}{l} \text{provided that} \\ b \neq -1 \end{array}$

(2) $h(x) = x^2 \qquad j(x) = x - 2 \qquad$ Then $h(4) = 16$ and $j(2) = 0$

Thus $h(4) + j(2) = 16 + 0 = 16$

$h(4) - j(2) = 16 - 0 = 16$

$h(4)j(2) = (16)(0) = 0$

$\dfrac{h(4)}{j(2)} = \dfrac{16}{0} \qquad$ undefined

Exercise 81:

Given that $f(x) = x^2 + 1$ and $g(x) = x - 3$, evaluate the following:

(1) $f(2) + g(1)$

(2) $f(-3) - g(5)$

(3) $f(0)g(6)$

(4) $\dfrac{f(-1)}{g(0)}$

(5) $f(-1) + g(-1)$

(6) $\dfrac{f(2)}{g(1)}$

(7) $f(-4)g(0)$

(8) $\dfrac{f(5)}{g(3)}$

(9) $f(t) + g(t)$

(10) $f(a)g(b)$

(11) $[f(-2)]^2$

(12) $[g(2)]^3$

Chapter Test 6

I. Graph the following sets in the same plane:

$A = \{(x,y) \mid 3x + 2y = 8\}$

$B = \{(x,y) \mid 5x - 3y = 7\}$

$A \cap B = ?$

II. Solve the following system of equations by addition and subtraction:

$2x + 5y = 11$

$4x - 3y = 9$

III. Solve the following system of equations by the method of substitution:
$$4x + y = 11$$
$$5x - 2y = 4$$

IV. Complete the following statements:
1. If f is a function of x, the value of f at x is denoted by _____.
2. f is a function of x if, for each x in the domain of f, there is _____
_____.

V. $g(x) = 2x - 3 \qquad f(x) = x^2 - 1$

Compute the following:

1. $g(1)f(2)$ 2. $g(3) + f(-3)$

3. $\dfrac{g(4)}{f(1)}$ 4. $[g(0)]^2$

5. $\dfrac{g(0)}{f(0)}$

VI. $f(x) = x^2 - 2x - 3 \qquad$ Domain of f: $x \in R$

1. Find the values of:
$f(-2), \ f(-1), \ f(0), \ f(1), \ f(2), \ f(3), \ f(4)$
2. Express the ordered pairs $(x, f(x))$ computed in part 1 as a set and graph the function.

VII. $A = \{-1,0,1,2\}, \qquad B = \{1,3,5\}, \qquad C = \{-2,-1,0,1\}$
1. $A \cup B =$ 2. $A \cap B =$
3. $C \cap \emptyset =$ 4. $A \cap C =$
5. $(A \cup B) \cap C =$ 6. $(A \cap C) \cup (B \cap C) =$

VIII. Construct a diagram showing the relationships of the following sets.
Q = the universal set
J = the set of integers
P = the set of primes

IX. A florist receives an order for a mixed arrangement of gladioli and lilies to cost $20.00. Lilies cost $1.50 each and gladioli cost $1.00 each. The florist decides that the size of the arrangement will require 15 flowers and $2.00 worth of greenery. How many of each type of flower should he use?

X. Mr. Delaney plans to invest a total sum of $22,500.00 in stocks and bonds. He has decided to buy bonds that pay 8% interest per year and a growth stock that costs $25.00 per share and pays $1.00 per share dividend per year. If he wants an income of $100.00 per month from the combined payments of interest and dividends, how should he divide the $22,500.00 between the two investments?

7
FIRST-DEGREE INEQUALITIES

7.1
The Order Postulates

The set of real numbers has another property besides the properties illustrated in the first four chapters that deserves our attention. This property is *order*. For example, if we contemplate at some length the set of integers and their positions on the number line, the intuitive notion of *order* in the successive positions of the integers is almost inescapable. (See Figure 7.1.) An elementary aspect of this is the fact that all of the negative integers

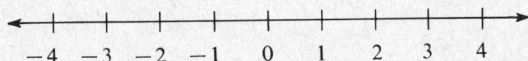

Figure 7.1

lie to the left of zero while all of the positive integers lie to the right of zero. If we select any integer and increase it by one (e.g., add $+1$ to the integer), we always get the next integer to the *right* of the one selected.

$$2 + 1 = 3$$
$$3 + 1 = 4$$
$$-7 + 1 = -6$$
$$-6 + 1 = -5$$
$$-1 + 1 = 0$$

These observations can also be applied to the set of real numbers: (1) all negative real numbers lie to the left of zero and all positive real numbers to the right of zero; (2) if we select any real number and add a *positive* number to it, we always get a real number which lies *to the right* of the original number selected.

Any two distinct real numbers, a and b, are represented by two distinct points on the number line, and one of them (we shall designate it by a) must lie to the left of the other, the point designated by b. Since the number a lies to the left of b, the number a would have to be *increased* by some positive number in order to "move to the right" or to equal b. We describe this situation by saying that a is "less than" b. This relationship is stated symbolically by a horizontal $\vee$ that *always points toward the smaller number:*

$$a < b \qquad \text{read "}a\text{ is less than } b\text{"}$$

Definition 1: For every $a,b \in R$, $a < b$ if and only if a positive number p exists such that $a + p = b$.

Examples:

(1) $2 < 6$	$2 + 4 = 6$	$p = 4$
(2) $4 < 7$	$4 + 3 = 7$	$p = 3$
(3) $-7 < -5$	$-7 + 2 = -5$	$p = 2$
(4) $-10 < -2$	$-10 + 8 = -2$	$p = 8$
(5) $\frac{1}{3} < \frac{1}{2}$	$\frac{1}{3} + \frac{1}{6} = \frac{1}{2}$	$p = \frac{1}{6}$

Conversely,

(6) $7 + 3 = 10$	$p = 3$	$7 < 10$
(7) $-5 + 2 = -3$	$p = 2$	$-5 < -3$
(8) $\frac{1}{2} + \frac{1}{4} = \frac{3}{4}$	$p = \frac{1}{4}$	$\frac{1}{2} < \frac{3}{4}$
(9) $-\frac{1}{3} + \frac{1}{3} = 0$	$p = \frac{1}{3}$	$-\frac{1}{3} < 0$
(10) $x + 5 = y$	$p = 5$	$x < y$

In every example above, the inequality sign points toward the smaller of the two numbers and that number always lies to the left of the other on the number line.

$a < b \qquad$ means "a is less than b, and a lies to the left of b on the number line"

Thus the statement $x < 0$ means that x lies to the left of zero on the number line; but any number that lies to the left of zero is a negative number. Therefore the statement $x < 0$ means that x is a negative number.

We can state the "less than" relationship in reverse by reversing the sense (or the order) of the inequality sign. If 2 is less than 6, then 6 is greater than 2; that is stated in the following manner:

$$2 < 6 \qquad \text{read "2 is less than 6"}$$
$$6 > 2 \qquad \text{read "6 is greater than 2"}$$

These two statements define the same relationship; note that the inequality sign continues to point toward the smaller number.

$$a < b \qquad \text{means } b > a$$

Then the statement $x > 0$ means that x lies to the *right* of zero on the number line and is therefore a positive number. In the statements below, the double-headed arrow is read "implies and is implied by."

Definition 2: $a < 0 \leftrightarrow a$ is a negative number.
Definition 3: $a > 0 \leftrightarrow a$ is a positive number.
Definition 4: $a \leq b \leftrightarrow a$ is less than or equal to b.
Definition 5: $a \geq b \leftrightarrow a$ is greater than or equal to b.

A statement of inequality may relate more than two quantities.

Examples:

$4 < 7$ two members
$2 < 5 < 8$ three members

The statement $2 < 5 < 8$ means the same as the statement $2 < 5$ *and* $5 < 8$. We shall refer to those parts of an inequality which are separated by inequality signs as the *members* of the inequality.

An essential characteristic of the property of order is expressed in an order postulate which asserts the truth of the *trichotomy* law. This law states, in effect, that for every real number, one and only one of *three* possibilities must occur. We shall call this postulate of the trichotomy law, O_t.

O_t For every $a \in R$, one and only one of the following is true: $a < 0$ or $a = 0$ or $a > 0$.

The second order postulate states the closure property of positive real numbers under addition and multiplication. We shall call this order postulate for closure, O_c.

O_c If a and b are positive real numbers, then $a + b$ is a positive number and ab is a positive number.

These order twins, "O_t and O_c," are both necessary and sufficient to establish the fact that a set is ordered. An example of an ordered set is the set of integers. Another example is the set of rational numbers.

7.2
First-Degree Inequalities in One Variable

We noted earlier that equations have great practical usefulness in solving many kinds of problems. But much human activity and considerable human interests are centered on quantities which are *not* equal: e.g., businessmen want their profits this year to be greater than last year; manufacturers seek ways to make production costs less than their competitors'; football teams devise plays to achieve a score greater than their opponents'; and home buyers look to the suburbs of cities for property on which taxes are less than city taxes. Thus our concern with unequal quantities, or in-equalities, pervades much of daily life. Another example: if C represents our im-mediate cash in hand, and, in an expansive mood, we suddenly decide to take x guests to dinner at a restaurant where E is the estimated cost per guest, then social comfort dictates that x must be chosen so that $C > Ex$.

In a more serious vein, suppose that a student with a quiz average of 93 is prepar-ing for a final exam which counts as one-third of his course grade, and he wishes to win an honor scholarship over a competitor who has already completed the course with a grade of 94. The student needs a grade, x, on the final exam such that

$$\frac{2}{3}(93) + \frac{1}{3}x > 94$$

Obviously, the greatest possible value of x is 100, and should that be the happy out-come, then the student's course grade would be $\frac{2}{3}(93) + \frac{1}{3}(100)$ or $95\frac{1}{3}$, which will clearly best the competitor. But what other values of x will produce a course grade greater than 94? More to the point, what is the *smallest* value of x that will produce a grade greater than 94? In other words, how do we find all possible values of x that will satisfy the inequality $\frac{2}{3}(93) + \frac{1}{3}x > 94$? The technique for doing this requires a knowledge of some of the mathematical properties of inequalities.

Inequalities have special properties that are in some ways similar to the special properties of equations and in other ways sharply different. For example, one of the equality axioms, E_a, states that we may add the same number to both sides of an equation and the truth of the equality is unchanged. Similarly, in the inequality $3 < 7$, if we add the same number, for example 2, to each side, we have $3 + 2 < 7 + 2$ or $5 < 9$, and the truth of the inequality is unchanged. On the other hand, we can reverse the two sides of any equation and the truth of the equation is unaltered (e.g. $4 + 2 = 6$ and $6 = 4 + 2$); but we *cannot* reverse two members of an in-equality: e.g. $3 < 7$ is true, but to say that $7 < 3$ is patently not true.

The properties of inequalities are explained and illustrated in the following dis-cussion and then restated in symbolic language. These symbolic statements are actually theorems about inequalities which can be proved using the definitions, the order postulates, and the properties of real numbers. Four important properties follow:

Property I

If the same number is added to the members of an inequality, the *order* of the inequality sign is unchanged.

Examples:

(1) Consider the inequality

$$2 < 5$$

Now add *any* number, for example the number 3, to both sides. On the left side we have $2 + 3$; on the right, $5 + 3$, and we note that

$$2 + 3 < 5 + 3$$

or

$$5 < 8$$

(2) $7 < 10$
Add -4 to each side.

$$7 - 4 < 10 - 4$$

or

$$3 < 6$$

(3) $-5 < -3$
Add 10 to both sides.

$$-5 + 10 < -3 + 10$$

or

$$5 < 7$$

(4) $2 < 6$
Add -6 to both sides.

$$2 - 6 < 6 - 6$$
$$-4 < 0$$

In general, stating Property I in symbolic language and labeling it Theorem O_1,* we have the following:

Theorem O_1:   Let $a, b, c, \in R$. If $a < b$, then $a + c < b + c$

---

*The proofs of the first four theorems stated here are given in the Appendix.

As noted earlier the above statement is similar to the addition axiom, E_a, for equations.

Property II

If three numbers are related in such a way that the first is less than the second and the second is less than the third, then the first is also less than the third.

Examples:

(1) Consider the numbers 3, 5, and 8. We note that

$$3 < 5 \quad \text{and} \quad 5 < 8$$
$$\text{then} \quad 3 < 8$$

(2) $-7 < -4$ and $-4 < -1$
then $-7 < -1$

(3) $-2 < 0$ and $0 < 3$
then $-2 < 3$

Stated in symbolic language, we have the following:

Theorem O_2:   Let $a,b,c, \in R$. If $a < b$ and $b < c$, then $a < c$

Note that the above statement is similar to the transitive axiom, E_t, for equations.

Property III

If the members of an inequality are multiplied or divided by the same *positive* number, then the *order* of the inequality sign is *unchanged*.

Examples:

(1) $3 < 5$
Multiply both sides by the positive number 4.

$$3 < 5$$
$$3(4) < 5(4)$$

or

$$12 < 20$$

(2) $-6 < -4$
Multiply both sides by the positive number 2.

$$-6 < -4$$
$$-6(2) < -4(2)$$

or

$$-12 < -8$$

(3) $10 < 15$
Divide both sides by the positive number 5.

$$10 < 15$$
$$\frac{10}{5} < \frac{15}{5}$$

or

$$2 < 3$$

In general, we have the following:

Theorem O_3:   Let $a, b, c, \in R$. If $a < b$ and $c > 0$, then $ac < bc$

Note: It can be proved that, if $c > 0$, then $\frac{1}{c} > 0$; thus, by Theorem O_3, if $a < b$,

then $a\left(\frac{1}{c}\right) < b\left(\frac{1}{c}\right)$ or $\frac{a}{c} < \frac{b}{c}$.

Property IV

If the members of an inequality are multiplied or divided by the same *negative* number, then the order of the inequality sign is *reversed*.

Examples:

(1) $-7 < -5$
Multiply each side by the *negative* number -1. On the left side, we have $-7(-1)$ and on the right, $-5(-1)$. We note that

$$-7(-1) > -5(-1)$$

or

$$7 > 5$$

Clearly, multiplying each side of $-7 < -5$ by -1 reverses the order of the inequality sign.

(2) $3 < 5$
Multiply both sides by the negative number -4.

$$3 < 5$$
$$3(-4) > 5(-4)$$

or

$$-12 > -20$$

(3) $-8 < -4$
Divide both sides by the negative number -2.

$$-8 < -4$$
$$\frac{-8}{-2} > \frac{-4}{-2}$$
$$4 > 2$$

In general, we have the following:

Theorem O₄: Let $a, b, c, \in R$. If $a < b$ and $c < 0$, then $ac > bc$

Note: It can be proved that if $c < 0$, then $\frac{1}{c} < 0$, and by Theorem O₄, if $a < b$, then $a\left(\frac{1}{c}\right) > b\left(\frac{1}{c}\right)$ or $\frac{a}{c} > \frac{b}{c}$.

If you look back over the foregoing discussion of properties of inequalities, you will find that, while the first three are similar to the properties of equations, Property IV is notably different. That difference should be examined carefully and remembered.

The following theorems state other useful properties of inequalities. In each theorem all literal symbols represent real numbers, and zero is excluded from all denominators.

Theorem O₅: If $a < b$, and a and b have the *same sign*, then $\frac{1}{a} > \frac{1}{b}$.
(Observe the reversal of the inequality symbol.)

Examples:

(1) $2 < 5$

but $\dfrac{1}{2} > \dfrac{1}{5}$

(2) $-4 < -2$

but $-\dfrac{1}{4} > -\dfrac{1}{2}$

In the statement of Theorem O_5, please note the emphasis placed on the phrase "same sign" and observe that this property *never* applies if a and b have *different* signs. For example, $-4 < 10$, but $-\dfrac{1}{4}$ is certainly not greater than $\dfrac{1}{10}$. But when two distinct numbers have the same sign, the order of the inequality sign between them is always reversed for their multiplicative inverses. In general

> If $\dfrac{a}{b}$ and $\dfrac{c}{d}$ have the same sign and $\dfrac{a}{b} < \dfrac{c}{d}$, then $\dfrac{b}{a} > \dfrac{d}{c}$.

> Theorem O_6: If $a < b$ and a and b are both *positive*, then
>
> $$a^2 < b^2$$

For example, $3 < 5$, and $3^2 < 5^2$ or $9 < 25$.

> Theorem O_7: If $a < b$ and a and b are both *negative*, then
>
> $$a^2 > b^2$$

For example, $-4 < -1$, but $(-4)^2 > (-1)^2$ or $16 > 1$.

> Theorem O_8: If $a < b$, then $a < \dfrac{a+b}{2}$ and $\dfrac{a+b}{2} < b$, or $a < \dfrac{a+b}{2} < b$

Examples:

(1) $4 < 6$

then $4 < \dfrac{4+6}{2} < 6$

or $4 < 5 < 6$

(2) $-5 < -3$

then $-5 < \dfrac{-5 + (-3)}{2} < -3$

or $-5 < -4 < -3$

A significant consequence of Theorem O_8 is illustrated in the following example:

(3)   Consider the inequality

$$4 < 8$$

Then $\qquad\qquad 4 < \dfrac{4 + 8}{2} < 8$

or $\qquad\qquad\qquad 4 < 6 < 8$

Continuing to apply Theorem O_8 to the above statement, we obtain

$$4 < \frac{4 + 6}{2} < 6 \quad \text{and} \quad 6 < \frac{6 + 8}{2} < 8$$

or $\qquad\qquad\quad 4 < 5 < 6 \quad \text{and} \quad 6 < 7 < 8$

Thus $\qquad\qquad 4 < 5 < 6 < 7 < 8 \qquad (\text{Theorem } O_2)$

We could get carried away and continue to apply Theorem O_8 to the last inequality, obtaining

$$4 < \frac{4 + 5}{2} < 5 < \frac{5 + 6}{2} < 6 < \frac{6 + 7}{2} < 7 < \frac{7 + 8}{2} < 8$$

or $\qquad\qquad 4 < 4\frac{1}{2} < 5 < 5\frac{1}{2} < 6 < 6\frac{1}{2} < 7 < 7\frac{1}{2} < 8$

Thus by beginning with $4 < 8$ and applying Theorem O_8 three successive times, we have, in the last statement above, found seven numbers, *all* of which lie between 4 and 8 on the number line. On the classic assumption that what has been done once (or twice or thrice) can always be repeated, we conclude that there are *infinitely many* real numbers which lie between 4 and 8.

In general, if you care to work out the arithmetic, by using Theorem O_8 you can show that, if $a < b$, then

$$a < \frac{a + b}{2} < \frac{a + 3b}{4} < \frac{a + 7b}{8} < \frac{a + 15b}{16} \cdots$$

where *each number we get* lies between a and b no matter how long the process is continued. From this we conclude that between *any* two distinct real numbers, there are infinitely many real numbers.

Exercise 82:

(1) The real numbers can be paired in one-to-one correspondence with the points on the number line. How many points (numbers) lie between any two distinct points (numbers) on the line?

(2) Given that $2 < 5$, state the inequality obtained when -10 is added to each side. (See Theorem O_1)

(3) Given that $-6 < -2$, state the inequality obtained when both sides are multiplied by 5. (See Theorem O_3)

(4) Given that $5 < 7$, state the inequality obtained when both sides are multiplied by -2. (See Theorem O_4)

(5) Given that $6 < 10$, state the inequality obtained when both sides are squared. (See Theorem O_6)

(6) Given that $-10 < -5$, state the inequality obtained when both sides are squared. (See Theorem O_7)

(7) Given that $\frac{7}{3} < \frac{5}{2}$, state the inequality obtained by replacing the number on each side with its multiplicative inverse. (See Theorem O_5)

(8) Given that $-6 < -3$, state the inequality obtained by replacing each number by its multiplicative inverse. (See Theorem O_5)

(9) Given that $9 < 12$, state the inequality obtained when both sides are divided by -3. (See Theorem O_4)

(10) Given that $4 < 5$, state the inequality obtained by replacing each number by its multiplicative inverse. (See Theorem O_5)

(11) Given that $5 < 8$ and $8 < x$, state the inequality relating 5 and x. (See Theorem O_2)

(12) Given that $1 < 4$, state the inequality obtained when both sides are multiplied by -6. (See Theorem O_4)

When an inequality contains a variable, or variables, the properties discussed above are applied to simplify the statement. For the present we shall consider only first-degree inequalities with one variable. In Section 7.3, solution sets for inequalities of this type are illustrated graphically and stated in set notation, but a preliminary problem with all such inequalities is to simplify each one as much as possible.

Suppose, for example, that we have a statement such as $x + 5 < 3$; this clearly imposes a limitation on the values of x that will make the statement true, and that limitation can be stated in a simpler manner by using Theorem O_1 to isolate x.

To illustrate:

$$x + 5 < 3$$

Add -5 to both sides (Theorem O_1).

$$x + 5 - 5 < 3 - 5$$

$$\text{or} \quad x < -2$$

Thus the statements

$$x + 5 < 3 \quad \text{and} \quad x < -2$$

impose exactly the same limitation on x, but the latter statement is much simpler than the first. Following are more examples of how we simplify such inequalities by using the properties of inequalities to isolate the variable.

Examples:

(1) $2x + 3 > x - 5$

Add -3 and $-x$ to both sides (Theorem O_1).

$$2x + 3 - 3 - x > x - 5 - 3 - x$$

Then combine the like terms to obtain

$$x > -8$$

which is a simplified form of the first statement.

(2) $3x < 12$

Here we apply Theorem O_3 and isolate x by dividing both sides of the inequality by the positive number 3.

$$3x < 12$$

$$\frac{\cancel{3}x}{\cancel{3}} < \frac{12}{3}$$

$$x < 4$$

(3) $4x - 6 < 7x + 12$

First add $-7x$ and 6 to both sides (Theorem O_1).

$$4x - 6 - 7x + 6 < 7x + 12 - 7x + 6$$

Combine similar terms

$$-3x < 18$$

Now, using Theorem O_4, divide both sides by the *negative* number -3.

$$-3x < 18$$

$$\frac{-3x}{-3} > \frac{18}{-3}$$

$$\boxed{\text{Note the } \textit{reversal} \text{ of the inequality symbol}}$$

$$x > 6$$

(4) $-7 < -2x + 3 < 9$

Add -3 to each member of the inequality.

$$-7 - 3 < -2x + 3 - 3 < 9 - 3$$

Combine similar terms.

$$-10 < -2x < 6$$

Using Theorem O$_4$, divide each member by -2.

$$-10 < -2x < 6$$

$$\frac{-10}{-2} > \frac{-2x}{-2} > \frac{6}{-2}$$

$$5 > x > -3$$

Note: This last inequality may also be written $-3 < x < 5$ because both

$$5 > x > -3$$

and

$$-3 < x < 5$$

say exactly the same thing: namely that x lies between -3 and 5. Observe that in both versions each inequality sign points toward the smaller number.

(5) $-\frac{2}{5} x < 6$

For convenience in computation, write $-\frac{2}{5}x$ as $\frac{-2x}{5}$ and multiply both sides by 5.

$$\frac{-2x}{\cancel{5}} \cdot \frac{\cancel{5}}{1} < 6 \cdot 5$$

$$-2x < 30$$

Divide both sides by -2.

$$\frac{-2x}{-2} > \frac{30}{-2}$$

$$x > -15$$

(6) $\frac{3x - 1}{2} < 10$

Multiply both sides by 2.

$$\frac{3x - 1}{\cancel{2}} \cdot \frac{\cancel{2}}{1} < 10 \cdot 2$$

$$3x - 1 < 20$$

Add 1 to both sides.

$$3x < 21$$

$$x < 7$$

Exercise 83:

Use the properties of inequalities to simplify the following statements:

(1) $x + 3 < 10$ (2) $x - 7 > 4$

(3) $4x > 9 - x$ (4) $-5x < 10$

(5) $-2x > -12$ (6) $-7x < 28$

(7) $\frac{1}{2}x < 3$ (8) $-\frac{1}{3}x > 4$

(9) $5x - 4 < 2x + 8$ (10) $3x + 5 > 5x - 1$

(11) $4x - 1 < 2x + 6$ (12) $3x - 8 > 7x + 2$

(13) $-2 < x - 4 < 7$ (See Example 4)   (14) $8 < 2 - x < 12$

(15) $3 < 2x - 5 < 9$ (16) $\frac{2x - 1}{3} < 1$ (See Example 6)

(17) $\frac{3x + 5}{2} < 4$ (18) $3 < \frac{2x + 5}{3} < 7$

7.3
Solution Sets of First-Degree Inequalities in One Variable

The solution set for an inequality is defined as the set of all numbers for which the inequality is a true statement. It is frequently helpful to graph the solution set on the number line before defining the set by rule.

Examples:

(1) $x > 5$

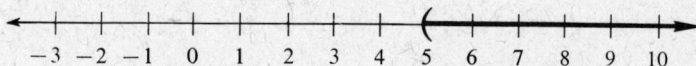

Figure 7.2

The curved line (half-parentheses) in Figure 7.2 means that the number 5 does not belong to the solution set. The solid bar shows that x can be any number which lies to the right of 5 on the number line.

The solution set is $\{x \mid x > 5\}$

(2) $x \leq -2$

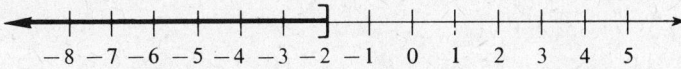

Figure 7.3

In Figure 7.3 the half-bracket (]) indicates that the number -2 is a member of the solution set; the solid bar shows that x can also be any number that lies to the left of -2 on the number line.

The solution set is $\{x \mid x \leq -2\}$

(3) $2x + 6 < x + 3$
$2x - x < 3 - 6$ Theorem O_1 (add -6 and $-x$ to both members)
$x < -3$ (See Figure 7.4)

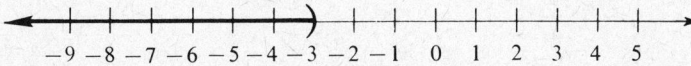

Figure 7.4

The solution set is $\{x \mid x < -3\}$

(4) $2x + 9 \leq 4x - 3$
$2x - 4x \leq -3 - 9$ Theorem O_1 (add $-4x$ and -9 to both members)
$-2x \leq -12$ Theorem O_4
$x \geq 6$ (multiply both members by $-1/2$)
             (See Figure 7.5)

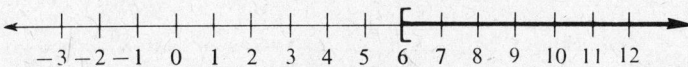

Figure 7.5

The solution set is $\{x \mid x \geq 6\}$

(5) $\dfrac{x}{3} - \dfrac{1}{2} > \dfrac{4x}{3} + \dfrac{3}{2}$

$\dfrac{6}{1}\cdot\dfrac{x}{3} - \dfrac{6}{1}\cdot\dfrac{1}{2} > \dfrac{6}{1}\cdot\dfrac{4x}{3} + \dfrac{6}{1}\cdot\dfrac{3}{2}$ Theorem O_3
                    (multiply both members by 6)

$2x - 3 > 8x + 9$
$2x - 8x > 9 + 3$ Theorem O_1
                   (add $-8x$ and 3 to both members)

$-6x > 12$ Theorem O_4
$x < -2$ (multiply both members by $-1/6$)

                   (See Figure 7.6)

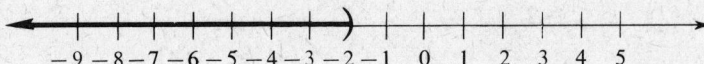

Figure 7.6

The solution set is $\{x \mid x < -2\}$

Exercise 84:

Graph and define by rule the solution set of each inequality.

(1) $5x - 1 < 2x + 8$ (2) $2x + 3 > 5x + 6$

(3) $x + 2 \leq 3x - 4$ (4) $2(x - 3) < 5(x - 3)$

(5) $4x + 3 \geq 2x - 7$ (6) $8x - 2 \leq 9x - 10$

(7) $\dfrac{x}{3} - \dfrac{1}{5} \leq \dfrac{x}{5} + 1$ (8) $\dfrac{3x}{4} - \dfrac{2}{3} > \dfrac{x}{2} + \dfrac{1}{3}$

(9) $\dfrac{x}{5} - \dfrac{x}{2} < 3$ (10) $3(x - 1) \leq 2(x + 3)$

(11) $\dfrac{x - 1}{3} < \dfrac{x + 2}{2}$ (12) $\dfrac{2x + 1}{3} + \dfrac{x + 1}{2} \geq \dfrac{x - 1}{2}$

(13) $\dfrac{2x - 3}{4} > \dfrac{3x + 4}{3}$ (14) $\dfrac{2x + 3}{3} < 3x - 4$

7.4
Inequalities Involving Absolute Values

The statement $2 < x < 5$ means that the number x must satisfy two conditions simultaneously; e.g., x must be less than 5 *and* also greater than 2. The solution set

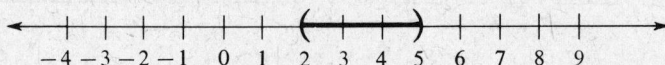

Figure 7.7

can be shown by a graph on the number line. The parentheses in Figure 7.7 indicate that 2 and 5 are not members of the solution set. This set is actually the *intersection* of *two sets*, the set of all numbers less than 5 and the set of all numbers greater than 2.

The solution set is $\{x \mid x > 2\} \cap \{x \mid x < 5\}$

Examples:

(1) $-2 \leq x \leq 2$

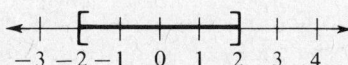

Figure 7.8

The solution set is $\{x \mid x \geq -2\} \cap \{x \mid x \leq 2\}$

(2) $-5 < x < 0$

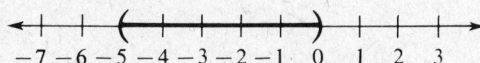

Figure 7.9

The solution set is $\{x \mid x > -5\} \cap \{x \mid x < 0\}$

We encounter the kind of situation illustrated above in working with inequalities that involve absolute values of quantities **less than** a given number.

Examples:

(1) $|x| < 4$
This statement means that x can be any number whose *distance from zero* (regardless of direction) is less than four units (see Section 1.4).

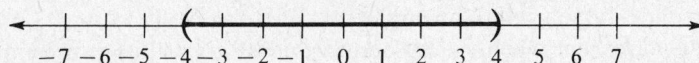

Figure 7.10

From Figure 7.10 it can be seen that the two statements $|x| < 4$ and $-4 < x < 4$ *mean exactly the same thing*. The solution set for each is $\{x \mid x > -4\} \cap \{x \mid x < 4\}$. In symbolic language:

Let $x, a \in R$ with $a > 0$. Then $|x| < a$ means $-a < x < a$. The solution set is $\{x \mid x > -a\} \cap \{x \mid x < a\}$. (See Figure 7.11.)

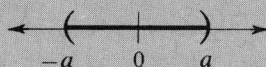

Figure 7.11

In like manner, $|x| \leq a$ means $-a \leq x \leq a$, and on the graph the solution set would be enclosed by brackets rather than parentheses.

(2) $|2x - 3| < 7$

Then $-7 < 2x - 3 < 7$.

Add 3 to all members of the inequality.

$$-7 + 3 < 2x - 3 + 3 < 7 + 3$$

or $\qquad -4 < 2x < 10$

Multiply all members by 1/2 to isolate x.

$$\frac{-4}{2} < \frac{2x}{2} < \frac{10}{2}$$

$$-2 < x < 5$$

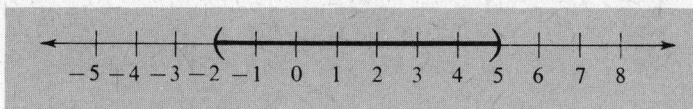

Figure 7.12

The solution set is $\{x \mid x > -2\} \cap \{x \mid x < 5\}$. (See Figure 7.12)

A comparison of the absolute-value *inequalities* discussed here with the absolute-value *equations* discussed in Section 5.7 can be informative in regard to similarities and differences. Note particularly that in both the equations and the inequalities the absolute-value symbol always means that *two* values (or sets of values) must be considered, except when the absolute value of a variable expression is less than, greater than, or equal to zero.

Compare

$$(1) \quad |x + 2| = 5$$

with $\qquad (2) \quad |x + 2| < 5$

(1) The equation $|x + 2| = 5$ means

$$x + 2 = 5 \quad \text{or} \quad x + 2 = -5$$

So $\qquad\qquad x = 5 - 2 \quad \text{or} \quad x = -5 - 2$

and $\qquad\qquad x = 3 \quad \text{or} \quad x = -7$

Hence the solution set is $\{3, -7\}$

(2)   The inequality $|x + 2| < 5$ means

$$x + 2 < 5 \quad \text{and} \quad x + 2 > -5$$

So $\qquad\qquad x < 5 - 2 \quad \text{and} \quad x > -5 - 2$

Thus $\qquad\qquad x < 3 \quad \text{and} \quad x > -7$

Consequently, the solution is the intersection of two overlapping sets, all numbers which are less than 3 and all the numbers greater than -7.

Hence the solution set is $\{x \mid x > -7\} \cap \{x \mid x < 3\}$

In the second type of absolute-value inequalities to be considered, we shall find that the two possibilities implied by the absolute-value sign lead to a solution that is the union of two disjoint sets.

Exercise 85:

Find the solution set of each of the following inequalities and show each set on a line graph:

(1) $|x| < 3$ (2) $|x| \leq 8$

(3) $|2x| < 6$ (4) $\left|\dfrac{5x}{2}\right| \leq 5$

(5) $|x + 3| < 4$ (6) $|x - 2| < 6$

(7) $|x - 8| < 12$ (8) $\left|x + \dfrac{2}{3}\right| < 5/3$

(9) $\left|x - \dfrac{1}{2}\right| < 3/2$ (10) $|x - 3/4| < 7/4$

(11) $|x - 1| < 3$ (12) $|x + 3| \leq 4$

(13) $|3x - 2| < 9$ (14) $\left|\dfrac{x - 2}{2}\right| < 1$

(15) $|2x + 5| < 9$ (16) $\left|\dfrac{2x + 1}{3}\right| < 3$

The statement $x > 5$ or $x < -2$ will be true for any value of x that is greater than 5 *or* less than -2. Figure 7.13 shows the two sets of numbers that will satisfy this inequality. The solution set for $x > 5$ or $x < -2$ is the *union* of the elements of the

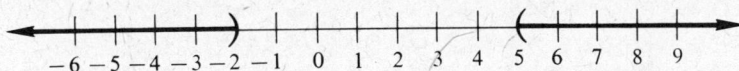

$$\xleftarrow{\hspace{2em}} \underset{-6\ -5\ -4\ -3\ -2\ -1 \quad 0 \quad 1 \quad 2 \quad 3 \quad 4 \quad 5 \quad 6 \quad 7 \quad 8 \quad 9}{\longrightarrow}$$

Figure 7.13

two sets shown on the graph. The solution set is $\{x \mid x < -2\} \cup \{x \mid x > 5\}$. Comparing this with the problems of Exercise 85, we see that *or* means "union" but *and* means "intersection."

Inequalities for which the solution set is the union of two sets are encountered in inequalities involving absolute values and the **"greater than"** relationship.

Example:

(1) $|x| > 3$

This statement means that x can be any number whose distance from zero (in either direction) is *greater than* three units. Figure 7.14 shows the two sets of numbers which satisfy this condition. From the graph it can be seen

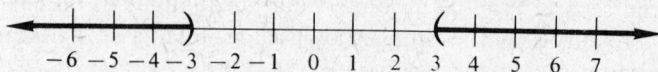

Figure 7.14

that $|x| > 3$ means the same thing as the statement $x < -3$ or $x > 3$. The solution set is $\{x \mid x < -3\} \cup \{x \mid x > 3\}$, which is the union of the elements of the two sets shown on the graph.

In symbolic language,

Let $x, a \in R$ with $a > 0$. Then $|x| > a$ means $x < -a$ or $x > a$. The solution set is $\{x \mid x < -a\} \cup \{x \mid x > a\}$. (See Figure 7.16.)

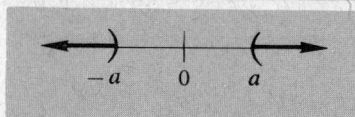

Figure 7.15

In like manner $|x| \geq a$, $a > 0$, means $x \leq -a$ or $x \geq a$ and on the graph the two sets would each be closed at one end point by a half-bracket.

(2) $|2x - 4| > 8$

$$\text{Then } 2x - 4 < -8 \quad or \quad 2x - 4 > 8$$
$$2x < -8 + 4 \quad or \quad 2x > 8 + 4$$
$$2x < -4 \quad or \quad 2x > 12$$
$$x < -2 \quad or \quad x > 6$$

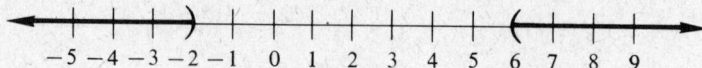

Figure 7.16

The solution set is $\{x \mid x < -2\} \cup \{x \mid x > 6\}$. (see Figure 7.16)

Exercise 86:

Find the solution sets of the following inequalities and graph each set on the number line:

(1) $|x| > 5$ (2) $|3x| > 12$

(3) $\left|\dfrac{2x}{3}\right| > 4$ (4) $|x - 4| > 5$

(5) $|x + 3| \geq 3$ (6) $|2x - 4| > 6$

(7) $|3x - 5| > 11$ (8) $\left|\dfrac{x - 3}{2}\right| \geq 3$

(9) $\left|\dfrac{3x - 2}{4}\right| > 2$ (10) $\left|\dfrac{2x - 5}{3}\right| \geq 3$

7.5
First-Degree Inequalities in Two Variables

Mathematical statements such as the following are examples of first-degree inequalities in two variables:

$$x - y < 2 \qquad x - y > 2$$
$$x - y \leq 2 \qquad x - y \geq 2$$

The best way to get acquainted with the inequalities shown above is to approach them by way of a very close relative, namely the *equality* $x - y = 2$. From Chapter 6 we recall that the equation $x - y = 2$ defines an infinite set of points in the plane which lie in a straight line. The graph of $x - y = 2$, shown in Figure 7.17, divides

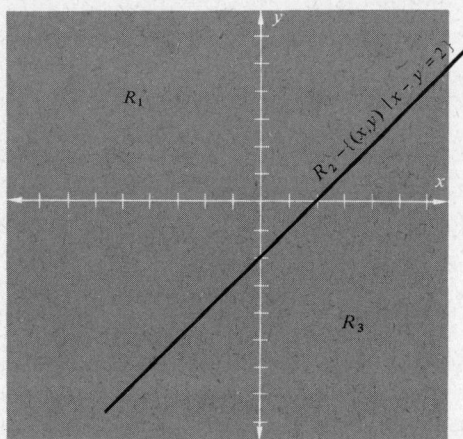

Figure 7.17

the coordinate plane into three distinct regions: the set of all points *above* the line (labeled R_1 on the graph), the set of all points *on* the line (R_2 on the graph) and the set of all points *below* the line (R_3 on the graph).

The straight line which contains all points in R_2 (i.e., all points whose coordinates satisfy the equation $x - y = 2$) divides the coordinate plane into two *half-planes*, which are labeled R_1 and R_3 in Figure 7.17. We could get an accurate premonition of things to come by pausing right here to observe that, since *all* points whose coordinates satisfy the equation $x - y = 2$ lie *on the line* (i.e. in R_2), then for *all* points in R_1 and R_3 one of two things must be true: either $x - y < 2$ or $x - y > 2$ at *every* point in these two regions.

To pursue that thought further, let us consider any one of the four inequalities previously listed as relatives of $x - y = 2$; for example,

$$x - y < 2$$

A *solution* of this inequality is an ordered pair of numbers (x_1, y_1) which satisfies the inequality; that is, $x_1 - y_1 < 2$. The solution *set* of this inequality is the set of *all* ordered pairs (x, y) which make the inequality true. We designate the set of all such pairs in the following manner:

$$\{(x, y) \mid x - y < 2\}$$

The *graph* of this inequality is the set of all points in the plane whose coordinates satisfy $x - y < 2$.

The relation of the solution sets of the inequalities $x - y < 2$ and $x - y > 2$ to the solution set of the equation $x - y = 2$ can be seen quite easily by writing all three in a slightly different form and then referring to the graph of the equation $x - y = 2$.

$$x - y < 2 \text{ is equivalent to } x < 2 + y$$

$$x - y = 2 \text{ is equivalent to } x = 2 + y$$

$$x - y > 2 \text{ is equivalent to } x > 2 + y$$

We shall select a point, P, on the line $x = 2 + y$ and through that point draw a line parallel to the x-axis. Note that at every point on the parallel line all values of y are the same. (See Figure 7.18.) Looking at the graph, we see that any point on the parallel line to the *left* of point P (at which $x = 2 + y$) will have an abscissa (x-value) *less than* $2 + y$, while any point on the parallel line to the *right* of point P will have an abscissa *greater than* $2 + y$. From the graph in Figure 7.18 we observe further that *no matter where* point P is located on the line $x = 2 + y$, any point to the left of P on a line parallel to the x-axis will lie in R_1 and have an abscissa *less than* $2 + y$; moreover any point to the right of P on a line parallel to the x-axis will lie in R_3 and have an abscissa *greater than* $2 + y$. Hence we see that all points in R_1 satisfy the inequality $x < 2 + y$ (or $x - y < 2$) and all points in R_3 satisfy the inequality $x > 2 + y$ (or $x - y > 2$).

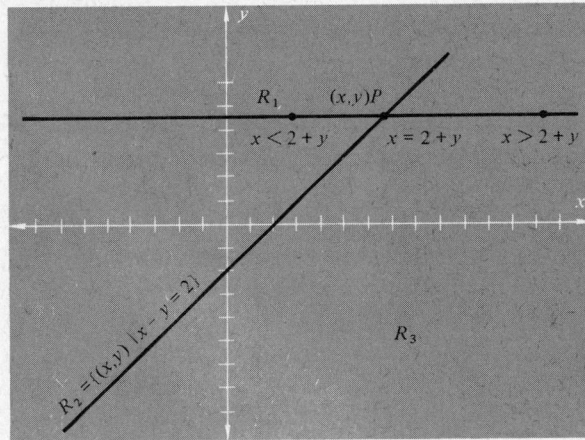

Figure 7.18

So the solution set of a first-degree inequality in two variables can be found by first graphing the line defined by the related *equation* and then deciding which one of the two half-planes separated by the line contains all the points whose coordinates satisfy the inequality. All such points lie on the *same* side of the line. Thus we have only to select one point that is *not* on the line and test its coordinates in the inequality. If the coordinates of the test point satisfy the inequality, then *all* points in that half-plane will also satisfy the inequality. If the coordinates of the test point do *not* satisfy the inequality, then all points in the other half-plane will satisfy it.

To graph the solution set of a particular inequality, we indicate the half-plane containing all points which satisfy the inequality by shading it with lines parallel or perpendicular to the line separating the half-planes. For example, let us graph the solution set of

$$2x - y < 3$$

1.   Graph the equation $2x - y = 3$ (Figure 7.19).

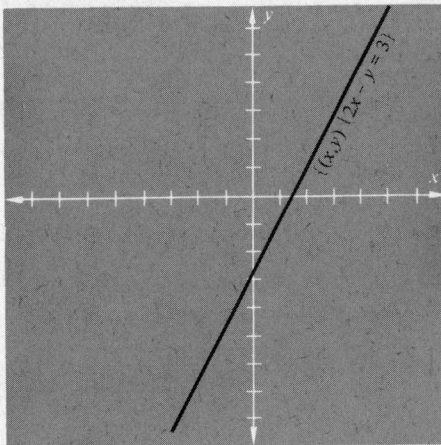

Figure 7.19

2. Select any point not on the line, for example (0,0), and test the coordinates in the inequality.

$$2x - y < 3$$

$$2(0) - 0 < 3$$

$$0 - 0 < 3 \qquad \text{True}$$

3. Then *all* points in the half-plane containing (0,0) satisfy the inequality, and we shade this region with lines to graph the solution set. (See Figure 7.20.)

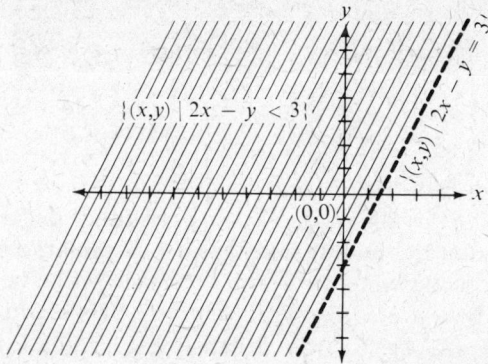

Note: The dotted line indicates that the points on the line, $2x - y = 3$, do not belong to the solution set of $2x - y < 3$.

Figure 7.20

Briefly summarizing the preceding discussion in symbolic language we have the following:

The straight line defined by a linear equation $ax + by = c$ divides the plane into two half-planes. One of the half-planes contains all points which satisfy the inequality $ax + by < c$, and the other half-plane contains all points which satisfy the inequality $ax + by > c$.

Note carefully: If either a or b equals zero in an equation of the type $ax + by = c$, the set of points indicated must be clearly stated in proper set language. Compare the graphs of the solution sets of the following inequalities:

A. $\{x \mid x > 3\}$ (See Section 7.3)

Figure 7.21

B. $\{(x,y) \mid x > 3\}$

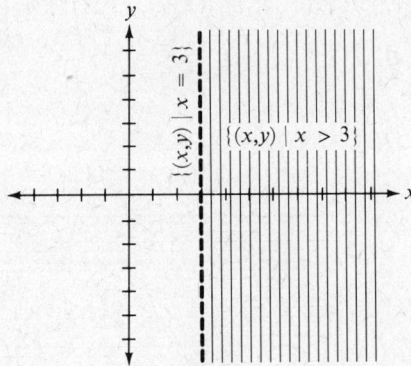

Figure 7.22

Examples:

(1)   Graph the solution set of $\{(x,y) \mid 3x + 2y < 6\}$ (see Figure 7.23).

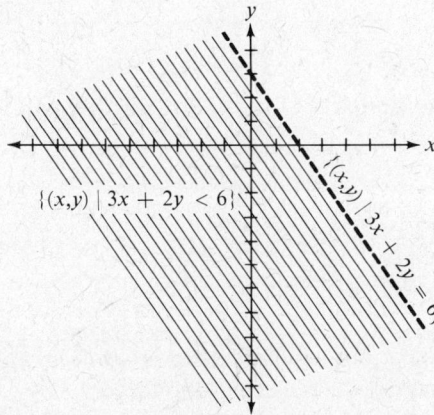

Figure 7.23

(2)   Graph the solution set of $\{(x,y) \mid 3x - y \geq 3\}$ (see Figure 7.24).

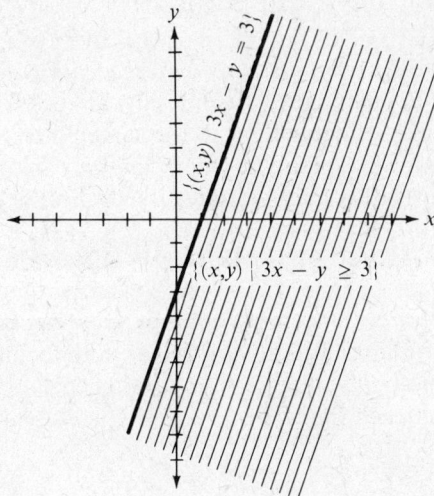

Figure 7.24

Note: The solid line indicates that the points on the line are part of the solution set of (x,y) $3x - y \geq 3$.

(3)   Graph the solution set of $\{(x,y) \mid y < 2\}$ (see Figure 7.25).

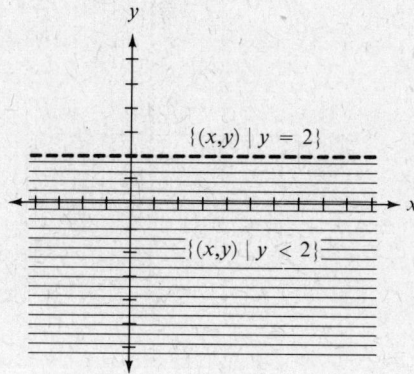

Figure 7.25

Exercise 87:

Graph the solution sets of the following inequalities:

(1) $\{(x,y) \mid x + y < 4\}$

(2) $\{(x,y) \mid 2x - y \geq 3\}$

(3) $\{(x,y) \mid x > -2\}$

(4) $\{(x,y) \mid x - 2y < 4\}$

(5) $\{(x,y) \mid 3x - y \leq 6\}$

(6) $\{(x,y) \mid 5x - y > 2\}$

(7) $\{(x,y) \mid y \geq -3\}$

(8) $\{(x,y) \mid 3x + 2y < 6\}$

In many applications of mathematics we frequently need the solution set of a *system* of inequalities such as the following:

$$3x - 2y < 6$$
$$5x + 2y > 20$$

To graph the solution set of the above system we must locate all points in the plane whose coordinates satisfy *both* inequalities. Such points are easily located by graphing the solution set of each of the inequalities in the same plane; the solution set of the system is then that part of the plane in which the two solution sets intersect. Thus, the solution set of the system may be stated symbolically as follows:

$$\{(x,y) \mid 3x - 2y < 6\} \cap \{(x,y) \mid 5x + 2y > 20\}$$

An accurate graph of the solution set can be found by graphing the related lines and shading the appropriate half-planes. It is advisable (for both facility and accuracy) to graph the related lines by finding the intersection point of the lines algebraically and then plot one other point on each line. This procedure is illustrated below.

Graph the solution set of the system

$$\{(x,y) \mid 3x - 2y < 6\}$$
$$\{(x,y) \mid 5x + 2y > 20\}$$

The related lines are

$$3x - 2y = 6$$
$$5x + 2y = 20$$

Solving these simultaneously, we obtain

$$3x - 2y = 6$$
$$\underline{5x + 2y = 20}$$
$$8x = 26$$
$$x = \frac{26}{8}$$
$$x = \frac{13}{4} \qquad \text{(the abscissa of the intersection point)}$$

Substituting $x = \frac{13}{4}$ in the first equation, we obtain

$$3\left(\frac{13}{4}\right) - 2y = 6$$
$$\frac{39}{4} - 2y = 6$$
$$39 - 8y = 24$$
$$-8y = 24 - 39$$
$$-8y = -15$$
$$y = \frac{-15}{-8} = \frac{15}{8} \qquad \text{(the ordinate of the intersection point)}$$

$$\{(x,y) \mid 3x - 2y = 6\} = \left\{\left(\frac{13}{4}, \frac{15}{8}\right), (0,-3)\dots\right\}$$
$$\{(x,y) \mid 5x + 2y = 20\} = \left\{\left(\frac{13}{4}, \frac{15}{8}\right), (4,0)\dots\right\}$$

The graph of the solution set is shown in Figure 7.26.

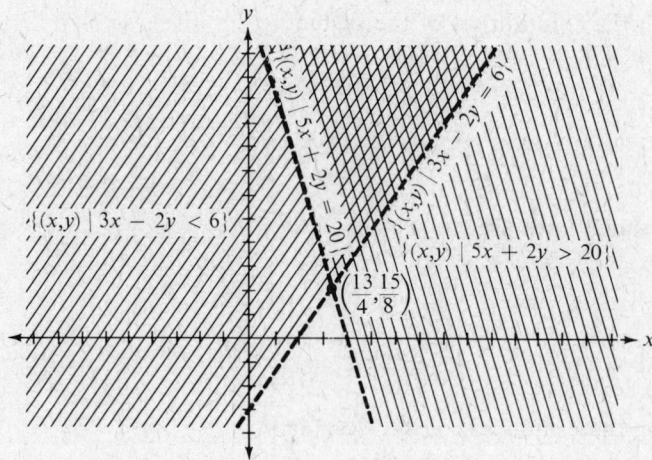

Figure 7.26

The solution set, $\{(x,y) \mid 3x - 2y < 6\} \cap \{(x,y) \mid 5x + 2y > 20\}$, is the set of points in the crosshatched section of the plane which lies between the two related lines above their intersection at the point $\left(\dfrac{13}{4}, \dfrac{15}{8}\right)$.

Examples:

(1) Graph the solution set of the system

$$\{(x,y) \mid y \leq 4\}$$
$$\{(x,y) \mid x + y > 0\}$$
(see Figure 7.27)

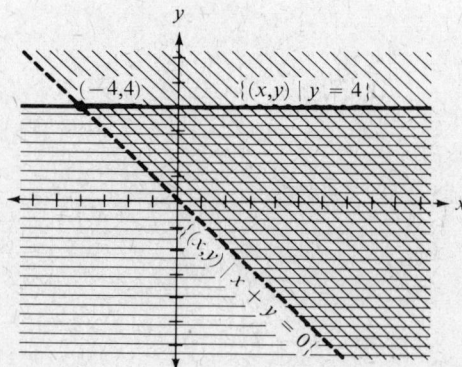

Figure 7.27

(2) Graph the solution set of the system

$$\{(x,y) \mid 2x - y > 3\}$$
$$\{(x,y) \mid x + 2y > 4\}$$
(see Figure 7.28)

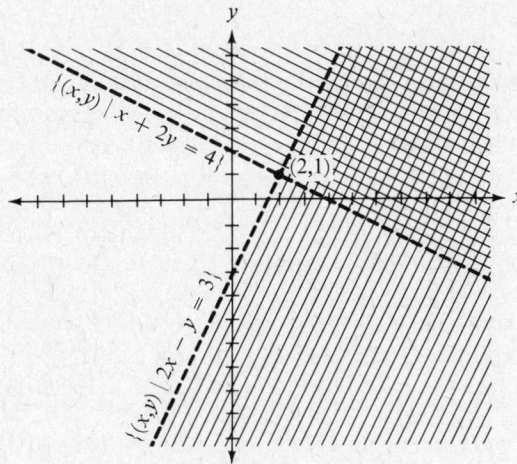

Figure 7.28

    (3)   Graph the solution set of the system

$$\{(x,y) \mid 2x - y \geq 3\}$$
$$\{(x,y) \mid x + 2y < 4\}$$
$$\{(x,y) \mid y > -3\}$$

(see Figure 7.29)

Note that the first two inequalities in this system have the same related lines as those in Example 2. On this graph we include the line $y = -3$ and then shade the appropriate half-planes. See Figure 7.29. The solution set of the system is the set of all points

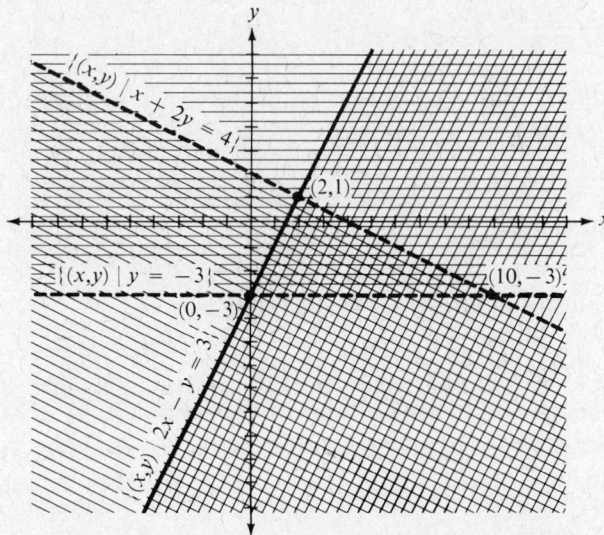

Figure 7.29

contained in the triangular section of the plane which has vertices at the points $(0,-3)$, $(2,1)$, and $(10,-3)$ and all points on the boundary of the triangle between $(0,-3)$ and $(2,1)$.

Exercise 88:

Graph the solution sets of the following systems of inequalities:

(1) $\{(x,y) \mid 2x + y < 8\}$
$\{(x,y) \mid y < 0\}$

(2) $\{(x,y) \mid x \le -2\}$
$\{(x,y) \mid 3x + 2y > 0\}$

(3) $\{(x,y) \mid x + y < 3\}$
$\{(x,y) \mid 2x - 3y > -4\}$

(4) $\{(x,y) \mid x + y \ge 4\}$
$\{(x,y) \mid 3x - y > 0\}$

(5) $\{(x,y) \mid 3x + 5y > 9\}$
$\{(x,y) \mid x - y < -5\}$

(6) $\{(x,y) \mid 3x + 5y \ge 9\}$
$\{(x,y) \mid x - y < -5\}$
$\{(x,y) \mid y \le 12\}$

(7) $\{(x,y) \mid x - y > 1\}$
$\{(x,y) \mid 3x + 2y \le 18\}$
$\{(x,y) \mid x < 4\}$

(8) $\{(x,y) \mid 3x - 2y > -6\}$
$\{(x,y) \mid 3x + y < 12\}$
$\{(x,y) \mid y \ge -12\}$

Chapter Test 7

I.   Graph and define by rule the solution set of each of the following inequalities:
1. $5x - 3 < 2x + 6$
2. $x + 4 \le 3x - 8$
3. $2(x - 1) > 3(x + 3)$
4. $\dfrac{x}{3} - \dfrac{5}{4} \ge \dfrac{3x}{4}$
5. $\dfrac{3x + 1}{2} - \dfrac{x + 1}{6} < \dfrac{x - 2}{3}$

II. Find the solution set of each of the following inequalities and show each set in a line graph:
1. $|x - 4| < 6$
2. $\left| x + \dfrac{1}{3} \right| \le \dfrac{4}{3}$
3. $\left| \dfrac{2x + 2}{5} \right| > 4$
4. $|3x - 2| \ge 7$

III. Graph the solution sets of the following:
1. $\{(x,y) \mid 2x - y < 6\}$
2. $\{(x,y) \mid x - 2y \ge 4\}$

IV. Graph the solution sets of the following systems of inequalities:
1. $\{(x,y) \mid 5x - 4y > -10\}$
$\{(x,y) \mid 5x + 3y < 25\}$
2. $\{(x,y) \mid 5x - 4y \le -10\}$
$\{(x,y) \mid 5x + 3y > 25\}$
$\{(x,y) \mid y < 10\}$

8
RATIO, PROPORTION, AND VARIATION

8.1
Ratio; The Metric System

When we speak of the "ratio" of one quantity to another, we are actually talking about a method of *comparing* two quantities. Such comparisons are useful in widely different ways that can range all the way from hunting a spouse to shopping for toothpaste, as the ensuing discussion will show.

Suppose, for example, that you are a coed who is trying to decide between two universities and that one of your objectives is an active social life involving men. How would you choose between University A with an enrollment of 2000 men and 1000 women and University B with an enrollment of 12,000 men and 10,000 women? Would you choose University B because it has 2000 more men than women rather than University A with only 1000 more men than women? Not if you think it over carefully. Instead of making a comparison by subtraction (i.e., the difference between the numbers of men and women), you would be much better off making a comparison by *division*: comparing men to women in each university by division, we have

$$\text{University } A: \frac{2000}{1000} = 2 \qquad \text{University } B: \frac{12,000}{10,000} = 1.2$$

Thus we see that at University A there are 2 men for each woman while at University B there are only 1.2 men (mathematically speaking) for each woman. Speaking more realistically, at University A there are 10 men for every 5 women while at University B there are 6 men for every 5 women. Obviously, the *thinking* coed with an active social life involving men in mind would head for University A.

Seriously, this way of comparing two numbers by division has many useful applications The resulting number is called a *ratio:*

> The ratio of a number a to a number b, $b \neq 0$, is the quotient $\dfrac{a}{b}$.

In other words, the ratio of two numbers is simply *the first number named divided by the second.*

$$\text{The } \textit{ratio} \text{ of 3 to 5 is } \frac{3}{5}$$

$$\text{The } \textit{ratio} \text{ of 5 to 3 is } \frac{5}{3}$$

First, let us consider the comparison of numbers of like nature. Following are some examples:

Examples:

(1)   A gasoline tank of a car having a capacity of 60 liters has 45 liters in it. Hence the tank is

$$\frac{45}{60} \text{ or } \frac{3}{4} \text{ full}$$

(2)   Suppose we wish to compute the ratio of one hour to 30 minutes. To do so, we must first express the quantities in the *same* units. Since

$$1 \text{ hour } = 60 \text{ minutes}$$

the ratio we wish to express is

$$\frac{60}{30} = 2$$

(3)   Suppose we wish to compute the ratio of one dime to one quarter. Since

$$1 \text{ dime } = 10 \text{ cents and } 1 \text{ quarter } = 25 \text{ cents}$$

the desired ratio is

$$\frac{10}{25} = \frac{2}{5}$$

The preceding examples illustrate the fact that, for simplicity and convenience in computation, we express ratios in their lowest terms. When mixed numbers are involved, such as 4 and $\frac{2}{5}$ inches, we write the number of inches as $4\frac{2}{5}$. Since $4\frac{2}{5}$ really means $4 + \frac{2}{5}$, it also means $\frac{20}{5} + \frac{2}{5}$ or $\frac{22}{5}$. The latter form, $\frac{22}{5}$, is the most convenient way to write this number in performing arithmetic computations. Note the following examples carefully:

(4)   Compute the ratio of $2\frac{3}{4}$ inches to $6\frac{1}{2}$ inches.

$$2\frac{3}{4} = 2 + \frac{3}{4} = \frac{8}{4} + \frac{3}{4} = \frac{11}{4}$$

$$6\frac{1}{2} = 6 + \frac{1}{2} = \frac{12}{2} + \frac{1}{2} = \frac{13}{2}$$

The ratio is

$$\frac{2\frac{3}{4}}{6\frac{1}{2}} = \frac{\frac{11}{4}}{\frac{13}{2}} = \frac{11}{4} \div \frac{13}{2} = \frac{11}{\cancel{4}_2} \cdot \frac{\cancel{2}}{13} = \frac{11}{26}$$

(5)   Compute the ratio of 3.2 ounces to 16 ounces.

$$1 \text{ pound } = 16 \text{ ounces}$$

The ratio is

$$\frac{3.2}{16} = \frac{3.2}{16} \cdot \frac{10}{10} = \frac{32}{160} = \frac{(\cancel{32})(1)}{(\cancel{32})(5)} = \frac{1}{5}$$

Here we shall digress briefly from the discussion of ratios to observe that the units compared in the last two examples — inches and ounces — belong to the English system of measurement, which is probably the system most familiar to you. However, in physics, chemistry, and related medical sciences, the *metric* system is used almost exclusively. While the metric system may not be quite so familiar to you, it is actually simpler because it is a system in which all units are expressed in multiples of ten.

For convenient reference, a short table of the metric system is included at the end of this section. Here we note briefly the basic units of the system and some of the more common prefixes used with those units.

1.   The unit of length is the *meter*. A meter is slightly longer than a yard; one meter = 39.37 in. The abbreviation for meter is the letter *m*.

2.   The unit of weight is the *gram*. A gram is much smaller than an ounce; one gram = .035 ounce. The abbreviation for gram is the letter *g*.

3.   The unit of volume is the *liter*. A liter and a quart are almost the same size; one liter = 1.057 quart. The abbreviation for liter is the letter *l*.

In the metric system these basic units are used with the same prefixes to denote divisions and multiplications by powers of ten. The most common prefixes are the following:

1.   *milli* (abbreviated by "m") means one-thousandth.

1 millimeter (1 mm) means $\frac{1}{1000}$ of a meter

1 milligram (1 mg) means $\frac{1}{1000}$ of a gram

1 milliliter (1 ml) means $\frac{1}{1000}$ of a liter

2. *centi* (abbreviated by "c") means one-hundredth.

1 centimeter (1 cm) means $\frac{1}{100}$ of a meter

1 centigram (1 cg) means $\frac{1}{100}$ of a gram

1 centiliter (1 cl) means $\frac{1}{100}$ of a liter

3. *kilo* (abbreviated by "k") means one thousand.
    1 kilometer (1 km) means 1000 meters
    1 kilogram (1 kg) means 1000 grams
    1 kiloliter (1 kl) means 1000 liters

When meters are multiplied by meters the result is expressed as meters squared or m². Similarly, centimeters times centimeters times centimeters produces cubic centimeters or centimeters cubed. It is useful to remember that one liter is equal to 1000 cubic centimeters; this is written as 1000 cm³ or 1000 cc. In this discussion we shall use the abbreviation cm³ for cubic centimeters.

Continuing our discussion of ratios, suppose that we wish to compute the ratio of 50 grams to one kilogram. We proceed as follows:

$$1 \text{ kilogram} = 1000 \text{ grams}$$

The ratio is

$$\frac{50}{1000} = \frac{5}{100}$$

Now, observe carefully that in *all* of the preceding examples of computing ratios, we compared *like* quantities in the *same* units; i.e., we compared liters to liters, minutes to minutes, cents to cents, inches to inches, ounces to ounces, and grams to grams. Such ratios produce pure or *dimensionless* numbers, and *the ratios remain the same* regardless of the type of units in which the quantities are expressed. To illustrate:

The ratio of 80 centimeters to 200 centimeters is

$$\frac{80}{200} = \frac{2\cancel{(40)}}{5\cancel{(40)}} = \frac{2}{5}$$

If we change centimeters to meters $\left(\text{one centimeter} = \frac{1}{100} \text{ of a meter}\right)$, we have

80 centimeters $= \frac{80}{100}$ or $\frac{4}{5}$ of a meter, and 200 centimeters $= \frac{200}{100}$ or 2 meters.

Comparing these same lengths in *meters*, we have $\dfrac{\frac{4}{5}}{2} = \frac{4}{5} \div 2 = \frac{\overset{2}{\cancel{4}}}{5} \cdot \frac{1}{\underset{1}{\cancel{2}}} = \frac{2}{5}$ which is precisely the ratio we obtained by comparing these lengths in centimeters.

Thus a ratio which produces a pure (dimensionless) number has the advantage of being valid for any system of measurement.

Exercise 89:

(1)   Express the following ratios as fractions in lowest terms:

  (a)   The ratio of 2 to 3

  (b)   The ratio of 3 to 2

  (c)   The ratio of 7 to 10

  (d)   The ratio of 12 to 5

  (e)   The ratio of $\frac{1}{2}$ to $\frac{3}{5}$

  (f)   The ratio of 1.7 to .43

(2)   Convert all quantities to the same units and express the following ratios as fractions in lowest terms:

  (a)   The ratio of 5 cents to 2 dimes

  (b)   The ratio of 40 seconds to 1 minute

  (c)   The ratio of 600 liters to 1 kiloliter

  (d)   The ratio of 75 centimeters to $1\frac{1}{2}$ meters

  (e)   The ratio of 1 meter and 30 centimeters to 3 meters

  (f)   The ratio of 500 grams to 2 kilograms

  (g)   The ratio of 85 centigrams to 40 grams

  (h)   The ratio of 750 meters to 2 kilometers

  (i)   The ratio of 16 minutes to $1\frac{1}{3}$ hours

  (j)   The ratio of 550 milliliters to one liter

As noted earlier, in the foregoing discussion of ratios we compared only like quantities in the same units. It is frequently useful, however, to compare quantities of a *different* nature. For example, if a car is driven 100 kilometers in 2 hours, the ratio of the distance traveled to the time elapsed in traveling is

$$\frac{100}{2} = 50 \text{ kilometers per hour}$$

Since there is no way that the unit "kilometer" can be expressed in terms of the unit "hour," we are stuck with the phrase, "kilometers per hour," which is called the *dimension* of the ratio.

Similarly, if one places 18 bananas in a cage with 3 monkeys, the ratio of bananas to monkeys *expressed in the proper dimension is*

$$\frac{18}{3} = 6 \text{ bananas per monkey}$$

The monkeys may not work out the distribution in exactly this way if any of them is overly aggressive, but all monkey business aside, the ratio given is mathematically accurate and correctly expressed in the dimension of "bananas per monkey." In general

---

If a and b, $b \neq 0$, represent numbers in different units, $\dfrac{a}{b}$ *has the dimension* of "units of a per units of b."

Examples:

1. A 50-gram can of pineapple juice costs 20 cents. The ratio of cost to volume of the pineapple juice is

$$\frac{20}{50} \text{ cents per gram} = \frac{2}{5} \text{ cents per gram} = .4 \text{ cents per gram}$$

Observe that the value of the ratio is the number of cents which each gram of the pineapple juice costs.

We use ratios of this type to express what is called the "density" of a substance. One would agree intuitively that molasses is "heavier" than gasoline; i.e., a cup filled with molasses would weigh more than the same cup filled with gasoline. Note that we are considering the *same volume* (namely, one cupful) of each liquid; obviously a barrel of gasoline would weigh more than a cupful of molasses. Thus, to describe the "heaviness" of different substances sensibly we take the *ratio of units of weight to units of volume* for each substance. This ratio is called the *density* of the substance, and, correctly, we should say that molasses is "denser" than gasoline.

For example, it can be demonstrated that 54.2 grams of mercury fill a 4-cm³ container; so the density of mercury is given by the ratio

$$\frac{54.2}{4} = 13.55 \text{ grams per cm}^3$$

Note that the value of the ratio is the number of grams which *each* cubic centimeter of mercury weighs. We shall henceforth use the letter g to represent grams.

Examples:

(1) 3.95 grams of alcohol fill a 5-cm³ container. What is the density of alcohol?

$$\frac{3.95}{5} = .79 \text{ g per cm}^3$$

(2)   There are 12.7 centimeters in 5 inches. The ratio of centimeters to inches is

$$\frac{12.7}{5} = 2.54 \text{ cm per in.}$$

(3)   In the game of bridge, 52 cards are dealt to 4 players. The ratio of cards to players is

$$\frac{52}{4} = 13 \text{ cards per player}$$

(4)   There are 11 pounds of candy in a 5-kilogram assortment. The ratio of pounds to kilogram is

$$\frac{11}{5} = 2.2 \text{ pounds per kilogram}$$

(5)   One and a half grams of boric acid are dissolved in 40 cm³ of water. The ratio of boric acid to water is

$$\frac{1\frac{1}{2}}{40} = \frac{\frac{3}{2}}{40} = \frac{3}{2}\cdot\frac{1}{40} = \frac{3}{80} = 0.0375 \text{ gram (of boric acid) per cm}^3 \text{ (of water)}$$

(6)   If, in the solution shown in Example 5, enough water is added to make a 50-cm³ solution, the ratio of boric acid to the total amount of *solution* is

$$\frac{1\frac{1}{2}}{50} = \frac{\frac{3}{2}}{50} = \frac{3}{2}\cdot\frac{1}{50} = \frac{3}{100} = 0.03 \text{ gram (of boric acid) per cm}^3 \text{ (of solution)}$$

(7)   The regular size of Super Sparkle toothpaste costs 67 cents while the giant economy size costs \$1.87. If the regular size contains $16\frac{3}{4}$ grams and the giant economy size contains $42\frac{1}{2}$ grams, which is the better buy?

Compute the ratio of cost to quantity in each case.

Regular size:

$$\frac{67}{16.75} = \frac{6700}{1675} = 4 \text{ cents per gram}$$

Giant economy size:

$$\frac{1.87}{42.5} \text{ dollars per gram} = \frac{187}{42.5} \text{ cents per gram}$$

$$= \frac{1870}{425} = 4.4 \text{ cents per gram}$$

Since both ratios were computed in terms of cents per gram, we can compare them to conclude that the best buy, at 4 cents per gram, is the regular size.

Exercise 90:

Express the following ratios as fractions in lowest terms with the proper dimension:

(1)   The ratio of 12 crackers to 4 parrots

(2)   The ratio of 28 people to 7 cars

(3)   The ratio of 80 cents to 16 grams

(4)   The ratio of 15 cents to 32 ounces

(5)   The ratio of $1.08 to 2.7 liters

(6)   The ratio of 94 kilometers to 2 hours

(7)   The ratio of 42 cents to 1.5 meters

(8)   The ratio of 81 kilometers to $1\frac{1}{2}$ hours

(9)   If a $24\frac{1}{4}$-gram tube of toothpaste costs 97 cents, how much does each gram of toothpaste in that tube cost?

(10)   If 40.4 liters of gasoline cost $110.10, what is the price of each liter of gasoline?

(11)   A tourist driving in Mexico observed a road sign giving the distance to the next town as 25 kilometers. He clocked the mileage on his American-made car and found that he drove $15\frac{1}{2}$ miles to reach the town. How many miles are there in each kilometer?

(12)   If 50 tablets of a vitamin preparation contain 12,500 mg of Vitamin C, how much Vitamin C is contained in each tablet?

(13)   3.3 grams of gasoline fill a 5cc container. What is the density of gasoline?

(14)   What is the density of a medication if 3 grams of it fill a 2cc syringe?

Table 8.1

Metric System	Conversion Factors
Length	
1 millimeter (mm) = .001 meter (m) 1 centimeter (cm) = .01 meter 1 decimeter (dm) = .1 meter 1 kilometer (km) = 1000 meters	1 meter = $\begin{cases} 39.37 \text{ inches} \\ 3.28 \text{ feet} \\ 1.09 \text{ yards} \end{cases}$ 1 kilometer = .62 mile

Weight

1 milligram (mg) = .001 gram (g)
1 centigram (cg) = .01 gram
1 decigram (dg) = .1 gram
1 kilogram (kg) = 1000 grams

$$1 \text{ gram} = \begin{cases} .035 \text{ ounces} \\ .0022 \text{ pounds} \end{cases}$$

1 kilogram = 2.2 pounds

Volume

1 milliliter (ml) = .001 liter (l)
1 centiliter (cl) = .01 liter
1 deciliter (dl) = .1 liter
1 kiloliter (kl) = 1000 liters

$$1 \text{ liter} = \begin{cases} 1.06 \text{ quarts} \\ .264 \text{ gallon} \end{cases}$$

$$(1 \text{ liter} = 1000 \text{ cm}^3 = 61.02 \text{ in.}^3 = .0353 \text{ ft}^3)$$

8.2
Proportion; Conversion of Units

In applications of mathematics, problems frequently arise which involve the *equality* of two ratios. For example, the statement

$$\frac{1}{2} = \frac{3}{6}$$

tells us that $\frac{1}{2}$ and $\frac{3}{6}$ represent the same quantity. Please note that it is the *ratios* of the numbers which are equal, not the individual numbers themselves. In like manner

$$\frac{3}{5} = \frac{21}{35}$$

and

$$\frac{2}{3} = \frac{8}{12}$$

In mathematics a statement that two ratios are equal is called a *proportion*. Thus a proportion is simply a special kind of *equation*, namely, an equation that has exactly one fraction on each side of the equal mark.

If a, b, c, d are numbers and $b \neq 0$, $d \neq 0$, then $\frac{a}{b} = \frac{c}{d}$ is a *proportion*.

In the proportion

$$\frac{a}{b} = \frac{c}{d}$$

the numbers a and d are called the *extremes* and the numbers b and c are called the *means* of the proportion. For example, in the proportion

$$\frac{3}{5} = \frac{9}{15},$$

3 and 15 are the extremes while 5 and 9 are the means. Note that

$$3 \cdot 15 = 45 \text{ and } 5 \cdot 9 = 45$$

so that the *product of the means is equal to the product of the extremes* in this example. That is not an accident but is true in general. For, if

$$\frac{a}{b} = \frac{c}{d} \qquad (b, d \neq 0)$$

then
$$\frac{a}{\not b} \cdot \frac{\not b d}{1} = \frac{c}{\not d} \cdot \frac{b \not d}{1} \qquad E_m$$

and
$$ad = bc$$

Thus in every proportion, the product of the means is equal to the product of the extremes. This fact provides us with a handy method of evaluating a variable in an equation which is a proportion. Consider the following examples:

(1) $\quad \dfrac{2}{x} = \dfrac{1}{5}$

$\qquad (2)(5) = (x)(1)$

$\qquad (x)(1) = (2)(5)$

$\qquad\quad\ x = 10$

(2) $\quad \dfrac{x}{8} = \dfrac{5}{2}$

$\qquad (2)(x) = (8)(5)$

$\qquad\quad 2x = 40$

$\qquad\quad\ x = 20$

(3) $\quad \dfrac{4}{9} = \dfrac{x}{36}$

$\qquad (4)(36) = (9)(x)$

$\qquad\quad 144 = 9x$

$\qquad\quad 9x = 144$

$\qquad\quad\ x = \dfrac{144}{9}$

$\qquad\quad\ x = 16$

The proportion

$$\frac{a}{b} = \frac{c}{d}$$

can be read, "*a* is to *b* as *c* is to *d*." So, since

$$\frac{3}{5} = \frac{9}{15}$$

we say, "3 is to 5 as 9 is to 15."

Examples:

(1) Find x if 33 is to x as 11 is to 3.

$$\frac{33}{x} = \frac{11}{3}$$
$$(3)(33) = 11x$$
$$11x = 99$$
$$x = 9$$

(2) Find x if $x + 2$ is to 4 as 5 is to 7.

$$\frac{x + 2}{4} = \frac{5}{7}$$
$$7(x + 2) = (4)(5)$$
$$7x + 14 = 20$$
$$7x = 6$$
$$x = \frac{6}{7}$$

Conversion of Units   Proportions provide us with a simple method of changing units of measurements (grams, ounces, meters, inches, liters, etc.) from one system to another one. Suppose that we wish to convert a measure of 8 pounds to its equivalent weight in kilograms. From Table 8.1 at the end of Section 8.1 we note that 1 kilogram = 2.2 pounds; let x kilograms = 8 pounds and then compare *kilograms* to *pounds* as follows:

  1 (kilogram) is to 2.2 (pounds) as x (kilograms) is to 8 (pounds)

or $$\frac{1}{2.2} = \frac{x}{8}$$

then $(2.2)(x) = (1)(8)$ (The product of the means = the product of the extremes)

or $2.2x = 8$

and $22x = 80$

$$x = \frac{80}{22}$$

$$x = 3.64 \text{ kilograms (in 8 pounds)}$$

Note: When converting units, we frequently encounter decimals which continue indefinitely. In such cases we cooperate with reality by "rounding off" the resulting number to that degree of accuracy compatible with the accuracy of the information we are using. The preceding example produced, for the value of x, the number $3.636\overline{363}$, and the number was rounded off to 3.64. In the examples and problems which follow, we shall round off all results to three figures, *not including* zeros whose only purpose is to place the decimal point in the number. (To do this, drop the remaining digits and increase the last digit in the result by 1 if the number which followed it was equal to or greater than 5.) This is a purely arbitrary arrangement and also a temporary one for our convenience. In actual practice, results of computations are always rounded off according to the accuracy of the numbers used in the computations.

Examples:

(1) Convert 6 ounces to grams.
In Table 8.1 we find that 1 gram = .035 ounces.
Let x grams = 6 ounces; then compare *grams* to *ounces* as follows:

1 (gram) is to .035 (ounces) as x (grams) is to 6 (ounces)

or $\qquad \dfrac{1}{.035} = \dfrac{x}{6}$

or $\qquad .035x = 6 \qquad$ (The product of the means = the product of the extremes.)

and $\qquad 35x = 6000$

$\qquad x = \dfrac{6000}{35}$

$\qquad x = 171$ grams (in 6 ounces)

(2) Convert 5 quarts to liters.
1 liter = 1.06 quarts   (From Table 8.1)
Let x liters = 5 quarts; then
1 is to 1.06 as x is to 5

or $\qquad \dfrac{1}{1.06} = \dfrac{x}{5}$

$\qquad 1.06x = 5$

$\qquad 106x = 500$

$\qquad x = \dfrac{500}{106}$

$\qquad x = 4.72$ liters (in 5 quarts)

(3) Convert 5 quarts to centiliters.
From the preceding example we have 5 quarts = 4.72 liters.
There are 100 centiliters in every liter, so

4.72 liters = (4.72)(100) centiliters

= 472 centiliters (in 5 quarts)

(4) Convert 5 miles to kilometers.

1 kilometer = .62 mile   (From Table 8.1)

Let x kilometers = 5 miles; then

1 is to .62 as x is to 5

or $\dfrac{1}{.62} = \dfrac{x}{5}$

$.62x = 5$

$62x = 500$

$x = 8.06$ kilometers (in 5 miles)

Observe that in the five examples preceding we have in every case converted units in the English system to their equivalent measures in the metric system. We may, however, wish to convert metric units to English units, and, using proportions, we can do that just as easily. Note the following examples:

(5) Convert 50 kilometers to miles.
Let x represent the number of miles in 50 kilometers; then

1 kilometer = .62 miles   (From Table 8.1)
50 kilometers = x miles

Comparing kilometers to miles, we have

1 is to .62 as 50 is to x

or $\dfrac{1}{.62} = \dfrac{50}{x}$

$(1)(x) = (.62)(50)$

$x = 31.0$ miles (in 50 kilometers)

(6) Convert 3 meters to feet.
Let x represent the number of feet in 3 meters; then

meter = 3.28 feet   (From Table 8.1)
3 meters = x feet

Comparing meters to feet, we have

1 is to 3.28 as 3 is to x

or $\qquad \dfrac{1}{3.28} = \dfrac{3}{x}$

$\qquad (1)(x) = 3(3.28)$

$\qquad x = 9.84$ feet (in 3 meters)

To understand fully the efficiency and generality of the use of proportions in conversion of units, you should pause here and reflect on the fact that this method of converting units from one system of measurement to another would be valid even if we were dealing with completely alien systems. To illustrate that fact, suppose that an intelligent form of life has been discovered on some distant planet named Ott, and that, for reasons best known to the Ottians, they measure lengths in two different types of units which they call glogs and snicks. Suppose further that we have unearthed (unotted?) the information that there are 15 glogs in every 6 snicks. How many glogs would equal 44 snicks?

We proceed as follows:

$$15 \text{ glogs} = 6 \text{ snicks}$$

$$\text{let } x \text{ glogs} = 44 \text{ snicks}$$

Comparing glogs to snicks we have

15 is to 6 as x is to 44, or

$\qquad \dfrac{15}{6} = \dfrac{x}{44} \qquad\qquad\qquad Note: \left(\dfrac{15}{6} = \dfrac{\cancel{3}(5)}{\cancel{3}(2)} = \dfrac{5}{2} \right)$

so $\qquad \dfrac{5}{2} = \dfrac{x}{44}$

and $\qquad 2x = 5(44)$

$\qquad 2x = 220$

$\qquad x = 110$ glogs in 44 snicks

Exercise 91:

(1) Find the value of x in each proportion

(a) x is to 12 as 5 is to 24

(b) 3 is to x as 5 is to 2

(c) 4 is to 1 as x is to 5

(d) 1 is to 2 as 4 is to x

(e) x is to 3 as $x + 2$ is to 6

(f) x is to $x + 1$ as 9 is to 12

(2) Convert the following measurements to the required units.

(a) 2.6 ounces equal how many grams? (See Example 1)

(b) 3 quarts equal how many liters? (See Example 2)

(c) 4 ounces equal how many centigrams? (See Example 2)

(d) 12 pounds equal how many kilograms? (See first paragraph under Section 8.2.)

(e) 2 quarts equal how many centiliters? (See Examples 2 and 3)

(f) 12 miles equal how many kilometers? (See Example 4)

(g) 5 pints equal how many liters?

(h) 9 ounces equal how many grams? How many kilograms? How many centigrams?

(i) 5 inches equal how many centimeters?

(j) 32.2 feet equal how many meters?

(k) 20 miles equal how many kilometers?

(l) 5 meters equal how many feet? (See Example 6)

(m) 35 kilometers equal how many miles? (See Example 5)

(n) 14 grams equal how many ounces?

Some types of functions appear so often in applications of mathematics that a special terminology has been developed for them. Three types which we shall consider are called *direct variation*, *inverse variation*, and *joint variation*.

8.3
Direct Variation

When two quantities are related so that one of them is always equal to a nonzero constant times the other one, we say that one *varies directly* as the other. If, for example, a car moves at a constant velocity of 50 kilometers an hour due north, the distance d in kilometers that the car travels in a specific number of hours t can be tabulated as follows:

t (hours)	1	2	3	4	5	6	etc.
d (kilometers)	50	100	150	200	250	300	etc.

The tabulated numbers immediately reflect the fact that as t increases, d also increases. Specifically, we should say that as t increases, the absolute value of d increases because, if this car were moving south instead of north (or started backing up at 50 kilometers an hour), the opposite directions are designated by positive and negative signs. The mathematically significant fact illustrated by the table is that d is always equal to 50 times the corresponding value of t, and the relation between d and t is defined by the function

$$d = 50t$$

where $f(t) = 50t$ and $d = f(t)$. We describe this situation by saying that d *varies directly* as t, and the number, 50, is called the *constant of variation*.

Functions of this type occur in abundance, not only in natural laws (such as the law of gravitation) but also in human activities that can include such widely different things as baking a cake and running a business. The gift-shop owner (see Section 6.7 who computed the selling price of an object as twice its cost was dealing in direct variation; i.e., the selling price varied directly as the cost and the constant of variation was 2. If you have ever worked by the hour, computed the circumference of a circle, invested money at simple interest, or even contemplated objects (rocks, apples, bottles, sky-divers) dropped from a height, then you have already encountered direct variation. Consider the following examples:

1. If a man works for $3.50 an hour, his wages, w, vary directly as the number of hours, t, that he works, or
    $$w = \$3.50t$$
    The constant of variation is 3.50.

2. The circumference, c, of a circle varies directly as the length of its diameter, d, or
    $$c = \pi d$$
    The constant of variation is π.

3. The interest, I, paid on a sum of money invested at 6% simple interest varies directly as the amount, P, invested, or
    $$I = .06P$$
    The constant of variation is .06.

4. It has been demonstrated that, neglecting air resistance, a rock dropped from the top of a cliff will fall with a velocity that increases at the rate of 9.8 meters per second each second, or
    $$V = 9.8t$$
    Thus the velocity varies directly as the time, and the constant of variation is 9.8.

5. The formula for computing the volume, V, of a sphere of radius, r, is

$$V = \frac{4}{3}\pi r^3$$

The volume varies directly as the cube of the radius, and the constant of variation is $\frac{4}{3}\pi$.

All of the above examples are alike in one way: in each one a quantity is equal to a nonzero *constant* times another quantity. Also these examples give rise to functions whose domains are determined by the nature of the problem. In Examples 2, 3, and 5, the domain of each function is the set of positive numbers, although we could take the view, in Example 3 for instance, that no money invested means that no interest is earned, or when $P = 0, I = 0$. Similarly in Examples 1 and 4, it would make sense for the variable, t, to equal zero; e.g., if a man works no time at all, then he collects no wages at all. From a purely technical viewpoint, *every* function defined in the foregoing examples is satisfied by the ordered pair (0,0), and we shall restrict our use of the terminology "direct variation" to have this meaning. That is, zero may or may not be in the domain of the function, but when zero *is* in the domain, the value of the function that corresponds to it is *always* zero also. Thus a function of the type

$$y = 5x$$

is an example of direct variation, but one of the type

$$y = 5x + 6$$

is not.

Problems involving variation occur frequently in the study of physics, chemistry, engineering, and business mathematics, in which cases the domain is obvious from the nature of the problem. We shall concentrate here on the terminology for such problems and generally leave the domains unspecified.

Summarizing the foregoing discussion of direct variation in symbolic language, we will choose the letter k to represent the constant of variation and the letters x and y to represent variables.

A variable y is said to *vary directly* as a variable x if there is a fixed, nonzero real number, k, such that
$$y = kx$$
k is called the *constant of variation*.

Similarly, if y varies directly as x^n, $n \in N$, there is a nonzero constant, k, such that $y = kx^n$.

For example, if $y = 3x$, then y varies directly as x and the constant of variation is 3. There is nothing magic about the symbols x and y. If, for instance, we know that

$$s = \sqrt{2}t$$

then s varies directly as t and the constant of variation is $\sqrt{2}$.

Note that if
$$s = \sqrt{2}\,t$$

then $\quad \sqrt{2}\,t = s$

and $\quad t = \dfrac{1}{\sqrt{2}}s$

So we could also say that t varies directly as s and the constant of variation is $\dfrac{1}{\sqrt{2}}$.

In general, *if y varies directly as x, then x varies directly as y* because, if
$$y = kx$$

then $\quad kx = y$

and $\quad x = \dfrac{1}{k}y$

where $\dfrac{1}{k}$ is certainly a nonzero constant if k is.

Now let us consider an interesting characteristic of the function $y = kx$. A subset of the ordered pairs defined by this function can be expressed symbolically by using subscripts; e.g., let x_1 (read "x sub-one") represent a specific nonzero value of x in $y = kx$, and let x_2 represent another specific nonzero value of x. The *corresponding values* of y in $y = kx$ can be labeled y_1 and y_2 respectively; i.e., when $x = x_1$, in $y = kx$, then $y = y_1$ and when $x = x_2$, $y = y_2$. Substituting these ordered pairs in the equation we obtain

1. $y_1 = kx_1$ and $y_2 = kx_2$
   Dividing the first equation by x_1 and the second by x_2, we have

2. $\dfrac{y_1}{x_1} = k$ and $\dfrac{y_2}{x_2} = k$

   Since $\dfrac{y_1}{x_1}$ and $\dfrac{y_2}{x_2}$ are *both* equal to k, by E_t, they are equal to each other

   Therefore

3. $\dfrac{y_1}{x_1} = \dfrac{y_2}{x_2}$

The last equation is the special type that is called a proportion. From Statement 3 we note that, since x_1 and x_2 represent *any* distinct nonzero values of x, the ordered pairs defined by $y = kx$ are pairs of numbers (x,y) in which the *ratios* of y to x in each pair $(x \neq 0)$ are *always equal*. When the ratio of two variables remains constant, we say the variables are *proportional*. The fact that these variables are proportional has given rise to the terminology that, if y varies directly as x (i.e., $y = kx$), then y is *directly proportional* to x, and the constant of variation is also called the *constant of proportionality*. Similarly, if y varies directly as x^n, then y is directly proportional to x^n. To illustrate:

$$y = 3x$$

Let $x_1 = 3$ $x_2 = 4$ $x_3 = 5$ $x_4 = 6$

Then $y_1 = 9$ $y_2 = 12$ $y_3 = 15$ $y_4 = 18$

$\dfrac{y_1}{x_1} = \dfrac{9}{3} = 3$ $\dfrac{y_2}{x_2} = \dfrac{12}{4} = 3$ $\dfrac{y_3}{x_3} = \dfrac{15}{5} = 3$ $\dfrac{y_4}{x_4} = \dfrac{18}{6} = 3$

Thus $\dfrac{y_1}{x_1} = \dfrac{y_2}{x_2} = \dfrac{y_3}{x_3} = \dfrac{y_4}{x_4} = 3$

The constant, 3, is called the constant of variation or the constant of proportionality.

Thus when two quantities are in direct variation, *the ratio of one to the other remains constant.* If $y = 7x$, then dividing both sides by x, $x \neq 0$, we obtain $\dfrac{y}{x} = 7$, which means that for all ordered pairs (x,y) satisfying this equation, the ratio of y to x is always 7. Similarly, if $y = 6x^2$, for $x \neq 0$, $\dfrac{y}{x^2} = 6$ for all ordered pairs (x,y) which satisfy the equation.

Following are examples of problems involving direct variation:

1. If y varies directly as x and the constant of variation is 5, what is y when $x = 7$?

    $y = kx$ and $k = 5$

    $y = 5x$ *or* $\dfrac{y}{x} = 5$

    When $x = 7$ When $x = 7$

    $y = 5 \cdot 7$ $\dfrac{y}{7} = 5$

    $y = 35$ $y = 35$

2. If u is directly proportional to v and the constant of proportionality is $-\dfrac{1}{2}$, what is v when $u = 3$?

    $u = kv$ and $k = -\dfrac{1}{2}$ *or* $\dfrac{u}{v} = -\dfrac{1}{2}$

    $u = -\dfrac{1}{2}v$ When $u = 3$

    $-\dfrac{1}{2}v = u$

    $v = -2u$ $\dfrac{3}{v} = -\dfrac{1}{2}$

    When $u = 3$ $-v = 6$ $\left(\begin{array}{l}\text{The product of the} \\ \text{means} = \text{the product} \\ \text{of the extremes}\end{array}\right)$

    $v = -2(3)$ $v = -6$

    $v = -6$

3. If r varies directly as s and $r = 3$ when $s = 2$, what is the constant of variation?

$$r = ks$$
$$3 = 2k$$
$$\frac{3}{2} = k$$
$$k = \frac{3}{2}$$

4. If y varies directly as x and $y = 2$ when $x = 5$, what is y when $x = 10$? *Note:* We must first compute the constant of variation.

$$y = kx$$
$$2 = 5k$$
$$\frac{2}{5} = k$$
$$k = \frac{2}{5}$$

So $y = \frac{2}{5}x$

When $x = 10$
$$y = \frac{2}{5} \cdot 10$$
$$y = 4$$

5. The distance that an object will fall freely from rest, neglecting air resistance, varies directly as the square of the time it falls. If an object falls 78.4 meters in 4 seconds, how far will it fall in 10 seconds?

First we choose symbols to represent the variables, distance and time.
Let s = the distance in meters
and t = the time in seconds
Since s varies directly as t^2, a constant, k, exists such that
$$s = kt^2$$
We compute k from the fact that $s = 78.4$ when $t = 4$.
$$78.4 = k(4)^2$$
$$78.4 = 16k$$
$$4.9 = k$$
So $s = 4.9t^2$
When $t = 10$
$$s = 4.9(10)^2$$
$$s = 490 \text{ meters} = \text{distance the object falls in 10 seconds}$$

6. The pressure at any point on the base of a container filled with liquid varies directly as the depth of the liquid. If the pressure exerted by a liquid on the base of a tank is 6 newtons per square meter when the depth is 2 meters, what is the pressure per square meter when the depth is 10 meters?*

*The newton is a unit of force in the metric system.

Let p = pressure and d = depth

$p = kd$

Since $p = 6$ when $d = 2$ we obtain

$6 = k(2)$ or $2k = 6$

$k = 3$

So $p = 3d$

When $d = 10$

$p = 3 \cdot 10$

$p = 30$ newtons per square meter

7. The amount of formaldehyde contained in a preservative solution called formalin is directly proportional to the amount of solution. If there are 30 grams of formaldehyde in 100 grams of formalin, how many grams of formalin would be needed to contain 75 grams of formaldehyde?

Let x grams = the amount of formalin

y grams = the amount of formaldehyde

$y = kx$

$30 = k(100)$

$.3 = k$

So $y = .3x$

When $y = 75$

$75 = .3x$

$750 = 3x$

$3x = 750$

$x = 250$ grams of formalin

Note: An alternate way to solve this problem is to express the relationship as a proportion.

$$\frac{y}{x} = \frac{30}{100}$$

When y is 75

$$\frac{75}{x} = \frac{30}{100}$$

The product of the means equals the product of the extremes

$30x = 7500$

$x = 250$ grams of formalin

Exercise 92:

(1) If y varies directly as x with the constant of variation $k = 10$, what is y when $x = 3$?

(2) If y is directly proportional to x^2 with the constant of proportionality $k = 2$, what is y when $x = 2$?

(3) If r varies directly as s and $r = 8$ when $s = 4$, what is the constant of variation?

(4) If y is directly proportional to x^3 and $y = 3$ when $x = 2$, what is the constant of proportionality?

(5) If y is directly proportional to x and $y = 2$ when $x = 3$, what is y when $x = 6$?

(6) If w varies directly as z^2 and $w = 18$ when $z = 3$, what is w when $z = 4$?

(7) The distance, d, required to brake a car to a stop varies directly as the *square* of the speed, s, of the car. If a car moving 64 kilometers an hour can be stopped in 27 meters, what distance is required to stop a car moving 96 kilometers an hour?

(8) The weight of a man, E, on the earth's surface is directly proportional to his weight, M, on the moon. If a man weighs 90 kilograms on the earth and 15 kilograms on the moon, what is the weight on the moon of a man who weighs 72 kilograms on the earth? What is your weight on the moon?

(9) The weight, w, of a fixed length of wire varies directly as the *square* of its diameter, d. If a coil of wire with a diameter of $\frac{1}{2}$ centimeter weighs 3 kilograms, what is the weight of a coil of wire of the same length with a diameter of $\frac{2}{3}$ centimeter?

(10) The number of cups of sugar, S, required to bake a layer cake varies directly as the number, n, of layers. If 5 cups of sugar are required for 2 layers, how much sugar is needed to bake a five-layer cake?

(11) The amount, D, of sulfur dioxide produced by burning sulfur in air is directly proportional to the amount of sulfur, S, burned. If 32 grams of sulfur produce 64 grams of sulfur dioxide, how many grams of sulfur dioxide will be produced from 70 grams of sulfur?

(12) At a fixed interest rate over a specific time, the present value, P, of an investment varies directly as the amount of money, A, originally invested. Mr. Jones found that he would have $5000 in five years by investing $3712.35 at a certain interest rate. His brother wants to have $8000 in five years by investing money at the same interest rate. How much should the brother invest?

(13) The resistance of a conductor to the flow of electricity is measured in units called ohms. For wire of fixed diameter, the resistance, R, is directly proportional to the length, L, of the wire. If the resistance of 30 meters of copper wire is .0641 ohms, what is the resistance of 260 meters of the wire?

(14) The pressure, p, on the hull of a submerged submarine varies directly as its depth, d, below the surface of the water. If the pressure is 146 kilograms per square meter at a depth of 15.2 meters, what is the pressure at a depth of 38 meters?

(15) The weight (in grams) of hydrogen, H, that will combine with oxygen varies directly as the weight (in grams) of the oxygen, O. If 3 grams of

hydrogen combine exactly with 24 grams of oxygen, how many grams of oxygen will combine with 16 grams of hydrogen?

(16) The number of machines, N, that a crew can assemble is directly proportional to the number of hours, t, that they work. If a crew can assemble 18 machines in 8 hours, how many hours must they work to assemble 100 machines?

(17) The air resistance, R, to an object moving at a high speed, S, varies directly as the square of the speed of the object. What is the ratio of the air resistance of a car moving 60 kilometers an hour to a car moving 80 kilometers an hour?

(18) The area, A, of a circle is directly proportional to the square of its radius, r. What is the ratio of the area of a circle with a radius of 4 centimeters to the area of a circle with a radius of 2 centimeters?

(19) Heat produced by the oxidation of protein in the human body is directly proportional to the amount of protein oxidized. If the oxidation of 8 grams of protein produces 32 Calories of heat, how much protein must be oxidized to produce 52 Calories of heat?

(20) In computing the deductions for his income tax return, Mr. Carter finds that in his income bracket, he is allowed to deduct $90.00 for his state sales tax. The state sales tax rate is 4%. If the city in which Mr. Carter lives charges 1% additional sales tax, what should be his total deduction for state and local sales tax?

(21) At a constant pressure the volume of a gas is directly proportional to its absolute temperature. If the temperature of 90 cm³ of gas is 273°, what will the volume of the gas be if the temperature is increased to 455° and the pressure remains the same?

(22) The amount of heat produced in the human body by the oxidation of fat varies directly as the amount of fat oxidized. If the oxidation of 5 grams of fat produces 45 Calories of heat, how much fat must be oxidized to produce 72 Calories of heat?

8.4
Inverse Variation; Joint Variation

Inverse Variation   In the discussion of direct variation we noted that, as the absolute value of the independent variable increased, the absolute value of the function also increased. In inverse variation the opposite is true; i.e., as the absolute value of the independent variable increases, the absolute value of the function decreases.

Consider, for example, some possible dimensions (length and width) of a rectangle whose area is 60 square meters. If the width of the rectangle is 1 meter, then the length is 60 meters; if the width is 2 meters, then the length is 30 meters. The following table shows corresponding values of such widths and lengths:

w (width in meters)	1	2	3	4	5	6
L (length in meters)	60	30	20	15	12	10

From the figures in the table it is clear that as w *increases*, L *decreases*. Furthermore every rectangle having the dimensions shown above will have an area of 60 square meters; i.e., in every ordered pair (w, L),

$$wL = 60$$

$$\text{or} \quad w = \frac{60}{L}$$

The last equation defines a function of L $\left(\text{namely, } \frac{60}{L}\right)$ in which the value of the function w is always equal to the *constant* 60 *divided by* L. We describe this situation by saying that w *varies inversely* as L, and the number 60 is called the constant of variation.

The same kind of relationship occurs when a volume of gas is subjected to increased pressure at the same temperature. When gas is enclosed in a sealed container, it distributes itself evenly about the interior and presses uniformly against the sides, top, and base of the container with equal force. This action is called the pressure exerted by the gas. It is measured in units of force per unit of area — such as pounds per square foot, dynes per square centimeter, or newtons per square meter.*

Figure 8.1 represents a sealed container with a top that can be moved up or down. Let us suppose that it contains 120 cubic meters of gas, which exerts a pressure against the container of 40 newtons per square meter.

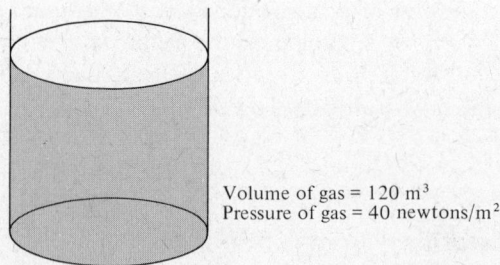

Volume of gas = 120 m³
Pressure of gas = 40 newtons/m²

Figure 8.1

Dynes and *newtons* are units of force in the metric system. A dyne is the force required to give a mass of one gram an acceleration of one centimeter per second per second. One newton equals 100,000 dynes; 4.45 newtons equal one pound of force.

Suppose now that the top is pushed down so that the volume is reduced; the gas is then confined in a smaller space and will exert a greater pressure against the interior surfaces of the container. Boyle's law states that the volume occupied by a gas varies inversely as the pressure, if the temperature remains constant. For example, if the volume of gas shown in Figure 8.1 is reduced from 120 cubic meters to 60 cubic meters, the pressure will rise from 40 newtons per square meter to 80 newtons per square meter, provided that the temperature of the gas remains the same. See Figure 8.2.

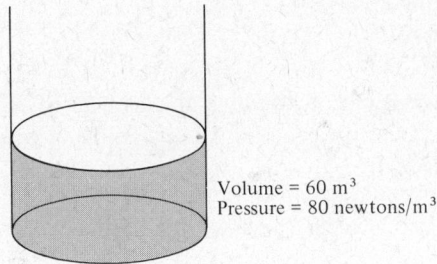

Volume = 60 m³
Pressure = 80 newtons/m³

Figure 8.2

The table below shows how the pressure changes as the volume is reduced.

V = volume of gas in cubic meters	120	60	40	30	20
p = pressure of gas in newtons per square meter	40	80	120	160	240

A thoughtful examination of the figures above shows not only that the pressure increases as the volume decreases but also that the *product* of the corresponding values of pressure and volume is always the *same number*; i.e.

$$pV = 4800$$

or

$$V = \frac{4800}{p} \qquad (p \neq 0)$$

Thus *V varies inversely* as *p*, and the constant of variation is 4800.

Summarizing this kind of relation between two variables in symbolic language, we have the following:

A variable y is said to *vary inversely* as a variable x if there is a nonzero constant, k, such that

$$y = \frac{k}{x} \qquad (x \neq 0)$$

k is called the constant of variation.

Similarly, y varies inversely as x^n ($x \neq 0$, $n \in N$) if there is a nonzero constant, k, such that $y = \dfrac{k}{x^n}$.

If y varies inversely as x, then we also say that y is *inversely proportional* to x and the constant of variation becomes the constant of proportionality just as in the case of direct variation. If

$$y = \frac{3}{x}$$

then y varies inversely as x and the constant of variation is 3. If

$$s = \frac{\sqrt{2}}{t}$$

then s is inversely proportional to t and the constant of proportionality is $\sqrt{2}$. *Note:* If

$$s = \frac{\sqrt{2}}{t}$$

$$\text{then} \qquad st = \sqrt{2}$$

$$\text{and} \qquad t = \frac{\sqrt{2}}{s}$$

So we could also say that t is inversely proportional to s with the same constant of proportionality, $\sqrt{2}$.

In general, *if y varies inversely as x then x varies inversely as y* with the *same* constant of variation because, if

$$y = \frac{k}{x} \qquad (x \neq 0)$$

$$\text{then} \qquad xy = k$$

$$\text{and} \qquad x = \frac{k}{y}$$

Following are five examples of problems involving inverse variation:

Example 1:

If y varies inversely as x and the constant of variation is 7, what is y when $x = 5$?

$$k = 7 \text{ and } y = \frac{k}{x}$$

$$y = \frac{7}{x}$$

$$y = \frac{7}{5}$$

Example 2:

If u is inversely proportional to v and $u = 6$ when $v = -3$, what is the constant of proportionality?

$$u = \frac{k}{v}$$

$$6 = \frac{k}{-3}$$

$$-18 = k$$

$$k = -18$$

Example 3:

If r varies inversely as s and $r = 2$ when $s = 5$, what is s when $r = 10$? First we must compute the constant of variation:

$$r = \frac{k}{s}$$

$$2 = \frac{k}{5}$$

$$10 = k$$

$$k = 10$$

So $\quad r = \dfrac{10}{s}$

$$10 = \frac{10}{s}$$

$$10s = 10$$

$$s = 1$$

Example 4:

At a constant temperature, the volume of a gas varies inversely as the pressure. If the pressure of a certain gas is 12 newtons per square meter when the volume is

300 cubic meters, what is the pressure when the volume is reduced to 180 cubic meters? First we must choose symbols to represent the variables.

Let V = the volume
and p = the pressure

V varies inversely as p, so a nonzero constant k exists such that

$$V = \frac{k}{p}$$

When $V = 300$ cubic meters, $p = 12$ newtons per square meter.

$$\text{So} \quad 300 = \frac{k}{12}$$

$$\text{and} \quad 3600 = k$$

Since $k = 3600$, the equation is

$$V = \frac{3600}{p}$$

When $V = 180$ cubic meters, we have

$$180 = \frac{3600}{p}$$

$$180p = 3600$$

$$p = \frac{3600}{180}$$

$$p = 20 \text{ newtons per square meter}$$

Example 5:

If t varies inversely as the square of w and $t = 9$ when $w = 2$, what is t when $w = 6$?

First we compute the constant of variation.

$$t = \frac{k}{w^2}$$

$$9 = \frac{k}{2^2}$$

$$9 = \frac{k}{4}$$

$$36 = k$$

$$k = 36$$

$$\text{So} \qquad t = \frac{36}{w^2}$$

$$t = \frac{36}{6^2}$$

$$t = \frac{36}{36}$$

$$t = 1$$

Inverse Variation Expressed in Terms of Direct Variation   In the discussion of inverse variation we observed that if $y = \dfrac{k}{x}$, then $x = \dfrac{k}{y}$. Thus neither x nor y can equal zero and each one has a multiplicative inverse (or reciprocal). The function

$$y = \frac{k}{x}$$

can also be written

$$y = k\left(\frac{1}{x}\right)$$

where $\dfrac{1}{x}$ is the reciprocal of x.

In the latter form, $y = k\left(\dfrac{1}{x}\right)$, we have the variable y equal to the constant, k, *times* $\dfrac{1}{x}$. This means that y varies directly as $\dfrac{1}{x}$. In general, if y varies *inversely* as x, then y varies *directly* as $\dfrac{1}{x}$.

Examples:

1. If w varies inversely as v, then w varies directly as $\dfrac{1}{v}$ and a nonzero constant, k, exists such that

$$w = k\left(\frac{1}{v}\right)$$

2. If t varies inversely as the square of u, then t varies directly as $\dfrac{1}{u^2}$, and

$$t = k\left(\frac{1}{u^2}\right)$$

Joint Variation   The fact that inverse variation can always be stated in terms of direct variation is helpful in dealing with *joint variation*. The statements "z varies *jointly* as x and y" and "z varies *directly* as x and y" both mean exactly the same thing. This fact, however, is hardly enlightening if one has no idea what either statement means, and the terminology is not much help. Specifically, z varies jointly as x and y if and only if z varies directly as x when y is constant and z varies directly as y when x is constant. All of this boils down to a surprisingly simple mathematical relation because it can be demonstrated that, under the stated conditions, z varies directly as the *product* of x and y.

A familiar example of joint variation is the relation of the area of a triangle to its base and height. If the base of a triangle remains fixed (constant) and the height increases, then the area of the triangle increases; or if the height remains fixed and the base increases, the area also increases. Furthermore, the area is equal to the constant $\frac{1}{2}$ times the *product* of the base and the height. Thus we say that the area of a triangle varies *jointly* as its base and height; e.g. if A is the area of a triangle having base, b, and height, h, then

$$A = \frac{1}{2}bh$$

To summarize, if z varies jointly as x and y, *then z varies directly as the product of x and y*, and a nonzero constant, k, exists such that

$$z = k(xy)$$
or
$$z = kxy$$

Similarly, if t varies jointly as w, u, and v, then t varies directly as the product of w, u, and v, and a nonzero constant, k, exists such that

$$t = k(wuv)$$
or
$$t = kwuv$$

Example:

t varies jointly as u and v. If $t = 12$ when $u = 3$ and $v = \frac{1}{2}$, what is t when $u = 4$ and $v = \frac{1}{4}$?

$$t = kuv$$

First we compute the value of k.

$$12 = k(3)\left(\frac{1}{2}\right)$$
$$12 = \frac{3}{2}k$$
$$24 = 3k$$
$$8 = k$$
$$k = 8$$

So $t = 8uv$
$$t = 8(4)\left(\frac{1}{4}\right)$$
$$t = 8$$

Now suppose we were told that y varies *directly* as x and *inversely* as z. This statement means that y varies *directly* as x and $\frac{1}{z}$, which in turn means that y varies directly as the product of x and $\frac{1}{z}$. Thus a nonzero constant, k, exists such that

$$y = k\left(x \cdot \frac{1}{z}\right)$$

or

$$y = \frac{kx}{z}$$

Example:

w varies directly as t and inversely as the square of v. If $w = 12$ when $t = 8$ and $v = 2$, what is w when $t = 12$ and $v = 3$?

w varies *directly* as t and $\dfrac{1}{v^2}$

Therefore a nonzero constant, k, exists such that

$$w = k\left(t \cdot \frac{1}{v^2}\right)$$

$$w = \frac{kt}{v^2}$$

First we compute the value of k.

$$12 = \frac{k(8)}{2^2}$$

$$12 = \frac{8k}{4}$$

$$12 = 2k$$

$$6 = k \qquad \text{E}_\text{m}$$

So $\qquad w = \dfrac{6t}{v^2}$

$$w = \frac{6(12)}{3^2}$$

$$w = \frac{72}{9}$$

$$w = 8$$

Various combinations of direct and inverse variation are possible in which several variables vary independently. The following examples illustrate some of these.

Examples:

1. The safe load, S, on a rectangular beam supported at both ends varies directly as its width and the square of its depth and inversely as its length. State a formula relating these quantities.

    First we must choose symbols to represent the variables involved.

$$S = \text{safe load} \qquad w = \text{width}$$
$$d = \text{depth} \qquad L = \text{length}$$

Then S varies *directly* as w and d^2 and $\dfrac{1}{L}$.

Thus a nonzero constant, k, exists such that

$$S = k\left(w \cdot d^2 \cdot \frac{1}{L}\right)$$

or

$$S = \frac{kwd^2}{L}$$

2. A beam 5 centimeters wide, 10 centimeters deep, and 275 centimeters long will support a load of 300 kilograms. What is the safe load for a beam of the same length and width but a depth of 15 centimeters?

$$S = \frac{kwd^2}{L}$$

We compute k from the facts that $S = 300$ when $w = 5$, $d = 10$, and $L = 275$.

$$300 = \frac{k(5)(10)^2}{275}$$
$$300 = \frac{500k}{275}$$
$$82{,}500 = 500k$$
$$165 = k$$

So

$$S = \frac{165wd^2}{L}$$

When $w = 5$, $d = 15$, and $L = 275$, we obtain

$$S = \frac{165(5)(15)^2}{275}$$
$$S = \frac{185{,}625}{275}$$
$$S = 675 \text{ kilograms}$$

3. y varies directly as the square of x and inversely as t and z. State a formula relating these quantities.

$$y \text{ varies } directly \text{ as } x^2, \frac{1}{t}, \text{ and } \frac{1}{z}$$

Thus a nonzero constant, k, exists such that

$$y = k\left(x^2 \cdot \frac{1}{t} \cdot \frac{1}{z}\right)$$

or

$$y = \frac{kx^2}{tz}$$

4.   Newton's law of gravitation states that the gravitational attraction between two bodies varies directly as the product of their masses and inversely as the square of the distance between them. State a formula relating these quantities.

Let G = the gravitational attraction
     m_1 = mass of one body
     m_2 = mass of the second body
     d = the distance between them

Then G varies *directly* as m_1, m_2, and $\frac{1}{d^2}$; therefore a nonzero constant, k, exists such that

$$G = k\left(m_1 \cdot m_2 \cdot \frac{1}{d^2}\right)$$

or

$$G = \frac{km_1m_2}{d^2}$$

Exercise 93:

(1)   If r varies inversely as s with the constant of variation $k = \frac{1}{2}$, what is r when $s = 5$?

(2)   If z varies jointly as u and v with the constant of variation $k = -2$, what is z when $u = 2$ and $v = 3$?

(3)   If y varies directly as x and inversely as w with the constant of variation $k = 3$, what is y when $x = 4$ and $w = 6$?

(4)   If r varies inversely as s and t with the constant of variation $k = \frac{1}{3}$, what is r when $s = 2$ and $t = \frac{1}{5}$?

(5)   If u is inversely proportional to t and s and $u = 3$ when $t = 2$ and $s = 3$, what is the constant of proportionality?

(6)   If z varies directly as x and inversely as y, and $z = 4$ when $x = 3$ and $y = 5$, what is z when $x = 4$ and $y = 6$?

(7)   If y varies directly as t and inversely as u and the square of v, and $y = \frac{1}{4}$ when $t = 6$, $u = 3$, and $v = 2$, what is y when $t = 9$, $u = 2$, and $v = 3$?

(8) The volume, V, of a cone varies jointly as its height, h, and the square of its radius, r. If a cone with a height of 8 centimeters and a radius of 2 centimeters has a volume of 33.5 cm^3, what is the volume of a cone with a height of 6 centimeters and a radius of 4 centimeters?

(9) The weight of an object varies inversely as the square of its distance from the center of the earth. If a man weighs 90 kilograms on the earth's surface, what will he weigh 2000 kilometers above the earth? 5000 kilometers above the earth? The radius of the earth is 6400 kilometers. (First, choose symbols to represent the two variables, weight and distance.)

(10) At a constant temperature, the volume of a gas varies inversely as the pressure. If the pressure of a certain gas is 40 newtons per square meter when the volume is 600 cubic meters, what will the pressure be when the volume is reduced to 240 cubic meters?

(11) The resistance of a conductor to the flow of electricity is measured in units called ohms. The resistance of wire to the flow of electricity varies directly as the length of the wire and inversely as the square of its diameter. If 30 meters of copper wire with a diameter of .65 centimeter has a resistance of .0159 ohm, what is the resistance of 300 meters of copper wire having a diameter of .4 centimeter?

(12) The maximum distance that a car can travel is directly proportional to the number of liters of gasoline in the tank and inversely proportional to the wind speed. If the maximum distance for a car with 76 liters of gasoline given a wind speed of 48 kilometers per hour is 480 kilometers, what is the maximum distance for that car with 38 liters given a 64-km/h wind?

(13) The intensity of illumination falling on a surface from a given source of light is inversely proportional to the square of the distance from the source of light. The unit for measuring the intensity of illumination is usually the *foot-candle*. If a given source of light gives an illumination of 1 foot-candle at a distance of 10 feet, what would the illumination be from the same source at a distance of 20 feet?

(14) Refer to Problem 13. What happens if the strength of the light source varies? It turns out that for a given distance, the illumination varies directly with the strength of the light source. Hence the intensity of illumination falling on a surface varies directly with the strength of the light source and inversely with the square of the distance from the light source. If a certain type of 100-watt bulb gives an illumination of 4.80 foot-candles at a distance of 5 feet, how much illumination would three lamps of the same type give at a distance of 10 feet?

(15) The volume occupied by a gas varies directly as its absolute temperature and inversely as the pressure it is under. At a temperature of 300° and under a pressure of 5.88 newtons per square centimeter, a sample of gas occupies a volume of 40 milliliters. What volume will this sample of gas occupy at a temperature of 360° and under a pressure of 8.82 newtons?

Chapter Test 8

I. Convert all quantities to the same units and express the following ratios as fractions in lowest terms:
    1. The ratio of 4 nickels to 6 dimes
    2. The ratio of 40 seconds to 3 minutes
    3. The ratio of 80 centimeters to 2 meters
    4. The ratio of 300 grams to 5 kilograms
    5. The ratio of 35 cents to $1.05

II. Express the ratio in lowest terms with the proper dimension.
    1. The ratio of 15 rats to 5 cats
    2. The ratio of 224 kilometers to $3\frac{1}{2}$ hours
    3. The ratio of $7.50 to 6 liters
    4. The ratio of 104 kilometers to 6 liters of gas
    5. The ratio of 240 cents to 5 meters
    6. The ratio of 18 dollars to 6 kilograms

III. 2.2 pounds = 1 kilogram. Use proportions to make the following conversions:
    1. Convert 18 pounds to kilograms
    2. Convert 11 kilograms to pounds.
    3. Convert $\frac{1}{4}$ pound to grams.

IV. A quantity, x, varies jointly as y and z and inversely as the square of t. If x is equal to 18 when $y = 3$, $z = 4$, and $t = 2$, what is x when $y = 4$, $z = 9$, and $t = 3$?

V. The weight of an object on the moon varies directly as its weight on earth. A sack of rocks brought from the moon weighed 6 kilograms on the moon and 36 kilograms on the earth. If a space crew can allow 210 kilograms additional earth weight for moon-rock samples, what should be the moon weight of the rocks gathered?

VI. The time required to assemble machines in a factory varies directly as the number of machines to be assembled and inversely as the number of crews working. If 6 crews can assemble 10 machines in 5 hours, how many crews will be needed to assemble 20 machines in 6 hours?

9
EXPONENTS

Exponents evolved from an expedient merger of convenience with common sense. As observed in Section 2.3, we frequently encounter repeated factors in mathematical expressions that produce terms such as the following:

$$xxx$$

$$yyyy$$

$$zzzzz$$

In order to condense such cumbersome expressions, inventive mathematicians of an earlier age devised the exponent, a number written to the right of and slightly above the factor that is to be repeated.

Thus, xxx is written x^3, and it means *x used as a factor three times*. In similar fashion:

$$yyyy \text{ is written } y^4$$

$$zzzzz \text{ is written } z^5$$

$$6xx \text{ is written } 6x^2$$

$$3 \cdot 3 \cdot yyy \text{ is written } 3^2y^3 \text{ or } 9y^3$$

By definition, x^1 means x. Therefore, when there is no exponent written for a factor, it is always understood to be *one*.

In general language, if $n \in N$ and $x \in R$, x^n means *x used as a factor n times*. Compare the following:

$$
\begin{array}{c}
x + x + x = 3x \\
x \cdot x \cdot x = x^3
\end{array}
$$

Please do not make a mystery of exponents. Their meaning is basically quite simple, and so, granted some necessary restrictions, is their behavior.

Examples:

$$5x^4 = 5xxxx$$
$$7y^6 = 7yyyyyy$$
$$4z^1 = 4z$$
$$3(a + b)^2 = 3(a + b)(a + b)$$
$$10(x - y)^3 = 10(x - y)(x - y)(x - y)$$

The expression x^4 is called a *power* and it is made up of two distinct parts called the *base* and the *exponent* (or degree). In the power x^4, x is the *base* of the power and 4 is the *exponent* of the power. In like manner:

$$y^6 \nearrow \text{exponent} \searrow \text{base}$$

$$2^5 \nearrow \text{exponent} \searrow \text{base}$$

$$(a + b)^3 \nearrow \text{exponent} \searrow \text{base}$$

Once created, these exponents developed a mathematical personality all their own in that they began to exhibit characteristics which belonged only to them. We call these characteristics that are peculiar to exponents "the laws of exponents."

We shall investigate these laws by observing the behavior of exponents which are natural numbers and prove some of the elementary theorems about them. In a more rigorous treatment it would be proved that these laws hold in a quite general context, in which the exponents are real numbers or even more general types of numbers. The purpose here is to proceed only as far as rational exponents. Proof of the laws of exponents will be left to more advanced treatises. From time to time, certain natural restrictions on the bases are unavoidable.

Accepting these laws without proof, we shall direct our investigation primarily toward operations with rational exponents and the motivation for the definitions given to them.

9.1
The First Law of Exponents

Let us consider what happens when two powers are multiplied together, *each power having the same base*.

> In the expression $x^2 \cdot x^3$, x^2 means xx and x^3 means xxx.
> Then $x^2 \cdot x^3$ means $xx \cdot xxx = xxxxx = x^5$

Therefore: $x^2 \cdot x^3 = x^5$
In similar fashion:

1. $y^2 \cdot y^4 = (y \cdot y)(y \cdot y \cdot y \cdot y) = yyyyyy = y^6$
   or $y^2 \cdot y^4 = y^6$

2. $3^2 \cdot 3^4 = (3 \cdot 3)(3 \cdot 3 \cdot 3 \cdot 3) = 3 \cdot 3 \cdot 3 \cdot 3 \cdot 3 \cdot 3 = 3^6$
   or $3^2 \cdot 3^4 = 3^6$

3. $x^2 \cdot x^3 \cdot x^4 = (x \cdot x)(x \cdot x \cdot x)(x \cdot x \cdot x \cdot x) = xxxxxxxxx = x^9$
   or $x^2 \cdot x^3 \cdot x^4 = x^9$

After considering this particular situation and its results, it becomes obvious that, in such products, the exponent in the result is always the *sum* of the exponents of the like bases. Thus

$$x^2 \cdot x^3 = x^{2+3} = x^5$$
$$y^2 \cdot y^4 = y^{2+4} = y^6$$
$$3^2 \cdot 3^4 = 3^{2+4} = 3^6$$

But if the bases in these products are *not* the same, what happens? Consider the term

$$x^2 \cdot y^3$$
$$x^2 = xx$$
$$\text{and } y^3 = yyy$$
$$\text{then } x^2 \cdot y^3 = xx \cdot yyy = x^2 y^3$$

Here the powers do not have the same bases, and the situation is completely different. Therefore

$$x^2 \cdot x^3 = x^{2+3} = x^5$$
$$x^2 \cdot y^3 = x^2 y^3$$

The product of two powers having the same base can be represented symbolically in the following manner:

$$\text{Let} \quad a,b \in N \quad \text{and} \quad x \in R$$

1. $x^a = xxxx \ldots x$ (x used as a factor a times)
2. $x^b = xxxx \ldots x$ (x used as a factor b times)
3. Then $x^a \cdot x^b = (xxxx \ldots x)(xxxx \ldots x)$

$$\underbrace{}_{a \text{ factors}} \quad \underbrace{}_{b \text{ factors}}$$

$x^a \cdot x^b = (xxxxxxxxx \ldots x)$ (x used as a factor $a + b$ times)
4. Therefore $x^a \cdot x^b = x^{a+b}$

The First Law of Exponents
The product of powers each having the same base.

$$a,b \in N \qquad x \in R$$
$$x^a \cdot x^b = x^{a+b}$$

This law can be restated in plain, simple English as follows: To multiply powers together having the same base, *leave the base alone and add the exponents.*

Using the symmetric axiom this law can also be stated

$$x^{a+b} = x^a \cdot x^b$$

a fact that should be noted carefully since it will sometimes be convenient to use this law in reverse.

In the following examples and exercises, all exponents are restricted to natural numbers.

Examples:

   (1) $2^2 \cdot 2^3 = 2^{2+3} = 2^5 = 32$
   (2) $(-2)^4(-2)^3 = (-2)^{4+3} = (-2)^7 = -128$
   (3) $(3)^4(3) = (3)^{4+1} = (3)^5 = 243$
   (4) $y^2 \cdot y^7 = y^{2+7} = y^9$
   (5) $x^n \cdot x = x^{n+1}$
   (6) $y^{a+b} \cdot y^{a+b} = y^{a+b+a+b} = y^{2a+2b}$
   (7) $(a + b)^5(a + b)^3 = (a + b)^{5+3} = (a + b)^8$
   (8) $(x - 2y)^c(x - 2y) = (x - 2y)^{c+1}$
   (9) $x^3 \cdot x^4 \cdot x^6 = x^{3+4+6} = x^{13}$
  (10) $(-3)^2(-3)^4(-3) = (-3)^{2+4+1} = (-3)^7 = -2187$
  (11) $3a^4 \cdot 5a^6 = 3 \cdot 5 \cdot a^4 \cdot a^6 = 15a^{4+6} = 15a^{10}$
  (12) $2x^3 \cdot 7x^n = 2 \cdot 7 \cdot x^3 \cdot x^n = 14x^{3+n}$

Exercise 94:

Multiply the following:

(1) $x^6 \cdot x^8$ (2) $y^3 \cdot y^{10}$

(3) $a^b \cdot a^2$ (4) $2^3 \cdot 2^4$

(5) $3x^3 \cdot 2x^8$ (6) $4y^b \cdot 2^y$

(7) $6x^a \cdot 3x^b$ (8) $4^3 \cdot 4^n$

(9) $(3z^7)(-5z^2)$ (10) $(-6)^3(-6)^4$

(11) $(x^{2m+n})(x^{m+n})$ (12) $y^3 \cdot y^5 \cdot y^2$

(13) $x^a \cdot x^6 \cdot x$ (14) $(-2)^3(-2)^2$

(15) $(y^6)(z^4)(z^6)$ (16) $(-4)^2(-4)^3$

(17) $(-4)^2(-2)^3$ (18) $(x - 2y)^3(x - 2y)^2$

(19) $(a + 4)^5(a + 4)^n$ (20) $(2x + 1)^3(2x + 1)^7(2x + 1)$

(21) $x^{a+2} \cdot x^{3-a}$ $(a < 3)$ (22) $3y^{b+1} \cdot 2y^{b+4}$

(23) $6x^{m+n} \cdot 3x^{n+1}$ (24) $(2a + 3)^4(2a + 3)^2$

(25) $(-7)^4(-7)^2$ (26) $(3y - x)^{n-2}(3y - x)^{n+2}$ $(n > 2)$

(27) $x^2 \cdot y^n \cdot y^3$ (28) $x^n \cdot 3x$

(29) $(2x^a)(-3x^a)(4x^b)$ (30) $(-5)^a(-5)^2$

9.2
The Second Law of Exponents

In order to arrive at the second law of exponents, let us consider what happens to exponents when two powers, each having the same base, are divided. In all cases we stipulate that the divisor cannot be zero.

> In the expression $\dfrac{x^5}{x^2}$
>
> $x^5 = xxxxx$ and $x^2 = xx$
>
> then $\dfrac{x^5}{x^2} = \dfrac{xxxxx}{xx} = \dfrac{\cancel{x}\cancel{x}xxx}{\cancel{x}\cancel{x}} = xxx = x^3$
>
> Therefore: $\dfrac{x^5}{x^2} = x^3$

In similar fashion:

> $\dfrac{2^6}{2^4} = \dfrac{(\cancel{2})(\cancel{2})(\cancel{2})(\cancel{2})(2)(2)}{(\cancel{2})(\cancel{2})(\cancel{2})(\cancel{2})} = (2)(2) = 2^2$
>
> Therefore: $\dfrac{2^6}{2^4} = 2^2$

$$\frac{y^8}{y^5} = \frac{yyyyyyyy}{yyyyy} = yyy = y^3$$

Therefore: $\frac{y^8}{y^5} = y^3$

Here again, a fairly obvious result keeps recurring. In the quotients above, the exponent in the result is always equal to the *difference* of the exponents in the dividend and the divisor. Furthermore, that difference is taken in the same manner: that is, the lower exponent is subtracted from the upper exponent. To do this, of course, we mean that we reverse the sign of the exponent in the divisor and add it to the exponent in the dividend. Consequently

1. $\dfrac{x^5}{x^2} = x^{5-2} = x^3$

2. $\dfrac{2^6}{2^4} = 2^{6-4} = 2^2$

3. $\dfrac{y^8}{y^5} = y^{8-5} = y^3$

But what happens if we have a quotient in which the bases are *not* the same? Consider the expression

4. $\dfrac{x^5}{y^2}$

    $x^5 = xxxxx$

    and $y^2 = yy$

    then $\dfrac{x^5}{y^2} = \dfrac{xxxxx}{yy} = \dfrac{x^5}{y^2}$

The situation described in the first three of the foregoing examples can be written symbolically

$$\frac{x^a}{x^b} = x^{a-b}$$

where, in each case, a and $b \in N$ and a is greater than b. We can prove that, for this particular case, x^a/x^b will always be equal to x^{a-b} in the following manner.

Proof of the second law of exponents:

$$a, b \in N \qquad (a > b)$$
$$x \in R \qquad (x \neq 0)$$

1. $-b + b = 0$
2. $a = a + 0$

3. Then $a = a - b + b$

4. And $x^a = x^{a-b+b}$

5. $\dfrac{x^a}{x^b} = \dfrac{x^{a-b+b}}{x^b} = \dfrac{x^{a-b} \cdot x^b}{x^b}$

6. $\dfrac{x^a}{x^b} = x^{a-b} \cdot \dfrac{x^b}{x^b} = x^{a-b} \cdot 1$

7. Therefore $\dfrac{x^a}{x^b} = x^{a-b}$

The second law of exponents for the case where $a,b \in N$ and $a > b$ can be stated in the following manner:

> **The Second Law of Exponents**
> **The quotient of two powers each having the same base.**
>
> $a,b \in N$ $x \in R$ $(x \neq 0)$
>
> $$\dfrac{x^a}{x^b} = x^{a-b}$$

In the following examples and exercises, all exponents are natural numbers and a is greater than b. Zero is excluded from all denominators.

Examples:

(1) $\dfrac{x^{10}}{x^3} = x^{10-3} = x^7$

(2) $\dfrac{y^7}{y^2} = y^{7-2} = y^5$

(3) $\dfrac{a^{3n}}{a} = a^{3n-1}$

(4) $\dfrac{x^{2a+b}}{x^{a-b}} = x^{2a+b-(a-b)} = x^{2a+b-a+b} = x^{a+2b}$

(5) $\dfrac{5^6}{5^2} = 5^{6-2} = 5^4$

(6) $\dfrac{(x+y)^6}{(x+y)^5} = (x+y)^{6-5} = (x+y)^1 = x+y$

(7) $\dfrac{(x+2)^4}{(x+2)} = \dfrac{(x+2)^4}{(x+2)^1} = (x+2)^{4-1} = (x+2)^3$

(8) $\dfrac{(-3)^5}{(-3)^2} = (-3)^{5-2} = (-3)^3 = -27$

(9) $\dfrac{6x^5}{3x^2} = \dfrac{6}{3} \cdot \dfrac{x^5}{x^2} = 2x^{5-2} = 2x^3$

$$(10) \quad \frac{x^{5n+2}y^{3n-1}}{x^{4n-3}y^{2n-1}} = \frac{x^{5n+2}}{x^{4n-3}} \cdot \frac{y^{3n-1}}{y^{2n-1}}$$
$$= x^{5n+2-4n+3}y^{3n-1-2n+1}$$
$$= x^{n+5}y^n$$

The process followed in each example above can be summarized as follows: to divide powers having the same base, *leave the base alone and subtract the exponent of the divisor from the exponent of the dividend.*

Exercise 95:

Simplify the following using the second law of exponents:

(1) $\dfrac{8^5}{8^2}$ 　　　　　　　　　　　　(2) $\dfrac{(-5)^3}{(-5)}$

(3) $\dfrac{(-2)^5}{(-2)^2}$ 　　　　　　　　　(4) $\dfrac{y^{2b}}{y}$

(5) $\dfrac{x^a}{x^2}$ 　　　　　　　　　　　　(6) $\dfrac{6x^4}{3x^2}$

(7) $\dfrac{y^{3b}}{y^{2b}}$ 　　　　　　　　　　(8) $\dfrac{28x^7}{8x^3}$

(9) $\dfrac{(-7)^5}{(-7)^3}$ 　　　　　　　　(10) $\dfrac{8x^n}{2x}$ 　$(n > 1)$

(11) $\dfrac{x^{a+5}}{x^{4+a}}$ 　　　　　　　　(12) $\dfrac{(-2x)^{2a}}{(-2x)^a}$

(13) $\dfrac{y^b}{y^2}$ 　$(b > 2)$ 　　　　(14) $\dfrac{y^{a+b}}{2y^{a-b}}$

(15) $\dfrac{6x^{4a}}{15x^2}$ 　　　　　　　　(16) $\dfrac{(2x-y)^{4a}}{(2x-y)^{a-b}}$

(17) $\dfrac{(x+y)^5}{(x+y)^2}$ 　　　　　　(18) $\dfrac{(2x-y)^{2n}}{(2x-y)}$

(19) $\dfrac{20x^8y^5}{12x^3y^2}$ 　　　　　　(20) $\dfrac{35x^4y^{10}}{14x^3y^8}$

(21) $\dfrac{(2y+3)^{3n-2}}{(2y+3)^{2n-1}}$ 　　　(22) $\dfrac{a^{b+2}c^{d+3}}{a^bc^{1+d}}$

(23) $\dfrac{b^{3n+2}c^{2n+1}}{b^{3n-1}c^{2n}}$ 　　　　(24) $\dfrac{x^{4a-1}y^{5a-2}}{x^{2a-1}y^{3a-2}}$

9.3
The Third Law of Exponents

The third law of exponents is a three-pronged affair. To arrive at it we ask what happens to the exponents if we have a power in which the base itself is also a power, or the product of several powers, or the quotient of two powers. Examples of these three situations are the following (again we make the condition that no divisor is equal to zero):

I.   A power in which the base is also a power:
     $(x^2)^3$ In this expression the *exponent* is 3, and the *base* is the power x^2.

II. A power in which the base is the product of two or more powers:
     $(x^2y^3)^4$ In this expression the *exponent* is 4, and the *base* is the product of the two powers x^2 and y^3.

III. A power in which the base is the quotient of two powers:
     $\left(\dfrac{x^2}{y^3}\right)^4$ In this expression the *exponent* is 4, and the *base* is the quotient of the two powers x^2 and y^3.

Let us consider the first of these three examples. Here we have $(x^2)^3$. This means that the base, x^2, must be used as a factor three times.

$$
\begin{aligned}
(x^2)^3 \text{ means } & x^2 \cdot x^2 \cdot x^2 \\
\text{but } x^2 \cdot x^2 \cdot x^2 &= (xx)(xx)(xx) \\
&= xxxxxx \\
&= x^6 \\
\text{Therefore } (x^2)^3 &= x^6
\end{aligned}
$$

In the same manner:

$$
\begin{aligned}
(y^3)^4 \text{ means } & y^3 \cdot y^3 \cdot y^3 \cdot y^3 \\
\text{but } y^3 \cdot y^3 \cdot y^3 \cdot y^3 &= (yyy)(yyy)(yyy)(yyy) \\
&= yyyyyyyyyyyy \\
&= y^{12} \\
\text{Therefore } (y^3)^4 &= y^{12}
\end{aligned}
$$

The foregoing results tell us that

$$
\begin{aligned}
(x^2)^3 &= x^6 \\
(y^3)^4 &= y^{12}
\end{aligned}
$$

In both cases the final exponent in the result of the operation is equal to the *product* of the two exponents in the power. This is a situation then in which, by *multiplying the exponents*, we can arrive quickly at the correct result.

Examples:

$$(x^2)^3 = x^{(2)(3)} = x^6$$
$$(y^3)^4 = y^{(3)(4)} = y^{12}$$
$$(x^3)^5 = x^{(3)(5)} = x^{15}$$

In symbolic language, if $x \in R$ and $a,b \in N$,

$$(x^a)^b = x^a x^a x^a x^a x^a \ldots x^a \qquad (x^a \text{ used as a factor } b \text{ times})$$
$$(x^a)^b = x^{a+a+a+a+a \ldots +a} \qquad (b \text{ terms})$$

When the number a is added to itself b times, the result is ab.

$$(x^a)^b = x^{ab}$$

The Third Law of Exponents: Part I
A power in which the base is also a power.

$$a,b \in N \qquad x \in R$$
$$(x^a)^b = x^{ab}$$
$$(x^b)^a = x^{ba} \quad \text{or} \quad x^{ab}$$

Since $(x^a)^b = x^{ab}$ and $(x^b)^a = x^{ab}$, by the symmetric axiom we have

$$x^{ab} = (x^a)^b \quad \text{or} \quad (x^b)^a$$

When we consider the examples given of the other two cases, we find the same thing happening. In the final results, the exponents are multiplied together.

The expression $(x^2 y^3)^4$ means that the base $x^2 y^3$ must be used as a factor four times.

$(x^2 y^3)^4$ means $x^2 y^3 \cdot x^2 y^3 \cdot x^2 y^3 \cdot x^2 \cdot y^3$
but this means $(xx)(yyy)(xx)(yyy)(xx)(yyy)(xx)(yyy)$
$$= (xx)(xx)(xx)(xx)(yyy)(yyy)(yyy)(yyy)$$
$$= (xxxxxxxx)(yyyyyyyyyyyy)$$
$$= x^8 y^{12}$$
Therefore $(x^2 y^3)^4 = x^8 y^{12}$

We can arrive at the same result by *multiplying the exponent of each separate factor in the base by the exponent of the power.* Thus

$$(x^2y^3)^4 = x^{(2)(4)}y^{(3)(4)} = x^8y^{12}$$

In like manner:

1. $(x^4y^2)^5 = x^{(4)(5)}y^{(2)(5)} = x^{20}y^{10}$
2. $(a^3b^4)^3 = a^{(3)(3)}b^{(4)(3)} = a^9b^{12}$
3. $(2x^3y)^6 = 2^{(1)(6)}x^{(3)(6)}y^{(1)(6)} = 2^6x^{18}y^6 = 64x^{18}y^6$

Putting this law in symbolic language we have

The Third Law of Exponents: Part II

$$a,b,c \in N \qquad x,y \in R$$
$$(x^ay^b)^c = x^{ac}y^{bc}$$
$$x^{ac}y^{bc} = (x^ay^b)^c$$

Examples:

(1) $(x^3y^5)^4 = x^{(3)(4)}y^{(5)(4)} = x^{12}y^{20}$
(2) $(2x^5)^3 = (2^1x^5)^3 = 2^{(1)(3)}x^{(5)(3)} = 2^3x^{15} = 8x^{15}$
(3) $(3x^2y^4)^4 = 3^{(1)(4)}x^{(2)(4)}y^{(4)(4)} = 3^4x^8y^{16} = 81x^8y^{16}$
(4) $(x^ay^3)^b = x^{ab}y^{3b}$

In the final part of the third law, we consider the type of expression given earlier as an example, $\left(\dfrac{x^2}{y^3}\right)^4$, or a power in which the base is itself the quotient of two other powers. This power, $\left(\dfrac{x^2}{y^3}\right)^4$, means that the base $\dfrac{x^2}{y^3}$ must be used as a factor four times.

$$\left(\frac{x^2}{y^3}\right)^4 = \frac{x^2}{y^3} \cdot \frac{x^2}{y^3} \cdot \frac{x^2}{y^3} \cdot \frac{x^2}{y^3}$$

$$= \frac{xx}{yyy} \cdot \frac{xx}{yyy} \cdot \frac{xx}{yyy} \cdot \frac{xx}{yyy}$$

$$= \frac{xxxxxxxx}{yyyyyyyyyyyy}$$

$$= \frac{x^8}{y^{12}}$$

Therefore $\left(\dfrac{x^2}{y^3}\right)^4 = \dfrac{x^8}{y^{12}}$

We can arrive at exactly the same result by *multiplying each exponent of the factors in the numerator and denominator by the exponent of the power*. Thus

$$\left(\frac{x^2}{y^3}\right)^4 = \frac{x^{(2)(4)}}{y^{(3)(4)}} = \frac{x^8}{y^{12}}$$

Stating this in general language we have the following:

The Third Law of Exponents: Part III
A power in which the base is the quotient of two other powers.

$a, b,$ and $c \in N$ $x, y \in R$ $y \neq 0$

$$\left(\frac{x^a}{y^b}\right)^c = \frac{x^{ac}}{y^{bc}}$$

or

$$\frac{x^{ac}}{y^{bc}} = \left(\frac{x^a}{y^b}\right)^c$$

In the examples and exercises which follow, all exponents are natural numbers, and zero is excluded from all denominators.

Examples:

(1) $\left(\dfrac{x^7}{y^2}\right)^2 = \dfrac{x^{(7)(2)}}{y^{(2)(2)}} = \dfrac{x^{14}}{y^4}$

(2) $\left(\dfrac{2y^3}{z^2}\right)^3 = \dfrac{2^{(1)(3)} y^{(3)(3)}}{z^{(2)(3)}} = \dfrac{2^3 y^9}{z^6} = \dfrac{8y^9}{z^6}$

(3) $\left(\dfrac{3x^3}{2y^4}\right)^3 = \left(\dfrac{3^1 x^3}{2^1 y^4}\right)^3 = \dfrac{3^{(1)(3)} x^{(3)(3)}}{2^{(1)(3)} y^{(4)(3)}} = \dfrac{3^3 x^9}{2^3 y^{12}} = \dfrac{27x^9}{8y^{12}}$

(4) $\left(\dfrac{x^{a+b}}{y^{a-b}}\right)^{a+b} = \dfrac{x^{(a+b)(a+b)}}{y^{(a-b)(a+b)}} = \dfrac{x^{a^2+2ab+b^2}}{y^{a^2-b^2}}$ (*a* is greater than *b*)

Please note that in every example given for all three parts of the third law, in each case we were dealing either with *factors* or with quotients of *factors*. The third law is never applied to exponents of separate *terms*.

Note the following *carefully:*

$$\left(\frac{a^2b^2}{c}\right)^2 = \frac{a^{(2)(2)}b^{(2)(2)}}{c^{(1)(2)}} = \frac{a^4b^4}{c^2}$$

$$\text{but} \quad \left(\frac{a^2+b^2}{c}\right)^2 = \frac{(a^2+b^2)^2}{c^2} = \frac{a^4+2a^2b^2+b^4}{c^2}$$

Exercise 96:

Remove the parentheses.

(1) $(x^3)^5$

(2) $(3x^2)^4$

(3) $(2x^3y^4)^3$

(4) $(5xy^3)^3$

(5) $\left(\dfrac{a}{b^2}\right)^4$

(6) $\left(\dfrac{2c}{d^5}\right)^5$

(7) $(x^2y^3)^5$

(8) $(a^3b^x)^y$

(9) $(3x^ay^4)^2$

(10) $(-2a^nb^3)^3$

(11) $(x^{a+b})^{a-b}$ $(a > b)$

(12) $(3a^mb^n)^4$

(13) $\left(\dfrac{2x^a}{y^b}\right)^5$ $(a > b)$

(14) $\left(\dfrac{3x^2y}{z^5}\right)^a$

(15) $\left(\dfrac{xy^b}{2z^3}\right)^4$

(16) $\left(\dfrac{-3x^4y^3}{2z^7}\right)^2$

(17) $\left(\dfrac{4a^2b^3}{3c^4}\right)^3$

(18) $\left(\dfrac{5^2x^4y^6}{2^4z^8}\right)^2$

(19) $\left(\dfrac{2x^3}{y^5}\right)^5$

(20) $\left(\dfrac{x^ay^{2c}}{z^b}\right)^c$

(21) $\left(\dfrac{x^{m-1}}{y^m}\right)^{m+1}$

(22) $\left(\dfrac{x^{3b}}{y^{2b}}\right)^b$

9.4
Zero Exponents and Negative Exponents

In the preceding sections the laws of exponents have been restricted to exponents which were natural numbers. If we extend these laws to include exponents that are any integers, we would immediately encounter such expressions as x^0 or x^{-4}. These expressions are meaningless in our former definition of exponents. How can x be used as a factor *zero times?* Or *minus four times?* In what way could these numbers be defined so that they would obey the laws of exponents?

We can find answers to these questions by asking some others. What mathematical situation, for instance, could produce a number such as x^0?

From the second law of exponents we know that

$$\frac{x^a}{x^b} = x^{a-b} \qquad \begin{array}{l} \text{if } x \neq 0 \\ a,b \in N \text{ and } a \text{ is greater than } b \end{array}$$

Consider this quantity:

$$\frac{x^2}{x^2} \qquad x \neq 0$$

If the second law of exponents *were* valid for $a = b = 2$, we would have

$$\frac{x^2}{x^2} = x^{2-2} = x^0$$

But plain common sense also tells us that

$$\frac{x^2}{x^2} = 1$$

since any number, except zero, divided by itself is always equal to one. So what could be more natural and plausible than to *define* x^0 in the following way:

$$x^0 = 1 \quad \text{if} \quad x \neq 0$$

When this is done, however, it must be recognized that for the exponent zero we have discarded the definition: "x^n equals x taken as a factor n times."

Summarizing, we have, for $x \neq 0$,

$$\boxed{\begin{array}{ll} \dfrac{x^2}{x^2} = x^0 & \text{if the second law were valid for } a = b = 2 \\[2ex] \dfrac{x^2}{x^2} = 1 \end{array}}$$

Thus if we wish to apply the second law here and assign a meaning to x^0 on that basis, it appears from the situation above that it should be the number *one* since the transitive axiom states that if two quantities are both equal to the same quantity, they are equal to each other.

We could also consider the following:

1. $\dfrac{x^a}{1} = x^a$ $\qquad\qquad \dfrac{x^a}{x^0} = x^{a-0} = x^a$

2. $x^a \cdot 1 = x^a$ $\qquad\quad x^a \cdot x^0 = x^{a+0} = x^a$

Here we see that the first and second laws of exponents are valid if $x^0 = 1$. Consequently, any number raised to the zero degree, except zero, is *defined* to be *one*.

The following examples provide further evidence in support of that definition.

Examples:

(1) $\dfrac{y^6}{y^6} = y^{6-6} = y^0$ and $\dfrac{y^6}{y^6} = 1$ $y \neq 0$

   $y^0 = 1$

(2) $\dfrac{5^2}{5^2} = 5^{2-2} = 5^0$ and $\dfrac{5^2}{5^2} = \dfrac{25}{25} = 1$

   $5^0 = 1$

(3) $\dfrac{(-2)^3}{(-2)^3} = (-2)^{3-3} = (-2)^0$ and $\dfrac{(-2)^3}{(-2)^3} = \dfrac{-8}{-8} = 1$

   $(-2)^0 = 1$

(4) $\dfrac{(a+b)^2}{(a+b)^2} + (a+b)^{2-2} = (a+b)^0$ and $\dfrac{(a+b)^2}{(a+b)^2} = 1$ $a+b \neq 0$

   $(a+b)^0 = 1$

By assigning the value *one* to powers with a zero exponent (provided the base is not zero), we can find a meaningful definition for exponents which are negative integers.

Consider the quantity

$$\frac{1}{x^6} \qquad x \neq 0$$

Since $x^0 = 1$, we can replace 1 by x^0 in the above expression.

$$\frac{1}{x^6} = \frac{x^0}{x^6}$$

Now if we assume the second law is valid for *a less than b* we will have

$$\frac{1}{x^6} = \frac{x^0}{x^6} = x^{0-6} = x^{-6}$$

which tells us that

$$x^{-6} = \frac{1}{x^6}$$

In general, if n is any nonnegative integer and $x \neq 0$,

$$\frac{1}{x^n} = \frac{x^0}{x^n} = x^{0-n} = x^{-n} \quad \text{or} \quad x^{-n} = \frac{1}{x^n}$$

provided the second law is valid for *a* less than *b*.

Consequently for $n \in N$ and $x \neq 0$, we *define* $x^{-n} = 1/x^n$, so that the second law will hold for a less than b.

In like manner:

(1) $\quad \dfrac{1}{y^7} = \dfrac{y^0}{y^7} = y^{0-7} = y^{-7} \quad$ or $\quad y^{-7} = \dfrac{1}{y^7}$

(2) $\quad \dfrac{1}{5^2} = \dfrac{5^0}{5^2} = 5^{0-2} = 5^{-2} \quad$ or $\quad 5^{-2} = \dfrac{1}{5^2}$

(3) $\quad \dfrac{1}{z^8} = \dfrac{z^0}{z^8} = z^{0-8} = z^{-8} \quad$ or $\quad z^{-8} = \dfrac{1}{z^8}$

Repeating this definition in symbolic language we have

$$n \in N \qquad x \neq 0$$
$$x^{-n} = \frac{1}{x^n}$$

In arriving at this definition we followed the same procedure as in the case of x^0; the definition was chosen to satisfy the *laws* of exponents without regard for the former definition of positive integral exponents. The expression x^{-n} does *not* mean "x used as a factor $-n$ times"; it means $1/x^n$.

We can now restate the second law of exponents for any integral exponents.

$$\frac{x^a}{x^b} = x^{a-b} \quad \text{if} \quad x \neq 0 \quad \text{and} \quad a,b \in J$$

Although we do not do so here, it is a simple matter, with the definitions we have given x^0 and x^{-n} ($x \neq 0$), to verify the laws of exponents under the assumption that the exponents are integers (positive, negative, or zero).

Using the definition that $x^{-n} = 1/x^n$ if $x \neq 0$, we can simplify expressions with negative exponents. In all examples we exclude zero from the denominator.

Examples:

(1) $\quad 6^{-2} = \dfrac{1}{6^2} = \dfrac{1}{36}$

(2) $\quad 2^{-3} = \dfrac{1}{2^3} = \dfrac{1}{8}$

(3) $\dfrac{1}{3^{-4}} = 3^4 = 81$

(4) $\dfrac{1}{2^{-5}} = 2^5 = 32$

If we have an expression such as

(5) $\dfrac{x^{-2}y^3}{a^4b^{-1}}$

and we wish to express the same value with all exponents positive, we proceed as follows:

$$x^{-2} = \frac{1}{x^2} \text{ and } b^{-1} = \frac{1}{b}$$

Then

$$\frac{x^{-2}y^3}{a^4b^{-1}} = \frac{\frac{1}{x^2}\cdot y^3}{a^4\cdot\frac{1}{b}} = \frac{\frac{y^3}{x^2}}{\frac{a^4}{b}}$$

$$= \frac{y^3}{x^2} \div \frac{a^4}{b} = \frac{y^3}{x^2}\cdot\frac{b}{a^4} = \frac{by^3}{a^4x^2}$$

$$\boxed{\text{Therefore } \frac{x^{-2}y^3}{a^4b^{-1}} = \frac{by^3}{a^4x^2}}$$

(6) $\dfrac{3a^{-3}b^2}{2^{-1}y^{-3}} = \dfrac{3\left(\frac{1}{a^3}\right)(b^2)}{\left(\frac{1}{2}\right)\left(\frac{1}{y^3}\right)} = \dfrac{\left(\frac{3b^2}{a^3}\right)}{\left(\frac{1}{2y^3}\right)} = \dfrac{3b^2}{a^3} \div \dfrac{1}{2y^3} = \dfrac{3b^2}{a^3}\cdot\dfrac{2y^3}{1} = \dfrac{6b^2y^3}{a^3}$

$$\boxed{\text{Therefore } \frac{3a^{-3}b^2}{2^{-1}y^{-3}} = \frac{6b^2y^3}{a^3}}$$

If you look *carefully* at the final results of the above two examples,

(5) $\dfrac{x^{-2}y^3}{a^4b^{-1}} = \dfrac{b^1y^3}{a^4x^2}$

(6) $\dfrac{3a^{-3}b^2}{2^{-1}y^{-3}} = \dfrac{3(2)^1b^2y^3}{a^3} = \dfrac{6b^2y^3}{a^3}$

it should eventually be clear that, in every case, all of the *factors* which had negative exponents changed position from the numerator to the denominator (or vice versa) in order to make their exponents positive, and all of the *factors* which had positive exponents stayed right where they were.

The word *factors* is deliberately emphasized in that last statement, because if these quantities had been *terms* instead of *factors*, *not one word of that statement would be true*. But as long as you are dealing with factors, not terms, you may move factors from the numerator to the denominator (or vice versa) of any fraction provided you change the sign of the exponent of each factor that you move.

If at this point in your mathematical progress, the tremendous operational differences in dealing with *terms* and *factors* has not been hammered home to you, now is the time to nail those differences down. As a brief review of your former encounters with terms and factors, *note the following situations carefully*:

	Factors	*Terms*
1.	$\dfrac{a\not b}{\not b} = a$	$\dfrac{a+b}{b} = \dfrac{a+b}{b}$
2.	$\dfrac{(3)(\not4)}{\not4} = 3$	$\dfrac{3+4}{4} = \dfrac{7}{4}$
3.	$(x^2y)^3 = x^6y^3$	$(x^2-y)^3 = x^6 - 3x^4y + 3x^2y^2 - y^3$

Now let us consider the following two situations, which may appear to be similar but which are actually worlds apart.

Examples:

(1) $\dfrac{x^{-2}y^{-1}}{z^{-2}}$ (In the numerator of this fraction there are two *factors*.)

(2) $\dfrac{x^{-2} - y^{-1}}{z^{-2}}$ (In the numerator of this fraction there are two *terms*.)

Example (1) is the same situation that has just been discussed. We can make the exponents positive by reversing the position of the *factors* in the numerator and denominator. Thus

$$\frac{x^{-2}y^{-1}}{z^{-2}} = \frac{z^2}{x^2y}$$

In Example (2) above, the quantities in the numerator are *not factors*. They are separated from each other by a minus sign, and this means that they are *terms*. A good working rule for basic mechanics in algebraic operations is simply this: *any operation that is true for factors is never going to be true for terms.*

To simplify Example (2) we make use of the definition we have for negative exponents.

$$x^{-n} = \frac{1}{x^n} \qquad x \neq 0$$

As a matter of fact, this definition and common sense will handle *any* situation with negative exponents provided that you can use the definition and your head at the same time. If you examine the laws of exponents carefully, you will see that each of them deals strictly with expressions involving *one term* only. Hence in dealing with expressions involving two or more terms, we may use these laws only to simplify *each term individually*.

In Example (2) we have

$$\frac{x^{-2} - y^{-1}}{z^{-2}} = \frac{\dfrac{1}{x^2} - \dfrac{1}{y}}{\dfrac{1}{z^2}}$$

This complex fraction can be simplified by either of the two methods discussed in Section 4.7.

1. The first method:

$$\frac{x^{-2} - y^{-1}}{z^{-2}} = \frac{\dfrac{1}{x^2} - \dfrac{1}{y}}{\dfrac{1}{z^2}} = \left(\frac{1}{x^2} - \frac{1}{y}\right) \div \frac{1}{z^2} = \frac{y - x^2}{x^2 y} \div \frac{1}{z^2}$$

$$= \frac{y - x^2}{x^2 y} \cdot \frac{z^2}{1} = \frac{z^2(y - x^2)}{x^2 y}$$

2. The second method:

$$\frac{x^{-2} - y^{-1}}{z^{-2}} = \frac{\left(\dfrac{1}{x^2} - \dfrac{1}{y}\right)}{\left(\dfrac{1}{z^2}\right)} \cdot \frac{x^2 y z^2}{x^2 y z^2} = \frac{y z^2 - x^2 z^2}{x^2 y} = \frac{z^2(y - x^2)}{x^2 y}$$

Consequently, we have arrived at this final result:

$$\boxed{\frac{x^{-2} - y^{-1}}{z^{-2}} = \frac{z^2(y - x^2)}{x^2 y}}$$

Please look over the above example and note the great difference in simplifying *terms* with negative exponents and *factors* with negative exponents. The *factors* merely reverse position in the numerator and denominator; *the terms do not*.

Exercise 97:

Make all of the exponents positive and simplify as much as possible.

(1) x^{-4}

(2) x^{-n}

(3) $x^0 y^{-1}$

(4) $x^3 y^{-2}$

(5) $\dfrac{x^{-2}}{y^3}$

(6) $\dfrac{x^{-2}}{y^{-3}}$

(7) $\dfrac{x^2}{y^{-3}}$

(8) $\dfrac{2x^{-2}}{y^0}$

(9) $x^{-3} y^2$

(10) $a^{-2} b$

(11) $2x^{-5} y^0$

(12) $6a^{-3} b^0$

(13) $\dfrac{x^2 y^{-4}}{z^{-5}}$

(14) $\dfrac{x^0}{y^{-3}}$

(15) $\dfrac{1}{x^{-8}}$

(16) $\dfrac{a^3 b^0}{c^{-2}}$

(17) $\dfrac{3a^{-2} b^{-1}}{x^{-3}}$

(18) $\dfrac{2x^{-3} y^2}{3^{-1} a b^{-4}}$

(19) $\left(\dfrac{2^{-1} a b^3}{3x^{-3} y}\right)^{-3}$

(20) $\left(\dfrac{5x^{-1} y^4}{2^{-1} x^{-3}}\right)^{-4}$

(21) $\left(\dfrac{3^{-2} a^4 b^{-1}}{7x^{-3} y^2}\right)^{-1}$

(22) $\dfrac{7^{-1} x^2 y^{-4}}{2^{-2} a^{-3} b}$

(23) $\dfrac{x^{-1} + x^{-2}}{y^{-3}}$

(24) $\dfrac{a^{-3} - b^{-1}}{b^{-2}}$

(25) $\dfrac{x^{-3} - x^{-1}}{x}$

(26) $\dfrac{(a+b)^0}{(a+b)^{-2}}$

(27) $\dfrac{(2x-y)^{-3}}{(2x-y)^0}$

(28) $\dfrac{x^{-3} + y^{-2}}{y^0}$

(29) $\dfrac{5^{-1} + 5^{-2}}{5^{-1}}$

(30) $\dfrac{x^{-1} + y^{-2}}{x^{-1} - y^{-2}}$

(31) $\left(\dfrac{3x^{-2}}{y}\right)^0$

(32) $\dfrac{a^{-2} + b^{-2}}{a^{-1}}$

(33) $\left(\dfrac{2x^{-1} y^{-2}}{3a^2 b^{-3}}\right)^{-2}$

(34) $\left(\dfrac{3^{-1} a^{-4} b}{2x^{-3} y}\right)^{-3}$

9.5
Fractional Exponents

Before examining the situation posed by powers with fractional exponents, the student should look again at a familiar process from arithmetic called the extraction of roots. The symbol used to indicate this operation is written $\sqrt[b]{}$ and is called a *radical*, where $b \in N$ and is called the *root index*.

In the expression $\sqrt[2]{9}$, or simply $\sqrt{9}$, the root index is 2, and the radical indicates that we find *one* of *two identical factors* of 9. The number under the radical, in this case 9, is called the *radicand*. If we have $\sqrt[3]{8}$, we want *one* of *three identical factors* of 8. In general, if $x \in R$,

$\sqrt[3]{x}$ means one of *three* identical factors whose product is x

$\sqrt[4]{x}$ means one of *four* identical factors whose product is x

$\sqrt[7]{x}$ means one of *seven* identical factors whose product is x

$\sqrt[n]{x}$ means one of n identical factors whose product is x where n is any positive integer.

In performing these operations with real numbers we run into an ambiguous situation when the root index is an *even* number, although this does not happen when the root index is odd and the results are restricted to real numbers. Consider the following examples *carefully*.

Examples:

(1) $\sqrt{64} = 8$ since $(8)(8) = 64$, but since $(-8)(-8)$ is also 64, there is the possibility of confusion as to the meaning of the symbol $\sqrt{64}$. That ambiguity will be eliminated in the discussion which follows.

(2) $\sqrt{-64}$ is not defined by *any* real number. (The numbers which equal $\sqrt{-64}$ will be discussed in Section 10.8.)

(3) $\sqrt[3]{64} = 4$ since $(4)(4)(4) = 64$

(4) $\sqrt[3]{-64} = -4$ since $(-4)(-4)(-4) = -64$

While the situation is perfectly clear when the root index is *odd*, it is rather murky when the root index is *even*. Every *positive* real number has *two* square roots, one positive and the other negative. A *negative* real number has no real square root at all.

In the latter case the situation is simply that the square root does not exist within the real number system. We will return to this point in Section 10.8. In the former case we have an ambiguous situation. To avoid this ambiguity we restrict all roots of real numbers to *principal roots*, which are defined as follows:

1. If a number has a positive root, that root is called the *principal root*.

$$\sqrt{64} = 8$$
$$\sqrt[3]{64} = 4$$

2. If a number does not have a positive root but does have a negative root, then the negative root is the *principal root*.

$$\sqrt[3]{-64} = -4$$
$$\sqrt[3]{-8} = -2$$

3. Even roots of negative real numbers are not defined in the real number system.

4. Zero has only one root of any order whatsoever, namely *zero*.

This takes care of all simple arithmetic situations. When the indicated root of a specific number is a real number, we have only to select the principal root in order to arrive at a unique result. The restriction to principal roots is desirable in the interest of eliminating ambiguities. Life becomes so complicated in the presence of such ambiguities that it was long ago agreed that principal values would *automatically* be signified unless an explicit, contrary agreement is signified.

Examples:

(1) $\sqrt{36} = 6$

(2) $\sqrt{81} = 9$

(3) $-\sqrt{81} = -(9) = -9$

(4) $\sqrt[3]{27} = 3$

(5) $\sqrt[3]{-27} = -3$

In dealing with powers and roots of literal numbers, which could represent either positive or negative numbers, we must find a means of stating the situation carefully and clearly.

1. The *equation* $x^2 = 4$ defines the set of all elements whose squares are 4. Here we are not limited to a *unique* value since the variable x can be any number whose square is 4. In this case the two values of x are 2 and -2 or

$$\{x \mid x^2 = 4\} = \{2, -2\}$$

2. But if we have the number

$$\sqrt{x^2} \qquad x \in R$$

we have a different situation altogether. Here we must choose the principal value of this square root. Although x^2 is positive (or zero), we have no way of knowing if x itself represents a positive or negative number. If x is *negative* and we say that $\sqrt{x^2} = x$, then we do *not* have the principal value. The only correct way to designate the principal value is the following:

$$\sqrt{x^2} = |x|$$

Then $|x| = x$ if x is positive and $|x| = -x$ if x is negative (see Section 1.4).

Examples:

(1) $x = 3$ and $x^2 = 9$

$\sqrt{9} = \sqrt{(3)^2} = |3| = 3$

(2) $x = -3$ and $x^2 = 9$

$\sqrt{9} = \sqrt{(-3)^2} = |-3| = -(-3) = 3$

In both cases above we had $\sqrt{9}$ and in both cases we arrived at the principal root.

Following are some examples of extracting roots. In every case $x \in R$.

Examples:

(1) $\sqrt[7]{128} = 2$

(2) $\sqrt{x^2} = |x|$ or $\left\{ \begin{array}{l} \sqrt{x^2} = x \text{ if } x \geq 0 \\ \sqrt{x^2} = -x \text{ if } x < 0 \end{array} \right\}$

(3) $\sqrt[3]{x^3} = x$

(4) $\sqrt[4]{x^4} = |x|$ or $\left\{ \begin{array}{l} \sqrt[4]{x^4} = x \text{ if } x \geq 0 \\ \sqrt[4]{x^4} = -x \text{ if } x < 0 \end{array} \right\}$

(5) $\sqrt[3]{x^6} = x^2$

(6) $\sqrt[4]{x^8} = x^2$ *Note:* $\sqrt[4]{x^8} = |x^2| = x^2$

(7) $\sqrt[4]{x^4 y^{12}} = |xy^3|$

(8) $\sqrt[4]{x^8 y^{16}} = x^2 y^4$

(9) $\sqrt[3]{y^9} = y^3$

(10) $\sqrt[6]{x^{18}} = |x^3|$

(11) $\sqrt[6]{x^{12}} = x^2$

(12) $\sqrt[4]{x^8 y^{20}} = x^2 |y^5|$

In order to illustrate the basic mechanics of operations with radicals in the simplest possible way, we shall for the present restrict all variables involving even roots to positive numbers unless otherwise stated. We shall find that there is one case where this restriction is a necessity rather than a temporary convenience, and that particular case will be pointed out when it is encountered. A sound proficiency in operations with radicals is eminently desirable for two reasons: (1) such operations are constantly encountered in applications at many levels, and (2) such proficiency, once acquired, can help the student gain a much clearer insight into a more generalized treatment of radicals.

Exercise 98:

Find the indicated root (assume all variables involving even roots are positive).

(1) $\sqrt{x^4}$ (2) $\sqrt{9x^4}$

(3) $\sqrt{36x^6}$ (4) $\sqrt[3]{-125}$

(5) $\sqrt[3]{y^6}$ (6) $\sqrt[3]{125y^6}$

(7) $\sqrt[4]{16}$ (8) $\sqrt[3]{-27}$

(9) $\sqrt[3]{-27x^9 y^{12}}$ (10) $\sqrt[4]{y^{16}}$

(11) $\sqrt[4]{16x^8 y^{16}}$ (12) $\sqrt[5]{-243}$

(13) $\sqrt[5]{-243a^5 b^{10}}$ (14) $\sqrt{81x^4 y^2}$

(15) $\sqrt[3]{-27x^9}$ (16) $\sqrt{64x^6 y^{12}}$

(17) $\sqrt{16y^{14}}$ (18) $\sqrt[3]{64x^6y^{12}}$

(19) $\sqrt[3]{125x^6y^{15}}$ (20) $\sqrt{121a^{10}b^8}$

In the foregoing sections, exponents that are either positive or negative integers or zero have been discussed and defined. Since we shall accept without proof the laws of exponents for all rational exponents, we need to consider the situation we have when an exponent is a fraction of the type a/b where $a,b \in J$ except that $b \neq 0$. We shall stipulate that all fractional exponents are to be expressed in their *lowest* terms; i.e., in any fractional exponent, a/b, a and b are relatively prime.

As an example, we need a meaning (or definition) for the expression

$$x^{1/3}$$

Let us consider the behavior of $x^{1/3}$ in an elementary operation with which we are familiar.

$$x^{1/3} \cdot x^{1/3} \cdot x^{1/3} = ?$$

From the first law of exponents we know that if a,b, and $c \in N$, then $x^a \cdot x^b \cdot x^c = x^{a+b+c}$. If we apply that law to this situation we would have

$$x^{1/3} \cdot x^{1/3} \cdot x^{1/3} = x^{(1/3)+(1/3)+(1/3)} = x^{3/3} = x$$

According to this, $x^{1/3}$ is *one* of *three identical factors whose product is x*. But we say that 2 is a *cube root* of 8 because 2 is one of three identical factors whose product is 8. Then $x^{1/3}$ could be the cube root of x for exactly the same reason.

Since $\sqrt[3]{8} = 2$ because $(2)(2)(2) = 8$
then $\sqrt[3]{x} = x^{1/3}$ because $(x^{1/3})(x^{1/3})(x^{1/3}) = x$

Since $\sqrt[5]{32} = 2$ because $(2)(2)(2)(2)(2) = 32$
then $\sqrt[5]{x} = x^{1/5}$ because $(x^{1/5})(x^{1/5})(x^{1/5})(x^{1/5})(x^{1/5}) = x$

From the above examples it appears that a fractional exponent, such as 1/3 or 1/5, should represent a radical if it is to satisfy the laws of exponents. To clarify, a fractional exponent means one of the form a/b where a and b are relatively prime integers and $b \in N$, $b \neq 1$.

Consequently we define

$$x^{1/b} = \sqrt[b]{x}$$

where $x \in R$, $b \in N$, $b \neq 1$, and $x \geqq 0$ when b is even. The rather numerous restrictions placed on this definition are designated to keep the quantities defined by it simultaneously extant, sensible, and restricted to the set of real numbers. The explanation of them follows:

1. If $b = 1$, then $x^{1/b}$ is equal to x^1 and is not a radical.
2. If $x < 0$ when b is *even*, then $\sqrt[b]{x}$ is not a real number. Example: $\sqrt{-4}$ does not exist within the real number system.

Now make this comparison:

$$\sqrt[3]{x^6} = x^2$$
$$\text{but} \quad x^{6/3} = x^2$$

Since both $\sqrt[3]{x^6}$ and $x^{6/3}$ are equal to the same quantity, x^2, they must be equal to each other.

$$\text{Therefore} \quad x^{6/3} = \sqrt[3]{x^6}$$
$$\sqrt[3]{x^6} = x^{6/3}$$

In this manner, we can rewrite fractional exponents as radicals or *vice versa*. In the above example, please note carefully how the fractional exponent shows up in the radical.

$$x^{6/3} = \sqrt[3]{x^6}$$

The *denominator* of the fractional exponent becomes the *root index* of the radical; the *numerator* of the fractional exponent remains the exponent of the base.

Examples:

(1) $x^{2/3} = \sqrt[3]{x^2}$

(2) $y^{4/5} = \sqrt[5]{y^4}$

(3) $z^{3/8} = \sqrt[8]{z^3}$ $\quad z > 0$

(4) $\sqrt[4]{a^3} = a^{3/4}$ $\quad a > 0$

Putting this situation in symbolic language, if x is the base and a/b is the exponent, we have the following:

$$a \in J \qquad x \in R$$
$$b \in N, b \neq 1$$
$$\text{and } x \geq 0 \text{ when } b \text{ is even}$$
$$x^{a/b} = \sqrt[b]{x^a}$$
$$\sqrt[b]{x^a} = x^{a/b}$$

Making use of this we can simplify the following expressions in the manner shown.

Examples:

(1) $9^{1/2} = \sqrt[2]{9^1} = \sqrt{9} = 3$

(2) $25^{1/2} = \sqrt{25} = 5$

(3) $8^{1/3} = \sqrt[3]{8} = 2$

(4) $16^{-1/2} = \dfrac{1}{16^{1/2}} = \dfrac{1}{\sqrt{16}} = \dfrac{1}{4}$

Note that we are using here, correctly, the definition $x^{-n} = \dfrac{1}{x^n}$ which was developed earlier for integral exponents.

(5) $81^{-1/4} = \dfrac{1}{81^{1/4}} = \dfrac{1}{\sqrt[4]{81}} = \dfrac{1}{3}$

(6) $(-8)^{1/3} = \sqrt[3]{-8} = -2$

(7) $(-32)^{1/5} = \sqrt[5]{-32} = -2$

Now consider this situation:

$$9^{3/2} = ?$$
$$\text{We note that} \quad 9^{3/2} = \sqrt[2]{9^3}$$

Then this question arises: what do we do first? Cube the 9 and then take the square root of the result, or take the square root of 9 and then cube the result?

We can answer this question by using the third law of exponents.

$$9^{3/2} = \sqrt{9^3}$$
$$9^{3/2} = (9^3)^{1/2} \quad \text{or} \quad (9^{1/2})^3$$
$$= (729)^{1/2} \quad \text{or} \quad (\sqrt{9})^3$$
$$= \sqrt{729} \quad \text{or} \quad (3)^3$$
$$= \underline{27} \quad \text{or} \quad \underline{27}$$

Consequently, when we have $\sqrt[2]{9^3}$ we may go either one of two ways: we may take the square root of 9 and cube the result, or we may get the cube of 9 and then take the square root of the result. Either way we arrive at the same answer.

If we put the above situation into symbolic language where $x \in R$ and $a,b \in J$, it is important to remember that, in this discussion, we have imposed the condition that x be positive when b is even. In cases such as $\sqrt[4]{x^4}$ this restriction has been a mere convenience designed to simplify an introduction to radicals. But that restriction ceases to be a temporary convenience and becomes a permanent necessity in the situation we are considering here.

To say that

$$(9^3)^{1/2} = (9^{1/2})^3$$

is quite correct. But to say that

$$(x^a)^{1/b} = (x^{1/b})^a$$

is *not* correct *unless x is positive when b is even.*

In the preceding discussion the necessary restrictions which must be placed in particular situations on negative bases with rational exponents have been examined. In the examples and exercises which follow, we shall limit all bases to positive numbers, and *within this limitation*, the laws of exponents and the definitions which have been given here can be summarized as follows:

Let x and y be positive real numbers, and let a and b be rational numbers. Then,

Law I $x^a \cdot x^b = x^{a+b}$

Law II $\dfrac{x^a}{x^b} = x^{a-b}$

Law III   (1) $(x^a)^b = x^{ab}$

(2) $(x^a y^b)^c = x^{ac} y^{bc}$

(3) $\left(\dfrac{x^a}{y^b}\right)^c = \dfrac{x^{ac}}{y^{bc}}$

Definition 1 $x^0 = 1$

Definition 2 $x^{-a} = \dfrac{1}{x^a}$

Definition 3 $x^{a/b} = (\sqrt[b]{x})^a = \sqrt[b]{x^a}$
$(a,b \in N \quad b \neq 1)$

The restriction that x and y be positive numbers may be relaxed on occasion as has been explained, but in order to illustrate operations with these laws in the simplest context, we shall allow only positive bases in the following examples and exercises.

Examples:

To express radicals as powers:

(1) $\sqrt[5]{x^3} = (x^3)^{1/5} = x^{3/5}$

(2) $\sqrt[5]{x^3y^2} = (x^3y^2)^{1/5} = x^{3/5}y^{2/5}$

(3) $\sqrt[4]{16x^8y^3} = (16x^8y^3)^{1/4} = 16^{1/4}x^{8/4}y^{3/4} = 2x^2y^{3/4}$

(4) $\sqrt[3]{-8x^2y^3} = (-8x^2y^3)^{1/3} = (-8)^{1/3}x^{2/3}y^{3/3} = -2x^{2/3}y$

To express powers as radicals:

(5) $64^{1/2} = \sqrt{64} = 8$

(6) $27^{2/3} = \sqrt[3]{(27)^2} = (3)^2 = 9$

(7) $16^{-1/2} = \dfrac{1}{16^{1/2}} = \dfrac{1}{\sqrt{16}} = \dfrac{1}{4}$

(8) $(81x^4)^{1/2} = \sqrt{81x^4} = 9x^2$

Exercise 99:

Write the following radicals in exponential form:

(1) $\sqrt[3]{x^2}$ (2) $\sqrt[5]{y^4}$

(3) $\sqrt[7]{a^2b^3}$ (4) $\sqrt[4]{16x^2y^3}$

(5) $\sqrt[12]{z^5}$ (6) $\sqrt[8]{a^5}$

(7) $\sqrt{25a^5b^7}$ (8) $\sqrt[3]{125yx^2}$

(9) $\sqrt[3]{-27xy^2}$ (10) $\sqrt[5]{-32x^3y}$

Write the following in radical form and simplify:

(11) $49^{1/2}$ (12) $121^{-1/2}$

(13) $8^{-2/3}$ (14) $27^{-1/3}$

(15) $(25x^2)^{1/2}$ (16) $36^{-1/2}$

(17) $(27x^3y^{12})^{1/3}$ (18) $(8a^6b^{15})^{2/3}$

(19) $(32x^{10})^{1/5}$ (20) $16^{3/2}$

Exercise 100:

Miscellaneous problems. Simplify the following using the laws of exponents; make all negative exponents positive by using the definition of negative exponents. All literal symbols represent positive numbers.

(1) $y^{-2} \cdot y^7$

(2) $x^n \cdot x$

(3) $(x - y)^{5/3}(x - y)^{1/3}$

(4) $y^{1/2} \cdot y^{1/3}$

(5) $(-3)^4(-3)^3$

(6) $(a + b)^{1/2}(a + b)^{1/2}$

(7) $x^{-3} \cdot x^{10}$

(8) $(6y^7)(3y^{-4})$

(9) $\dfrac{x^{5/2}}{x^{1/2}}$

(10) $\dfrac{y^3}{y^{-5}}$

(11) $\dfrac{x^4}{x^7}$

(12) $\dfrac{x^{2/3}}{x^{1/2}}$

(13) $\dfrac{(3a - b)^{7/2}}{(3a - b)^{1/2}}$

(14) $\dfrac{x^{-8}}{x^{10}}$

(15) $\dfrac{x^3 y^{-4}}{z^{-1}}$

(16) $\dfrac{3a^3 b^0}{c^{-2}}$

(17) $\left(\dfrac{2^{-1}xy^{-3}}{3a^{-2}b}\right)^2$

(18) $\left(\dfrac{5x^{-1/2}y^{1/4}}{2^{-1/2}x^{-3}}\right)^{-4}$

(19) $\dfrac{x^{-1} + y^{-2}}{y^{-1}}$

(20) $\dfrac{a^{-2} - b^{-1}}{b^{-2}}$

(21) $64^{-1/3}$

(22) $(36a^4 b^6)^{1/2}$

(23) $(25x^2 y^8)^{-1/2}$

(24) $\sqrt[3]{27x^9 y^{15}}$

(25) $\sqrt[5]{32a^5 b^{10}}$

(26) $(2x^{-1/2}y^0)^{-4}$

(27) $9^{-1/2}x^{-3}y^0$

(28) $(4x^2 y^{-4})^{-1/2}$

(29) $\dfrac{1}{8^{-1/3}}$

(30) $\dfrac{x + x^{-1}}{y^{-2}}$

Chapter Test 9

I. Write the following in exponential form:

1. $\sqrt[3]{y^2}$

2. $\sqrt[5]{x}$

3. $\sqrt[7]{x^2 y^3}$

4. $\sqrt[4]{16ab^2}$

II. Write the following in radical form and simplify:

1. $(x^4)^{1/2}$

2. $(y^3)^{1/3}$

3. $(25a^2)^{1/2}$

4. $(-8b^6)^{1/3}$

III. Write each of the following in a simpler form having no negative, zero, or fractional exponents:

1. 7^{-1}

2. $\left(\dfrac{2}{3}\right)^{-1}$

3. $\dfrac{2^{-1}}{3}$

4. $\left(\dfrac{4}{5}\right)^0$

5. $\dfrac{4^0}{5}$

6. $64^{1/2}$

7. $64^{1/3}$

8. $(2^{-4})(5^0)$

9. $(3^{-2})(3^0)$

10. $81^{1/2}$

11. $81^{-1/2}$

12. $27^{-2/3}$

IV. Simplify the following. Make all negative exponents positive by using the definition of negative exponents. All bases are positive numbers and zero is excluded from all denominators.

1. $x^a \cdot x$

2. $y^{1/2} \cdot y^{1/3} \cdot y^{7/6}$

3. $\dfrac{2x^{-4}}{y^{-3}}$

4. $\dfrac{2a^{-3}b^2}{3^{-1}x^2y^{-4}}$

5. $\left(\dfrac{4a^4b^{-6}}{9^{-1}x^2y^8}\right)^{-1/2}$

6. $\dfrac{a^{-1} + b^{-2}}{a^{-1}}$

7. $\dfrac{y^{-4}}{y^{-9}}$

8. $\left(\dfrac{x^{2/3}y^2}{a^4b^{-3}}\right)^3$

9. $8^{-2/3}$

10. $\dfrac{3^{-3} + 3^{-1}}{3^{-1} + 3^{-2}}$

10
RADICALS; COMPLEX NUMBERS

Many applications of algebra lead to equations whose solution sets involve numbers such as $\sqrt{2}$, $\sqrt[3]{5}$, $1 - \sqrt{3}$, or $2 + \sqrt{7}$. These numbers are called *irrational numbers* (meaning that they cannot be expressed as ratios of integers). We took a rather brief look at the set of irrational numbers in Section 4.3, but before we go into operations with them, we need to take a closer look at the numbers themselves.

10.1
Irrational Numbers Revisited

In Sections 1.1, 1.2, and 1.4 we used numbers of the real number system to locate points on a straight line, each number designating a particular point and representing that point's distance and direction from zero. We concluded that the rational numbers can be matched one to one with infinitely many points on the line. All such numbers are really non-ending, repeating decimals; each one may be expressed as the quotient of an integer and a nonzero integer; they have been named the rational numbers from the word *ratio*. The question was asked then and is repeated here: Do the rational numbers account for the location of *all* the points on the number line? By a very simple geometric construction we can show that the answer is no. To do so, let us look back to the numbers of arithmetic, specifically to the process for finding the square root of a number.

$$\sqrt{81} = 9 \text{ because } 9 \times 9 = 81$$
$$\sqrt{289} = 17 \text{ because } 17 \times 17 = 289$$

Now, what happened when you encountered $\sqrt{2}$? Since 2 is not a perfect square, you may have been introduced to a process for approximating its square root to any desired degree of accuracy.

$$\sqrt{2} = \sqrt{2.0000000000\overline{0}\ldots}$$

This value can be computed to any required number of decimal places by the method illustrated below. If you do not recall this procedure for computing square roots, relax and let the matter rest in happy oblivion. The method is shown here for illustrative purposes *only*. For future reference, the square roots of all integers from 1 to 100 are listed in the table on page 483.

$$
\begin{array}{r|l}
\sqrt{2.} \;00\;00\;00\;00\;00\ldots & 1.41421\ldots \\
1 & \\
\hline
24\;|\;1.\;00 & \\
\times 4\;96 & \\
\hline
281|\;4\;00 & \\
\times 1\;2\;81 & \\
\hline
2824|\;1\;19\;00 & \\
\times 4\;1\;12\;96 & \\
\hline
28282|\;6\;04\;00 & \\
\times 2\;5\;65\;64 & \\
\hline
282841|\;38\;36\;00 & \\
\times 1\;28\;28\;41 & \\
\hline
10\;07\;59\;00 &
\end{array}
$$

If curiosity should lure us into continuing this laborious process to twenty figures, we would wind up looking at the following:

$$\sqrt{2} = 1.4142135623730950488\ldots$$

at which point, with any sense at all, we would quit or at least start seeking a friendly computer.

Thus $\sqrt{2} = 1.41421\ldots$ (where the three dots indicate that the decimal goes on forever). This decimal, like the decimal equivalents of rational numbers, is non-ending, but no matter how long you carry out the above process for computing $\sqrt{2}$, the decimal never repeats. Think this over. *It never repeats*. If this decimal is non-repeating, then no *rational* number can ever represent it, because every rational number is equal to a repeating decimal. Here, then, is a number which *cannot* be written as the quotient of an integer and a nonzero integer. Furthermore, while we cannot get an exact value for $\sqrt{2}$ in decimals, we can represent $\sqrt{2}$ exactly as a distance on a straight line.

Example:

1. Consider that part of the number line between 0 and 2 (Figure 10.1).

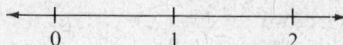

Figure 10.1

2. At zero construct a square whose sides are 1 unit in length (Figure 10.2).

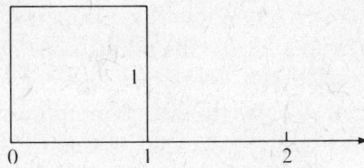

Figure 10.2

3. Now draw the diagonal of that square from the point 0 to the opposite corner (Figure 10.3). Label this point P.

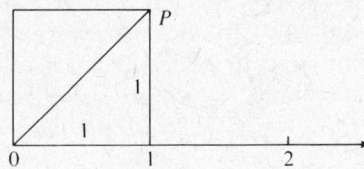

Figure 10.3

4. The hypotenuse of the right triangle with vertices at 0, P, and 1 is $0P$. The lengths of the two sides of this right triangle are each equal to 1. From the theorem of Pythagoras we know that in every right triangle, the sum of the squares of the two sides will always equal the square of the hypotenuse. Then

$$(0P)^2 = 1^2 + 1^2$$
$$(0P)^2 = 1 + 1$$
$$(0P)^2 = 2$$
$$\sqrt{(0P)^2} = \sqrt{2}$$
$$0P = \sqrt{2}$$

5. Now take a compass and mark off the distance $0P$ on the number line (Figure 10.4). Label the point R.

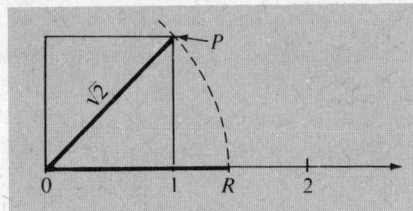

Figure 10.4

The point R is a point on the number line whose distance from 0 is exactly equal to $\sqrt{2}$. Therefore $\sqrt{2}$ is the only number which accurately describes the location of this point. Then $\sqrt{2}$ is a real number, since the real number system is the set of numbers which represent *all* of the points on a straight line.

By similar constructions we can show that there are infinitely many points on the number line whose distance and direction from zero cannot be designated by rational numbers but must be represented by numbers such as $\sqrt{3}, \sqrt{7}, \sqrt[3]{2}, \sqrt[3]{7}, \sqrt{15}, \sqrt[4]{11}$, etc. All of these numbers are non-ending, *non-repeating* decimals. The set of all such numbers, i.e., the set of *irrational* numbers, comprises a very important part of the real number system. An ancient example of an irrational number is $\pi = 3.14159\ldots$, a non-ending, non-repeating decimal. As noted earlier, the set of irrational numbers is designated throughout this text by the letter H.

An irrational number represents an *exact* distance on the number line, just as a rational number does. The difference is that any rational number can be written as the quotient of an integer and a nonzero integer while irrational numbers cannot be so written.

The sum of a rational and an irrational number is irrational. For example, $\sqrt{5}$ is a non-ending, non-repeating decimal; if we add a rational number, such as the number 3, to this decimal, we will still have a non-ending, non-repeating decimal. Thus, the number $3 + \sqrt{5}$ is an irrational number.

Exercise 101:

Classify the following numbers as rational or irrational:

(1) 7

(2) $-.3232\overline{23}\ldots$

(3) $4 + \sqrt{3}$

(4) $.604040\overline{04}\ldots$

(5) $-\dfrac{5}{11}$

(6) 165,782

(7) $\sqrt{13}$

(8) $\sqrt[3]{8}$

(9) $-\sqrt{3}$

(10) $6 - \sqrt{25}$

(11) $8 + \sqrt{7}$

(12) $-.183183183\overline{183}\ldots$

(13) $\dfrac{115}{221}$

(14) $.13113111311113111113\ldots$

10.2
The Multiplication of Monomials Containing Radicals

In order to work with irrational numbers with ease and accuracy, it is necessary to learn the arithmetic of such numbers. For example, the solution set of the equation $x^2 - 2x = 2$ is $\{1 + \sqrt{3}, 1 - \sqrt{3}\}$. The check of either of these two solutions, for instance $1 + \sqrt{3}$, requires the substitution of this number for x in the equation.

$$x^2 - 2x = 2$$
$$(1 + \sqrt{3})^2 - 2(1 + \sqrt{3}) = 2$$

When this is done we are faced with verifying the arithmetic identity above. Even before the solution, $1 + \sqrt{3}$, is found, it frequently appears in a disguised form such as $(2 + \sqrt{12})/2$, and a preliminary problem is to reduce this fraction to its simplest form. Consequently, before approaching algebraic problems which involve irrational numbers, the student needs first to learn the arithmetic of radicals: how to simplify them; how to add and subtract them; how to multiply and divide them; and finally, how to reduce fractions containing them. In this chapter we shall deal only with radicals in which *the literal symbols represent positive numbers* and *the radicand is positive or zero*.

The process of multiplying together two radicals having the same root index can be illustrated by using the laws of exponents.

If $(ab)^{1/2} = a^{1/2}b^{1/2}$
then $a^{1/2}b^{1/2} = (ab)^{1/2}$
and $\sqrt{2}\sqrt{3} = 2^{1/2} \cdot 3^{1/2} = (2 \cdot 3)^{1/2} = (6)^{1/2} = \sqrt{6}$
Therefore $\sqrt{2}\sqrt{3} = \sqrt{6}$

All radicals which represent real numbers and which have the *same* index can be multiplied together in the same fashion. If the root indices are *not* the same, they must be made the same before multiplication can be performed.

Examples:

$(1) \quad \sqrt{7}\sqrt{5} = \sqrt{35}$

$(2) \quad \sqrt{6}\sqrt{3} = \sqrt{18} = \sqrt{(9)(2)} = 3\sqrt{2}$

$(3) \quad \sqrt[3]{4}\sqrt[3]{5} = \sqrt[3]{20}$

$(4) \quad 2\sqrt{7}\sqrt{3} = 2\sqrt{21}$

$(5) \quad (2\sqrt{7})(3\sqrt{3}) = (2)(3)(\sqrt{7})(\sqrt{3}) = 6\sqrt{21}$

$(6) \quad \sqrt{2}\sqrt[3]{3} = (2)^{1/2}(3)^{1/3} = (2)^{3/6}(3)^{2/6} = (2^3)^{1/6}(3^2)^{1/6} = (8)^{1/6}(9)^{1/6}$
$\qquad\qquad = \sqrt[6]{8}\sqrt[6]{9} = \sqrt[6]{72}$

$(7) \quad \sqrt{5}\sqrt{5} = \sqrt{25} = 5$

$(8) \quad \sqrt{7}\sqrt{7} = \sqrt{49} = 7$

$(9) \quad \sqrt{3}\sqrt{3} = \sqrt{9} = 3$

$(10) \quad \sqrt{x}\sqrt{x} = \sqrt{x^2} = x$

Exercise 102:

Multiply the following:

(1) $\sqrt{3}\sqrt{5}$ (2) $\sqrt{2a}\sqrt{3b}$

(3) $(3\sqrt{7})(2\sqrt{5})$ (4) $(\sqrt{11})(\sqrt{11})$

(5) $(\sqrt{6})(\sqrt{6})$ (6) $\sqrt{7a}\sqrt{7a}$

(7) $\sqrt[3]{7a}\sqrt[3]{7a}$ (8) $\sqrt{3x}\sqrt{3x}$

(9) $\sqrt[3]{3x}\sqrt[3]{3x}$ (10) $(2\sqrt{5})(3\sqrt{5})$

(11) $(3\sqrt{2})(5\sqrt{2})$ (12) $(3\sqrt{2})(2\sqrt{5})$

(13) $(5\sqrt{2})(\sqrt{3})$ (14) $(\sqrt{5})(4\sqrt{2})$

(15) $(3\sqrt{2})(4\sqrt{3})$ (16) $(5\sqrt{7})(2\sqrt{3})$

(17) $(7\sqrt{11})(2\sqrt{5})$ (18) $(3\sqrt{8})(2\sqrt{2})$

(19) $(5\sqrt{3})^2$ (20) $(2\sqrt{7})^2$

(21) $(3\sqrt{5})^2$ (22) $(6\sqrt{6})^2$

(23) $(\sqrt[3]{5})(\sqrt{2})$ (24) $\sqrt[4]{a}\sqrt[3]{a}$

(25) $\sqrt[4]{x^3}\sqrt{x}$ (26) $\sqrt{2ab}\sqrt{2ab}$

(27) $\sqrt{10y}\sqrt{10y}$ (28) $\sqrt{5xy}\sqrt{5xy}$

10.3
The Simplification of Radicals

Skill in simplifying radicals is essential in numerous operations encountered in algebra, trigonometry, analytic geometry, and the calculus. It is also a skill that appears to be a source of unwarranted confusion to a great many students. For that reason the techniques of simplifying radicals will be presented here with an emphasis on the usual procedures that will amount to a deliberate over-emphasis. We shall explore these particular processes from a "never do this" viewpoint, and sum them up in three statements all beginning with the word *never*. But these three "Nevers" are not absolute. There are situations in some applications where either one of the first two could be more conveniently ignored than applied, but that decision should always be made from the vantage point of knowing the usual procedures *thoroughly*. Our philosophy here is that of the sea captain in the comic opera who claims *never* to have been sick at sea. "What, *never?*" the chorus chants, eliciting from the captain the admission, "Well, *hardly* ever." Thus these procedures are presented rather dogmatically, and the word "never" is to be translated "never in the exercises given in this chapter" or, more generally, "never without good reason."

If one needs the simplest possible radicand, for instance, he should never leave under a radical any factor for which the indicated root can be found. And that statement is actually the first of the three "Nevers," two of which will be discussed in this section and the remaining one in Section 10.8.

The First Never Never leave under any radical any *factor* for which the indicated root can be found.

Please note the language of that statement carefully. It states that a certain type of *factor* must not be left under a radical. It says nothing about removing *terms*. For example, in the expression (9)(16), 9 and 16 are *factors*. In the expression 9 + 16, 9 and 16 are *terms*.

Consider the following situation:

$$\sqrt{(9)(16)}$$

By the laws of exponents,

$$\sqrt{(9)(16)} = (9 \cdot 16)^{1/2} = 9^{1/2} \cdot 16^{1/2} = (\sqrt{9})(\sqrt{16})$$

Therefore

$$\sqrt{(9)(16)} = \sqrt{9}\sqrt{16} = (3)(4) = 12$$
$$or \ \sqrt{(9)(16)} = \sqrt{144} = 12$$

But $\sqrt{9 + 16}$ *does not equal* $\sqrt{9} + \sqrt{16}$

since $\sqrt{9 + 16} = \sqrt{25} = \underline{5}$

and $\sqrt{9} + \sqrt{16} = 3 + 4 = \underline{7}$

We cannot operate with *terms* separately under a radical, only *factors*. The only sensible thing to do with an expression under a radical involving more than one term is to factor the expression if possible. If not, let it alone.

Thus

$$\sqrt{a^2b^2} = \sqrt{a^2}\sqrt{b^2} = ab$$
$$\sqrt{a^2 + a^2b^2} = \sqrt{a^2(1 + b^2)} = \sqrt{a^2}\sqrt{1 + b^2} = a\sqrt{1 + b^2}$$

But

$$\sqrt{a^2 + b^2} = \sqrt{a^2 + b^2} \quad \textbf{(Hands off!)}$$

Have the good sense *to leave* $\sqrt{a^2 + b^2}$ *alone*. It cannot be made any simpler than it is because $a^2 + b^2$ is not a perfect square. In similar fashion, the exact value of $\sqrt{7}$ is written as $\sqrt{7}$ and left at that. It cannot be made any simpler because 7 is not a perfect square.

To illustrate the first Never, consider the quantity $\sqrt{24}$. The *indicated root* is the square root. The question is this: does the number 24 contain a *factor* for which the square root can be found?

$$24 = (8)(3)$$
$$24 = (2)(12)$$
$$24 = (4)(6)$$

The number 24 contains the factor 4, which is a perfect square. Therefore

$$\sqrt{24} = \sqrt{(4)(6)} = \sqrt{4}\sqrt{6} = \underline{2\sqrt{6}}$$

Consider $\sqrt[3]{24}$. Here the indicated root is the cube root. Does 24 contain a factor for which the *cube* root can be found? Yes, since $24 = (8)(3)$. Therefore:

$$\sqrt[3]{24} = \sqrt[3]{(8)(3)} = \sqrt[3]{8}\sqrt[3]{3} = \underline{2\sqrt[3]{3}}$$

Examples:

(1) $\sqrt{162} = \sqrt{(81)(2)} = \sqrt{81}\sqrt{2} = 9\sqrt{2}$

(2) $\sqrt[3]{48} = \sqrt[3]{(8)(6)} = \sqrt[3]{8}\sqrt[3]{6} = 2\sqrt[3]{6}$

(3) $\sqrt{36 + 4} = \sqrt{40} = \sqrt{4(10)} = \sqrt{4}\sqrt{10} = 2\sqrt{10}$

(4) $\sqrt{64 - 4} = \sqrt{60} = \sqrt{(4)(15)} = \sqrt{4}\sqrt{15} = 2\sqrt{15}$

(5) $\sqrt{75x^3} = \sqrt{(25)(3)(x^2)(x)}$
$\qquad = \sqrt{25}\sqrt{3}\sqrt{x^2}\sqrt{x}$
$\qquad = \sqrt{25}\sqrt{x^2}\sqrt{3}\sqrt{x}$
$\qquad\quad\ \downarrow\quad\downarrow\quad\downarrow\ \downarrow$
$\qquad = \ \ 5\ \ \ x\ \ \sqrt{3}\sqrt{x}$
$\qquad = 5x\sqrt{3x}$

(6) $\sqrt{18x^3y^2} = \sqrt{(9)(2)(x^2)(x)(y^2)}$
$\qquad\quad = \sqrt{9}\sqrt{2}\sqrt{x^2}\sqrt{x}\sqrt{y^2}$
$\qquad\quad = \sqrt{9}\sqrt{x^2}\sqrt{y^2}\sqrt{2}\sqrt{x}$
$\qquad\qquad\ \downarrow\quad\downarrow\quad\ \downarrow\quad\downarrow\ \downarrow$
$\qquad\quad = \ \ 3\ \ \ x\ \ \ y\ \ \sqrt{2}\sqrt{x}$
$\qquad\quad = 3xy\sqrt{2x}$

Exercise 103:

Simplify the following radicals:

(1) $\sqrt{32}$ (2) $\sqrt[3]{32}$

(3) $\sqrt[4]{32}$ (4) $-\sqrt{108}$

(5) $\sqrt[3]{54}$ (6) $-\sqrt{x^3}$

(7) $\sqrt{75x^3}$ (8) $\sqrt{64 + 16}$

(9) $\sqrt{36 - 16}$ (10) $\sqrt{x^6}$

(11) $\sqrt{8x^6}$ (12) $-\sqrt{27x^9}$

(13) $\sqrt{49x^2 + 9x^2}$ (14) $\sqrt{16x^2 - 4x^2}$

(15) $3\sqrt{12}$ (16) $-2\sqrt[3]{250}$

(17) $3x\sqrt{x^3}$ (18) $2\sqrt[3]{16y^9}$

(19) $-\sqrt{4x^3}$ (20) $5\sqrt[3]{8y^5}$

(21) $\sqrt{72a^3b^5}$ (22) $4\sqrt{27x^4y}$

(23) $-\sqrt{8x^6y^3z^4}$ (24) $\sqrt[3]{16x^6}$

(25) $\sqrt{98a^5b^7}$ (26) $\sqrt[3]{40x^7}$

(27) $\sqrt{300x^4y^5z^9}$ (28) $\sqrt{a^5b^7c^8}$

(29) $\sqrt[3]{54x^3y^5}$ (30) $\sqrt[4]{32a^5b^2}$

(31) $3y\sqrt{12xy^3}$ (32) $-2a\sqrt{18ab^5}$

(33) $\sqrt{25x^2 - 4x^2}$ (34) $\sqrt{4a^2 - 4}$

(35) $\sqrt{25x^2 + 50}$ (36) $\sqrt{x^2y^2 - x^2}$

(37) $\sqrt{4a^2 + 20}$ (38) $\sqrt{2a^2 - 4ab + 2b^2}$

(39) $\sqrt{64a^2 + 4a^2}$ (40) $\sqrt{x^3 + 2x^2 + x}$

It is sometimes possible to reduce the index (also called the *order*) of a radical. This situation, when it exists, is generally easier to see if the radical is rewritten in exponential form.

Examples:

(1) $\sqrt[4]{9x^2y^6} = \sqrt[4]{3^2x^2y^6} = (3^2x^2y^6)^{1/4}$
$$= 3^{2/4}x^{2/4}y^{6/4}$$
$$= 3^{1/2}x^{1/2}y^{3/2} = 3^{1/2}x^{1/2}(y^3)^{1/2}$$
$$= (3xy^3)^{1/2}$$
$$= \sqrt{3xy^3} = \sqrt{3xy^2y} = y\sqrt{3xy}$$

(2) $\sqrt[6]{4x^2y^8} = \sqrt[6]{2^2x^2y^8} = (2^2x^2y^8)^{1/6}$
$$= 2^{2/6}x^{2/6}y^{8/6}$$
$$= 2^{1/3}x^{1/3}y^{4/3}$$
$$= (2xy^4)^{1/3}$$
$$= \sqrt[3]{2xy^4} = y\sqrt[3]{2xy}$$

Exercise 104:

Reduce the order of the following radicals and simplify:

(1) $\sqrt[6]{9x^6z^4}$ (2) $\sqrt[8]{a^4b^6}$

(3) $\sqrt[10]{4x^2y^8}$ (4) $\sqrt[16]{x^4y^{12}}$

(5) $\sqrt[12]{a^6b^3}$ (6) $\sqrt[14]{x^2y^6}$

(7) $\sqrt[6]{a^4b^2c^8}$ (8) $\sqrt[4]{16a^{10}b^6}$

The Second Never   Never leave a *fraction* under a radical, and never leave a radical in the *denominator* of a fraction.

The application of this advice effectively removes radicals from the denominator of a fraction and thereby makes the denominator a rational number. For that reason the technique discussed here is called "rationalizing the denominator," a process that is extremely useful in simplifying many operations with radicals.

We shall consider the two parts of this second Never separately. First, *do not leave a fraction under a radical.*

Suppose we have the $\sqrt{3/5}$. Clearly, we have a fraction under the radical. By the laws of exponents,

$$\sqrt{\frac{3}{5}} = \left(\frac{3}{5}\right)^{1/2} = \frac{3^{1/2}}{5^{1/2}} = \frac{\sqrt{3}}{\sqrt{5}}$$

Therefore, $\sqrt{3/5} = \sqrt{3}/\sqrt{5}$, and now there is no longer a fraction under the radical. But we are not really any better off, because the second part of this Never says, *do not leave a radical in the denominator of any fraction.* And in writing that $\sqrt{3/5} = \sqrt{3}/\sqrt{5}$, we have done just that.

However, we can multiply any number by *one,* and still have the same number. Furthermore, $\sqrt{5}/\sqrt{5}$ is most certainly equal to one.

$$\sqrt{\frac{3}{5}} = \frac{\sqrt{3}}{\sqrt{5}} = \frac{\sqrt{3}}{\sqrt{5}} \cdot \frac{\sqrt{5}}{\sqrt{5}} = \frac{\sqrt{15}}{5}$$

Examples:

(1) $\sqrt{\dfrac{2}{3}} = \dfrac{\sqrt{2}}{\sqrt{3}} = \dfrac{\sqrt{2}}{\sqrt{3}} \cdot \dfrac{\sqrt{3}}{\sqrt{3}} = \dfrac{\sqrt{6}}{3}$

(2) $\sqrt{\dfrac{2a}{3b}} = \dfrac{\sqrt{2a}}{\sqrt{3b}} = \dfrac{\sqrt{2a}}{\sqrt{3b}} \cdot \dfrac{\sqrt{3b}}{\sqrt{3b}} = \dfrac{\sqrt{6ab}}{3b}$

(3) $\sqrt[3]{\dfrac{x}{y}} = \dfrac{\sqrt[3]{x}}{\sqrt[3]{y}} = \dfrac{\sqrt[3]{x}}{\sqrt[3]{y}} \cdot \dfrac{\sqrt[3]{y^2}}{\sqrt[3]{y^2}} = \dfrac{\sqrt[3]{xy^2}}{\sqrt[3]{y^3}} = \dfrac{\sqrt[3]{xy^2}}{y}$

(4) $\sqrt{\dfrac{2}{3}x^3} = \sqrt{\dfrac{2x^3}{3}} = \dfrac{\sqrt{2x^3}}{\sqrt{3}} \cdot \dfrac{\sqrt{3}}{\sqrt{3}} = \dfrac{\sqrt{6x^3}}{3} = \dfrac{x\sqrt{6x}}{3}$

(5) $\dfrac{4}{\sqrt{3}} = \dfrac{4}{\sqrt{3}} \cdot \dfrac{\sqrt{3}}{\sqrt{3}} = \dfrac{4\sqrt{3}}{3}$

(6) $\dfrac{a}{\sqrt{b}} = \dfrac{a}{\sqrt{b}} \cdot \dfrac{\sqrt{b}}{\sqrt{b}} = \dfrac{a\sqrt{b}}{b}$

(7) $\sqrt{\dfrac{3}{8}} = \dfrac{\sqrt{3}}{\sqrt{8}} = \dfrac{\sqrt{3}}{\sqrt{8}} \cdot \dfrac{\sqrt{2}}{\sqrt{2}} = \dfrac{\sqrt{6}}{\sqrt{16}} = \dfrac{\sqrt{6}}{4}$

Exercise 105:

Rationalize the denominators and simplify as much as possible.

(1) $\sqrt{\dfrac{3}{7}}$ (2) $\sqrt{\dfrac{5}{8}}$

(3) $\sqrt[3]{\dfrac{3}{4}}$ (4) $\sqrt[3]{\dfrac{2}{9}}$

(5) $\dfrac{2}{\sqrt{5}}$ (6) $\dfrac{5}{\sqrt[3]{2}}$

(7) $\sqrt{\dfrac{3}{x}}$ (8) $\dfrac{3}{\sqrt{x}}$

(9) $\sqrt[3]{\dfrac{2}{x^2}}$ (10) $\dfrac{2}{\sqrt[3]{x^2}}$

(11) $\sqrt{\dfrac{2x}{3}}$ (12) $\dfrac{5}{\sqrt{7x}}$

(13) $\dfrac{3a}{\sqrt{a}}$ (14) $\dfrac{2}{\sqrt{2}}$

(15) $\dfrac{x}{\sqrt{x}}$ (16) $\sqrt{\dfrac{3x}{5y}}$

(17) $\dfrac{3x}{\sqrt{5y}}$ (18) $\sqrt[3]{\dfrac{3a}{2b}}$

(19) $\sqrt{\dfrac{3}{5}x^3y}$ (20) $\sqrt{\dfrac{7a^2b^3}{3x}}$

(21) $\sqrt{\dfrac{2x^5y^2z}{7ab^2}}$ (22) $\sqrt{\dfrac{64x^2}{9y}}$

(23) $\sqrt[3]{\dfrac{64x^2}{9y}}$ (24) $\sqrt{\dfrac{8a^5}{3b^7}}$

(25) $\sqrt[3]{\dfrac{2b^2}{125x^9}}$ (26) $\sqrt[3]{\dfrac{3x^2}{4y}}$

(27) $\sqrt{\dfrac{5a}{7b^2}}$ (28) $\sqrt[3]{\dfrac{8a^3}{27b^2}}$

(29) $\dfrac{x}{\sqrt[3]{x}}$ (30) $\dfrac{3a}{\sqrt[3]{9a^2}}$

10.4
The Addition of Radicals

Addition means the same thing here that it has always meant, but the student could possibly use a reminder that when quantities are added, the only terms that can be combined are those which are exactly the same kinds of things.

Examples:

(1) $2x + 4x = 6x$ for all $x \in R$

since $2x + 4x = x(2 + 4)$
$= x(6)$
$= 6x$

but $2 + 4x$ *does not equal* $6x$ except in the special case where $x = 1$.

(2) $2\sqrt{3} + 4\sqrt{3} = 6\sqrt{3}$
because $2\sqrt{3} + 4\sqrt{3} = \sqrt{3}(2 + 4) = \sqrt{3}(6) = 6\sqrt{3}$

but $2 + 4\sqrt{3}$ **does not equal** $6\sqrt{3}$.

If this is not immediately and abundantly clear to you, perhaps you had better look over carefully the following examples of addition taken from different levels of learning. A good look should convince you that addition is beautifully consistent and what was not right in grade school is also not right in college.

Grade School	$\dfrac{2}{7} + \dfrac{3}{7} = \dfrac{5}{7}$
High School	$2x + 3x = 5x$
	$2\sqrt{7} + 3\sqrt{7} = 5\sqrt{7}$
College	$2\dfrac{dy}{dx} + 3\dfrac{dy}{dx} = 5\dfrac{dy}{dx}$

However,

Grade School	$\dfrac{2}{7} + \dfrac{3}{5} = \dfrac{2}{7} + \dfrac{3}{5}$
High School	$2x + 3y = 2x + 3y$
	$2 + 3\sqrt{7} = 2 + 3\sqrt{7}$
College	$2\dfrac{dy}{dx} + 3\dfrac{dx}{dt} = 2\dfrac{dy}{dx} + 3\dfrac{dx}{dt}$

To paraphrase a famous line, addition is addition is addition, no matter where you are in the learning process. In the addition of radicals, there is one added precaution: *all radicals should be simplified first.*

Example:

$$\sqrt{18} - 3\sqrt{8} + 5\sqrt{\frac{1}{2}}$$

The three radicals involved are $\sqrt{18}$, $\sqrt{8}$, and $\sqrt{1/2}$. Each one should be simplified before anything else is done.

$$\sqrt{18} = \sqrt{(9)(2)} = 3\sqrt{2}$$
$$\sqrt{8} = \sqrt{(4)(2)} = 2\sqrt{2}$$
$$\sqrt{\frac{1}{2}} = \frac{\sqrt{1}}{\sqrt{2}} = \frac{1}{\sqrt{2}} \cdot \frac{\sqrt{2}}{\sqrt{2}} = \frac{\sqrt{2}}{2}$$

Therefore

$$\sqrt{18} - 3\sqrt{8} + 5\sqrt{\frac{1}{2}} = 3\sqrt{2} - 3(2\sqrt{2}) + 5\frac{\sqrt{2}}{2}$$

$$= 3\sqrt{2} - 6\sqrt{2} + \frac{5\sqrt{2}}{2}$$

Note: The lowest common denominator is 2.

$$= \frac{6\sqrt{2} - 12\sqrt{2} + 5\sqrt{2}}{2}$$

$$= \frac{11\sqrt{2} - 12\sqrt{2}}{2}$$

$$= \frac{-\sqrt{2}}{2}$$

Example:

$$\sqrt{27a^3} + 2\sqrt{12a^3} - 3\sqrt{3a}$$
$$\sqrt{(9)(3)(a^2)(a)} + 2\sqrt{(4)(3)(a^2)(a)} - 3\sqrt{3a}$$
$$3a\sqrt{3a} + 2(2)a\sqrt{3a} - 3\sqrt{3a}$$
$$3a\sqrt{3a} + 4a\sqrt{3a} - 3\sqrt{3a}$$
$$7a\sqrt{3a} - 3\sqrt{3a}$$
$$\sqrt{3a}(7a - 3)$$

Exercise 106:

Simplify the following:

(1) $\sqrt{50} + 2\sqrt{8}$ (2) $\sqrt{63} - 2\sqrt{7} + \sqrt{27}$

(3) $2\sqrt[3]{2} - \sqrt[3]{54} + \sqrt[3]{250}$ (4) $x\sqrt{2x} + 4\sqrt{18x^3}$

(5) $\sqrt{\dfrac{1}{2}} + \sqrt{8}$ (6) $\sqrt{48} - 2\sqrt{\dfrac{1}{3}}$

(7) $\sqrt{75} + 4\sqrt{18} + 2\sqrt{12} - 2\sqrt{8}$ (8) $3\sqrt{2a^3} + a\sqrt{18a} - 2\sqrt{8a^3}$

(9) $\sqrt{125} + 2\sqrt{27} - \sqrt{20} + 3\sqrt{12}$ (10) $2\sqrt{12x} - 3\sqrt{\dfrac{1}{3}x}$

(11) $\sqrt{3a} + 5\sqrt{27a^3} - a\sqrt{3a}$ (12) $\sqrt{\dfrac{1}{3}} + 3\sqrt{27} - 2\sqrt{12}$

(13) $\sqrt{\dfrac{3}{5}} + 5\sqrt{60} - 3\sqrt{15}$ (14) $3\sqrt{50} - 4\sqrt{8} + \sqrt{27} - \sqrt{3}$

(15) $\sqrt{x^4y} - x\sqrt{9x^2y} + x^2\sqrt{16y}$ (16) $a^2\sqrt{8a^3b} - 2a^3\sqrt{18ab} + 3a\sqrt{50a^5b}$

(17) $\sqrt{3x^2y^3} - \sqrt{12x^3y} + \sqrt{27x^5y} - \sqrt{75y}$

(18) $\dfrac{\sqrt{25x}}{x} + \sqrt{\dfrac{9}{x}} - \dfrac{5}{\sqrt{x}}$

(19) $\sqrt{16ab^3} - \sqrt{9a^3b} + \sqrt{25a^3b^3}$ (20) $\sqrt{\dfrac{9ab^2}{2}} - \sqrt{\dfrac{a^4b}{3}} + \sqrt{\dfrac{a^3}{8}} + \sqrt{\dfrac{3b}{a^2}}$

(21) $\sqrt{\dfrac{a}{2}} - \sqrt{\dfrac{3a}{2}} + \sqrt{\dfrac{5a}{2}}$ (22) $\dfrac{x\sqrt{8x}}{y} - \dfrac{\sqrt{18x^3}}{y} + \dfrac{11x^2}{y\sqrt{2x}}$

10.5
Multiplication and Division of Radicals

When multiplying multinomials containing radicals, you should always make use of the special products of mathematics. In fact, this is one of the many reasons why you were asked to memorize the special products in the first place.

> **The Square of a Binomial**
>
> Since $(a + b)^2 = a^2 + 2ab + b^2$
> then $(3 + \sqrt{5})^2 = 9 + 6\sqrt{5} + 5 = 14 + 6\sqrt{5}$
> and $(2 + \sqrt{x})^2 = 4 + 4\sqrt{x} + x$

> **The Product of Two Binomials**
>
> Since $(a + 2b)(a - 3b) = a^2 - ab - 6b^2$
> then $(3 + 2\sqrt{5})(3 - 3\sqrt{5}) = 9 - 3\sqrt{5} - 6(5) = 9 - 3\sqrt{5} - 30 = -21 - 3\sqrt{5}$
> and $(3 + \sqrt{x})(4 - \sqrt{x}) = 12 + \sqrt{x} - x$

> **The Product of the Sum and the Difference of the Same Two Numbers**
>
> Since $(a + b)(a - b) = a^2 - b^2$
>
> then $(4 + \sqrt{7})(4 - \sqrt{7}) = (4)^2 - (\sqrt{7})^2 = 16 - 7 = 9$
>
> and $(3 + \sqrt{a})(3 - \sqrt{a}) = (3)^2 - (\sqrt{a})^2 = 9 - a$

Exercise 107:

Multiply the following:

(1) $(\sqrt{7} - 1)^2$

(2) $(3 + \sqrt{2})^2$

(3) $(\sqrt{5} + 1)(\sqrt{5} - 1)$

(4) $(\sqrt{11} + 3)(\sqrt{11} - 5)$

(5) $(\sqrt{3} + \sqrt{7})^2$

(6) $(\sqrt{10} + \sqrt{6})(\sqrt{10} - \sqrt{6})$

(7) $(2\sqrt{5} + 3)(3\sqrt{5} - 4)$

(8) $(4 + 2\sqrt{7})(4 - 2\sqrt{7})$

(9) $(\sqrt{3} + 5)(\sqrt{3} - 5)$

(10) $(\sqrt{7} + 4)^2$

(11) $(2\sqrt{3} - 4)(2\sqrt{3} + 4)$

(12) $(5\sqrt{3} + 1)^2$

(13) $(2\sqrt{7} - 1)(3\sqrt{7} + 2)$

(14) $(\sqrt{7} + \sqrt{3})(3\sqrt{7} - 2\sqrt{3})$

(15) $(3\sqrt{5} - 2\sqrt{7})^2$

(16) $(2\sqrt{3} - 4\sqrt{5})(2\sqrt{3} + 4\sqrt{5})$

(17) $(a + \sqrt{b})(a - \sqrt{b})$

(18) $(2\sqrt{7} + 3\sqrt{2})(\sqrt{7} - 5\sqrt{2})$

(19) $(a + \sqrt{b})(a + \sqrt{b})$

(20) $(\sqrt{a} + \sqrt{b})(\sqrt{a} - \sqrt{b})$

(21) $(x - \sqrt{3y})(x + \sqrt{3y})$

(22) $(\sqrt{x} + 5)(\sqrt{x} - 5)$

(23) $(\sqrt{x} + \sqrt{y})^2$

(24) $(3x + \sqrt{x})(3x - \sqrt{x})$

(25) $(\sqrt{a} - \sqrt{b})^2$

(26) $(2\sqrt{x} - \sqrt{y})(2\sqrt{x} + \sqrt{y})$

(27) $(3\sqrt{a} + 4)(3\sqrt{a} - 4)$

(28) $(\sqrt{5x} + 3)(\sqrt{5x} - 3)$

With quotients involving radicals we are not generally concerned with the division process itself other than to divide out common factors when they occur. But we are concerned with the second Never, since we can generally get a simpler form of such a quotient by making the radicals vanish from the denominator. In the following quotients, therefore, our purpose is to rationalize the denominator.

Examples:

(1) $\dfrac{\sqrt{3}}{\sqrt{7}} = \dfrac{\sqrt{3}}{\sqrt{7}} \cdot \dfrac{\sqrt{7}}{\sqrt{7}} = \dfrac{\sqrt{21}}{7}$

(2) $\dfrac{2}{\sqrt{5}} = \dfrac{2}{\sqrt{5}} \cdot \dfrac{\sqrt{5}}{\sqrt{5}} = \dfrac{2\sqrt{5}}{5}$

(3) $\dfrac{3}{1 + \sqrt{2}} = \dfrac{(3)}{(1 + \sqrt{2})} \cdot \dfrac{(1 - \sqrt{2})}{(1 - \sqrt{2})} = \dfrac{3 - 3\sqrt{2}}{1 - 2} = \dfrac{3 - 3\sqrt{2}}{-1} = -3 + 3\sqrt{2}$

(4) $\dfrac{2 + \sqrt{3}}{4 - \sqrt{3}} = \dfrac{(2 + \sqrt{3})}{(4 - \sqrt{3})} \cdot \dfrac{(4 + \sqrt{3})}{(4 + \sqrt{3})} = \dfrac{8 + 6\sqrt{3} + 3}{16 - 3} = \dfrac{11 + 6\sqrt{3}}{13}$

(5) $\dfrac{1 + \sqrt{x}}{1 - \sqrt{x}} = \dfrac{(1 + \sqrt{x})}{(1 - \sqrt{x})} \cdot \dfrac{(1 + \sqrt{x})}{(1 + \sqrt{x})} = \dfrac{1 + 2\sqrt{x} + x}{1 - x}$

Exercise 108:

Rationalize the denominators.

(1) $\dfrac{7}{\sqrt{3}}$ 　　(2) $\dfrac{5}{\sqrt{5}}$ 　　(3) $\dfrac{\sqrt{11}}{\sqrt{6}}$

(4) $\dfrac{1}{1 + \sqrt{2}}$ 　　(5) $\dfrac{2}{\sqrt{3} - 1}$ 　　(6) $\dfrac{3 + \sqrt{3}}{\sqrt{3} - 1}$

(7) $\dfrac{2 + 2\sqrt{3}}{3 - 2\sqrt{3}}$ 　　(8) $\dfrac{\sqrt{14} - 2}{\sqrt{7} - \sqrt{2}}$ 　　(9) $\dfrac{3 + \sqrt{6}}{\sqrt{3} + \sqrt{2}}$

(10) $\dfrac{2 + \sqrt{10}}{\sqrt{2} + \sqrt{5}}$ 　　(11) $\dfrac{\sqrt{35} - 5}{\sqrt{7} - \sqrt{5}}$ 　　(12) $\dfrac{\sqrt{ab} - a}{\sqrt{b} - \sqrt{a}}$

(13) $\dfrac{\sqrt{6} + \sqrt{3}}{\sqrt{6} - \sqrt{3}}$ 　　(14) $\dfrac{\sqrt{14} - \sqrt{7}}{\sqrt{14} + \sqrt{7}}$ 　　(15) $\dfrac{2 + \sqrt{10}}{\sqrt{2} + \sqrt{5}}$

(16) $\dfrac{a + \sqrt{ab}}{\sqrt{a} + \sqrt{b}}$ 　　(17) $\dfrac{\sqrt{10} - \sqrt{5}}{\sqrt{10} + \sqrt{5}}$ 　　(18) $\dfrac{\sqrt{8} - \sqrt{2}}{\sqrt{8} + \sqrt{2}}$

(19) $\dfrac{3\sqrt{2} - \sqrt{5}}{3 + \sqrt{5}}$ 　　(20) $\dfrac{2 + \sqrt{6}}{2 + \sqrt{3}}$ 　　(21) $\dfrac{\sqrt{15} - 3}{\sqrt{5} - \sqrt{3}}$

(22) $\dfrac{2\sqrt{5} + \sqrt{3}}{1 - \sqrt{3}}$ 　　(23) $\dfrac{\sqrt{x}}{1 + \sqrt{x}}$ 　　(24) $\dfrac{a - \sqrt{b}}{a + \sqrt{b}}$

(25) $\dfrac{2\sqrt{x} - 1}{2\sqrt{x} + 3}$ 　　(26) $\dfrac{a + b}{\sqrt{a} + \sqrt{b}}$ 　　(27) $\dfrac{\sqrt{x} + \sqrt{y}}{\sqrt{x} - \sqrt{y}}$

10.6
The Reduction of Fractions Containing Radicals

In Section 4.4 we discussed the process of reducing a fraction to its simplest terms. As we observed then, a fraction can be reduced to a simpler form if the same factor appears in both the numerator and the denominator. While this is always true, in fractions containing radicals, the successful search for common factors is frequently dependent upon the correct simplification of the radical. Consequently, in order to be sure of finding the simplest form, *always simplify the radical first.*

Example 1:

Reduce the fraction $\dfrac{2 + \sqrt{12}}{2}$.

(1) Simplify the radical first.

$$\sqrt{12} = \sqrt{(4)(3)} = 2\sqrt{3}$$

(2) $\dfrac{2 + \sqrt{12}}{2} = \dfrac{2 + 2\sqrt{3}}{2}$

(3) Factor the numerator.

$$\frac{2 + 2\sqrt{3}}{2} = \frac{2(1 + \sqrt{3})}{2} = 1 + \sqrt{3}$$

Therefore $\dfrac{2 + \sqrt{12}}{2} = 1 + \sqrt{3}$.

Example 2:

Reduce the fraction $\dfrac{10 - \sqrt{150 - 25}}{5}$.

(1) Simplify the radical first.

$$\sqrt{150 - 25} = \sqrt{125} = \sqrt{5(25)} = 5\sqrt{5}$$

(2) $\dfrac{10 - \sqrt{150 - 25}}{5} = \dfrac{10 - 5\sqrt{5}}{5}$

(3) Factor the numerator.

$$\frac{10 - 5\sqrt{5}}{5} = \frac{5(2 - \sqrt{5})}{5} = 2 - \sqrt{5}$$

Therefore $\dfrac{10 - \sqrt{150 - 25}}{5} = 2 - \sqrt{5}$.

Exercise 109:

Reduce the fractions.

(1) $\dfrac{6 + \sqrt{27}}{3}$

(2) $\dfrac{4 + \sqrt{8}}{4}$

(3) $\dfrac{5 - \sqrt{49 + 1}}{10}$

(4) $\dfrac{2 + \sqrt{64 - 4}}{8}$

(5) $\dfrac{6 - \sqrt{50 - 5}}{6}$

(6) $\dfrac{3 - \sqrt{9 + 40}}{4}$

(7) $\dfrac{4 + \sqrt{16 + 48}}{6}$ (8) $\dfrac{-5 + \sqrt{25 + 96}}{8}$

(9) $\dfrac{6 - \sqrt{30 - 2}}{2}$ (10) $\dfrac{6 - \sqrt{24 - 4}}{2}$

(11) $\dfrac{-11 - \sqrt{25 - 16}}{7}$ (12) $\dfrac{7 + \sqrt{100 - 2}}{7}$

(13) $\dfrac{6 + \sqrt{49 - 4}}{3}$ (14) $\dfrac{4 - \sqrt{100 + 28}}{8}$

10.7
Comments on "the simplest form"

The notion of "simplifying" an algebraic expression is one of the most baffling problems a student of elementary algebra faces. The term "simplify" is just too vague, which is one reason that the usual procedures for simplifying radicals have been rather arbitrarily spelled out in the preceding sections. We have concentrated primarily on the "simplest radical form," which is *generally* the most helpful in handling algebraic expressions, but not *always*. For example, if we wish to add

$$\frac{1}{\sqrt{a}} + \frac{\sqrt{b}}{a}$$

it is easier to rationalize the first denominator and write

$$\frac{\sqrt{a}}{a} + \frac{\sqrt{b}}{a} = \frac{\sqrt{a} + \sqrt{b}}{a}$$

But if we wish to multiply

$$\frac{1}{\sqrt{2}} \cdot \frac{1}{\sqrt{3}} \cdot \frac{5}{\sqrt{6}}$$

we should go right ahead and multiply

$$\frac{1}{\sqrt{2}} \cdot \frac{1}{\sqrt{3}} \cdot \frac{5}{\sqrt{6}} = \frac{5}{\sqrt{36}} = \frac{5}{6}$$

It would be ridiculous to write

$$\frac{1}{\sqrt{2}} \cdot \frac{1}{\sqrt{3}} \cdot \frac{5}{\sqrt{6}} = \frac{\sqrt{2}}{2} \cdot \frac{\sqrt{3}}{3} \cdot \frac{5\sqrt{6}}{6} = \frac{5\sqrt{36}}{36}$$

$$= \frac{5(\cancel{6})}{\cancel{36}}_{6}$$

$$= \frac{5}{6}$$

As for the so-called "simplest form" we may specify "simplest radical form" or "simplest exponential form." We might even ask for "simplest computational form" from the point of view of hand computation. Or on occasion we might demand the "simplest typographical form"! To illustrate:

(1) $\dfrac{1}{\sqrt[5]{2}+1} = \dfrac{\sqrt[5]{16} - \sqrt[5]{8} + \sqrt[5]{4} - \sqrt[5]{2} + 1}{3}$

The right side is in "simplest radical form," but for most purposes the left side would be easier to deal with typographically or computationally.

(2) $16\sqrt{2} = 2^{9/2}$

The left side is in "simplest radical form," the right side in "simplest exponential form," and they are of comparable complexity. But note that if we wished to compute this value to several decimal places by using the method of extracting a square root taught in grade school, we could just as easily use the form $\sqrt{512}$ as $16\sqrt{2}$. There would be no computational advantage in finding the $\sqrt{2}$ to several decimal places and then having to multiply the result by 16.

Now consider

(3) $27\sqrt[20]{2654208} = 2^{3/4} \cdot 3^{16/5}$

The left side is in "simplest radical form" and the right side in "simplest exponential form." No one in his right mind would maintain that the left side is really simpler than the right.

The foregoing remarks and examples all illustrate a fact mentioned earlier: that, while the first two Nevers are excellent ground rules for the beginner, they are far from absolute. After you have mastered these elementary techniques and understand them *fully*, you will be in a position to appreciate the fact that "the simplest form" will ultimately depend on the situation and purpose at hand plus your own good judgment in the matter.

COMPLEX NUMBERS

10.8
The Symbol *i*

In the sixteenth century, mathematicians who were busily pursuing their art began to run into problems whose solutions required the square roots of negative numbers, such as $\sqrt{-16}$. They had actually collided with this nemesis before the sixteenth century, but it was not until then that they began to take their present view of this source of puzzlement.

Since the square root of any number means one of its two *identical* factors, the $\sqrt{-16}$ has no meaning as a real number. Think it over. What two numbers, exactly alike, will give -16 when multiplied together? Certainly no *real* number could supply an answer. And that brought on another difficulty. In trying to express the $\sqrt{-16}$, the mathematicians had encountered an idea *for which no one had yet devised a generally accepted definition or symbol.* For instance, the symbols we readily recognize, such as "7" or "8," were not found lying under some ancient stone. The symbols "7" and "8" were deliberately invented by the human race to represent specific quantities.

Now it had become necessary to invent a symbol for numbers *whose squares were negative numbers*. Rather than devise some new squiggle, such as ℘ or ℑ, the mathematicians took a ready-made symbol that had no mathematical meaning at all and gave it one. They borrowed the letter "*i*" from the alphabet and defined i^2 to equal -1. The Greeks had done much the same thing in selecting a symbol for the universal constant $3.14159\ldots$. They took a letter pi (π) from their alphabet and gave it that value.

Thus i^2 was endowed with a mathematical meaning.

$$i^2 = -1$$

and in all mathematics (except that of radio and electronics where the letter "*j*" is used) i^2 *always* means -1.

$$\begin{aligned} \text{Since } & i^2 = -1 \\ \text{then } & \sqrt{i^2} = \sqrt{-1} \\ \text{and } & i = \sqrt{-1} \end{aligned}$$

Great care should be exercised in using these symbols correctly. Please note the following:

i^2 is always written as -1
$\sqrt{-1}$ is always written as i

After the numbers whose squares are negative had been given a symbolic name, it became a simple matter to write them in terms of that symbol.

Examples:

(1) $\sqrt{-16} = \sqrt{16(-1)} = \sqrt{16}\sqrt{-1} = 4i$
     $\underline{\sqrt{-16} = 4i}$

To show that $4i$ is one of two identical factors of -16,

$$(4i)(4i) = (4)(4)(i)(i) = 16i^2 = 16(-1) = -16$$

We should also note that

$$(-4i)(-4i) = (-4)(-4)(i)(i) = 16i^2 = 16(-1) = -16$$

Thus, since $\sqrt{-16}$ could possibly have two meanings, either $4i$ or $-4i$, we agree to let $\sqrt{-16} = 4i$ and $-\sqrt{-16} = -4i$.

$$\boxed{\begin{array}{c} \sqrt{-16} = 4i \\ -\sqrt{-16} = -4i \end{array}}$$

(2) $\sqrt{-4} = \sqrt{(4)(-1)} = \sqrt{4}\sqrt{-1} = 2i$
     Thus $\sqrt{-4} = 2i$
     because $(2i)(2i) = (2)(2)(i)(i) = 4i^2 = 4(-1) = -4$

(3) $\sqrt{-18} = \sqrt{(18)(-1)} = \sqrt{(9)(2)(-1)} = \sqrt{9}\sqrt{2}\sqrt{-1}$
     $$\qquad\qquad\qquad\qquad\qquad\qquad\quad \downarrow\ \ \downarrow\ \ \downarrow$$
     $$= \quad 3\sqrt{2}\ \ i$$
     so $\sqrt{-18} = 3i\sqrt{2}$

Note: In an expression such as $3i\sqrt{2}$, we prefer to write the factor i in *front* of the radical to emphasize the fact that it does not belong *under* the radical.

$$(3i\sqrt{2})(3i\sqrt{2}) = (3)(3)(i)(i)(\sqrt{2})(\sqrt{2}) = (9)(i^2)(2)$$
$$= (9)(-1)(2) = -18$$

(4) $\sqrt{-7} = \sqrt{(7)(-1)} = \sqrt{7}\sqrt{-1} = \sqrt{7}\,i = i\sqrt{7}$
     because $(i\sqrt{7})(i\sqrt{7}) = (i)(i)(\sqrt{7})(\sqrt{7}) = i^2(7) = (-1)(7) = -7$

(5) $\sqrt{-24a^3} = \sqrt{(4)(6)(-1)(a^2)(a)} = \sqrt{4}\sqrt{6}\sqrt{-1}\sqrt{a^2}\sqrt{a}$ $(a > 0)$
     $$= \sqrt{4}\sqrt{a^2}\sqrt{-1}\sqrt{6}\sqrt{a}$$
     $$\qquad\quad \downarrow\ \ \downarrow\ \ \downarrow\ \ \downarrow\ \ \downarrow$$
     $$= \ 2\ \ a\ \ i\ \ \sqrt{6}\sqrt{a}$$
     so $\sqrt{-24a^3} = 2ai\sqrt{6a}$

because $(2ai\sqrt{6a})(2ai\sqrt{6a}) = 4a^2i^2(6a) = 24a^3i^2 = 24a^3(-1) = -24a^3$

In the fundamental operations of addition, subtraction, multiplication, and division, the symbol i behaves as any other literal symbol such as x or y.

$$\text{Since } 4x + 2x = 6x$$
$$\text{then } 4i + 2i = 6i$$

$$\text{Since } 5 + 2y = 5 + 2y$$
$$\text{then } 5 + 2i = 5 + 2i$$

$$\text{Since } \frac{6x^3}{x} = \frac{6xx\cancel{x}}{\cancel{x}} = 6x^2$$

$$\text{then } \frac{6i^3}{i} = \frac{6iii}{i} = 6i^2$$

$$\text{but } i^2 = -1$$

$$\text{then } \frac{6i^3}{i} = 6i^2 = 6(-1) = -6$$

Because the mathematicians of the sixteenth century were mildly distrustful of this new kind of number, they gave the numbers containing i the unfortunate and misleading name of *imaginary numbers*. The name somehow stuck, and we are stuck with it. Just do not let it fool you. Every symbol we use for *any* quantity is nothing more than a figment of our creative imaginations, and "$4i$" is no more imaginary than "7" or "8" is imaginary in the literal sense of the word.

The adoption of the symbol i to represent $\sqrt{-1}$ leads us to the third *Never* in simplifying radicals, where the word *never*, as explained in Section 10.3, is still an emphatic version of, "Well, *hardly* ever."

The Third Never Never leave a *negative* (numerical) *factor* under a square root radical.

This *Never* comes from the fact that $\sqrt{-1}$ is always written as i. In the following examples and exercises all literal symbols represent positive numbers.

Examples:

(1) $\sqrt{-25} = \sqrt{25(-1)} = \sqrt{25}\sqrt{-1} = 5i$

(2) $\sqrt{-27} = \sqrt{(9)(3)(-1)} = \sqrt{9}\sqrt{-1}\sqrt{3} = 3i\sqrt{3}$

(3) $\sqrt{\dfrac{-2a^3}{3}} = \sqrt{\dfrac{-2a^3}{3}} = \dfrac{\sqrt{(2)(-1)(a^2)(a)}}{\sqrt{3}} = \dfrac{ai\sqrt{2a}}{\sqrt{3}} = \dfrac{ai\sqrt{2a}}{\sqrt{3}} \cdot \dfrac{\sqrt{3}}{\sqrt{3}} = \dfrac{ai\sqrt{6a}}{3}$

(4) $\sqrt{9x^2 - 25x^2} = \sqrt{-16x^2} = \sqrt{(16)(-1)(x^2)} = 4ix$

(5) $\dfrac{5 - \sqrt{6 - 31}}{10} = \dfrac{5 - \sqrt{-25}}{10} = \dfrac{5 - \sqrt{(25)(-1)}}{10}$

$$= \dfrac{5 - 5i}{10} = \dfrac{\cancel{5}(1 - i)}{\underset{2}{\cancel{10}}} = \dfrac{1 - i}{2}$$

Exercise 110:

Simplify as much as possible.

(1) $\sqrt{-36}$ (2) $\sqrt{-100}$

(3) $\sqrt{4 - 20}$ (4) $\sqrt{9 - 16}$

(5) $\sqrt{-50}$ (6) $\sqrt{-24}$

(7) $\sqrt{-49a^2}$ (8) $\sqrt{16x^2 - 25x^2}$

(9) $\sqrt{-8a^3}$ (10) $\sqrt{-81a^3b^2}$

(11) $\sqrt{-75x^7y^5}$ (12) $\sqrt{-\dfrac{3}{5}x^3y^2}$

(13) $\sqrt{\dfrac{-18x^5y}{a^2b^3}}$ (14) $\sqrt{36a^2 - 100a^2}$

(15) $\sqrt{-\dfrac{2}{3}a^3b}$ (16) $\sqrt{\dfrac{-20xy^2}{3ab^2}}$

(17) $\dfrac{6 + \sqrt{3 - 12}}{3}$ (18) $\dfrac{2 - \sqrt{4 - 40}}{2}$

(19) $\dfrac{8 - \sqrt{8 - 12}}{4}$ (20) $\dfrac{a^2 - \sqrt{7a^3 - 16a^3}}{a}$

10.9
Complex Numbers

Heretofore, in the algebraic operations which we have discussed, all literal symbols represented real numbers only. But with the addition of the symbol i to the realm of numbers, we have extended the number system to include numbers such as $6i$, $-4i$, $i\sqrt{3}$, $0i$, and $2/3i$, none of which are real numbers *except* $0i$ since $0i = 0$. These numbers all have the form bi, where b is a real number and i is the square root of -1. The elements of this set of numbers are called the *pure imaginary numbers*, and we shall henceforth refer to this set by the letter I.

$$I = \text{the set of all pure imaginary numbers}$$

or

$$I = \{bi \mid b \in R \quad \text{and} \quad i = \sqrt{-1}\}^*$$

This set of numbers not only gives us a means of expressing the square roots of negative numbers. It also provides us with solutions to algebraic equations such as $x^2 + 9 = 0$.

Examples:

(1) $\{x \mid x^2 + 9 = 0\} = \{3i, -3i\}$ $3i$ and $-3i \in I$

(2) $\{x \mid x^2 + 5 = 0\} = \{i\sqrt{5}, -i\sqrt{5}\}$ $i\sqrt{5}$ and $-i\sqrt{5} \in I$

Compare these with

(3) $\{x \mid x^2 - 4 = 0\} = \{2, -2\}$ 2 and $-2 \in R$

Techniques for finding these solution sets are discussed later, but here it might be instructive for the student to verify that the numbers given actually are solutions.

In the examples above *all* the equations are "polynomial equations," meaning that the left side is a polynomial when the right side is zero. In mathematics, the phrase *algebraic number* means a number that can be a root of a polynomial equation with rational coefficients, such as $x^2 + 9 = 0$ or $x^2 - 4 = 0$. Consequently, *both* of the sets I and R contain algebraic numbers. So the set of pure "imaginary" numbers is a logical extension of the number system and, indeed, a necessary one if we are to attempt to express the square roots of the negative numbers or solve equations such as $x^2 + 9 = 0$. The student should note carefully a close relationship of the elements of I and R: the square of any element of I produces an element of R.

1. $(2i)^2 = (2i)(2i) = 4i^2 = 4(-1) = -4$
            $2i \in I$ $-4 \in R$

2. $(i\sqrt{5})^2 = (i\sqrt{5})(i\sqrt{5}) = 5i^2 = 5(-1) = -5$
            $i\sqrt{5} \in I$ $-5 \in R$

Figure 10.5 is the diagram of the real numbers shown in Chapter 4.

---

*Some authors exclude $0 = 0i$ from the set of pure imaginaries. The definition given here adopts the view of algebraists who retain the zero so as to preserve more of the algebraic structure. Thus, if the zero is retained, the set $I = \{bi \mid b \in R\}$ forms an Abelian group under addition. Also, over the reals I forms a one-dimensional vector space with basis $\{i\}$, and the complex plane is the Cartesian product of R and I. None of these statements is possible with the zero excluded from I.

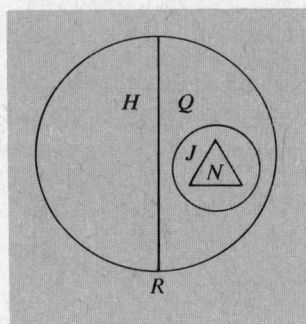

Figure 10.5

Now let us consider two operations with the two sets R and I.

$R \cup I =$ the set of all real and all pure imaginary numbers

$R \cap I = \{0\}$

In Figure 10.6 the elements of these different sets are represented by the areas enclosed by the plane figures.

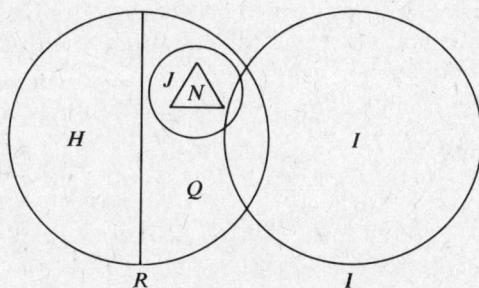

$R \cap I = \{0\}$

Figure 10.6

Figure 10.7 is a diagram representing only the two sets R and I.

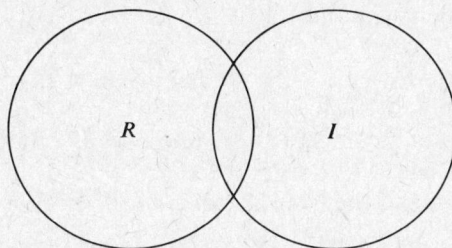

$R \cap I = \{0\}$

Figure 10.7

Now let us take Figure 10.7 and enclose all of it in an area C. Would C represent some universal set? Is there a set of numbers to which *all* of the elements of R and I belong? (See Figure 10.8.)

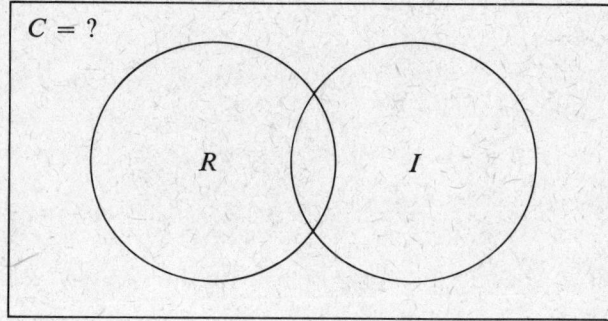

Figure 10.8

The answer to that question is yes. There is a set of numbers, C, which contains *both* R and I as subsets.

Not only can we *find* numbers which belong to C, we literally cannot avoid them when we attempt to find solutions for equations such as $x^2 - 2x + 10 = 0$. By substituting either number for x in this equation, we can prove that the solution set for $x^2 - 2x + 10 = 0$ is $\{1 + 3i, 1 - 3i\}$. Substituting the number $1 + 3i$,

$$x^2 - 2x + 10 = 0$$
$$(1 + 3i)^2 - 2(1 + 3i) + 10 = 0$$
$$1 + 6i + 9i^2 - 2 - 6i + 10 = 0$$
$$1 + 6i + 9(-1) - 2 - 6i + 10 = 0$$
$$1 + 6i - 9 - 2 - 6i + 10 = 0$$
$$1 + 10 - 9 - 2 + 6i - 6i = 0$$
$$11 - 11 + 6i - 6i = 0$$
$$0 = 0$$

The student may show that $1 - 3i$ is a solution of this equation in the same manner. Let us consider these two numbers:

$$1 + 3i \quad \text{and} \quad 1 - 3i$$

They are algebraic numbers *since they are the roots of a polynomial equation with rational coefficients, but they do not belong to R because they cannot be expressed without the symbol i, and they do not belong to I because neither one of them is of

---

*The fact that some of the elements of a set are algebraic numbers does not mean that all of the elements must be algebraic numbers. Here we are merely showing that the numbers $1 + 3i$ and $1 - 3i$ have a common bond with some of the elements of I and R.

the form bi, where $b \in R$ and $i = \sqrt{-1}$. The way to symbolize the form of these numbers is

$$a + bi$$

where $a, b \in R$ and $i = \sqrt{-1}$.

We could find infinitely many more examples of such numbers by letting a and b equal *any real numbers*.

Examples:

$2 + 5i$	$a = 2$	$b = 5$
$6 - 10i$	$a = 6$	$b = -10$
$-4 + 8i$	$a = -4$	$b = 8$
$-\sqrt{3} - 7i$	$a = -\sqrt{3}$	$b = -7$
$\frac{3}{4} - \frac{1}{2}i$	$a = \frac{3}{4}$	$b = -\frac{1}{2}$

Remember that, in a number of the type $a + bi$, a and b can be any real numbers.

I.   Consider now what happens to these numbers when $a = 0$.

   1. $0 + 7i$ $[a = 0, b = 7]$
       but $0 + 7i = 7i$ and $7i \in I$

   2. $0 - 4i$ $[a = 0, b = -4]$
       $0 - 4i = -4i$ and $-4i \in I$

   3. $0 - i\sqrt{5}$ $[a = 0, b = -\sqrt{5}]$
       $0 - i\sqrt{5} = -i\sqrt{5}$ and $-i\sqrt{5} \in I$

Therefore, *when $a = 0$ in a number of the form $a + bi$, the number has the form bi, which is an element of I.*

II.   Now let $b = 0$ in a number of the form $a + bi$.

   1. $4 + 0i$ $[a = 4, b = 0]$
       but $4 + 0i = 4$ and $4 \in R$

   2. $-3 - 0i$ $[a = -3, b = 0]$
       $-3 - 0i = -3$ and $-3 \in R$

   3. $\sqrt{7} + 0i$ $[a = \sqrt{7}, b = 0]$
       $\sqrt{7} + 0i = \sqrt{7}$ and $\sqrt{7} \in R$

Numbers of the form $a + bi$ are called *complex numbers*, and this set of numbers actually is the universal set of our number system. As we have shown, the set of real numbers, R. and the set of pure imaginary numbers, I, are both subsets of the set of complex numbers. To emphasize these points in your mind, study the following statements carefully:

Every complex number has the form $a + bi$.

Every real number is a *complex number* in which $b = 0$.

Every pure imaginary number is a *complex number* in which $a = 0$.

For future reference, all complex numbers that are also real numbers will continue to be called real numbers; all *other* complex numbers (i.e., those in which the co-efficient of i is not zero) will be called imaginary numbers.

We shall call the set of complex numbers C

$$C = \{a + bi \mid a,b \in R \quad \text{and} \quad i = \sqrt{-1}\}$$

Figures 10.9 and 10.10 are two diagrams of the entire number system, the second showing the relationships of some significant subsets of R.

$C = $ the set of complex numbers
$R = $ the set of real numbers
$I = $ the set of pure imaginary numbers
$Q = $ the set of rational numbers
$H = $ the set of irrational numbers
$J = $ the set of integers
$N = $ the set of natural numbers

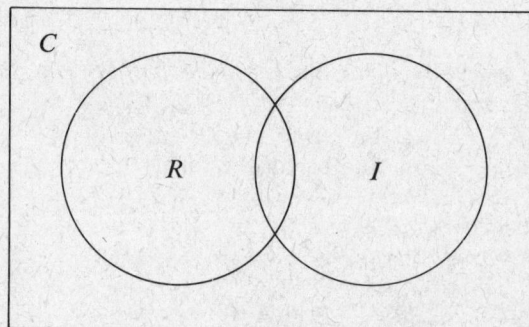

Figure 10.9

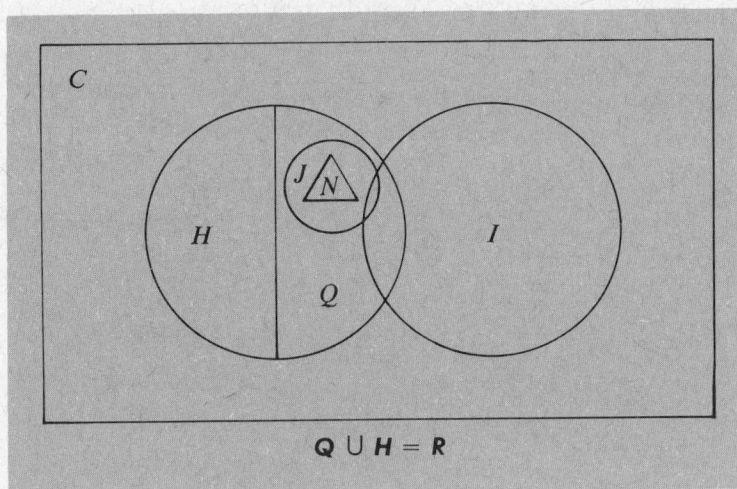

$$Q \cup H = R$$

Figure 10.10

Exercise 111:

Assign each of the following numbers to the correct sets C, R, and I:

(1) 6 $6 \in C$ and $6 \in R$ (2) $-6i$ $-6i \in C$ and $-6i \in I$

(3) $2 + 7i$ $2 + 7i \in C$ (4) $\sqrt{3}$

(5) $i\sqrt{3}$ (6) $4 - i\sqrt{5}$

(7) 10,285   (8) $\sqrt{7} + 2i$

(9) $\sqrt{7} + 2$ (10) $-i + 24$

(11) $4i + 7$ (12) $\sqrt{13} - 2$

(13) $i(\sqrt{7} + 1)$ (14) $-16 - 20i$

(15) What is an algebraic number?

10.10
Operations with Complex Numbers

It can be demonstrated, although we shall not do so here, that in the four fundamental operations — addition, subtraction, multiplication, and division — the complex numbers obey the same laws as the elements of one of its subsets, the real numbers. We observed in Section 10.8 that the symbol i behaves like any other literal symbol in the fundamental operations; e.g., $2i + 3i = 5i$, but $2 + 3i = 2 + 3i$. Following are illustrations of the fundamental operations with numbers of the form $a + bi$:

I. *Addition:*

Add $3 + 4i$ and $2 + 7i$.

$$(3 + 4i) + (2 + 7i) = 3 + 4i + 2 + 7i$$
$$= 3 + 2 + 4i + 7i$$
$$= (3 + 2) + (4i + 7i)$$
$$= 5 + i(4 + 7)$$
$$= 5 + i(11)$$
$$= 5 + 11i$$

II. *Subtraction:*

Subtract $2 - 5i$ from $4 + 3i$.

$$(4 + 3i) - (2 - 5i) = 4 + 3i - 2 + 5i$$
$$= 4 - 2 + 3i + 5i$$
$$= (4 - 2) + (3i + 5i)$$
$$= 2 + i(3 + 5)$$
$$= 2 + i(8)$$
$$= 2 + 8i$$

III. *Multiplication:*

Note: In the product of two complex numbers, we always replace i^2 by its equal, -1.

1. $(5 + 3i)^2 = 5^2 + 2(5)(3i) + (3i)^2$
$$= 25 + 30i + 9i^2$$
$$= 25 + 30i + 9(-1)$$
$$= 25 + 30i - 9$$
$$= 16 + 30i$$

2. $(3 + i)(2 + 3i) = 6 + 11i + 3i^2$
$$= 6 + 11i + 3(-1)$$
$$= 6 + 11i - 3$$
$$= 3 + 11i$$

3. $(4 + 3i)(4 - 3i) = 4^2 - (3i)^2$

$$= 16 - 9i^2$$
$$= 16 - 9(-1)$$
$$= 16 + 9$$
$$= 25$$

IV. *Division:*

Note: In dividing complex numbers we proceed as in the division of radicals; i.e., we make the denominator a rational (and therefore a *real*) number.

1. $\dfrac{2}{3 - i} = \dfrac{(2)}{(3 - i)} \cdot \dfrac{(3 + i)}{(3 + i)} = \dfrac{6 + 2i}{9 - i^2} = \dfrac{6 + 2i}{9 - (-1)}$

$$= \dfrac{6 + 2i}{9 + 1}$$
$$= \dfrac{6 + 2i}{10}$$
$$= \dfrac{\cancel{2}(3 + i)}{\cancel{10}_{\,5}}$$
$$= \dfrac{3 + i}{5}$$

2. $\dfrac{2 + i}{2 - i} = \dfrac{(2 + i)}{(2 - i)} \cdot \dfrac{(2 + i)}{(2 + i)} = \dfrac{4 + 4i + i^2}{4 - i^2}$

$$= \dfrac{4 + 4i + (-1)}{4 - (-1)}$$
$$= \dfrac{4 + 4i - 1}{4 + 1}$$
$$= \dfrac{3 + 4i}{5}$$

3. $\dfrac{4 - 2i}{3 + 2i} = \dfrac{(4 - 2i)}{(3 + 2i)} \cdot \dfrac{(3 - 2i)}{(3 - 2i)} = \dfrac{12 - 14i + 4i^2}{9 - 4i^2}$

$$= \dfrac{12 - 14i + 4(-1)}{8 - 4(-1)}$$
$$= \dfrac{12 - 14i - 4}{9 + 4}$$
$$= \dfrac{8 - 14i}{13}$$

We reduce fractions involving complex numbers exactly the same way that we reduce fractions involving real numbers; namely, by dividing out common *factors*. Look over the following examples carefully:

(1) $\dfrac{2 + \sqrt{4 - 20}}{2} = \dfrac{2 + \sqrt{-16}}{2}$

$= \dfrac{2 + 4i}{2}$

$= \dfrac{\cancel{2}(1 + 2i)}{\cancel{2}}$

$= 1 + 2i$

(2) $\dfrac{3 - \sqrt{4 - 40}}{6} = \dfrac{3 - \sqrt{-36}}{6}$

$= \dfrac{3 - 6i}{6}$

$= \dfrac{\cancel{3}(1 - 2i)}{\cancel{6}\,2}$

$= \dfrac{1 - 2i}{2}$

(3) $\dfrac{4 + \sqrt{9 - 21}}{4} = \dfrac{4 + \sqrt{-12}}{4}$

$= \dfrac{4 + \sqrt{(-1)(4)(3)}}{4}$

$= \dfrac{4 + 2i\sqrt{3}}{4}$

$= \dfrac{\cancel{2}(2 + i\sqrt{3})}{\cancel{4}\,2}$

$= \dfrac{2 + i\sqrt{3}}{2}$

Exercise 112:

Perform the indicated operations.

(1) $(3 + 2i) + (4 - 5i)$ (2) $(-7 + 9i) + (2 - 10i)$

(3) $(5 - 6i) - (3 - 2i)$ (4) $(8 + 5i) - (2 + 3i)$

(5) $(2 + 3i\sqrt{7}) + (4 - 5i\sqrt{7})$ (6) $(4 - 2i\sqrt{5}) - (3 + 6i\sqrt{5})$

(7) $(3 - 2i)^2$ (8) $(4 + i)^2$

(9) $(2 + i\sqrt{3})^2$ (10) $(5 + 2i\sqrt{2})^2$

(11) $(2 + 4i)(3 - 5i)$ (12) $(1 + 2i)(6 + 3i)$

(13) $(4 - 5i)(2 - i)$ (14) $(3 + i\sqrt{2})(4 + 3i\sqrt{2})$

(15) $(7 + i)(7 - i)$ (16) $(3 + 4i)(3 - 4i)$

(17) $(6 + 2i)(6 - 2i)$ (18) $\dfrac{6}{1 - 2i}$

(19) $\dfrac{1 + i}{2 - i}$ (20) $\dfrac{4 + 3i}{2 - 3i}$

(21) $\dfrac{6 + i}{6 - i}$ (22) $\dfrac{1 - 4i}{2 + 4i}$

(23) $\dfrac{5 + i\sqrt{3}}{5 - i\sqrt{3}}$ (24) $\dfrac{2 + i\sqrt{5}}{1 - i\sqrt{5}}$

Reduce each of the following fractions:

(25) $\dfrac{5 + \sqrt{25 - 50}}{10}$ (26) $\dfrac{4 - \sqrt{16 - 24}}{2}$

(27) $\dfrac{3 + \sqrt{9 - 45}}{6}$ (28) $\dfrac{-2 - \sqrt{4 - 20}}{4}$

(29) $\dfrac{5 - \sqrt{25 - 100}}{10}$ (30) $\dfrac{2 + \sqrt{4 - 24}}{4}$

(31) Show that $1 + i$ is a solution of the equation $x^2 - 2x + 2 = 0$.

(32) Show that $2 - 3i$ is a solution of the equation $x^2 - 4x + 13 = 0$.

Chapter Test 10

I.   Simplify the following. Express each with the simplest radicand and rational denominators. All literal symbols represent positive numbers.

1. $\sqrt{24a^3 b}$ 2. $\sqrt{-12x^3}$

3. $\sqrt[3]{\dfrac{2a^2}{3b^2}}$ 4. $\sqrt{64x^3 - 4x^3}$

5. $\sqrt{\dfrac{5a}{2b}}$ 6. $\sqrt{\dfrac{3x^3 y^5}{2a^3}}$

7. $\sqrt{\dfrac{-32x^3}{y}}$ 8. $\sqrt[3]{\dfrac{-8x^5}{y^2}}$

II. Multiply the following and simplify:

1. $(2 + \sqrt{5})(3 - 2\sqrt{5})$
2. $(3 - \sqrt{7})^2$
3. $(4 + 2i)^2$
4. $(2\sqrt{5} + 3)(2\sqrt{5} - 3)$
5. $(3 - 2i)(4 + 3i)$
6. $(\sqrt{10} + 2i)(\sqrt{10} - 2i)$

III. Rationalize the denominators and simplify as much as possible.

1. $\dfrac{\sqrt{21} + 3}{\sqrt{7} - \sqrt{3}}$
2. $\dfrac{2\sqrt{5} - 1}{\sqrt{5} + 2}$
3. $\dfrac{3 + 2i}{3 - 2i}$
4. $\dfrac{5}{6 + 2i}$

IV. Reduce the following:

1. $\dfrac{4 + \sqrt{20} + 4}{2}$
2. $\dfrac{-6 - \sqrt{9 - 36}}{3}$
3. $\dfrac{10 + \sqrt{100 - 75}}{5}$
4. $\dfrac{2 + \sqrt{4 - 16}}{2}$

V. Show that $1 + i\sqrt{5}$ is a root of the equation $x^2 - 2x + 6 = 0$.

11
QUADRATIC EQUATIONS IN ONE VARIABLE

Before discussing quadratic equations in one variable, we are going to backtrack and repeat, for emphasis as well as for other good reasons, a definition given in Section 1.8 and again in Section 3.7. In both of those sections a *polynomial* was defined to be a monomial or multinomial in which the variable (or variables) has only *nonnegative integral exponents*. In a real polynomial, the coefficients of the variables are real numbers; a rational polynomial is one in which all coefficients are rational numbers.

There is one class of polynomials, those involving *one variable only*, which are especially important, and we are going to consider here the general form of such polynomials.

Examples:

Polynomials in one variable:

(1) $3x^4 - 6x^3 + 2x^2 - x + 4$ a polynomial in x of degree 4
(2) $3y^2 - 7y + 2$ a polynomial in y of degree 2
(3) $6z^3 + 3z^2 + 2z + 5$ a polynomial in z of degree 3

The general form of such a polynomial can be represented as follows:

Let x be the variable.
Let n be the greatest exponent of x; n is a nonnegative integer.
Let $a_0, a_1, a_2, a_3, a_4 \ldots$ (in general, a_i) represent the coefficients of the terms; all a_i's $\in R$.
Then a *polynomial in x*, which we shall call $P(x)$, has the following form:

$$P(x) = a_0 x^n + a_1 x^{n-1} + a_2 x^{n-2} + a_3 x^{n-3} + \cdots + a_{n-1} x + a_n$$

where n is the degree of the polynomial provided that $a_0 \neq 0$.

The variable x in the above form is, of course, a stand-in for any single variable. If we use the notation $P(x)$ to mean always a polynomial in x, then $P(y)$ and $P(z)$ would mean respectively a polynomial in y and a polynomial in z.

An *equation* of the type $P(x) = 0$ means any equation in which the left side is a polynomial in a single variable when the right side is zero. We would also include in this class of equations any equation which could be put in the form $P(x) = 0$ after clearing all parentheses, combining like terms, and using the addition axiom to make one side of the equation zero.

Examples:

Equations of the type $P(x) = 0$:

(1) $3x^4 - 5x^3 + 2x^2 - 7x + 2 = 0$

(2) $4x^2 = 5x + 7$ or $4x^2 - 5x - 7 = 0$

(3) $x(x^2 - 3x) = 2(x + 1)$ or $x^3 - 3x^2 = 2x + 2$ or $x^3 - 3x^2 - 2x - 2 = 0$

(4) $5x^3 - 2x^2 + 7x = 4x^2 + 2x - 1$ or $5x^3 - 6x^2 + 5x + 1 = 0$

An equation of this type is considered to be in its simplest form when one side is zero and the terms of the polynomial, $P(x)$, on the other side are arranged in descending order of the exponents. The *degree* of such an equation is the degree of $P(x)$.

A quadratic equation in one variable can always be written in the form $P(x) = 0$. The word *quadratic* means "square" or "to the second degree," and a quadratic equation in one variable is one that can be written in the form $P(x) = 0$, where $P(x)$ is a *polynomial of degree two*. In this chapter we shall consider only quadratic equations which have rational coefficients.

Examples:

Quadratic equations in one variable:

(1) $3x^2 = 5x - 2$ or $3x^2 - 5x + 2 = 0$

(2) $5z^2 = 6$ or $5z^2 - 6 = 0$

(3) $10 = 3y - y^2$ or $y^2 - 3y + 10 = 0$

(4) $6t^2 = 0$

(5) $6u^2 = 5u$ or $6u^2 - 5u = 0$

(6) $(x - 3)(x - 1) = 4$ or $x^2 - 4x + 3 = 4$ or $x^2 - 4x - 1 = 0$

(7) $5y^2 - 6y = 3 + 2y^2 - y$ or $3y^2 - 5y - 3 = 0$

In each equation above there is only one variable, and, when each one is put in its simplest form, the left side is a rational polynomial of degree two.

This precise manner of classifying equations of the form $P(x) = 0$ by the degree of the polynomial, $P(x)$, is not simply a matter of observing the mathematical niceties. The methods by which we solve such equations increase sharply in complexity as the *degree* of the equation advances. The method which worked so well for solving first-degree equations in one variable (Chapter 5) is useless for solving quadratic equations in one variable. And the general method which we shall derive for solving these quadratic equations will solve *only* these equations. Consequently one should learn early to be acutely aware of the *degree* of all such equations, since that degree will dictate what general methods of solution are available.

11.1
The Standard Form of a Quadratic Equation in One Variable

From the preceding discussion, we can conclude that the three *essential* ingredients of a quadratic equation in one variable are the following:

1. One variable only — any variety, such as x, y, z, t, $\sin \Theta$, $\tan \alpha$, y' or y''.
2. The variable may have only nonnegative integral exponents of which the greatest must be *two*.
3. And, to have an equation at all, one equal mark.

Look over the following statements carefully:

(1) $3y^2 - 7x = 2$ (2) $x^2 - 3x = x^3 + 4$

(3) $2(\sin \Theta)^2 = 1$ (4) $z^2 = 0$

(5) $2x - 7 = 3x + 4$ (6) $x^2 - 5x + 4$

(7) $3y^2 - 7y = 3y - 4$ (8) $3x^2y - x = 10$

(9) $4z^2 - 7z + 2$ (10) $x^2 + y^2 = 25$

(11) $y^2 - 3y = 2$ (12) $3z - 7z = 4z^2 - 8z + 5$

Give the reasons why, in the statements above, the only quadratic equations in one variable are numbers (3), (4), (7), (11), and (12).

The easiest way to settle the matter of recognizing a quadratic equation in one variable is to set up a universal pattern or standard form into which all of them will fit after all parentheses have been cleared and the like terms have been combined.

To establish such a pattern we shall let x represent the variable. That variable must appear to the second degree; so the general pattern will have a term involving x^2 with any numerical coefficient except zero. The only other *nonnegative integral* powers of x available, with exponents not greater than two, are x^1 and x^0, both of which could have any numerical coefficients including zero. To symbolize these terms we shall let the first letters of the alphabet (a, b, and c) represent the numerical coefficients of the terms and choose the letter a to represent the coefficient of x^2, with the provision that $a \neq 0$. The letters b and c will represent the coefficients of x^1 and x^0 respectively.

$$ax^2 = ax^2$$
$$bx^1 = bx$$
$$cx^0 = c(1) = c$$

Putting these terms on the left side of the equal mark and zero on the right side, we have the following:

$$ax^2 + bx + c = 0 \qquad a \neq 0$$

In the box above we have the universal pattern or the standard form of *every quadratic equation* in one variable, where x is a stand-in for *any* variable, and a, b, and c represent *any* constants except that $a \neq 0$. Therefore, if an equation is a quadratic equation in one variable, it can be made to fit this form. And when such an equation is put in this standard form, we can read off the specific values of a, b, and c in that particular equation.

Examples:

(1) $3x^2 = 7 - 2x$ the variable is x
$3x^2 + 2x - 7 = 0$ standard form

Compare this equation with

$ax^2 + bx + c = 0$
$a = 3, b = 2, c = -7$

(2) $8 = 5y^2 - 4y$ the variable is y
$-5y^2 + 4y + 8 = 0$ standard form

Compare this equation with

$ay^2 + by + c = 0$
$a = -5, b = 4, c = 8$

(3) $2z^2 = 11$ the variable is z
$2z^2 - 11 = 0$
      or
$2z^2 + 0z - 11 = 0$ standard form

Compare this equation with

$az^2 + bz + c = 0$
$a = 2, b = 0, c = -11$

(4) $3t^2 = 4t$ the variable is t
$3t^2 - 4t = 0$
      or
$3t^2 - 4t + 0 = 0$ standard form

Compare this equation with

$$at^2 + bt + c = 0$$
$$a = 3, b = -4, c = 0$$

(5) $(2x - 1)(x + 3) = 6$ the variable is x
     $2x^2 + 5x - 3 = 6$
     $2x^2 + 5x - 9 = 0$ standard form
     $a = 2, b = 5, c = -9$

Exercise 113:

Put the following quadratic equations in standard form and identify the values of a, b, and c:

(1) $3x^2 - 5x + 7 = 0$ (2) $2y^2 = 3y - 8$
(3) $10 = 2t - 5t^2$ (4) $4x^2 = 7x$
(5) $8z^2 = 9$ (6) $3 = 2y - 8y^2$
(7) $24x - 5x^2 = 0$ (8) $7y^2 - 4 = 0$
(9) $z^2 = z - 1$ (10) $3w = 2w - 7w^2$
(11) $6x^2 - 7x = 3x^2 + 2x - 4$ (12) $9y^2 + 6 = 3y^2 - 2y - 5$
(13) $4(x^2 + 2x) = -3 + 3x$ (14) $(x + 2)(3x - 1) = 14$
(15) $(x - 5)(2x + 1) = 10x$ (16) $(3x + 2)(2x - 1) = x^2 - 4$

11.2
Solving a Quadratic Equation by Completing the Square

The method of solving a quadratic equation by completing a square is somewhat tedious, but it has one tremendous virtue: once mastered, you may use it to solve *any* quadratic equation in one variable. It is called the "method of completing the square" because we are going to do exactly that. Remember that a perfect square is any number that has two identical factors.

Examples:

Perfect squares:

(1) $49 = 7^2$
(2) $81 = 9^2$
(3) $a^2 - 6a + 9 = (a - 3)^2$
(4) $x^2 + 10x + 25 = (x + 5)^2$
(5) $y^2 - \dfrac{1}{2}y + \dfrac{1}{16} = \left(y - \dfrac{1}{4}\right)^2$

The preceding comments might well prompt you to ask, "What exactly is meant by completing the square?" and "Why do we want to do it?" The phrase "completing the square" means to write the equation so that the variable is contained in a squared expression on one side of the equal mark and only a constant (i.e., *no variable at all*) is on the other side. We want to do this because, once that feat is accomplished, these equations are almost ridiculously easy to solve.

We shall consider first quadratic equations that are already in this happy form or else are easily put into the desired form. As an example, suppose we have the equation

$$x^2 - 9 = 0$$

Write the equation in the equivalent form

$$x^2 = 9$$

and note the following:

1. The left side of the equation is a variable squared and the right side is a constant.
2. To find the solution set of this equation, we have only to find all the numbers whose square is 9.
3. Clearly, there are exactly two such numbers, 3 and -3, since $(3)^2 = 9$ and $(-3)^2 = 9$.
4. So, if $x^2 = 9$, then $x = 3$ or $x = -3$.

Similarly, for $y^2 - 49 = 0$, we write

$$y^2 = 49$$

So　　　　　　　　　　　$y = 7$　　or　　$y = -7$

And in like manner,

$$2x^2 - 162 = 0$$
$$2x^2 = 162$$
$$x^2 = 81$$

So　　　　　　　　　　　$x = 9$　　or　　$x = -9$

Now suppose we had the equation

$$x^2 - 5 = 0$$

We would write

$$x^2 = 5$$

and the two possible values for x are

$$x = \sqrt{5}　　\text{or}　　x = -\sqrt{5}$$

Note that when we are seeking all possible values of a variable that has been squared, we must take into account that there are always two real numbers which have the same square, namely the number and its negative.

As a further example, if we have

$$t^2 + 36 = 0$$

we write $$t^2 = -36$$

So $$t = \sqrt{-36} \quad \text{or} \quad t = -\sqrt{-36}$$

Hence, $$t = 6i \quad \text{or} \quad t = -6i$$

All of the preceding examples illustrate the same technique, which is to arrange the equation in the following form:

A variable squared = a constant

This same technique may be applied to any equation in which a *variable expression squared* on one side equals a *constant* on the other. Note the following examples of this carefully:

1. $(x - 1)^2 = 9$

   In this equation the left side is the square of the binomial $(x - 1)$, and the *square* of that binomial must equal 9. Hence, the two possible values of $(x - 1)$ are 3 and -3. Therefore, we write

   $$x - 1 = 3 \quad \text{or} \quad x - 1 = -3$$

   Then we solve the two resulting *first-degree* equations for x.

   $$
   \begin{array}{ll}
   x - 1 = 3 & \quad \text{or} \quad x - 1 = -3 \\
   x = 1 + 3 & \quad\quad\quad\quad x = 1 - 3 \\
   \text{So} \quad x = 4 & \quad \text{or} \quad x = -2
   \end{array}
   $$

   Thus the solution set of $(x - 1)^2 = 9$ is $\{4, -2\}$. We check these results in the original equation as follows:

   $$
   \begin{array}{ll}
   \text{For } x = 4 & \quad\quad \text{For } x = -2 \\
   (x - 1)^2 = 9 & \quad\quad (x - 1)^2 = 9 \\
   (4 - 1)^2 = 9 & \quad\quad (-2 - 1)^2 = 9 \\
   (3)^2 = 9 & \quad\quad (-3)^2 = 9
   \end{array}
   $$

2. Similarly, $(x - 1)^2 = 5$.

   $$x - 1 = \sqrt{5} \quad \text{or} \quad x - 1 = -\sqrt{5}$$

So $\qquad x = 1 + \sqrt{5} \qquad$ or $\qquad x = 1 - \sqrt{5}$

Following are other examples of solving a quadratic equation in which one side is a square:

3. $(2x - 1)^2 = 25$

$$
\begin{array}{ll}
2x - 1 = 5 & \text{or} \qquad 2x - 1 = -5 \\
2x = 1 + 5 & \qquad\qquad 2x = 1 - 5 \\
2x = 6 & \qquad\qquad 2x = -4 \\
\end{array}
$$

So $\qquad x = 3 \qquad$ or $\qquad x = -2$

4. $(t - 2)^2 = 11$

$$
t - 2 = \sqrt{11} \qquad \text{or} \qquad t - 2 = -\sqrt{11}
$$

So $\qquad t = 2 + \sqrt{11} \qquad$ or $\qquad t = 2 - \sqrt{11}$

5. $(x + 3)^2 = -4$

$$
\begin{array}{ll}
x + 3 = \sqrt{-4} & \text{or} \qquad x + 3 = -\sqrt{-4} \\
x + 3 = 2i & \qquad\qquad x + 3 = -2i \\
\end{array}
$$

So $\qquad x = -3 + 2i \qquad$ or $\qquad x = -3 - 2i$

6. $\left(x - \dfrac{1}{2}\right)^2 = \dfrac{3}{4}$

$$
x - \frac{1}{2} = \sqrt{\frac{3}{4}} \qquad \text{or} \qquad x - \frac{1}{2} = -\sqrt{\frac{3}{4}}
$$

$$
x - \frac{1}{2} = \frac{\sqrt{3}}{2} \qquad\qquad x - \frac{1}{2} = \frac{-\sqrt{3}}{2}
$$

$$
x = \frac{1}{2} + \frac{\sqrt{3}}{2} \qquad\qquad x = \frac{1}{2} - \frac{\sqrt{3}}{2}
$$

So $\qquad x = \dfrac{1 + \sqrt{3}}{2} \qquad$ or $\qquad x = \dfrac{1 - \sqrt{3}}{2}$

7. $\left(y + \dfrac{2}{3}\right)^2 = \dfrac{-4}{9}$

$$
y + \frac{2}{3} = \sqrt{\frac{-4}{9}} \qquad \text{or} \qquad y + \frac{2}{3} = -\sqrt{\frac{-4}{9}}
$$

$$
y + \frac{2}{3} = \frac{2i}{3} \qquad\qquad y + \frac{2}{3} = \frac{-2i}{3}
$$

$$y = \frac{-2}{3} + \frac{2i}{3} \qquad\qquad y = \frac{-2}{3} - \frac{2i}{3}$$

So $\qquad\qquad y = \frac{-2 + 2i}{3} \qquad$ or $\qquad y = \frac{-2 - 2i}{3}$

Exercise 114:

Find the solution set of the following:

(1) $x^2 = 16$	(2) $y^2 = 8$	(3) $2x^2 = -18$
(4) $x^2 = -3$	(5) $x^2 - 25 = 0$	(6) $x^2 - 24 = 0$
(7) $x^2 + 100 = 0$	(8) $x^2 + 40 = 0$	(9) $(x - 2)^2 = 4$
(10) $(t + 4)^2 = 7$	(11) $(y - 3)^2 = -25$	(12) $(x + 1)^2 = -2$
(13) $\left(x - \frac{1}{6}\right)^2 = \frac{1}{36}$	(14) $\left(y + \frac{1}{5}\right)^2 = \frac{4}{25}$	(15) $\left(t - \frac{1}{2}\right)^2 = \frac{-1}{4}$
(16) $\left(x - \frac{1}{3}\right)^2 = \frac{-2}{9}$	(17) $(x + 3)^2 = \frac{1}{2}$	(18) $\left(x - \frac{1}{4}\right)^2 = \frac{-3}{16}$

When we encounter quadratic equations that are not in the convenient form of those in Exercise 114 (an equation, for example, such as $x^2 - 5x + 6 = 0$), we use a procedure to rewrite the equations in the desired form. There are seven fixed steps in this routine procedure, and the first four of them, when completed, are guaranteed to make the left side of the equation a perfect square. The seven steps in the procedure are outlined as follows:

Example 1:

Solve $3x^2 = 2x + 5$.

(1) Put the equation in standard form.

$$3x^2 - 2x - 5 = 0$$

(2) Using E_a, remove the constant term from the left side. The *constant term* is the term which does not contain the variable, x, and therefore always remains the same regardless of the value of x.

$$3x^2 - 2x = 5$$

(3) Using E_m, divide both sides of the equation by the *coefficient* of the *second-degree term* (in this equation the second-degree term is $3x^2$ and therefore we shall divide both sides by 3).

$$\frac{3x^2}{3} - \frac{2x}{3} = \frac{5}{3}$$

or $\quad x^2 - \frac{2}{3}x = \frac{5}{3}$

(4)   Add to each side of the equation the square of one-half of the coefficient of the *first-degree term* (in this equation the first-degree term is $-2/3x$).

$$-\frac{2}{3} = \text{the } \textit{coefficient} \text{ of } x$$

$$-\frac{\cancel{2}}{3}\cdot\frac{1}{\cancel{2}} = -\frac{1}{3} = \textit{one-half} \text{ of the coefficient of } x$$

$$-\frac{1}{3}\cdot-\frac{1}{3} = \frac{1}{9} = \text{the } \textit{square} \text{ of one-half of the coefficient of } x$$

$$\left[\text{add } \frac{1}{9} \text{ to each side}\right]$$

$$x^2 - \frac{2}{3}x = \frac{5}{3}$$

$$x^2 - \frac{2}{3}x + \frac{1}{9} = \frac{5}{3} + \frac{1}{9}$$

When Step 4 is completed, the left side of the equation will always be a perfect square. Since it is a perfect square, it can be written in factored form as a quantity squared.

(5)   Write the left side of the equation in factored form, and then combine the terms on the right side.

$$x^2 - \frac{2}{3}x + \frac{1}{9} = \frac{5}{3} + \frac{1}{9}$$

$$\left(x - \frac{1}{3}\right)^2 = \frac{15}{9} + \frac{1}{9}$$

$$\left(x - \frac{1}{3}\right)^2 = \frac{16}{9}$$

(6)   List *all possible values* of the second-degree expression, $(x - 1/3)^2$.

If $\left(x - \frac{1}{3}\right)^2 = \frac{16}{9}$

then $x - \frac{1}{3} = \sqrt{\frac{16}{9}}$ or $x - \frac{1}{3} = -\sqrt{\frac{16}{9}}$

Therefore $x - \dfrac{1}{3} = \dfrac{4}{3}$ or $x - \dfrac{1}{3} = -\dfrac{4}{3}$

(7) To compute the values of the variable, x, consider the two possibilities from the previous step.

$$x - \frac{1}{3} = \frac{4}{3} \quad \text{or} \quad x - \frac{1}{3} = -\frac{4}{3}$$

Using E_a, solve the two first-degree equations.

$$x = \frac{4}{3} + \frac{1}{3} \quad \text{or} \quad x = -\frac{4}{3} + \frac{1}{3}$$

$$x = \frac{5}{3} \quad \text{or} \quad x = -\frac{3}{3}$$

$$x = -1$$

The solution set is $\left\{\dfrac{5}{3}, -1\right\}$.

Following are the same seven steps repeated in abbreviated form:

Solve $3x^2 = 2x + 5$.

Step 1 $3x^2 - 2x - 5 = 0$

Step 2 $3x^2 - 2x = 5$

Step 3 $x^2 - \dfrac{2}{3}x = \dfrac{5}{3}$

Step 4 $x^2 - \dfrac{2}{3}x + \dfrac{1}{9} = \dfrac{5}{3} + \dfrac{1}{9}$

Step 5 $\left(x - \dfrac{1}{3}\right)^2 = \dfrac{16}{9}$

Step 6 $x - \dfrac{1}{3} = \sqrt{\dfrac{16}{9}}$ or $x - \dfrac{1}{3} = -\sqrt{\dfrac{16}{9}}$

$x - \dfrac{1}{3} = \dfrac{4}{3}$ or $x - \dfrac{1}{3} = -\dfrac{4}{3}$

Step 7 $x - \dfrac{1}{3} = \dfrac{4}{3}$ or $x - \dfrac{1}{3} = -\dfrac{4}{3}$

$x = \dfrac{4}{3} + \dfrac{1}{3}$ or $x = -\dfrac{4}{3} + \dfrac{1}{3}$

$x = \dfrac{5}{3}$ or $x = -\dfrac{3}{3}$

$= -1$

Check for $x = \dfrac{5}{3}$

$3x^2 = 2x + 5$

$3\left(\dfrac{5}{3}\right)^2 = 2\left(\dfrac{5}{3}\right) + 5$

$3\left(\dfrac{25}{\underset{3}{9}}\right) = \dfrac{10}{3} + 5$

$\dfrac{25}{3} = \dfrac{10}{3} + \dfrac{15}{3}$

$\dfrac{25}{3} = \dfrac{25}{3}$

Check for $x = -1$

$3x^2 = 2x + 5$

$3(-1)^2 = 2(-1) + 5$

$3(1) = -2 + 5$

$3 = -2 + 5$

$3 = 3$

While this method may look tortuous at first, it grows less so with repetition. The same seven steps are repeated in the same order each time a quadratic equation is solved by this method. Trace these steps in the example which follows.

Example 2:

Solve $10 = 2x - x^2$.

Step 1 $x^2 - 2x + 10 = 0$ Standard form

Step 2 $x^2 - 2x = -10$ Add (-10) to both sides of the equation.

Step 3 $x^2 - 2x = -10$ Divide both sides by the coefficient of x^2 which is 1.

Step 4 $x^2 - 2x + 1 = -10 + 1$ Add to each side the square of 1/2 of the coefficient of x.

Step 5 $(x - 1)^2 = -9$ Write the left side in factored form and combine the terms on the right side.

Step 6 $x - 1 = \sqrt{-9}$ or   List all possible values of the second-degree expression from Step 5.
 $x - 1 = -\sqrt{-9}$

so $x - 1 = 3i$ or
 $x - 1 = -3i$

Step 7 $x - 1 = 3i$ or   Compute the values of x from the two possibilities in Step 6.
 $x - 1 = -3i$

so $x = 1 + 3i$ or
 $x = 1 - 3i$

The solution set is $\{1 + 3i, 1 - 3i\}$.

The check for these two solutions was discussed on page 384.

Example 3:

Solve $y^2 = 2y + 2$.

Step 1 $y^2 - 2y - 2 = 0$

Step 2 $y^2 - 2y = 2$

Step 3 $y^2 - 2y = 2$

Step 4 $y^2 - 2y + 1 = 2 + 1$

Step 5 $(y - 1)^2 = 3$

Step 6 $y - 1 = \sqrt{3}$ or $y - 1 = -\sqrt{3}$

Step 7 $y = 1 + \sqrt{3}$ or $y = 1 - \sqrt{3}$

The solution set is $\{1 + \sqrt{3},\, 1 - \sqrt{3}\}$.

Check for $y = 1 + \sqrt{3}$

$y^2 = 2y + 2$

$(1 + \sqrt{3})^2 = 2(1 + \sqrt{3}) + 2$

$1 + 2\sqrt{3} + 3 = 2 + 2\sqrt{3} + 2$

$4 + 2\sqrt{3} = 4 + 2\sqrt{3}$

Check for $y = 1 - \sqrt{3}$

$y^2 = 2y + 2$

$(1 - \sqrt{3})^2 = 2(1 - \sqrt{3}) + 2$

$1 - 2\sqrt{3} + 3 = 2 - 2\sqrt{3} + 2$

$4 - 2\sqrt{3} = 4 - 2\sqrt{3}$

Example 4:

Solve $2 = 3t - 2t^2$.

Step 1 $2t^2 - 3t + 2 = 0$

Step 2 $2t^2 - 3t = -2$

Step 3 $t^2 - \dfrac{3}{2}t = -1$

Step 4 $t^2 - \dfrac{3}{2}t + \dfrac{9}{16} = -1 + \dfrac{9}{16}$

Step 5 $\left(t - \dfrac{3}{4}\right)^2 = -\dfrac{16}{16} + \dfrac{9}{16}$

$\left(t - \dfrac{3}{4}\right)^2 = -\dfrac{7}{16}$

Step 6 $t - \dfrac{3}{4} = \sqrt{-\dfrac{7}{16}}$ or $t - \dfrac{3}{4} = -\sqrt{-\dfrac{7}{16}}$

$t - \dfrac{3}{4} = \dfrac{i\sqrt{7}}{4}$ or $t - \dfrac{3}{4} = -\dfrac{i\sqrt{7}}{4}$

Step 7 $\quad t = \dfrac{3}{4} + \dfrac{i\sqrt{7}}{4}$ or $t = \dfrac{3}{4} - \dfrac{i\sqrt{7}}{4}$

$\qquad\quad t = \dfrac{3 + i\sqrt{7}}{4}$ or $t = \dfrac{3 - i\sqrt{7}}{4}$

The solution set is $\left\{\dfrac{3 + i\sqrt{7}}{4}, \dfrac{3 - i\sqrt{7}}{4}\right\}$.

Example 5:

Solve $x^2 = 6x$

Step 1 $\quad x^2 - 6x = 0$

Step 2 $\quad x^2 - 6x = 0 \qquad$ (The constant term is 0)

Step 3 $\quad x^2 - 6x = 0$

Step 4 $\quad x^2 - 6x + 9 = 0 + 9$

Step 5 $\quad (x - 3)^2 = 9$

Step 6 $\quad x - 3 = \sqrt{9}$ or $x - 3 = -\sqrt{9}$

Step 7 $\quad x = 3 + 3 \qquad$ or $\quad x = 3 - 3$

$\qquad\quad x = 6 \qquad\quad$ or $\quad x = 0$

Exercise 115:

Solve the following equations by completing the square:

(1) $x^2 = 2x + 15$ 　　(2) $2y^2 = 7y - 3$

(3) $t^2 = 8 - 2t$ 　　(4) $7x = 3 - 6x^2$

(5) $4x^2 - 9x + 2 = 0$ 　　(6) $z^2 - 2z = 35$

(7) $2y^2 + y - 15 = 0$ 　　(8) $2w^2 = 2w + 1$

(9) $9x^2 = 12x - 2$ 　　(10) $t^2 = 2t + 4$

(11) $y^2 - 4y + 1 = 0$ 　　(12) $8x^2 = 4x + 1$

(13) $3 = 4x - x^2$ 　　(14) $z^2 - 8z + 25 = 0$

(15) $4t^2 = 4t - 5$ 　　(16) $9x^2 = 24x - 17$

(17) $y^2 - 12y + 37 = 0$ 　　(18) $7 = 4x - x^2$

(19) $3 = 4t - 4t^2$ 　　(20) $3x^2 - 4x + 3 = 0$

(21) $x^2 = 4x - 2$ 　　(22) $6r = 3 - r^2$

(23) $4 = 9x - 2x^2$ 　　(24) $2x^2 = 1 - x$

(25) $3x^2 - 2x + 2 = 0$ 　　(26) $10y^2 = 2y + 5$

(27) $6x^2 - 2 = 0$ 　　(28) $2t^2 - 2t = 1$

(29) $4x^2 - 4x = 1$ 　　(30) $5y^2 - 9y = 0$ 　(See Example 5)

(31) $3t^2 - 16t = 0$ 　　(32) $9x^2 = 12x - 1$

11.3
Solution of Quadratic Equations by Formula

Here we shall repeat two statements from the foregoing section and invite your contemplation of the two together.

1. The standard form

$$ax^2 + bx + c = 0 \qquad a \neq 0$$

represents *every* quadratic equation in one variable.

2. The method of completing the square will solve *every* quadratic equation in one variable.

If you think that over for a bit, it should become clear to you that we can *solve* the equation $ax^2 + bx + c = 0$ by completing the square. By doing so we would be solving an equation that represents all such equations, and therefore the solution set of it would be the automatic solution set of every quadratic equation in one variable. That solution set is called the *quadratic formula*, and we shall find it by solving $ax^2 + bx + c = 0$, using the method of completing the square.

Step 1 $ax^2 + bx + c = 0$ Standard form

Step 2 $ax + bx = -c$ Add $(-c)$ to both sides.

Step 3 $x^2 + \dfrac{b}{a}x = -\dfrac{c}{a}$ Divide both sides of the equation by a.

Step 4

$$\frac{b}{a} = \text{the coefficient of } x$$

$$\frac{b}{a} \cdot \frac{1}{2} = \frac{b}{2a} = \text{one-half the coefficient of } x$$

$$\frac{b}{2a} \cdot \frac{b}{2a} = \frac{b^2}{4a^2} = \text{the square of one-half of the coefficient of } x$$

$$x^2 + \frac{b}{a}x + \frac{b^2}{4a^2} = \frac{b^2}{4a^2} - \frac{c}{a} \qquad \text{Add } \frac{b^2}{4a^2} \text{ to both sides of the equation.}$$

Step 5 $\left(x + \dfrac{b}{2a}\right)^2 = \dfrac{b^2}{4a^2} - \dfrac{4ac}{4a^2}$ Write the left side in factored form and combine the terms on the right side.

$$\left(x + \frac{b}{2a}\right)^2 = \frac{b^2 - 4ac}{4a^2}$$

Step 6   If $\left(x + \dfrac{b}{2a}\right)^2 = \dfrac{b^2 - 4ac}{4a^2}$, then   List all possible values of the second-degree expression, $\left(x + \dfrac{b}{2a}\right)^2$.

$$x + \frac{b}{2a} = \sqrt{\frac{b^2 - 4ac}{4a^2}} \qquad \text{or} \qquad x + \frac{b}{2a} = -\sqrt{\frac{b^2 - 4ac}{4a^2}}$$

Step 7 $x + \dfrac{b}{2a} = \dfrac{\sqrt{b^2 - 4ac}}{2a}$ Compute the values of x from the two possibilities in Step 6.

$$\text{or } x + \frac{b}{2a} = -\frac{\sqrt{b^2 - 4ac}}{2a}$$

$$x = -\frac{b}{2a} + \frac{\sqrt{b^2 - 4ac}}{2a} \qquad \text{or} \qquad x = -\frac{b}{2a} - \frac{\sqrt{b^2 - 4ac}}{2a}$$

$$x = \frac{-b + \sqrt{b^2 - 4ac}}{2a} \qquad \text{or} \qquad x = \frac{-b - \sqrt{b^2 - 4ac}}{2a}$$

Therefore, for any equation of the type

$$ax^2 + bx + c = 0 \qquad a \neq 0$$

the solution set is $\left\{ \dfrac{-b + \sqrt{b^2 - 4ac}}{2a}, \dfrac{-b - \sqrt{b^2 - 4ac}}{2a} \right\}$.

This solution set is called the *quadratic formula*.

As illustrated by the above results, the solutions of a quadratic equation are arithmetic combinations of the constants a, b, and c in the equation. Thus, if we can correctly identify the values of a, b, and c in a specific quadratic equation, we have only to substitute those values in the quadratic formula to get the solution set for that particular equation.

Examples:

(1) $3x^2 = 2x + 5$

$3x^2 - 2x - 5 = 0$

$ax^2 + bx + c = 0$

$a = 3,\ b = -2,\ c = -5$

$x = \dfrac{-b \pm \sqrt{b^2 - 4ac}}{2a}$

$x = \dfrac{-(-2) \pm \sqrt{(-2)^2 - 4(3)(-5)}}{2(3)}$

$x = \dfrac{2 \pm \sqrt{4 + 60}}{6} = \dfrac{2 \pm \sqrt{64}}{6}$

$x = \dfrac{2 \pm 8}{6}$

$$x = \frac{2 + 8}{6} \quad \text{or} \quad x = \frac{2 - 8}{6}$$

$$x = \frac{10}{6} \qquad x = \frac{-6}{6}$$

$$x = \frac{5}{3} \quad \text{or} \quad x = -1$$

The solution set is $\left\{ \frac{5}{3}, -1 \right\}$.

(2) $\quad x^2 = 2x - 10$

$\quad x^2 - 2x + 10 = 0$

$\quad a = 1, b = -2, c = 10$

$$x = \frac{-b \pm \sqrt{b^2 - 4ac}}{2a}$$

$$x = \frac{-(-2) \pm \sqrt{(-2)^2 - 4(1)(10)}}{2(1)}$$

$$x = \frac{2 \pm \sqrt{4 - 40}}{2} = \frac{2 \pm \sqrt{-36}}{2}$$

$$x = \frac{2 \pm 6i}{2}$$

$$x = \frac{2 + 6i}{2} \quad \text{or} \quad x = \frac{2 - 6i}{2}$$

$$x = \frac{\cancel{2}(1 + 3i)}{\cancel{2}} \quad \text{or} \quad x = \frac{\cancel{2}(1 - 3i)}{\cancel{2}}$$

$\quad x = 1 + 3i \quad \text{or} \quad x = 1 - 3i$

The solution set is $\{1 + 3i, 1 - 3i\}$.

Check this with Example 2 in Section 11.2.

(3) $\quad 3y^2 = 7$

$\quad 3y^2 - 7 = 0$

$\quad$ or $3y^2 + 0y - 7 = 0$

$\quad a = 3, b = 0, c = -7$

$$y = \frac{-b \pm \sqrt{b^2 - 4ac}}{2a}$$

$$y = \frac{-0 \pm \sqrt{0^2 - 4(3)(-7)}}{2(3)}$$

$$y = \frac{\pm\sqrt{84}}{6} = \frac{\pm\sqrt{4(21)}}{6} = \frac{\pm 2\sqrt{21}}{6}$$

$$y = \frac{\pm \cancel{2}\sqrt{21}}{\underset{3}{\cancel{6}}} = \frac{\pm\sqrt{21}}{3}$$

The solution set is $\left\{ \frac{\sqrt{21}}{3}, \frac{-\sqrt{21}}{3} \right\}$.

(4) $5t^2 = 4t$

$5t^2 - 4t = 0$

$5t^2 - 4t + 0 = 0$

$a = 5, b = -4, c = 0$

$$t = \frac{-b \pm \sqrt{b^2 - 4ac}}{2a}$$

$$t = \frac{-(-4) \pm \sqrt{(-4)^2 - 4(5)(0)}}{2(5)}$$

$$t = \frac{4 \pm \sqrt{16 - 0}}{10}$$

$$t = \frac{4 \pm 4}{10}$$

$$t = \frac{4 + 4}{10} \quad \text{or} \quad t = \frac{4 - 4}{10}$$

$$t = \frac{8}{10} = \frac{4}{5} \quad \text{or} \quad t = \frac{0}{10} = 0$$

The solution set is $\left\{\frac{4}{5}, 0\right\}$.

(5) $x^2 - 6x + 9 = 0$

$a = 1, b = -6, c = 9$

$$x = \frac{-(-6) \pm \sqrt{(-6)^2 - 4(1)(9)}}{2(1)}$$

$$x = \frac{6 \pm \sqrt{36 - 36}}{2} = \frac{6 \pm \sqrt{0}}{2}$$

$$x = \frac{6 + 0}{2} \quad \text{or} \quad x = \frac{6 - 0}{2}$$

$$x = \frac{6}{2} = 3 \quad \text{or} \quad x = \frac{6}{2} = 3$$

The solution set is $\{3\}$.

Exercise 116:

Solve the following equations by formula:

(1) $2x^2 = x + 3$

(2) $y^2 + y - 12 = 0$

(3) $x^2 - 6x = 0$

(4) $t^2 - 7 = 0$

(5) $3y^2 - 11y = 4$

(6) $x^2 - 8x + 16 = 0$

(7) $z^2 + 9 = 0$

(8) $3x^2 - 5x = 0$

(9) $y^2 - 7y - 8 = 0$

(10) $4t^2 = 12t - 9$

(11) $x^2 - x - 1 = 0$

(12) $2y^2 - y - 2 = 0$

(13) $w^2 - 8w + 25 = 0$

(14) $9t^2 - 24t + 16 = 0$

(15) $2x^2 + 7x = 0$ (16) $x^2 - 16 = 0$

(17) $9y^2 = 12y - 1$ (18) $25z^2 - 10z + 2 = 0$

(19) $8x^2 + 4x = 1$ (20) $y^2 + 12 = 0$

(21) $t^2 - 2t = 15$ (22) $3 = 7x - 2x^2$

(23) $x^2 + 2x = 8$ (24) $7y = 3 - 7y^2$

(25) $4r^2 = 9r - 2$ (26) $x^2 - 35 = 2x$

(27) $2y^2 = 15 - y$ (28) $2y^2 - 2y - 1 = 0$

(29) $2 = 12x - 9x^2$ (30) $x^2 - 2x - 4 = 0$

(31) $x^2 = 4x - 1$ (32) $4y + 1 = 8y^2$

(33) $t^2 = 4t - 3$ (34) $x^2 + 25 = 8x$

(35) $4x^2 + 5 = 4x$ (36) $9y^2 - 24y + 17 = 0$

(37) $x^2 = 12x - 37$ (38) $4t - t^2 = 7$

(39) $4x^2 - 4x + 3 = 0$ (40) $3y^2 = 4y - 3$

(41) $3t^2 = 2t - 2$ (42) $10x^2 - 2x = 5$

(43) $6r^2 - 2 = 0$ (44) $2y^2 = 2y + 1$

(45) $4x^2 = 4x + 1$ (46) $5x^2 - 9x = 0$

(47) $3y^2 - 16y = 0$ (48) $9y^2 - 12y + 1 = 0$

(49) $2t^2 + 4 = 9t$ (50) $2x^2 + x = 1$

11.4
Solution of Quadratic Equations by Factoring

The foregoing methods of solving quadratic equations can be used to solve all quadratic equations in one variable. Some quadratic equations can be solved quickly and easily by the method of *factoring*. This method is handy to know when it is available, but unfortunately it is not generally available unless the equation has solutions, or roots, which are rational numbers. But because it is a quick, easy method of solution when the roots are rational, it is highly convenient to know.

The method of solving a quadratic equation by factoring is based on a significant fact about any product *that is equal to zero*. A premonition of that fact may be aroused by considering a product that is *not* equal to zero. For instance, if you are told that x and y are numbers whose product is 12, what information could you get about x or y by knowing that $xy = 12$? The answer is, "Not much," because x (or y) could be *any* number except zero. For example,

$$(1)(12) \qquad (2)(6) \qquad (3)(4) \qquad \left(\frac{1}{2}\right)(24) \qquad (5)\left(\frac{12}{5}\right) \qquad (9)\left(\frac{4}{3}\right) \qquad \text{etc.}$$

are *all* products that equal 12, and there is obviously no end to the possibilities.

But suppose you were told that x and y are two numbers whose product is 0. If we set up a table of some possibilities, it would look like the following:

Value of x	1	-3	$\frac{1}{2}$	275	$-\frac{7}{8}$	0	0	0	0	0	etc.
Value of y	0	0	0	0	0	13	$\sqrt{2}$	-26	$\frac{4}{5}$	382	
The product xy	0	0	0	0	0	0	0	0	0	0	

A thoughtful look at this table should prod the suspicion that if a product is equal to *zero*, then at least one of its factors must *be* zero. This is certainly not the case when a product is equal to any number *except* zero; e.g., the fact that $xy = 12$ does not require that either x or y *be* 12 — far from it! The unique property of products which are equal to zero is that one of the factors must be zero. In short, if $(a)(b) = 0$, then either $a = 0$ or $b = 0$ or both. This fact is the basis of the solution of quadratic equations by factoring. Thus if we have an equation equal to zero, and if we can *factor* the other side of the equation, then we shall have a *product equal to zero*. This means that the equation will be true if and only if one or both of the factors are equal to zero.

Examples:

(1) $x^2 = 5x - 6$
$x^2 - 5x + 6 = 0$

Factor the left side.

$$(x - 3)(x - 2) = 0$$

Since the product $(x - 3)(x - 2)$ is equal to *zero*, this equation will be true only when $x - 3$ or $x - 2$ is equal to zero.

$$x - 3 = 0 \quad \text{or} \quad x - 2 = 0$$
$$x = 3 \quad \text{or} \qquad x = 2$$

Since the numbers 3 and 2 are the only numbers that will make the product equal to zero, they are the only numbers that will make the equation true.

Repeating the steps we have the following:

$x^2 = 5x - 6$
$x^2 - 5x + 6 = 0$
$(x - 3)(x - 2) = 0$
$x - 3 = 0 \quad \text{or} \quad x - 2 = 0$
$x = 3 \quad \text{or} \quad x = 2$

The solution set is $\{3, 2\}$.

(2) $3x^2 - 16x = -5$
       $3x^2 - 16x + 5 = 0$
       $(3x - 1)(x - 5) = 0$
       $3x - 1 = 0$ or $x - 5 = 0$
       $3x = 1$ or $x = 5$
       $x = \dfrac{1}{3}$ or $x = 5$

The solution set is $\left\{ \dfrac{1}{3}, 5 \right\}$.

(3) $x^2 = 25$
       $x^2 - 25 = 0$
       $(x + 5)(x - 5) = 0$
       $x + 5 = 0$ or $x - 5 = 0$
       $x = -5$ or $x = 5$

The solution set is $\{-5, 5\}$.

(4) $3x^2 = 12x$
       $3x^2 - 12x = 0$
       $3x(x - 4) = 0$
       $3x = 0$ or $x - 4 = 0$
       $x = 0$ or $x = 4$

The solution set is $\{0, 4\}$.

(5) $x^2 = 8x - 16$
       $x^2 - 8x + 16 = 0$
       $(x - 4)(x - 4) = 0$
       $x - 4 = 0$ or $x - 4 = 0$
       $x = 4$ or $x = 4$

The solution set is $\{4\}$.

(6) $4(x^2 + 2x) = -3$
       $4x^2 + 8x = -3$
       $4x^2 + 8x + 3 = 0$
       $(2x + 3)(2x + 1) = 0$
       $2x + 3 = 0$ or $2x + 1 = 0$
       $2x = -3$ or $2x = -1$
       $x = -\dfrac{3}{2}$ or $x = -\dfrac{1}{2}$

The solution set is $\left\{ -\dfrac{3}{2}, -\dfrac{1}{2} \right\}$.

Exercise 117:

Solve the following equations by the method of factoring:

(1) $x^2 - 9x = 0$ (2) $y^2 - 49 = 0$

(3) $z^2 = z + 12$ (4) $2x^2 - 3x = 5$

(5) $15y^2 + 7y - 2 = 0$ (6) $x^2 + 4x + 4 = 0$

(7) $t^2 - 36 = 0$ (8) $2w^2 = 9w + 5$

(9) $4x^2 = 28x - 49$ (10) $6z^2 - 3z = 0$

(11) $3y^2 = 19y + 14$ (12) $8x^2 = 26x - 15$

(13) $15x^2 = 20x$ (14) $9w^2 + 6w + 1 = 0$

(15) $5 = 16z - 12z^2$ (16) $3x^2 = 14x - 15$

(17) $t^2 = 64$ (18) $16x^2 = 1$

(19) $6 = 13y - 6y^2$ (20) $3x^2 - x = 0$

(21) $3(3y^2 + 4y) = -4$ (22) $2(4x^2 + 7x) = 15$

(23) $3(3x^2 - 10x) = -25$ (24) $2(4z^2 + 5z) = -3$

(25) $2(4t^2 - 3t) = 9$ (26) $4x^2 - 3x = 0$

(27) $3x^2 - 14x + 8 = 0$ (28) $12t^2 = 5t + 3$

(29) $15y^2 - 7y = 4$ (30) $6x^2 + 43x = 40$

(31) $x^2 - 30x + 200 = 0$ (32) $6y^2 - 29y + 30 = 0$

(33) $4y^2 - 9 = 0$ (34) $36t^2 - 25 = 0$

(35) $9r^2 - 49 = 0$ (36) $2x^2 - 8 = 0$

(37) $3x^2 - 5x = 0$ (38) $2y^2 + 7y = 39$

(39) $12x^2 - 8x + 1 = 0$ (40) $12x^2 - 7x = 12$

11.5
Fractional Equations Which Can Be Reduced to Quadratic Equations

As observed in Section 5.6, when an equation contains fractions, we multiply both sides of the equation by the lowest common denominator of all the fractions in the equation. When the multiplier (the L.C.D.) contains the variable, there is always the possibility that the equation derived by this process does not have the same solution set as the original equation. Therefore it is always necessary to check the original equation for any solution obtained in this manner (see Section 5.6).

Examples:

(1) $\dfrac{x}{x + 1} - \dfrac{3}{x + 4} = \dfrac{9}{(x + 4)(x + 1)}$

The L.C.D. is $(x + 4)(x + 1)$.

Using E_m, multiply both sides of the equation by this number.

$$\frac{x}{\cancel{(x+1)}} \cdot \frac{(x+4)\cancel{(x+1)}}{1} - \frac{3}{\cancel{(x+4)}} \cdot \frac{\cancel{(x+4)}(x+1)}{1}$$

$$= \frac{9}{\cancel{(x+4)}\cancel{(x+1)}} \cdot \frac{\cancel{(x+4)}\cancel{(x+1)}}{1}$$

$$x(x+4) - 3(x+1) = 9$$
$$x^2 + 4x - 3x - 3 = 9$$
$$x^2 + x - 12 = 0$$
$$(x+4)(x-3) = 0$$
$$x + 4 = 0 \quad \text{or} \quad x - 3 = 0$$
$$x = -4 \quad \text{or} \quad x = 3$$

Check for $x = -4$

$$\frac{x}{x+1} - \frac{3}{x+4} = \frac{9}{(x+4)(x+1)}$$

$$\frac{-4}{-4+1} - \frac{4}{-4+4}$$

$$= \frac{9}{(-4+4)(-4+1)}$$

$$\frac{-4}{-3} - \frac{3}{0} = \frac{9}{(0)(-3)}$$

$$\frac{4}{3} - \frac{3}{0} = \frac{9}{0} \quad \text{(undefined)}$$

[-4 cannot be a root of the original equation because $\frac{4}{3} - \frac{3}{0} = \frac{9}{0}$ is meaningless]

Check for $x = 3$

$$\frac{x}{x+1} - \frac{3}{x+4} = \frac{9}{(x+4)(x+1)}$$

$$\frac{3}{3+1} - \frac{3}{3+4} = \frac{9}{(3+4)(3+1)}$$

$$\frac{3}{4} - \frac{3}{7} = \frac{9}{(7)(4)}$$

$$\frac{3}{4} - \frac{3}{7} = \frac{9}{28}$$

$$\frac{21}{28} - \frac{12}{28} = \frac{9}{28}$$

$$\frac{9}{28} = \frac{9}{28}$$

The solution set is $\{3\}$.

(2) $\dfrac{x}{x-1} - \dfrac{1}{x-2} = \dfrac{11}{(x-2)(x-1)}$

$$\frac{(x-2)\cancel{(x-1)}}{1} \cdot \frac{x}{\cancel{(x-1)}} - \frac{\cancel{(x-2)}(x-1)}{1} \cdot \frac{1}{\cancel{(x-2)}}$$

$$= \frac{\cancel{(x-2)}\cancel{(x-1)}}{1} \cdot \frac{11}{\cancel{(x-2)}\cancel{(x-1)}}$$

$$x(x-2) - 1(x-1) = 11$$
$$x^2 - 2x - x + 1 = 11$$
$$x^2 - 3x - 10 = 0$$
$$(x-5)(x+2) = 0$$
$$x - 5 = 0 \text{ or } x + 2 = 0$$
$$x = 5 \text{ or } x = -2$$

Check for $x = 5$

$$\frac{x}{x-1} - \frac{1}{x-2} = \frac{11}{(x-2)(x-1)}$$

$$\frac{5}{5-1} - \frac{1}{5-2} = \frac{11}{(5-2)(5-1)}$$

$$\frac{5}{4} - \frac{1}{3} = \frac{11}{(3)(4)}$$

$$\frac{15}{12} - \frac{4}{12} = \frac{11}{12}$$

$$\frac{11}{12} = \frac{11}{12}$$

Check for $x = -2$

$$\frac{x}{x-1} - \frac{1}{x-2} = \frac{11}{(x-2)(x-1)}$$

$$\frac{-2}{-2-1} - \frac{1}{-2-2}$$

$$= \frac{11}{(-2-2)(-2-1)}$$

$$\frac{-2}{-3} - \frac{1}{-4} = \frac{11}{(-4)(-3)}$$

$$\frac{2}{3} + \frac{1}{4} = \frac{11}{12}$$

$$\frac{8}{12} + \frac{3}{12} = \frac{11}{12}$$

$$\frac{11}{12} = \frac{11}{12}$$

The solution set is $\{5, -2\}$.

Exercise 118:

Solve the following equations:

(1) $\quad x - 2 = \dfrac{8}{x}$

(2) $\quad 3x - \dfrac{1}{2} - \dfrac{1}{x} = 0$

(3) $\quad 2x - \dfrac{5x}{x+3} = \dfrac{15}{x+3}$

(4) $\quad x + 1 = \dfrac{4}{x+1}$

(5) $\quad x - 1 = \dfrac{12}{x-2}$

(6) $\quad x + \dfrac{20}{x-4} = \dfrac{5x}{x-4} - 2$

(7) $\quad \dfrac{5x+5}{x+2} + 3x = \dfrac{x^2}{x+2}$

(8) $\quad \dfrac{x^2+6}{x-1} + \dfrac{x-2}{x-1} = 2x$

(9) $\quad \dfrac{x^2+3x}{x-3} + 1 = \dfrac{12}{x-3} - x$

(10) $\quad x + \dfrac{6}{x-3} = \dfrac{2x}{x-3}$

(11) $\quad \dfrac{x}{x+2} + \dfrac{1}{x+1} = \dfrac{6}{(x+1)(x+2)}$

(12) $\quad \dfrac{2x}{x+2} + \dfrac{x+1}{(x+3)(x+2)} = \dfrac{2}{x+3}$

(13) $\quad \dfrac{x}{x+2} + \dfrac{4}{x-1} = \dfrac{x+11}{x^2+x-2}$

(14) $\quad \dfrac{x}{x-1} + \dfrac{2}{x-4} = \dfrac{6}{x^2-5x+4}$

(15) $\quad \dfrac{1}{x+2} + \dfrac{x}{x-3} = \dfrac{2x+9}{x^2-x-6}$

11.6
Equations Containing Radicals

In various topics and applications of mathematics we are frequently confronted with equations which contain square-root radicals such as the following:

$$\sqrt{x + 1} = 2$$
$$\sqrt{3x + 1} = 2 + \sqrt{x + 1}$$

The most expedient way to solve such an equation is to eliminate the radicals, and, as noted in Chapter 10, the most expedient way to eliminate a square-root radical is to square it.

However, when we square both sides of an equation, the equation we get may have solutions that do not belong to the original one. For example, consider the equation

$$x + \sqrt{6 - 5x} = 0$$

If we isolate the term containing the radical in this equation, we have

$$x = -\sqrt{6 - 5x}$$

Then squaring both sides, we obtain

$$(x)^2 = (-\sqrt{6 - 5x})^2$$
$$x^2 = 6 - 5x$$

Putting this quadratic equation in standard form and solving by factoring, we have

$$x^2 + 5x - 6 = 0$$
$$(x + 6)(x - 1) = 0$$
$$x + 6 = 0 \quad \text{or} \quad x - 1 = 0$$
$$x = -6 \quad \text{or} \quad x = 1$$

Check for $x = -6$
in the *original* equation

Check for $x = 1$
in the *original* equation

$x + \sqrt{6 - 5x} = 0$

$-6 + \sqrt{6 - 5(-6)} = 0$

$-6 + \sqrt{6 + 30} = 0$

$-6 + \sqrt{36} = 0$

$-6 + 6 = 0$

True

$x + \sqrt{6 - 5x} = 0$

$1 + \sqrt{6 - 5(1)} = 0$

$1 + \sqrt{6 - 5} = 0$

$1 + \sqrt{1} = 0$

$1 + 1 = 0$

Not true

The solution set of $x + \sqrt{6 - 5x} = 0$ is $\{-6\}$.

In the preceding illustration, we note that -6 is a solution of both the radical equation and the quadratic equation obtained by squaring both sides of the radical equation; but the other solution of the derived quadratic equation, the number 1, is *not* a solution of $x + \sqrt{6 - 5x} = 0$. Roots that satisfy the derived equation but not the original equation are called *extraneous* roots, and they can be identified by checking all solutions in the original equation.

We should also observe that the solution set of the radical equation is a subset of the solution set of the derived quadratic equation. In general, it is true that the roots of a radical equation will always be a subset of the roots of the equation we obtain by squaring both sides, but remember that a subset of a given set may be *all* of the elements of the set, or *some* of the elements of the set, or *none* of the elements of the set, i.e., the empty set. Indeed, it sometimes happens that a radical equation (like some fractional equations) has no solution. Obviously, all roots of the derived equation must be checked in the *original* equation in order to determine whether or not they are extraneous roots.

To see why extraneous roots can occur when both sides of an equation are squared, examine the following illustration:

In the equation

$$x = 3$$

the solution set is $\{3\}$.

But, if we square both sides of $x = 3$ we obtain the following:

$$(x)^2 = (3)^2$$

or

$$x^2 = 9$$

Solving this last equation by factoring, we have

$$x^2 - 9 = 0$$
$$(x + 3)(x - 3) = 0$$
$$x + 3 = 0 \quad \text{or} \quad x - 3 = 0$$
$$x = -3 \quad \text{or} \quad x = 3$$

So the solution set of $x^2 = 9$ is $\{3, -3\}$.

Note that while $\{3, -3\}$ is the solution set of $x^2 = 9$, only the element 3 is a solution of the *original* equation, $x = 3$. The number -3 is an extraneous root that resulted from squaring both sides of $x = 3$. In short, the fact that $a^2 = b^2$ does not guarantee that $a = b$ because a could also equal $-b$, and the statement, $a^2 = b^2$, would still be true.

In the preceding discussion, frequent references have been made to the process of "squaring both sides of an equation." In that phrase the word *sides* should probably be up in lights as a warning to unwary students who tend to fall into the grievous error of squaring all *terms* in an equation. Note the following *carefully:*

$$4 + 3 = 7$$

but

$$4^2 + 3^2 \neq 7^2 \qquad \text{(i.e. } 16 + 9 \neq 49)$$

However,

$$4 + 3 = 7$$

and

$$(4 + 3)^2 = 7^2$$

or

$$4^2 + 2(4)(3) + 3^2 = 7^2$$
$$16 + 24 + 9 = 49$$
$$49 = 49$$

In the work that follows, it is absolutely necessary to remember that the square of a binomial is a trinomial. For example

$$(a + b)^2 = a^2 + 2ab + b^2$$

and
$$(\sqrt{x + 1} + 3)^2 = (\sqrt{x + 1})^2 + 2(3)\sqrt{x + 1} + 3^2$$
$$= x + 1 + 6\sqrt{x + 1} + 9$$
$$= x + 6\sqrt{x + 1} + 10$$

In squaring binomials which contain radicals, the entire radical is a single term. Be sure also that you distinguish between a binomial and a monomial which contains two factors. Compare

$$(xy)^2 = x^2 y^2$$
$$(x + y)^2 = x^2 + 2xy + y^2$$

with

$$(\sqrt{a + 1}\sqrt{a - 2})^2 = (\sqrt{a + 1})^2(\sqrt{a - 2})^2 = (a + 1)(a - 2) \quad \text{or} \quad a^2 - a - 2$$

$$(\sqrt{a + 1} + \sqrt{a - 2})^2 = (\sqrt{a + 1})^2 + 2\sqrt{a + 1}\sqrt{a - 2} + (\sqrt{a - 2})^2$$
$$= a + 1 + 2\sqrt{a + 1}\sqrt{a - 2} + a - 2$$
$$= 2a - 1 + 2\sqrt{a + 1}\sqrt{a - 2}$$

Before we begin to solve radical equations, practice squaring the expressions in the following exercise.

Exercise 119:

Simplify the following:

(1) $(ab)^2$ (2) $(a + b)^2$
(3) $(\sqrt{x + 2})^2$ (4) $(\sqrt{x} + 2)^2$
(5) $(3\sqrt{x - 4})^2$ (6) $(3 - \sqrt{x - 4})^2$
(7) $(2\sqrt{x + 1})^2$ (8) $(2\sqrt{x + 1} + 3)^2$
(9) $(\sqrt{2x - 1})\sqrt{3x + 1})^2$ (10) $(\sqrt{2x - 1} + \sqrt{3x + 1})^2$
(11) $(3\sqrt{x} + 3\sqrt{2x - 1})^2$ (12) $(3\sqrt{x + 3} - \sqrt{2x - 1})^2$

Following is an example of solving a radical equation by eliminating the radicals. Check each step and the accompanying explanation carefully.

$$\sqrt{x + 7} + x = 5$$

1. If we square both sides of this equation as it is written above we would have

$$(\sqrt{x + 7} + x)^2 = 5^2$$
$$x + 7 + 2x\sqrt{x + 7} + x^2 = 25$$

in which we *still* have the radical $\sqrt{x + 7}$. We have done nothing wrong, but neither are we making any progress.

2. However, the equation

$$\sqrt{x + 7} + x = 5$$

may be written in the form

$$\sqrt{x + 7} = 5 - x$$

in which the term containing the radical is isolated.

3. Squaring both sides of the equation written with the radical isolated we obtain

$$(\sqrt{x + 7})^2 = (5 - x)^2$$
$$x + 7 = 25 - 10x + x^2$$

or

$$x^2 - 10x + 25 = x + 7$$

4. Putting this quadratic equation in standard form and solving by factoring, we obtain

$$x^2 - 10x + 25 - x - 7 = 0$$
$$x^2 - 11x + 18 = 0$$
$$(x - 9)(x - 2) = 0$$
$$x - 9 = 0 \quad \text{or} \quad x - 2 = 0$$
$$x = 9 \qquad \text{or} \qquad x = 2$$

The solution set of $x^2 - 11x + 18$ is $\{9, 2\}$.

5. Checking the possible solutions in the *original* equation we have

$$
\begin{array}{cc}
x = 9 & x = 2 \\
\sqrt{x + 7} + x = 5 & \sqrt{x + 7} + x = 5 \\
\sqrt{9 + 7} + 9 = 5 & \sqrt{2 + 7} + 2 = 5 \\
\sqrt{16} + 9 = 5 & \sqrt{9} + 2 = 5 \\
4 + 9 \neq 5 & 3 + 2 = 5
\end{array}
$$

Thus the solution set of $\sqrt{x + 7} + x = 5$ is $\{2\}$.

If you will look over the preceding illustration carefully, particularly steps 2 and 3, it should become obvious to you that the most efficient way to eliminate a radical is to *isolate the radical* first. Hence our procedure in solving such equations can be stated as follows:

---

1. Isolate the radical if possible.
2. Then square both sides of the equation.

When it is not possible to isolate the radical at the outset, we look for the possibility of doing so in one of the successive steps, continuing to square both sides until we have eliminated all radicals. In this situation, our first step listed above should be revised to read as follows:

---

1. Isolate the radical as much as possible, and *continue* to isolate radicals occurring in the successive steps of solving the equation.

Study the following examples carefully:

(1) $\sqrt{2x + 10} - 4 = 0$

$\sqrt{2x + 10} = 4$

$(\sqrt{2x + 10})^2 = (4)^2$

$2x + 10 = 16$

$2x = 16 - 10$

$2x = 6$

$x = 3$

Check for $x = 3$

$\sqrt{2x + 10} - 4 = 0$

$\sqrt{2(3) + 10} - 4 = 0$

$\sqrt{6 + 10} - 4 = 0$

$\sqrt{16} - 4 = 0$

$4 - 4 = 0$

The solution set of $\sqrt{2x + 10} - 4 = 0$ is $\{3\}$.

(2) $\sqrt{2x + 1} + x = 7$

$\sqrt{2x + 1} = 7 - x$

$(\sqrt{2x + 1})^2 = (7 - x)^2$

$2x + 1 = 49 - 14x + x^2$

or

$x^2 - 14x + 49 = 2x + 1$

$x^2 - 14x + 49 - 2x - 1 = 0$

$x^2 - 16x + 48 = 0$

$(x - 4)(x - 12) = 0$

$x - 4 = 0$ or $x - 12 = 0$

$x = 4$ or $x = 12$

Check for $x = 4$

$\sqrt{2x + 1} + x = 7$

$\sqrt{2(4) + 1} + 4 = 7$

$\sqrt{8 + 1} + 4 = 7$

$\sqrt{9} + 4 = 7$

$3 + 4 = 7$

Check for $x = 12$

$\sqrt{2x + 1} + x = 7$

$\sqrt{2(12) + 1} + 12 = 7$

$\sqrt{24 + 1} + 12 = 7$

$\sqrt{25} + 12 = 7$

$5 + 12 \neq 7$

The solution set of $\sqrt{2x + 1} + x = 7$ is $\{4\}$.

(3) $\sqrt{2x + 1} + \sqrt{2x - 4} = 1$

Note: Since we cannot isolate both radicals, we settle for isolating one of them.

$$\sqrt{2x + 1} = 1 - \sqrt{2x - 4}$$

Squaring both sides we obtain

$$(\sqrt{2x + 1})^2 = (1 - \sqrt{2x - 4})^2$$

$$2x + 1 = 1 - 2\sqrt{2x - 4} + 2x - 4$$

At this point, isolate the term $-2\sqrt{2x-4}$ and combine similar terms.

$$2x + 1 - 1 - 2x + 4 = -2\sqrt{2x-4}$$
$$4 = -2\sqrt{2x-4}$$

Simplify this equation by dividing both sides by 2.

$$2 = -\sqrt{2x-4}$$

Now square both sides.

$$(2)^2 = (-\sqrt{2x-4})^2$$
$$4 = 2x - 4$$

or

$$2x - 4 = 4$$
$$2x = 8$$
$$x = 4$$

Check the possible solution, 4, in the original equation.

$$\sqrt{2x+1} + \sqrt{2x-4} = 1$$
$$\sqrt{2(4)+1} + \sqrt{2(4)-4} = 1$$
$$\sqrt{9} + \sqrt{4} = 1$$
$$3 + 2 \neq 1$$

The solution set of $\sqrt{2x+1} + \sqrt{2x-4} = 1$ is $\varnothing$.

(4) $\sqrt{6x+2} + \sqrt{2x+6} = \sqrt{15x+17}$

There is no way to isolate either of the radicals on the left, so we simply start squaring.

$$(\sqrt{6x+2} + \sqrt{2x+6})^2 = (\sqrt{15x+17})^2$$
$$6x + 2 + 2\sqrt{6x+2}\sqrt{2x+6} + 2x + 6 = 15x + 17$$

Isolate the term $2\sqrt{6x+2}\sqrt{2x+6}$ and combine similar terms.

$$2\sqrt{6x+2}\sqrt{2x+6} = 15x + 17 - 6x - 2 - 2x - 6$$
$$2\sqrt{6x+2}\sqrt{2x+6} = 7x + 9$$

Square both sides of the above equation.

$$(2\sqrt{6x+2}\sqrt{2x+6})^2 = (7x+9)^2$$
$$4(6x+2)(2x+6) = 49x^2 + 126x + 81$$
$$4(12x^2+40x+12) = 49x^2 + 126x + 81$$
$$48x^2 + 160x + 48 - 49x^2 - 126x - 81 = 0$$
$$-x^2 + 34x - 33 = 0$$

or

$$x^2 - 34x + 33 = 0$$
$$(x-33)(x-1) = 0$$
$$x - 33 = 0 \quad \text{or} \quad x - 1 = 0$$
$$x = 33 \quad \text{or} \quad x = 1$$

Check for $x = 33$

$$\sqrt{6(33)+2} + \sqrt{2(33)+6} = \sqrt{15(33)+17}$$
$$\sqrt{198+2} + \sqrt{66+6} = \sqrt{495+17}$$
$$\sqrt{200} + \sqrt{72} = \sqrt{512}$$
$$\sqrt{(100)(2)} + \sqrt{(36)(2)} = \sqrt{(256)2}$$
$$10\sqrt{2} + 6\sqrt{2} = 16\sqrt{2}$$
$$16\sqrt{2} = 16\sqrt{2}$$

Check for $x = 1$

$$\sqrt{6(1)+2} + \sqrt{2(1)+6} = \sqrt{15(1)+17}$$
$$\sqrt{6+2} + \sqrt{2+6} = \sqrt{15+17}$$
$$\sqrt{8} + \sqrt{8} = \sqrt{32}$$
$$\sqrt{(4)(2)} + \sqrt{(4)(2)} = \sqrt{(16)(2)}$$
$$2\sqrt{2} + 2\sqrt{2} = 4\sqrt{2}$$
$$4\sqrt{2} = 4\sqrt{2}$$

The solution set of $\sqrt{6x+2} + \sqrt{2x+6} = \sqrt{15x+17}$ is $\{1, 33\}$.

Exercise 120:

Find the solution sets of the following equations:

(1) $\sqrt{7x+2} = 4$

(2) $\sqrt{2x+3} - 3 = 0$

(3) $\sqrt{3x+12} - 6 = 0$

(4) $\sqrt{5x+1} - 4 = 0$

(5) $\sqrt{3x - 8} - \sqrt{x} = 0$ (6) $\sqrt{6x - 5} - x = 0$

(7) $\sqrt{5x + 1} - 1 = x$ (8) $\sqrt{x + 2} - \sqrt{x} = 2$

(9) $x + \sqrt{4x + 1} = 5$ (10) $3 + x = \sqrt{6x + 13}$

(11) $\sqrt{5x + 1} - \sqrt{4x + 4} = 0$ (12) $x - 1 = \sqrt{7 - x}$

(13) $\sqrt{6x - 8} - \sqrt{3x + 4} = 0$ (14) $\sqrt{3 - 3x} - 1 = 2x$

(15) $\sqrt{2x + 1} - \sqrt{x} = 1$ (16) $\sqrt{2x + 2} = 3 + \sqrt{2x - 1}$

(17) $\sqrt{3x - 2} - \sqrt{x} = 2$ (18) $\sqrt{4x + 5} - \sqrt{x + 4} = 2$

(19) $\sqrt{3x + 1} = 4 + \sqrt{x + 3}$ (20) $\sqrt{3x + 4} - \sqrt{x + 2} = 2$

(21) $\sqrt{x + 2} + 5 = \sqrt{3x + 3}$ (22) $\sqrt{2x + 4} - \sqrt{x + 3} = 1$

(23) $\sqrt{4x + 4} - \sqrt{x - 2} = \sqrt{2x + 3}$

(24) $\sqrt{3x + 4} - \sqrt{2x + 1} = \sqrt{x - 3}$

(25) $\sqrt{4x + 1} + \sqrt{x - 1} = \sqrt{7x + 2}$

(26) $\sqrt{2x - 1} + \sqrt{x - 1} = x$

11.7
The Solution of Stated Problems Leading to Quadratic Equations

When the facts of a stated problem are translated into symbolic language to set up an equation, it often happens that the resulting equation is quadratic. The process of arriving at the equation is exactly the same as stated in Section 5.8, and you should review that discussion at this point. Below is a repetition of the summary given in Chapter 5, which lists the techniques involved in solving all stated problems.

Summary

1. Read the problem carefully and repeatedly.
2. Decide what quantity you are looking for and give it a name, such as x.
3. Draw a figure representing the conditions given in the problem whenever possible.
4. Tabulate all facts known about x in terms of x.
5. Look over the tabulated facts and find two equal quantities.
6. Write the equation and solve it.

When a stated problem leads to a quadratic equation, the equation has *two* solutions (which could be two equal numbers), while the conditions of the problem generally admit only *one* solution. In solving these problems, then, it will be necessary to look over the solutions carefully and select the one that not only satisfies the equation, but that also satisfies the conditions stated in the problem.

Example 1:

If two positive consecutive integers are squared and added, the result is one more than twelve times the smaller integer. What are the numbers?

(1) Let x = the smaller of the two integers
(2) Then $x + 1$ = the next consecutive integer
(3) x^2 = the square of the first integer
(4) $(x + 1)^2$ = the square of the second integer
(5) $x^2 + (x + 1)^2$ = the sum of the squares of the two integers
(6) $12x$ = twelve times the smaller integer
(7) $12x + 1$ = one more than twelve times the smaller integer

From (5) and (7)

$$x^2 + (x + 1)^2 = 12x + 1$$
$$x^2 + x^2 + 2x + 1 = 12x + 1$$
$$2x^2 + 2x - 12x + 1 - 1 = 0$$
$$2x^2 - 10x = 0$$
$$2x(x - 5) = 0$$
$$2x = 0 \quad \text{or} \quad x - 5 = 0$$
$$x = 0 \quad \text{or} \quad x = 5$$

Since zero is not a positive integer, the only solution that satisfies the conditions of the problem is 5.

$$5 = \text{the smaller of the two integers}$$
$$6 = \text{the next consecutive integer}$$

Example 2:

The area of a rectangle is 60 square meters. If the length of the rectangle is 4 meters longer than the width, find the dimensions of the rectangle.

(1) Let x meters = the width of the rectangle
(2) Then $x + 4$ meters = the length of the rectangle
(3) And $x(x + 4)$ square meters = the area of the rectangle
(4) 60 square meters = the area of the rectangle
     From (3) and (4)

$$x(x + 4) = 60$$
$$x^2 + 4x - 60 = 0$$
$$(x + 10)(x - 6) = 0$$
$$x = -10 \quad \text{or} \quad x = 6$$

Since x is the width of the rectangle, it cannot be a negative number; so we take the positive value, 6.

$$x = 6 \text{ meters} = \text{the width}$$
$$x + 4 = 10 \text{ meters} = \text{the length}$$

Example 3:

A man rows upstream a distance of 2 kilometers and returns. If the rate of the current is 2 km/h and the round trip required 1 hour and 20 minutes, how fast was the man rowing?

Note: (rate)(time) = distance; therefore, time $= \dfrac{\text{distance}}{\text{rate}}$

$\xrightarrow{\text{upstream}}$ rate $= x - 2$ km/h; distance $= 2$ kilometers

$\xleftarrow{\text{downstream}}$ rate $= x + 2$ km/h; distance $= 2$ kilometers

(1) Let x km/h $=$ the rate at which the man rows

(2) 2 km/h $=$ the rate of the current

(3) Then $x - 2$ km/h $=$ the rate he traveled *upstream*

(4) 2 kilometers $=$ the distance he traveled *upstream*

(5) Then $\dfrac{2}{x - 2} = \dfrac{\text{distance}}{\text{rate}} = $ time he traveled upstream

(6) $x + 2$ km/h $=$ the rate he traveled *downstream*

(7) 2 kilometers $=$ the distance he traveled *downstream*

(8) Then $\dfrac{2}{x + 2} = \dfrac{\text{distance}}{\text{rate}} = $ time he traveled downstream

(9) From (5) and (8)

$\dfrac{2}{x - 2} + \dfrac{2}{x + 2} = $ time (in hours) required for the round trip

(10) $1\dfrac{1}{3}$ or $\dfrac{4}{3}$ hours $=$ 1 hour and 20 minutes $=$ time required for the round trip

From (9) and (10)

$$\frac{2}{x - 2} + \frac{2}{x + 2} = \frac{4}{3}$$
$$3(2)(x + 2) + 3(2)(x - 2) = 4(x + 2)(x - 2)$$
$$6x + 12 + 6x - 12 = 4x^2 - 16$$
$$12x = 4x^2 - 16$$

$$3x = x^2 - 4$$
$$x^2 - 3x - 4 = 0$$
$$(x - 4)(x + 1) = 0$$
$$x - 4 = 0 \text{ or } x + 1 = 0$$
$$x = 4 \text{ or } x = -1$$

Since x equals the rate at which the man rows, x must be positive.

$$x \text{ km/h} = 4 \text{ km/h} = \text{the rate the man rows}$$

Example 4:

A packing company has an order for cardboard boxes with clear plastic tops for packaging matches. The boxes are required to be 3 centimeters deep, twice as long as they are wide, and each must have a volume of 150 cm³. The company plans to make the cardboard boxes by cutting 3-centimeter squares out of the corners of flat rectangular pieces of cardboard and turning up the sides. What should be the dimensions of the original pieces of cardboard from which the boxes are made?

(1) Let x centimeters = the length of the original piece of cardboard (Figure 11.1).

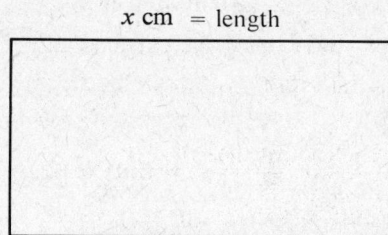

x cm = length

Figure 11.1

(2) Cut 3-centimeter squares from each corner. The overall length will be reduced by 6 centimeters, and the new width is required to be one-half of the new length, or $\dfrac{x - 6}{2}$ centimeters (Figure 11.2).

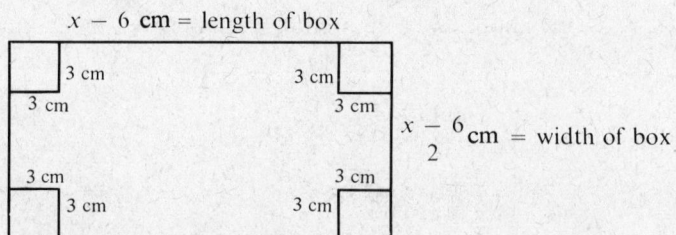

$x - 6$ **cm** = length of box

3 cm   3 cm
3 cm   3 cm

$\dfrac{x - 6}{2}$ cm = width of box

3 cm   3 cm
3 cm   3 cm

Figure 11.2

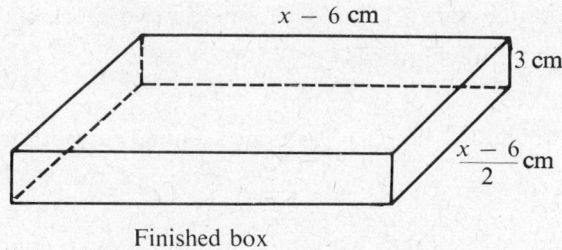

Finished box

Figure 11.3

(3) $3(x - 6)\left(\dfrac{x - 6}{2}\right)$ cm³ = the volume of the box

(4) 150 cm³ = the volume of the box

Therefore

$$3(x - 6)\left(\frac{x - 6}{2}\right) = 150$$
$$3(x - 6)(x - 6) = 300$$
$$(x - 6)(x - 6) = 100$$
$$x^2 - 12x + 36 = 100$$
$$x^2 - 12x - 64 = 0$$
$$(x - 16)(x + 4) = 0$$
$$x - 16 = 0 \quad x + 4 = 0$$
$$x = 16 \quad \text{or} \quad x = -4$$

Since x must be a positive number, the length of the original piece of cardboard is 16 centimeters.

Original length = 16 centimeters

Original width = $\dfrac{x - 6}{2} + 6 = \dfrac{10}{2} + 6 = 5 + 6 = 11$ centimeters

Example 5:

Jack can wash and wax the family car in one hour less than Bob can. The two working together can complete the job in $1\dfrac{1}{5}$ hours. How much time would each require working alone?

(1) Let x hours = the time Jack needs to wax the car

(2) Then in *one* hour, Jack can complete $\dfrac{1}{x}$ of the job

(3) $x + 1$ hours = the time Bob needs to wax the car

(4) Then in *one* hour, Bob can complete $\dfrac{1}{x + 1}$ of the job

(5) From (2) and (4), in *one* hour, the two working together can complete
$\dfrac{1}{x} + \dfrac{1}{x+1}$ of the job.

(6) $1\dfrac{1}{5}$ hours or $\dfrac{6}{5}$ hours = the time required for both working together to wax the car

(7) Then, in *one* hour, working together they can complete $\dfrac{1}{\frac{6}{5}}$ or $\dfrac{5}{6}$ of the job

From (5) and (7)

$$\frac{1}{x} + \frac{1}{x+1} = \frac{5}{6}$$
$$\frac{5}{6} = \frac{1}{x} + \frac{1}{x+1}$$
$$5(x)(x+1) = 6(x+1) + 6(x)$$
$$5x(x+1) = 6x + 6 + 6x$$
$$5x^2 + 5x = 12x + 6$$
$$5x^2 - 7x - 6 = 0$$
$$(x-2)(5x+3) = 0$$
$$x - 2 = 0 \quad \text{or} \quad 5x + 3 = 0$$
$$x = 2 \quad \text{or} \quad x = -\frac{3}{5}$$

Since the time is positive, $x = 2$ hours (for Jack to wax the car) and $x + 1 = 3$ hours (for Bob to wax the car).

Exercise 121:

Solve the following stated problems:

(1) One positive integer exceeds another by 2. If their product is 24, what are the numbers?

(2) If two positive consecutive integers are squared and then added, the result is 5 more than the square of the next consecutive integer. What are the numbers?

(3) In a two-digit number, the unit's digit is 2 more than the ten's digit. The sum of the squares of the digits is 18 more than the ten's digit. Find the numbers.

(4) If a positive number is added to nine times its multiplicative inverse, the result is twice the original number. What is the number?

(5) When five more than a number is divided by one less than the number, the quotient is equal to the divisor. Find the set of real numbers that satisfy this property.

(6) When seven more than a number is divided by two less than the number, the quotient is equal to the divisor. Find the set of real numbers that satisfies this property.

(7) When one and one-half less than a number is divided by one more than the number, the quotient is equal to the divisor. Find the set of complex numbers which satisfy this property.

(8) The area of a triangle is 6 square meters. The altitude of the triangle is 3 meters less than one-third of the base. Find the length of the altitude.

(9) A number greater than 5 is decreased by 5 and squared. The result is 3 less than the original number. What is the number?

(10) A rectangle is twice as long as it is wide. If the length is decreased by 4 meters and the width decreased by 2 meters, the area is 72 square meters. What were the original dimensions?

(11) The area of a rectangle is 48 square meters. If the length is 2 meters longer than the width, find the dimensions of the rectangle.

(12) In a right triangle, one side is 2 centimeters longer than the other side. The hypotenuse is 10 centimeters long. What are the lengths of the sides?

(13) A rectangle is inscribed in a circle of diameter 13 centimeters. If the length of the rectangle is 7 centimeters longer than the width, what are its dimensions?

(14) Find the base of a triangle with area 6 square meters if the altitude of the triangle is 4 meters shorter than the base.

(15) A rectangular box is twice as long as it is wide. The height is 2 centimeters longer than the width. If the surface area of the box is 208 square centimeters, what are the dimensions of the box?

(16) The height of a cylinder is 8 centimeters and its surface area is 130π square centimeters. What is the radius of the base? *Hint:* Remove the top and base of the cylinder and then mentally slice open the hollow shell and spread it flat. The sum of the areas of the three surfaces (the top, the base, and the flattened shell) will equal the surface area.

(17) When a sum of money, P, is invested at an interest rate, r, and the interest is compounded annually for a period of n years, the original investment at the end of n years has increased to an amount, A, that is computed from the equation $A = P(1 + r)^n$. Certificates of deposit (money invested for a fixed period of years) pay higher interest rates than ordinary savings accounts.

On his graduation from high school John received gifts of money from parents and relatives that totalled $3600.00. He plans to spend a summer in Europe after two years in college and needs $4200. for expenses. If John buys a 2-year certificate of deposit with the $3600. now, what interest rate does he need in order to have the desired sum of money at the end of two years?

(18) A landscape architect is designing a rectangular flower bed to be bordered with 28 plants that are placed 1 meter apart. He needs an inner rectangular space in the center for plants that must be 1 meter from the border of the bed and that require 24 square meters for planting. What should the overall dimensions of the flower bed be?

(19) A homeowner has a backyard containing 2880 square meters and wants to build a pool with a surrounding tile walk that will occupy no more than one-third of the space, or 960 square meters. He finds a pool company that offers special low prices on rectangular pools. If he decides to build a rectangular pool that is twice as long as it is wide with a 4-meter tiled area surrounding it, what should the dimensions of the pool be?

(21) A packaging firm plans to make rectangular boxes that are 2 centimeters deep, 4 centimeters longer than they are wide, and with a volume of 280 cm³ per box. The boxes are to be made from flat rectangles of cardboard by cutting 2-centimeter squares from each corner and turning up the sides. What should be the dimensions of the flat pieces of cardboard? (See Example 4.)

(22) A factory tests the road performance of new model cars by driving them at two different rates of speed for at least 100 kilometers at each rate. The speed rates range from 50 to 70 km/h in the lower range and from 70 to 90 km/h in the higher range. A driver plans to test a car on an available speedway by driving it for 120 kilometers at a speed in the lower range and then driving 120 kilometers at a rate that is 20 km/h faster. At what rates should he drive if he plans to complete the test in $3\frac{1}{2}$ hours?

(23) Bill's father can paint a room in two hours less than Bill can paint it. Working together they can complete the job in two hours and twenty-four minutes. How much time would each require working alone?

(24) Of two inlet pipes, the larger one can fill a swimming pool 4 hours faster than the smaller pipe. When both pipes are open, the pool is filled in three hours and forty-five minutes. If only the smaller pipe is open, how many hours are required to fill the pool?

(25) A train traveled 240 kilometers at a certain rate of speed. When the engine was replaced by an improved model, the speed was increased by 20 km/h and the travel time for the trip was decreased by 1 hour. What was the rate of each engine?

(26) The rate of the current in a stream is 3 km/h. A man rowed upstream for 3 kilometers and returned. The round trip required 1 hour and 20 minutes. How fast was he rowing?

(27) A pilot flying at a constant rate against a headwind of 50 km/h flew for 750 kilometers, then reversed direction and returned to his starting point. He completed the round trip in 8 hours. What was the speed of the plane?

(28) A school wishes to build a rectangular parking lot against the side of a building and enclose the other three sides with a brick wall. The budget will allow only enough money (for labor and material) to build a 100-meter length of brick wall to enclose this lot. There is to be a 4-meter opening across the center front of the lot for ingress and egress. It has been determined that the greatest area that can be enclosed under these conditions is 1352 square meters. What should the dimensions of the lot be in order to enclose this maximum area?

(29) A container company manufactures cans by joining the sides of flat rectangular pieces of metal to form hollow cylinders and then adding bases and tops. They have orders for a can that has a height of 6 centimeters, and a volume of 54π cm^3. What should be the dimensions of the flat rectangle of metal used to form the hollow cylinder for the can?

(30) Two drivers are testing the same model car at speeds that differ by 20 km/h. The one driving at the slower rate drives 70 kilometers down a speedway and returns by the same route. The one driving at the faster rate drives 76 kilometers on the speedway and returns by the same route. They left at the same time and the fast car returned $\frac{1}{2}$ hour earlier than the slower car.

At what rates were the cars driven?

Chapter Test 11

  I. Solve by completing the square.

$$2x^2 = 4x + 3$$

 II. Solve by formula.
      1. $x^2 - 2x + 4 = 0$
      2. $2x^2 = 5x - 3$

III. Solve by factoring.
      1. $x^2 = 16$
      2. $3x^2 + 5x = 0$
      3. $6x^2 = 4 - 5x$
      4. $4 - 11x = 3x^2$

 IV. Solve for x.
      1. $\dfrac{x}{x-2} - \dfrac{x-6}{x^2 - 5x + 6} = \dfrac{3}{x-3}$
      2. $\sqrt{2x+4} - x = 2$
      3. $\sqrt{2x+3} - \sqrt{x-2} = 2$

V. A rectangular pool surrounded by a 3-meter boardwalk is to be built in a park. The space available requires that the outside perimeter of the walk be limited to 190 meters. If the surface area of the pool is to be 1590 square meters, what should be the dimensions of the pool?

VI. A driver wants to test a car at two rates of speed which vary by 20 km/h. On each test he plans to drive a 175-kilometer route. At what rates should he drive the car on the two runs if he plans to complete both tests in 6 hours?

12
LOGARITHMS

The algebraic operations discussed in the first ten chapters of this text can be summarized briefly as the addition, the subtraction, the multiplication, the division, the involution (raising to powers), and the evolution (extraction of roots) of real numbers. In working with equations, we have repeatedly employed these operations, together with the equality axioms, to "isolate" or to "solve for" one particular quantity in an equation.

For example, the equation $4 + 2 = 6$ defines a relationship involving three numbers, 4, 2, and 6. We can state the relationship *explicitly* of any *one* of these numbers to the other two in the following manner:

$$6 = 4 + 2$$
$$4 = 6 - 2$$
$$2 = 6 - 4$$

Another example is the equation $A = \pi r^2$ in which the three numbers involved are A, π, and r. To state the explicit value of π or r in this equation means that we wish to *isolate* π or r on one side of the equal mark. By using the equality axioms and the previously defined operations we can show these explicit values in the following manner:

$$A = \pi r^2$$

$$\pi = \frac{A}{r^2}$$

$$r = \sqrt{\frac{A}{\pi}} \quad \text{or} \quad \frac{\sqrt{\pi A}}{\pi}$$

In the equation $3^2 = 9$, again we have a relationship involving three numbers, 3, 2, and 9. If we should seek to express each of these relationships explicitly, that is, if we should ask, what exactly is the relationship of any *one* of these numbers to the other two, we would have difficulty in finding the explicit value of the exponent, 2.

It is clear in the equation $3^2 = 9$ that

$$9 = 3^2$$
$$\text{and} \quad 3 = \sqrt[2]{9}$$
$$\textit{but} \quad 2 = ?$$

The algebraic operations which have previously been so useful cannot answer this question $2 = ?$ in the equation $3^2 = 9$.

12.1
Definition of a Logarithm

The question asked above, i.e., $2 = ?$ in the equation $3^2 = 9$, needs to be answered, because we shall frequently want to find the explicit value of the exponent in equations of this type. For example, in the equation $2^x = 7$, $x = ?$ For the equation $3^2 = 9$ we can describe the relationship of the exponent 2 to the numbers 3 and 9 in simple English by stating that

"2 is the *exponent* indicating the power to which the *base*, 3, must be raised in order to produce the *number* 9."

This statement can be translated into mathematical language by using the word *logarithm*. The statement

"2 is the *logarithm* of the *number* 9 to the *base* 3"

is by definition the same as the preceding statement in quotation marks.

In the latter statement, if we replace the word *logarithm* by its common abbreviation, *log*, then write the base 3 as a subscript to the word log, and finally replace the verb *is* by an equal mark, instead of the statement

"2 is the logarithm of the number 9 to the base 3"

we have the equation

$$2 = \log_3 9$$

and it defines *exactly the same relationship* as the equation $3^2 = 9$.

$$2 = \log_3 9 \quad \text{means} \quad 3^2 = 9$$

Examples:

(1) $3 = \log_2 8$ means $2^3 = 8$

(2) $5 = \log_2 32$ means $2^5 = 32$

(3) $4 = \log_5 625$ means $5^4 = 625$

(4) $6 = \log_{10} 1{,}000{,}000$ means $10^6 = 1{,}000{,}000$

(5) $\dfrac{1}{2} = \log_{36} 6$ means $36^{1/2} = 6$

(6) $-\dfrac{1}{3} = \log_{27} \dfrac{1}{3}$ means $27^{-1/3} = \dfrac{1}{3}$

By the symmetric axiom, the above equations can be written in reverse. For example, the first three equations with the sides reversed would be

(1) $\log_2 8 = 3$ means $2^3 = 8$

(2) $\log_2 32 = 5$ means $2^5 = 32$

(3) $\log_5 625 = 4$ means $5^4 = 625$

Note carefully that in each of these equations, the value of the *logarithm* in the equation on the left is the *exponent* in the equation on the right. One can get off to a fine start in understanding logarithmic notation by repeating numerous times "*a logarithm is an exponent*." By using this notation we may write exponential statements in logarithmic form and thereby state the explicit value of the exponent. In the following examples the double-headed arrow is read "implies and is implied by."

Examples:

(1) $3^2 = 9 \longleftrightarrow \log_3 9 = 2$

(2) $4^3 = 64 \longleftrightarrow \log_4 64 = 3$

(3) $49^{1/2} = 7 \longleftrightarrow \log_{49} 7 = \dfrac{1}{2}$

(4) $8^{-1/3} = \dfrac{1}{2} \longleftrightarrow \log_8 \dfrac{1}{2} = -\dfrac{1}{3}$

(5) $2^x = 7 \longleftrightarrow \log_2 7 = x$

Conversely, any logarithmic statement may be expressed in exponential form.

Examples:

(1) $\log_4 16 = 2 \longleftrightarrow 4^2 = 16$

(2) $\log_2 8 = 3 \longleftrightarrow 2^3 = 8$

(3) $\log_{81} 9 = \dfrac{1}{2} \longleftrightarrow 81^{1/2} = 9$

(4) $\log_{125} \dfrac{1}{5} = -\dfrac{1}{3} \longleftrightarrow 125^{-1/3} = \dfrac{1}{5}$

(5) $\log_6 x \;\; = 2 \longleftrightarrow 6^2 = x \;\;$ or $\;\; 36 = x$

Exercise 122:

Write the following exponential statements in logarithmic form:

(1) $3^4 = 81$

(2) $2^5 = 32$

(3) $10^3 = 1,000$

(4) $8^2 = 64$

(5) $5^3 = 125$

(6) $64^{1/2} = 8$

(7) $25^{1/2} = 5$

(8) $27^{1/3} = 3$

(9) $7^1 = 7$

(10) $23^1 = 23$

(11) $5^1 = 5$

(12) $12^0 = 1$

(13) $10^0 = 1$

(14) $\left(\dfrac{2}{3}\right)^0 = 1$

(15) $9^{-1/2} = \dfrac{1}{3}$

(16) $8^{-1/3} = \dfrac{1}{2}$

(17) $25^{-1/2} = \dfrac{1}{5}$

(18) $64^{-1/3} = \dfrac{1}{4}$

(19) $32^{-1/5} = \dfrac{1}{2}$

(20) $100^{-1/2} = \dfrac{1}{10}$

Exercise 123:

Write the following logarithms in exponential form:

(1) $\log_5 5 = 1$

(2) $\log_7 7 = 1$

(3) $\log_{10} 10 = 1$

(4) $\log_b b = 1$

(5) $\log_8 1 = 0$

(6) $\log_{25} 1 = 0$

(7) $\log_{10} 1 = 0$

(8) $\log_{5/3} 1 = 0$

(9) $\log_5 25 = 2$

(10) $\log_3 81 = 4$

(11) $\log_{10} 100 = 2$

(12) $\log_{10} 1,000 = 3$

(13) $\log_{10} 10,000 = 4$

(14) $\log_{10} \dfrac{1}{10} = -1$

(15) $\log_{27} \dfrac{1}{3} = -\dfrac{1}{3}$

(16) $\log_{36} \dfrac{1}{6} = -\dfrac{1}{2}$

(17) $\log_{49} \dfrac{1}{7} = -\dfrac{1}{2}$

(18) $\log_{81} \dfrac{1}{3} = -\dfrac{1}{4}$

(19) $\log_{64} \dfrac{1}{8} = -\dfrac{1}{2}$

(20) $\log_{10} \dfrac{1}{100} = -2$

(21) $\log_2 32 = 5$ (22) $\log_2 64 = 6$

(23) $\log_2 128 = 7$ (24) $\log_4 64 = 3$

(25) $\log_8 64 = 2$

Exercise 124:

Put the following statements in exponential form and find the value of x:

(1) $\log_{10} x = 2$ (2) $\log_5 x = 3$

(3) $\log_{16} x = \dfrac{1}{2}$ (4) $\log_{27} x = -\dfrac{1}{3}$

(5) $\log_{10} x = -4$ (6) $\log_{10} x = 3$

(7) $\log_x 25 = 2$ (8) $\log_x 8 = 3$

(9) $\log_x \dfrac{1}{6} = \dfrac{1}{2}$ (10) $\log_x \dfrac{1}{3} = -\dfrac{1}{2}$

(11) $\log_x \dfrac{1}{10} = -1$ (12) $\log_x .1 = -1$

(13) $\log_x .01 = -2$ (14) $\log_x .001 = -3$

(15) $\log_7 49 = x$ (16) $\log_4 64 = x$

(17) $\log_5 125 = x$ (18) $\log_{10} 10 = x$

(19) $\log_6 6 = x$ (20) $\log_{13} 13 = x$

(21) $\log_{15} 1 = x$ (22) $\log_{10} 1 = x$

(23) $\log_{45} 1 = x$ (24) $\log_{1/2} 1 = x$

(25) $\log_{64} \dfrac{1}{8} = x$ (26) $\log_{125} \dfrac{1}{5} = x$

To put a logarithmic statement in symbolic language we shall designate the following symbols for the three numbers involved:

b = base
x = the exponent
N = the number produced when b is raised to the power x

Then

$$\log_b N = x \longleftrightarrow b^x = N$$
$$(b, x, N \in R \qquad b > 0 \text{ and } b \neq 1)$$

The statement $\log_b N = x$ is read, x *is the exponent indicating the power to which the base,* b, *must be raised in order to produce the number* N. The number one is excluded as a base because one raised to any power can produce only the number one.

In this discussion we shall consider the logarithms of *positive numbers only*. If we should continue to restrict exponents to rational numbers and require also that the base, b, of a logarithm be a positive number, then in the expression, $\log_b N = x$, N will always be a *positive number*. As an illustration of this fact, consider the positive base 9 and the succession of rational exponents $-2, -1, -1/2, 0, 1/2, 1, 2$.

$$9^{-2} = \frac{1}{9^2} = \frac{1}{81} \qquad 9^0 = 1 \qquad 9^{1/2} = 3$$

$$9^{-1} = \frac{1}{9^1} = \frac{1}{9} \qquad\qquad\qquad 9^1 = 9$$

$$9^{-1/2} = \frac{1}{9^{1/2}} = \frac{1}{3} \qquad\qquad\qquad 9^2 = 81$$

In every case the number N (obtained by raising the base to the indicated power) is a positive number. Actually, it is true that for any positive base, b, the number N will always be positive if the exponent, x, is any real number. In more advanced work it is shown that logarithms of negative numbers with positive bases are imaginary numbers.

Restating the definition of a logarithm we have the following:

$$\log_b N = x \longleftrightarrow b^x = N$$
$$(x, b \in R \qquad b > 0, b \neq 1)$$

In the equation on the right, $b^x = N$, we may substitute for x the value which is *equal to x* in the equation on the left. Thus we have

$$b^x = N$$

or

$$\boxed{b^{\log_b N} = N}$$

The equation in the box above is an important identity which should be memorized.

Examples:

(1) $10^{\log_{10} 5} = 5$ (2) $10^{\log_{10} 8} = 8$

(3) $10^{\log_{10} x} = x$ (4) $b^{\log_b y} = y$

(5) $a^{\log_a y} = y$ (6) $e^{\log_e x} = x$

12.2
Two Properties of Logarithms; Logarithmic Equations

Since a logarithm is the explicit value of an exponent, we find that logarithms have some special properties which can be derived from the laws of exponents. The first property which we shall consider, Property 1, can be proved in the following manner.

Proof of Property 1   Consider two different powers each having the same base b with real exponents represented by x and y: (1) $b^x = M$, and (2) $b^y = N$, where M and N represent the unique numbers produced by raising the base b to the respective powers x and y. By writing each of the above equations in logarithmic form, we can state the explicit values of the exponents x and y.

1. $b^x = M \longleftrightarrow \log_b M = x$

2. $b^y = N \longleftrightarrow \log_b N = y$

3. In Equation (2)

$$b^y = N$$

Multiply both sides by b^x.

$$b^x b^y = b^x N$$

4. Then $b^{x+y} = b^x \, N$ (first law of exponents)

5. From Statement (1) $b^x = M$; in Equation (4), replace b^x by M and we have

$$b^{x+y} = MN$$

6. Expressing this last equation in logarithmic form, we have

$$\log_b MN = x + y$$

In this equation, we replace x and y by the values of each given in Statements (1) and (2).

$$x = \log_b M \quad \text{and} \quad y = \log_b N$$

7. Therefore, from Statement (6)

$$\log_b MN = x + y$$

we have

$$\boxed{\log_b MN = \log_b M + \log_b N}$$

The equation boxed in above states the first property of logarithms, which we shall henceforth call Property 1. By the symmetric axiom this property is true in reverse, and it should be thoroughly *understood* and *memorized* both ways. Restating Property 1 literally forwards and backwards, we have the following:

Property 1

A. $\log_b MN = \log_b M + \log_b N$
The logarithm of the product of two positive numbers is equal to the *sum* of the logarithms of the numbers.

B. $\log_b M + \log_b N = \log_b MN$
The *sum* of the logarithms of two positive numbers is equal to the logarithm of the *product* of the two numbers.

Property 1 is equally true for the logarithm of the product of more than two factors, e.g., $\log_b MNS = \log_b M + \log_b N + \log_b S$.

Illustration of Property 1:   Given that

$$\log_2 4 = 2 \qquad \log_2 8 = 3 \qquad \log_2 16 = 4 \qquad \log_2 32 = 5 \qquad \log_2 64 = 6$$

Property 1

A. $\boxed{\log_2 (4)(8) = \log_2 4 + \log_2 8}$

or $\log_2 32 = \log_2 4 + \log_2 8$
$$5 \;=\; 2 + \quad 3$$

B. $\boxed{\log_2 16 + \log_2 4 = \log_2 (16)(4)}$

or $\log_2 16 + \log_2 4 = \log_2 64$
$$4 + \quad 2 = \quad 6$$

The second property of logarithms which we shall consider can be derived by referring to the same equations given in Statements (1) and (2) for the proof of Property 1.

Proof of Property 2

1. $b^x = M \longleftrightarrow \log_b M = x$
2. $b^y = N \longleftrightarrow \log_b N = y$
3. In equation (1)

$$b^x = M$$

Divide both sides by b^y.

$$\frac{b^x}{b^y} = \frac{M}{b^y}$$

4. Then $b^{x-y} = M/b^y$ (second law of exponents).
5. From statement (2) $b^y = N$; replacing b^y by N in Equation (4) we have

$$b^{x-y} = \frac{M}{N}$$

6. Putting this last equation in logarithmic form we have

$$\log_b \frac{M}{N} = x - y$$

7. From (1) and (2) replace x and y in (6) by their explicit values: $x = \log_b M$ and $y = \log_b N$. Therefore from

$$\log_b \frac{M}{N} = x - y$$

we have

$$\boxed{\log_b \frac{M}{N} = \log_b M - \log_b N}$$

The last equation states the second property of logarithms, which we shall call Property 2. Like Property 1, it must be learned backwards and forwards since it is equally useful both ways.

Property 2

A. $\log_b \dfrac{M}{N} = \log_b M - \log_b N$

The logarithm of the quotient of two positive numbers is equal to the logarithm of the numerator minus the logarithm of the denominator.

B. $\log_b M - \log_b N = \log_b \dfrac{M}{N}$

The *difference* of the *logarithms* of two positive numbers is equal to the *logarithm* of the quotient of the two numbers.

Note the following *carefully:*

1. $\log_b \dfrac{M}{N}$ is the *logarithm* of a quotient.

2. $\dfrac{\log_b M}{\log_b N}$ is *not* the logarithm of a quotient; it is the quotient of two logarithms which is an entirely different situation.

Property 2 is *true only* for the logarithm of a quotient.

Illustration of Property 2:   Given that

$$\log_2 4 = 2 \qquad \log_2 8 = 3 \qquad \log_2 16 = 4 \qquad \log_2 32 = 5 \qquad \log_2 64 = 6$$

Property 2

A. $$\boxed{\log_2 \frac{32}{8} = \log_2 32 - \log_2 8}$$

or $\log_2 4 = \log_2 32 - \log_2 8$
$$2 = \qquad 5 - \qquad 3$$

B. $$\boxed{\log_2 64 - \log_2 16 = \log_2 \frac{64}{16}}$$

or $\log_2 64 - \log_2 16 = \log_2 4$
$$6 - \qquad 4 = \qquad 2$$

When two positive numbers, A and C, are equal, then the values of their logarithms to the same base are also equal. If

$$A = C \qquad (A \in R \qquad A > 0)$$

then $$\log_b A = \log_b C \qquad (b > 0, b \neq 1)$$

Under the conditions stated, the converse of this is also true. If

$$\log_b A = \log_b C$$

then $$A = C$$

Consequently, if we have an equation such as $\log_b x = \log_b 7$, we can conclude that $x = 7$. The intelligent use of this fact along with the first two properties of logarithms can be exceedingly helpful in serving certain equations. Consider the following examples, in all of which $x \in R^+$:

Examples:

(1) $\log_4 x = \log_4 5 + \log_4 3$
$\log_4 x = \log_4 (5)(3)$ (Property 1)
$\log_4 x = \log_4 15$
$x = 15$

(2) $\log_7 x = \log_7 18 - \log_7 2$
$\log_7 x = \log_7 \dfrac{18}{2}$ (Property 2)
$\log_7 x = \log_7 9$
$x = 9$

(3) $\log_8 6 = \log_8 x - \log_8 4$
$\log_8 6 + \log_8 4 = \log_8 x$
$\log_8 (6)(4) = \log_8 x$ (Property 1)
$\log_8 24 = \log_8 x$
$24 = x$

(4) $\log_5 7x = \log_5 28 - \log_5 2$
$\log_5 7x = \log_5 \dfrac{28}{2}$ (Property 2)
$\log_5 7x = \log_5 14$
$7x = 14$
$x = 2$

(5) $\log_3 (2x + 1) = \log_3 9 + \log_3 7$
$\log_3 (2x + 1) = \log_3 (9)(7)$ (Property 1)
$\log_3 (2x + 1) = \log_3 63$
$2x + 1 = 63$
$2x = 63 - 1$
$2x = 62$
$x = 31$

(6) $\log_7 x^2 = \log_7 5 + \log_7 x + \log_7 2$
$\log_7 x^2 - \log_7 x = \log_7 5 + \log_7 2$
$\log_7 \dfrac{x^2}{x} = \log_7 (5)(2)$ $\left(\begin{array}{l}\text{Property 2}\\\text{Property 1}\end{array}\right)$
$\log_7 x = \log_7 10$
$x = 10$

(7) $\log_{10} \sqrt{3x + 2} + \log_{10} 4 = \log_{10} 56 - \log_{10} \sqrt{3x + 2}$

$\log_{10} \sqrt{3x + 2} + \log_{10} \sqrt{3x + 2} = \log_{10} 56 - \log_{10} 4$

$\log_{10} (\sqrt{3x + 2})(\sqrt{3x + 2}) = \log_{10} \dfrac{56}{4}$ $\qquad \left(\begin{array}{l} \text{Property 1} \\ \text{Property 2} \end{array} \right)$

$\log_{10} (3x + 2) = \log_{10} 14$

$3x + 2 = 14$

$3x = 12$

$x = 4$

In the examples given thus far, every term in each equation is expressed as the logarithm of a number. But suppose we had an equation such as the following:

$$\log_3 4x = \log_3 8 + 1$$

We cannot apply Property 1 to the sum of the two terms on the right unless we have the sum of two *logarithms*. To achieve this aim, we can write the number 1 as a logarithm *to base 3* in the following way:

$$1 = \log_3 3 \longleftrightarrow 3^1 = 3$$

and then substitute $\log_3 3$ for 1 in the equation:

$$\log_3 4x = \log_3 8 + 1$$
$$\downarrow \qquad\qquad \downarrow \qquad \downarrow$$
$$\log_3 4x = \log_3 8 + \log_3 3$$

Then, applying Property 1 on the right, we obtain

$$\log_3 4x = \log_3 (8)(3)$$
$$\log_3 4x = \log_3 24$$
$$4x = 24$$
$$x = 6$$

Now suppose the numerical term in the preceding equation had been 2 or 7 or 9. We can write all of these numbers as logarithms to base 3 in the following way:

$$2 = \log_3 3^2 \longleftrightarrow 3^2 = 3^2$$
$$7 = \log_3 3^7 \longleftrightarrow 3^7 = 3^7$$
$$9 = \log_3 3^9 \longleftrightarrow 3^9 = 3^9$$

This handy notion permits us to express any real number as a logarithm to any positive base. For example,

$$5 = \log_4 4^5 = \log_6 6^5 = \log_{13} 13^5 \text{ etc.}$$

In general, letting R^+ represent the set of *positive* real numbers,

> If $x \in R$ and $a \in R^+$, then
> $$x = \log_a a^x$$

Note the use of this fact in the remaining examples of logarithmic equations.

(8) $\log_4 8x = \log_4 5 + 2$
$\log_4 8x = \log_4 5 + \log_4 4^2$
$\log_4 8x = \log_4 5 + \log_4 16$
$\log_4 8x = \log_4 (5)(16) \quad$ (Property 1)
$\log_4 8x = \log_4 80$
$ 8x = 80$
$ x = 10$

(9) $\log_3 (x + 1) = \log_3 7 + 4$
$\log_3 (x + 1) = \log_3 7 + \log_3 3^4$
$\log_3 (x + 1) = \log_3 7 + \log_3 81$
$\log_3 (x + 1) = \log_3 (7)(81) \quad$ (Property 1)
$\log_3 (x + 1) = \log_3 567$
$ x + 1 = 567$
$ x = 567 - 1$
$ x = 566$

Exercise 125:

Solve the following equations for x:

(1) $\log_5 x = \log_5 24$
(2) $\log_6 2x = \log_6 26$
(3) $\log_3 (x + 1) = \log_3 16$
(4) $\log_8 (2x - 1) = \log_8 15$
(5) $\log_7 x = \log_7 8 + \log_7 6$

(6) $\log_2 x = \log_2 12 - \log_2 4$

(7) $\log_6 x - \log_6 3 = \log_6 7$

(8) $\log_4 3x = \log_4 12$

(9) $\log_8 6 = \log_8 x - \log_8 4$

(10) $\log_5 5x - \log_5 3 = \log_5 10$

(11) $\log_9 2x = \log_9 3 + \log_9 4$

(12) $\log_4 (x + 1) - \log_4 7 = \log_4 17$

(13) $\log_6 \sqrt{2x} + \log_6 2 = \log_6 24 - \log_6 \sqrt{2x}$

(14) $\log_2 x + \log_2 5 = \log_2 30$

(15) $\log_{10} 4x^2 - \log_{10} 4 = \log_{10} 2x + \log_{10} 24$

(16) $\log_7 x^2 + \log_7 3 = \log_7 2x + \log_7 18$

(17) $\log_{10} 3x + \log_{10} 2x - \log_{10} 5 = \log_{10} x + \log_{10} 6 + \log_{10} 2$

(18) $\log_3 (x^2 - 5x + 6) - \log_3 (x - 2) = \log_3 7 + \log_3 2$

(19) $\log_9 \sqrt{5x + 1} - \log_9 2 = \log_9 13 - \log_9 \sqrt{5x + 1}$

(20) $\log_2 8x = \log_2 3 + 5$ (See Example 8)

(21) $\log_3 4x = \log_3 72 - 2$

(22) $\log_2 (3x + 1) = \log_2 5 + 3$

(23) $\log_4 \sqrt{x + 1} + 2 = \log_4 48 - \log_4 \sqrt{x + 1}$

(24) $\log_3 (x^2 - 6x + 8) - \log_3 6 = \log_3 (x - 4) + 2$

12.3
Common Logarithms

In the preceding exercise the bases of the logarithms were all positive numbers such as 3, 4, 5, 8, 10, etc. Since any positive number except the number one can serve as a base for logarithms, the human race long ago decided to simplify working with logarithms by agreeing to use the same base. In fact, this has been agreed on *twice*, so that we now have available to us tables of logarithms with values computed to two different bases: the *natural* logarithms for which the base is a positive irrational number symbolized by the letter e ($e = 2.71828\ldots$), and the *common* logarithms for which the base is always the number ten. Since this discussion is limited to logarithms of positive numbers only, common logarithms will serve our purposes adequately, and we shall confine our attention to this system. Because the base is always ten, it has become customary to write "$\log_{10} a$" simply as "$\log a$", where the omission of the base means that base 10 is understood. Any base *other than ten* must be written in the usual manner. For emphasis, we shall continue to write the base 10 in the following discussion.

Consider the following logarithms to base 10 carefully, and ask yourself what numbers could correctly replace the question marks:

1. $\log_{10} .01 = ?$ or $10^? = .01$

2. $\log_{10} .1 = ?$ or $10^? = .1$

3. $\log_{10} 1 = ?$ or $10^? = 1$
4. $\log_{10} 10 = ?$ or $10^? = 10$
5. $\log_{10} 100 = ?$ or $10^? = 100$

By looking at the column on the *right* we can easily see that

$$10^{-2} = .01 \qquad \left[10^{-2} = \frac{1}{10^2} = \frac{1}{100} = .01 \right]$$

$$10^{-1} = .1 \qquad \left[10^{-1} = \frac{1}{10} = .1 \right]$$

$$10^0 \; = 1$$

$$10^1 \; = 10$$

$$10^2 \; = 100$$

Replacing the question marks in the column on the left by the correct values of exponents given above, we have the following:

(1) $\log_{10} .01 \; = -2$
(2) $\log_{10} .1 \;\; = -1$
(3) $\log_{10} 1 \;\;\; = 0$
(4) $\log_{10} 10 \;\; = 1$
(5) $\log_{10} 100 = 2$

If we wished to expand this brief table, we could begin by noting that $\log_{10} 1 = 0$ and $\log_{10} 10 = 1$. Then the common logarithms of numbers which lie between 1 and 10 (such as $\log_{10} 2, \log_{10} 3, \log_{10} 5.8, \log_{10} 9.3$) should all have values that lie *between* 0 and 1 if we make the reasonable assumption that $a < b \rightarrow \log a < \log b$. For example, $\log_{10} 2$ must be a number that is *greater than zero* (since $10^0 = 1$), but at the same time *less than one* (since $10^1 = 10$).

$$\log_{10} 2 = x \quad (\text{or } 10^x = 2) \qquad \begin{array}{l}(x \text{ must be greater than 0} \\ \text{but less than 1})\end{array}$$

The actual computation of this logarithm involves operations beyond the scope of this text, but these values have been computed and arranged in tables where they may be found for any number within the range of the tables. In these tables, for example, by reading the *number* in the number column on the left and the *logarithm of the number* in the proper column to the right of the number, we would find the following values correct to five decimal places:

(1) $\log_{10} 2 = .3010$ or $10^{.3010} = 2$
(2) $\log_{10} 3 = .4771$ or $10^{.4771} = 3$
(3) $\log_{10} 5 = .6990$ or $10^{.6990} = 5$
(4) $\log_{10} 8 = .9031$ or $10^{.9031} = 8$

With these values, we could, if we wished, use the first property of logarithms to compute many others.

Examples:

(1) $\log_{10} 6 = \log_{10} (3)(2) = \log_{10} 3 + \log_{10} 2$
$$= .4771 + .3010$$
$$= .7781$$

$$\boxed{\log_{10} 6 = .7781 \quad \text{or} \quad 10^{.7781} = 6}$$

(2) $\log_{10} 16 = \log_{10} (8)(2) = \log_{10} 8 + \log_{10} 2$
$$= .9031 + .3010$$
$$= 1.2041$$

$$\boxed{\log_{10} 16 = 1.2041 \quad \text{or} \quad 10^{1.2041} = 16}$$

Now let us consider $\log_{10} 3.25$. Since 3.25 lies between 1 and 10, its logarithm to base 10 must lie between 0 and 1. The tables would show that

$$\log_{10} 3.25 = 0.5119$$

Using this, we shall compute the following logarithms:

1. $\log_{10} .0325 = \log_{10} \dfrac{3.25}{100}$

$\qquad\qquad\quad = \log_{10} 3.25 - \log_{10} 100 \qquad$ (Property 2)
$\qquad\qquad\quad = 0.5119 - 2$
$\qquad\qquad\quad = \underline{-2 + .5119}$

2. $\log_{10} .325 \; = \log_{10} \dfrac{3.25}{10}$

$\qquad\qquad\quad = \log_{10} 3.25 - \log_{10} 10 \qquad$ (Property 2)
$\qquad\qquad\quad = 0.5119 - 1$
$\qquad\qquad\quad = \underline{-1 + .5119}$

3. $\log_{10} 32.5 \; = \log_{10} (3.25)(10)$

$\qquad\qquad\quad = \log_{10} 3.25 + \log_{10} 10 \qquad$ (Property 1)
$\qquad\qquad\quad = 0.5119 + 1$
$\qquad\qquad\quad = \underline{1 + .5119}$

4. $\log_{10} 325 = \log_{10} (3.25)(100)$
    $= \log_{10} 3.25 + \log_{10} 100$ (Property 1)
    $= 0.5119 + 2$
    $= \underline{2 + .5119}$

Listing these results in abbreviated form we have the following:

(1) $\log_{10} .0325 = -2 + .5119$
(2) $\log_{10} .325 = -1 + .5119$
(3) $\log_{10} 3.25 = 0 + .5119$
(4) $\log_{10} 32.5 = 1 + .5119$
(5) $\log_{10} 325 = 2 + .5119$

In the list above, each logarithm consists of two distinct terms. The *first* term in each is an integer and is called the *characteristic*. The second term in each one is the positive decimal fraction .5119, and it is called the *mantissa*. From these results the following observations should be made:

1. The common logarithms of numbers which are the same except for the position of the decimal point will always have the *same mantissa*.

2. Any *mantissa* different from zero is always a *positive decimal fraction*.

3. The characteristic is always an *integer*. It will be *positive* for any number greater than or equal to 10, *negative* for any number less that 1 but greater than zero, and *zero* for any number less than 10 but greater than or equal to 1.

The mantissas listed in tables of logarithms are always positive, and, in order to use the tables, we shall write *all* logarithms so that the decimal parts are positive. Referring again to the logarithms listed for .0325, .325, 3.25, etc., we find that there is no problem about keeping the mantissa positive in Examples (3), (4), and (5) because the characteristic in each case is either zero or a positive integer.

(3) $\log_{10} 3.25 = 0 + .5119 = 0.5119$
(4) $\log_{10} 32.5 = 1 + .5119 = 1.5119$
(5) $\log_{10} 325 = 2 + .5119 = 2.5119$

But in Examples (1) and (2)

(1) $\log_{10} .0325 = -2 + .5119$
(2) $\log_{10} .325 = -1 + .5119$

we cannot combine the negative characteristic and the positive mantissa without arriving at a negative decimal.

It is perfectly true that

$$\log_{10} .0325 = -2 + .5119 = -1.4881$$
$$= -1 - .4881$$

but the decimal $-.4881$ is *not* the mantissa for the number .0325.

Because we need a positive decimal to use the logarithmic tables, we write all negative characteristics in a binomial form such as the following:

$$-1 = 9 - 10$$
$$-2 = 8 - 10$$
$$-3 = 7 - 10$$
$$-4 = 6 - 10$$

Then we add the mantissa to the *positive term* of the characteristic and leave the logarithm in binomial form.

Examples:

(1) $\log_{10} .0325 = -2 + .5119$
$$= 8 - 10 + .5119$$
$$= 8 + .5119 - 10$$
$$= \underline{8.5119 - 10}$$

(2) $\log_{10} .325 = -1 + .5119$
$$= 9 - 10 + .5119$$
$$= 9 + 5.119 - 10$$
$$= \underline{9.5119 - 10}$$

The tables of common logarithms list *only* the mantissa for any given sequence of figures in a number, a sensible procedure when one pauses to reflect that numbers which are the same except for the position of the decimal point will all have the same mantissa. The characteristic for the logarithm of a given number must be computed from the position of the decimal point in the number.

The relation of the characteristic to the position of the decimal point in the number may be illustrated in the following way: Any positive real number may be expressed as the product of a number between 1 and 10 and *some power of ten*; written in this form, a number is said to be in scientific notation.

Examples:

(1) $.0325 = (3.25)(10^{-2})$
   and log $.0325 = -2 + .5119$

(2) $.325 = (3.25)(10^{-1})$
   and log $.325 = -1 + .5119$

(3) $3.25 = (3.25)(10^0)$
   and log $3.25 = 0 + .5119$

(4) $32.5 = (3.25)(10^1)$
   and log $3.25 = 1 + .5119$

(5) $325 = (3.25)(10^2)$
   and log $325 = 2 + .5119$

Since the *exponent of 10* and the *characteristic* of the logarithm are always the *same integer*, the characteristic for the logarithm of any number may be computed by writing the number (literally or mentally) in this form.

Examples:

Number	Number in Scientific Notation	Characteristic
.063	$(6.3)(10^{-2})$	-2
00.063	$(6.3)(10^{-2})$	-2
.0063	$(6.3)(10^{-3})$	-3
.00637	$(6.37)(10^{-3})$	-3

Following are more examples of numbers and the characteristics of their common logarithms.

Example:

Number	Characteristic	Number	Characteristic
6	0	4820	3
4.5	0	8796.408	3
7.34	0	.5	-1 or $9 - 10$
32	1	.732	-1 or $9 - 10$
65.7	1	.8297	-1 or $9 - 10$
87.45	1	.04	-2 or $8 - 10$
245	2	.0336	-2 or $8 - 10$
562.3	2	.008	-3 or $7 - 10$
784.59	2	.007326	-3 or $7 - 10$

Exercise 126:

Find the characteristics.

(1)	4.602	(2)	3589	(3)	.884
(4)	265,000	(5)	.000377	(6)	3
(7)	20	(8)	16.48	(9)	3729.45
(10)	1.066	(11)	.054	(12)	7.35
(13)	.09832	(14)	.0043	(15)	38,000
(16)	49.73	(17)	2.8645	(18)	584.36
(19)	.0009843	(20)	.7	(21)	.654
(22)	.0059	(23)	37.9832	(24)	6.7032

Exercise 127:

Using four-place tables, find the logarithms of the following numbers:

(1)	36.2	(2)	.00971	(3)	432
(4)	685,000	(5)	7.84	(6)	.9
(7)	.0223	(8)	75.8	(9)	891
(10)	.0886	(11)	2.68	(12)	765
(13)	8000	(14)	.000389	(15)	84.4
(16)	23,600	(17)	.665	(18)	.0473
(19)	64.3	(20)	.00718	(21)	189,000,000
(22)	.00000835	(23)	6.28	(24)	342

It is frequently necessary to reverse the procedure of the foregoing exercise and find a number for which the logarithm is known. Consider that we know, for example, that the logarithm of a certain number is $8.7101 - 10$ and we wish to find the number. First, we should state the problem in the proper form. Let N represent the number to be found. Then

$$\log_{10} N = 8.7101 - 10$$
$$N = ?$$

The *two* things known about the number N are the following:

1. First, in the tables, we look *in the columns listing the mantissas* until we find .7101. The table shows that this is the mantissa for *any* number having the following sequence of figures: 513. Therefore the figures in the number N must be 513.

2. Besides being composed of this sequence of figures, the logarithm of the number N must *also* have a characteristic of -2.

We have previously observed that when a number is written as the product of a number between one and ten and some power of ten, the *characteristic* of the *logarithm* and the *exponent of ten* are the same integer. Hence, since the characteristic is -2, we may write

$$N = (5.13)(10^{-2})$$
$$\text{or} \quad N = .0513$$

Example:

$$\log N = 2.8116$$
$$N = ?$$

(1) The mantissa for N is .8116. In the tables the columns of mantissas show that the digits in the number N must be 648.

(2) The characteristic is $+2$. Therefore

$$N = (6.48)(10^2)$$
$$\text{or} \quad N = 648$$

Exercise 128:

Find the numbers which have the given logarithms.

(1) $\log N = 1.5366$ (2) $\log N = 9.4425 - 10$
(3) $\log N = 0.3032$ (4) $\log N = 3.6981$
(5) $\log N = 7.0212 - 10$ (6) $\log N = 2.2648$
(7) $\log N = 4.1761$ (8) $\log N = 8.4298 - 10$
(9) $\log N = 1.4409$ (10) $\log N = 2.6335$
(11) $\log N = 9.7760 - 10$ (12) $\log N = 0.8943$
(13) $\log N = 6.5966 - 10$ (14) $\log N = 2.3502$
(15) $\log N = 7.6513 - 10$ (16) $\log N = 5.7235$
(17) $\log N = 5.0864 - 10$ (18) $\log N = 9.3979 - 10$

12.4
Interpolation

Because of the obvious restrictions of space in the compiling of logarithmic tables, the numbers listed are limited in the number of digits given. By a process called "linear interpolation" we can calculate the mantissa, with reasonable accuracy, for the logarithm of a number containing one figure more than those given by the tables. We shall limit interpolation to one additional digit only, which will give us sufficient accuracy for our purposes.

If we wish to find, for example, the logarithm of 46.6689 and the numbers listed in the tables are limited to three digits, we would proceed in the following manner:

1. Round off the number 46.6689 to four digits, that is, to 46.67.

2. The number 46.67 lies between 46.60 and 46.70. In linear interpolation we assume that the graph of the logarithm for numbers between 46.60 and 46.70 (that is, for 46.61, 46.62, 46.63, etc.) is a straight line. While this assumption is not true, over this restricted range of numbers, it gives a result that is sufficiently accurate.

3. List the numbers and their logarithms in ascending order as follows:

$$\begin{matrix} \text{total} \\ \text{difference} \\ = .10 \end{matrix} \begin{cases} \begin{matrix} \text{partial} \\ \text{difference} \\ = .07 \end{matrix} \begin{cases} \log 46.60 = 1.6684 \\ \log 46.67 = 1. \quad ? \\ \log 46.70 = 1.6693 \end{cases} \end{cases} \begin{matrix} \text{total} \\ \text{difference} \\ = .0009 \end{matrix}$$

The figures on the left show that the number 46.67 includes .07 of the .10 difference between 46.60 and 46.70. Therefore we shall take .07/.10 = 7/10 = .7 of the difference between the two mantissas, and *increase* the smaller mantissa by that amount.

$$\begin{matrix} .0009 \\ \underline{\quad .7} \\ .00063 \end{matrix}$$

The correction is .00063; *rounded off to four places*, this is .0006. Adding this to the smaller mantissa, we have

$$\begin{matrix} .6684 = \text{mantissa for } 46.60 \\ \underline{.0006} = \text{correction} \\ .6690 = \text{mantissa for } 46.67 \end{matrix}$$

Therefore log 46.67 = 1.6690.

Two facts should be noted *carefully* about this interpolation process.

1. All corrections for mantissas are rounded off to the same number of places as the values given in the tables. We cannot interpolate tables to any greater degree of accuracy than that of the figures with which we are working.

2. Neither the decimal point in the number nor the characteristic of its logarithm is affected by the interpolation process. Consequently, we may ignore both and *work only with the mantissas.*

Example:

Find log 235.3.

$$10 \left\{ {}_3\!\left\{ \begin{array}{l} 2350 \\ 2353 \\ 2360 \end{array} \right. \begin{array}{l} \text{mantissa is .3711} \\ \text{mantissa is\quad ?} \\ \text{mantissa is .3729} \end{array} \right\} .0018$$

$$(.0018)\left(\frac{3}{10}\right) = (.0018)(.3) = .00054$$

The correction for the mantissa is .0005.

$$\begin{array}{ll} .3711 & \text{mantissa for 2350} \\ +.0005 & \\ \hline .3716 & \text{mantissa for 2353} \end{array}$$

The characteristic for 235.3 is 2.
Therefore log 235.3 = 2.3716.

This same process can be applied to find a number whose logarithm has a mantissa which is not given in the tables. If we know that

$$\log N = 3.4413$$

and we wish to find N, we should proceed in the following manner: In the table of mantissas find the mantissa just smaller and the one just greater than the mantissa for N.

$$.4413 = \text{mantissa for } N$$

From the columns of mantissas

$$\begin{array}{l} \text{total} \\ \text{difference} \\ = .0016 \end{array} \left\{ \begin{array}{l} \text{partial} \\ \text{difference} \\ = .0004 \end{array} \left\{ \begin{array}{l} .4409 = \text{mantissa for 2760} \\ .4413 = \text{mantissa for } N \\ .4425 = \text{mantissa for 2770} \end{array} \right\} \begin{array}{l} \text{total} \\ \text{difference} \\ = 10 \end{array} \right.$$

$$\frac{.0004}{.0016} = \frac{4}{16} = \frac{1}{4} \quad \text{and} \quad \frac{1}{4}(10) = 2.5$$

The correction for the last figure in the number N is 3. Therefore the sequence of digits for N is 2763.

Since log $N = 3.4413$, the characteristic is 3.

$$N = (2.763)(10^3)$$
$$N = 2763$$

Exercise 129:

Find the logarithms of the following numbers:

(1)   823.67   (2)   48.832
(3)   2.7849   (4)   .0010452
(5)   632,460   (6)   .2003573
(7)   .0932519   (8)   6.3487
(9)   .0038054   (10)   4682.5

Exercise 130:

Find the numbers which have the given logarithms.

(1) $\log N = 0.3862$ (2) $\log N = 8.5978 - 10$
(3) $\log N = 2.6741$ (4) $\log N = 1.8123$
(5) $\log N = 9.0384 - 10$ (6) $\log N = 7.2752 - 10$
(7) $\log N = 1.1654$ (8) $\log N = 3.2928$
(9) $\log N = 9.4446$ (10) $\log N = 0.7317$

12.5
Additional Properties of Logarithms; Exponential Equations

A most useful property of logarithms is one which can be proved for the logarithm of a power. Let b represent a positive base ($b \neq 1$), x and y real exponents, and N the number produced when b is raised to the power indicated by x.

1. $b^x = N \longleftrightarrow \log_b N = x$

2.   Then $(b^x)^y = N^y$
     or $b^{xy} = N^y$ (third law of exponents)

3.   Since $b^{xy} = N^y$

$$\log_b N^y = xy$$

4.   From statement (1), $x = \log_b N$

5.   Replacing x by $\log_b N$ in the equation

$$\log_b N^y = xy$$

we have

$$\log_b N^y = (\log_b N)y$$

or $\boxed{\log_b N^y = y \log_b N}$

This last equation states the third property of logarithms and, like the first two, it should be learned and understood thoroughly forwards and backwards.

Property 3

A. $\log_b N^y = y \log_b N$
The logarithm of a number raised to a power, y, is equal to the exponent, y, *multiplied* by the logarithm of the number.
B. $y \log_b N = \log_b N^y$
The product of a number, y, and the logarithm of a number, N, is equal to the logarithm of N^y.

Illustration of Property 3:   Given that

$$\log_2 2 = 1 \qquad \log_2 4 = 2 \qquad \log_2 32 = 5 \qquad \log_2 64 = 6$$

Property 3

A.
$$\boxed{\log_2 4^3 = 3 \log_2 4}$$

or $\quad \log_2 64 = 3 \log_2 4$
$$6 = 3 \ (2)$$

B.
$$\boxed{5 \log_2 2 = \log_2 2^5}$$

or $\quad 5 \log_2 2 = \log_2 32$
$$5 \ (1) = 5$$

This property can be used to raise numbers to powers or to extract roots. *You can avoid many careless errors if you will label the quantity to be computed by some symbol, such as N, and then state the problem as an equation to be solved for N.*

Examples:

(1) Find the value of 3^8.
The value of 3^8 is some number which we shall call N.
$$N = 3^8$$

Then $\log N = \log 3^8$

$\log N = 8 \log 3$ (Property 3)

$\log N = 8(.4771)$

$\log N = 3.8168$

$N = 6560$

(2) Find $\sqrt[5]{3240}$.

$\sqrt[5]{3240} = 3240^{1/5}$

$N = 3240^{1/5}$

$\log N = \log 3240^{1/5}$

$\log N = \frac{1}{5}(\log 3240)$ (Property 3)

$\log N = \frac{1}{5}(3.5105)$

$\log N = 0.7021$

$N = 5.036$

(3) In extracting roots of numbers whose logarithms have *negative* characteristics, we can avoid arithmetic complications by expressing the characteristic in a binomial form for which the *last term* is *evenly* divisible by the denominator of the power. This is necessary because the characteristic of the resulting logarithm must be an integer.

Find $\sqrt[3]{.0565}$.

$\sqrt[3]{.0565} = .0565^{1/3}$ Let N represent this number.

$N = .0565^{1/3}$

$\log N = \log .0565^{1/3}$

$\log N = \frac{1}{3}(\log .0565)$

$\log N = \frac{1}{3}(8.7520 - 10)$

The last term, 10, is *not* divisible by 3; consequently, the characteristic, -2, is written $28 - 30$ instead of $8 - 10$.

$$\log N = \frac{1}{3}(28.7520 - 30)$$

$$\log N = 9.5840 - 10$$

N is a number whose logarithm has a *mantissa* $= .5840$ and a *characteristic* $= 9 - 10$ or -1.

$$N = .3837$$

Exercise 131:

Compute the following values:

    (1) 2^{15} (2) 4^9

    (3) $\sqrt{33}$ (4) $\sqrt[3]{52}$

    (5) $\sqrt[3]{.287}$ (6) $\sqrt[5]{.0173}$

    (7) $\sqrt[4]{628,000}$ (8) $\sqrt{.4}$

    (9) $\sqrt{.9}$ (10) $\sqrt[4]{.386}$

The third property is also extremely useful in solving certain types of equations. Study the following examples carefully.

Examples:

    (1) $3^x = 5$

First, we apply common sense. Since $3^1 = 3$ and $3^2 = 9$, we can begin in the certain knowledge that, if $3^x = 5$, then *x must lie between 1 and 2*. Taking the logarithm of both sides and applying Property 3, we have

$$3^x = 5$$
$$\log 3^x = \log 5$$
$$x \log 3 = \log 5$$
$$x = \frac{\log 5}{\log 3}$$

Caution! Do *not* use Property 2 since $\frac{\log 5}{\log 3}$ is *not* the logarithm of a quotient. See Section 12.2.

$$x = \frac{.6990}{.4771}$$
$$\underline{x = 1.4651}$$

    (2) $x^4 = 20$

Trying common sense first, we know that

$$1^4 = 1$$
$$2^4 = 16$$
$$3^4 = 81$$

Therefore, since $x^4 = 20$, *x must lie between 2 and 3*.

$$x^4 = 20$$
$$\log x^4 = \log 20$$
$$4 \log x = \log 20$$

$$\log x = \frac{\log 20}{4}$$

$$\log x = \frac{1.3010}{4}$$

$$\log x = 0.3253$$

$$x = 2.115$$

Exercise 132:

Solve the following equations for x:

(1) $5^x = 30$ (2) $2^x = 12$

(3) $6^{x+1} = 40$ (4) $3^{x-2} = 25$

(5) $x^3 = 31$ (6) $x^5 = .0863$

(7) $x^{2.3} = 50$ (8) $x^{4.2} = 90$

(9) $2^{x-1} = 20$ (10) $4^{.3x} = 6$

(11) $7^{1/2x} = 35$ (12) $8^{1.6x} = 50$

Another useful property of logarithms is one which permits us to change the base of a given logarithm. Consider the equation

1. $b^x = N \longleftrightarrow \log_b N = x$ $(b, x \in R \quad b > 0 \quad b \neq 1)$

2. Let c be any positive base greater than 1.

$$b^x = N$$

$$\text{Then} \qquad \log_c b^x = \log_c N$$

3. $x \log_c b = \log_c N$

4. $x = \dfrac{\log_c N}{\log_c b}$

5. From Statement (1), $x = \log_b N$.

Replacing x by $\log_b N$ in Equation (4) we have

Property 4

$$\log_b N = \frac{\log_c N}{\log_c b}$$

This equation states the fourth property of logarithms and provides a method of changing a logarithm from a given base b to some other base c. An interesting result that can be derived from this property is shown in the following discussion:

Change the base of $\log_b a$ from b to a.

$$\log_b a = \frac{\log_a a}{\log_a b} \qquad \text{(Property 4)}$$

But $\log_a a = 1$. Therefore

$$\log_b a = \frac{\log_a a}{\log_a b} = \frac{1}{\log_a b}$$

or

Property 4A

$$\log_b a = \frac{1}{\log_a b}$$

Property 4A gives us a way of writing Property 4 which involves multiplication rather than the more tedious process of division. To illustrate:

$$\log_b N = \frac{\log_c N}{\log_c b} = \log_c N \frac{1}{\log_c b} = \log_c N \log_b c$$

$$\log_b N = \frac{\log_c N}{\log_c b} = \log_c N \log_b c \qquad \text{Alternate form of Property 4}$$

Illustration of Property 4:   Given that

$$\log_4 64 = 3 \qquad \log_2 64 = 6 \qquad \log_2 4 = 2$$

Property 4

$$\log_4 64 = \frac{\log_2 64}{\log_2 4}$$

or

$$\log_4 64 = \frac{\log_2 64}{\log_2 4}$$
$$3 = \frac{6}{2}$$

Examples:

(1) Find $\log_4 7$.

$$\log_4 7 = \frac{\log_{10} 7}{\log_{10} 4} \qquad \text{(Property 4)}$$

$$= \frac{.8451}{.6021}$$

$$\log_4 7 = 1.4036 \longleftrightarrow 4^{1.4036} = 7$$

(2) Find $\log_2 10$.

$$\log_2 10 = \frac{1}{\log_{10} 2} \qquad \text{(Property 4A)}$$

$$= \frac{1}{.3010}$$

$$\log_2 10 = 3.3223 \longleftrightarrow 2^{3.3223} = 10$$

Exercise 133:

Find the following logarithms:

(1) $\log_3 16$ (2) $\log_2 20$

(3) $\log_5 100$ (4) $\log_3 17$

(5) $\log_4 3860$ (6) $\log_2 3.65$

(7) $\log_{3.6} 25.7$ (8) $\log_{5.3} 424$

(9) $\log_7 10$ (10) $\log_{24} 10$

12.6

Computation with Logarithms

Logarithms are wonderfully handy for simplifying elongated and involved computations because they permit us to substitute addition for multiplication, subtraction for division, and division for extractions of roots. Before the days of electronic computers, they provided the most efficient technique available for computation. Any human computer, using logarithms, will profit by heeding the advice given earlier: label the quantity to be computed as N, and then, using the properties of logarithms, solve the resulting equation for N. Consider the following problem:

1. $\dfrac{432.8}{528,000}$

This operation is bound to produce some number; call it N.

$$N = \frac{432.8}{528,000}$$

$$\log N = \log \frac{432.8}{528,000}$$

$$\log N = \log 432.8 - \log 528,000$$

$$\log N = 2.6363 - 5.7226$$

Stop. If we combine the numbers above, we shall have a negative decimal which will *not* be the mantissa for log N. To avoid this we add *zero* to the *first* number in the form of $10 - 10$.

$$2.6363 + 10 - 10 = 12.6363 - 10$$

$$\log N = 12.6363 - 10 - 5.7226$$
$$\log N = 12.6363 - 5.7226 - 10$$

Combining the first two terms, we have

$$\log N = 6.9137 - 10$$

Therefore, N is a number whose logarithm has a *mantissa* = .9137 and a *characteristic* = 6 − 10 or −4.

$$N = .0008198$$

2. $\dfrac{2375}{.006984}$

$$N = \frac{2375}{.006984}$$

$$\log N = \log \frac{2375}{.006984}$$

$$\log N = \log 2375 - \log .006984$$

$$\log N = 3.3757 \quad - \quad (7.8441 - 10)$$

Note the *parentheses* enclosing the log .006984; this number is a binomial and the minus sign in front of it applies to both of the terms.

$$\log N = 3.3757 - 7.8441 + 10$$
$$\log N = 13.3757 - 7.8441$$
$$\log N = 5.5316$$
$$N = 340,100$$

3. $\dfrac{.428\sqrt[3]{72}}{6328\sqrt{.665}}$

$$N = \frac{.428\sqrt[3]{72}}{6328\sqrt{.665}}$$

$$\log N = \log \frac{.428(72)^{1/3}}{6328(.665)^{1/2}}$$

$\log N = \log .428 \ (72)^{1/3} - \log 6328 \ (.665)^{1/2}$ Property 2

$\log N = \log .428 + \log 72^{1/3} - (\log 6328 + \log .665^{1/2})$ Property 1

$\log N = \log .428 + \dfrac{1}{3} \log 72 - \log 6328 - \dfrac{1}{2} \log .665$ Property 3

$\log N = 9.6314 - 10 + \dfrac{1}{3}(1.8573) - 3.8013 - \dfrac{1}{2}(19.8228 - 20)$

$\log N = 9.6314 - 10 + .6191 - 3.8013 - 9.9114 + 10$

$\log N = 9.6314 + .6191 - 3.8013 - 9.9114$

$\log N = 10.2505 - 13.7127$

$$[10.2505 = 20.2505 - 10]$$

$\log N = 20.2505 - 13.7127 - 10$

$\log N = 6.5378 - 10$

$N = \underline{.0003450}$

Exercise 134:

Compute the values of the following:

(1) $(286{,}000)(.000487)$

(2) $\dfrac{43.7}{6.93}$

(3) $\dfrac{26.8}{4850}$

(4) $\dfrac{(8650)(.0732)}{(48.72)(.249)}$

(5) $\dfrac{\sqrt[3]{672}}{.0378}$

(6) $\dfrac{56.3\sqrt{954}}{.682}$

(7) $\dfrac{(.0036)(8659)}{24.83\sqrt[3]{.743}}$

(8) $(5287)(\sqrt{4265})(\sqrt[3]{2.895})$

(9) $\dfrac{387.6}{(2639)(.168)}$

(10) $\dfrac{472\sqrt[3]{387}}{9.54\sqrt[4]{.5645}}$

(11) $\dfrac{428{,}500}{\sqrt[3]{683}\sqrt[4]{.0893}}$

(12) $\sqrt[3]{\dfrac{40.86}{7.42\sqrt{359}}}$

Chapter Test 12

I. Write the following statements in logarithmic form:

(1) $9^{1/2} = 3$

(2) $2^5 = 32$

(3) $2^{-3} = \dfrac{1}{8}$

(4) $14^0 = 1$

(5) $10^{-2} = \dfrac{1}{100}$

(6) $8^{\log_8 x} = x$

II. Write the following statements in exponential form:

(1) $\log_3 81 = 4$

(2) $\log_{25} \dfrac{1}{5} = -\dfrac{1}{2}$

(3) $\log_{17} 17 = 1$

(4) $\log_a a^x = x$

(5) $\log_{35} 1 = 0$

(6) $\log_{10} .001 = -3$

III. Solve the following equations for x $(x \in R^+)$:

(1) $\log_3 (2x + 1) = \log_3 9 + \log_3 7$

(2) $\log_4 \sqrt{x + 1} + \log_4 5 = \log_4 20 - \log_4 \sqrt{x + 1}$

(3) $\log_2 (x - 1) = \log_2 6 + 3$

(4) $5^x = 30$

(5) $3^{2x+1} = 90$

(6) $x^{2.3} = 25$

IV. Compute the values of the following:

(1) $\log_5 15$

(2) $\sqrt[3]{.27}$ (Be *careful;* $(.3)^3 = .027$)

(3) $\dfrac{(.00432)\sqrt[4]{2440}}{(.0083)^2}$

Appendix

THE
FIELD
POSTULATES
AND
ELEMENTARY
THEOREMS

The algebraic operations considered in the foregoing chapters were approached and developed on an intuitive basis, i.e., by applying the "rules of arithmetic" to numbers symbolized by letters. But algebra had an even less sophisticated beginning; it began in a manner that can be described as *rhetorical* algebra. The ancient Egyptians, for instance, sometimes referred to an unknown number simply as "a heap." *Symbolic* algebra, the use of letters as symbols for numbers, is a fairly modern refinement, being only three or four hundred years old. But for thousands of years before, algebraic techniques for solving specific types of problems had slowly evolved: like Topsy in *Uncle Tom's Cabin*, they "just growed." Consequently, by the beginning of the nineteenth century the subject encompassed a variety of topics dealing with techniques, operations, and applications. There was little to suggest any structure.

Yet basic algebra, like geometry, can and should be based upon a set of postulates. Unfortunately, however, these postulates, necessarily couched in the symbolic language of algebra, are not understandable to the novice. Indeed, they were not even perceptible to the human race through several thousand years of tinkering with algebraic techniques. But after the habit of working with symbols of symbols is acquired, and the language necessary for doing so is learned, then (and only then) the mind is equipped with the experience and knowledge necessary to perceive and understand the postulates of algebra. It is our purpose in this appendix to state a set of postulates which form, in part, the foundations of algebra, and to study some of the logical consequences of these postulates.

A.1
The Closure Property

The real number system is made up of integers, quotients of two integers, and non-ending, non-repeating decimals. Let us confine our attention for the moment to the integers only. When we say "the set of integers" we mean *all* of the integers: the natural numbers, the negatives of the natural numbers, and zero.

In some arithmetic operations the set of integers has a property, or characteristic, called "closure." The word *closure* indicates a sort of exclusive mathematical club. When we say that the set of integers is *closed under addition*, we mean that, when this operation is used on integers, none but *integers* are qualified for membership, every other kind of number being excluded. For example, take any bunch of integers and *add* them.

$$4 + 5 = \underline{9}$$
$$3 + 4 + 6 = \underline{13}$$
$$8 + 2 + 7 + 1 = \underline{18}$$

The result is *always* another *integer*. In other words, we are saying that no matter how many integers you add together, or no matter how often you add different bunches of integers, the only kind of answer you can get is *always another integer*. To describe this situation, we say that the set of integers is *closed* under this operation (in this case, addition).

A set is *closed* under an operation provided that *only elements of that same set* are produced when the operation is applied to elements of the set.

Although closure may appear exceedingly obvious to you, it is a property that is fundamental in advanced work. A student would be well advised to give the matter sufficient time and thought to insure that this property will be clearly understood and remembered. The preceding discussion dealt with the closure of the set of integers under addition and multiplication. Now let us consider this property of closure with other sets of numbers under a designated operation.

We say, for example, that the set of real numbers is closed under addition, which means that the sum of two real numbers is *always* another real number. As another example, the set of rational numbers is closed under multiplication. On the other hand, the set of irrational numbers is *not* closed under multiplication: for example, $\sqrt{8}$ and $\sqrt{2}$ are both irrational, but $(\sqrt{8})(\sqrt{2}) = \sqrt{(8)(2)} = \sqrt{16} = 4$, and the product, 4, is rational. Consider the following questions about closure carefully and remember that *one* counterexample is sufficient to disprove the closure of any set under a specified operation.

A.2
Binary Operations

The four fundamental operations with real numbers in arithmetic and algebra were discussed in the first five chapters of this text. These operations are similar in one significant way: Each one designates a procedure for performing an operation with two numbers to produce another number. Furthermore, two of the fundamental operations, subtraction and division, are ambiguous unless the operations are performed on two numbers in a specified order. If the two numbers, for example, are 8 and 2, then the four fundamental operations applied to these numbers *in the order in which they are named* are as follows:

Addition: $\qquad\qquad$ $8 + 2 = 10$
Subtraction: $\qquad\quad$ $8 - 2 = 6$
Multiplication: $\qquad$ $(8)(2) = 16$
Division: $\qquad\qquad$ $8 \div 2 = 4$

Any such operation applied to two real numbers, given in a specific order, that produces a unique real number is called a *binary operation*. In general terms, a binary operation is defined as follows:

> **Binary operation** — Any rule or procedure of combination which, when applied to two elements of a set, given in a fixed order, produces a unique element of a set.

Note the following comparison of three binary operations:

1. *Addition*

$$\left. \begin{array}{l} 2 + 5 = 7 \\ 5 + 2 = 7 \end{array} \right\} 2 + 5 = 5 + 2$$

2. *Multiplication*

$$\left. \begin{array}{l} (2)(5) = 10 \\ (5)(2) = 10 \end{array} \right\} (2)(5) = (5)(2)$$

3. *Division*

$$\left. \begin{array}{l} 2 \div 5 = \dfrac{2}{5} \\[2mm] 5 \div 2 = \dfrac{5}{2} \end{array} \right\} \quad \dfrac{2}{5} \neq \dfrac{5}{2}$$

Observe that in all three examples above we have performed an operation on two numbers given in a specified order to produce a third number. The purpose of all this is twofold.

1. To show clearly what a binary operation is.
2. To emphasize the fact that a binary operation is *not* necessarily commutative.

If you mull over the foregoing discussion a bit, you may acquire a degree of instant mathematical maturity by realizing that the definition of a binary operation provides considerable leeway for creative activity: that is, the definition allows many possibilities of inventing procedures which, when applied to two real numbers in a specified order, will produce a unique real number. Thus a binary operation is basically a matching device and may be defined as follows: *a binary operation defines a procedure for matching an ordered pair of elements of a set with a unique element of a set.* Observing that any convenient symbol can be used to denote a binary operation, we can invent such operations with abandon. With that in mind, consider the binary operations defined in the following exercises and work out the problems.

Exercise A1:

1.　　For every $a, b \in R$, $a \text{ A } b = \dfrac{a + b}{2}$.

　　Evaluate:

　　(a)　6 A 2　　　　　　　(b)　13 A 7　　　　　(c)　-6 A 4

　　(d)　0 A -10　　　　　(e)　-8 A -4　　　(f)　$\dfrac{1}{2}$ A $\dfrac{3}{2}$

　　(g)　Is a A b commutative?

2.　　For every $a, b \in R$, $a \boxed{\text{X}} b = a^2 + b$.
　　Evaluate:

　　(a)　$2\boxed{\text{X}}4$　　　　(b)　$-2\boxed{\text{X}}4$　　　(c)　$5\boxed{\text{X}}3$

　　(d)　$\dfrac{1}{2}\boxed{\text{X}}\dfrac{3}{4}$　　　　(e)　$7\boxed{\text{X}}-3$　　　(f)　$-5\boxed{\text{X}}-25$

　　(g)　Is $a\boxed{\text{X}}b$ commutative?

3.　　For every $a, b \in R$ except $a \neq 0$, $a \; \square \; b = \dfrac{1}{a} + b$.

　　Evaluate:

　　(a)　$3\square\dfrac{2}{3}$　　　　(b)　$\dfrac{1}{2}\square-4$　　　(c)　$\dfrac{4}{5}\square\dfrac{3}{4}$

　　(d)　$-8\square-2$　　　　(e)　$-1\square-1$　　　(f)　$\dfrac{3}{8}\square\dfrac{1}{8}$

　　(g)　Is $a \; \square \; b$ commutative?

A.3
The Field Postulates

The problems in Exercise A1 show that binary operations are frequently not commutative. Consequently in Chapter 1, Sections 1.5 and 1.7, we paused to observe that the binary operations of multiplication and addition with real numbers *are* commutative, since this is actually a rather special kind of property.

　　Following is a summary of some special features which were noted in multiplication and addition with real numbers:

The sum or product of two real numbers is always another real number.	The closure property
$6 + 0 = 6$	The additive identity
$(6)(1) = 6$	The multiplicative identity
$6 + (-6) = 0$	The additive inverse for a real number
$3 + 2 = 2 + 3$ $(3)(2) = (2)(3)$	The commutative law for addition and multiplication

$(3 + 5) + 2 = 3 + (5 + 2)$ $(3 \cdot 5)(2) = (3)(5 \cdot 2)$	The associative law for addition and multiplication
$4(6 + 2) = 4(6) + 4(2)$	The distributive axiom
$(7)\left(\dfrac{1}{7}\right) = 1$	The multiplicative inverse of a real number

These special properties of real numbers under the binary operations of addition and multiplication are illustrations of the *field properties*, and any set whose elements exhibit the same special properties under two binary operations is called a *field*. Put in general language, these particular properties are called the *field postulates*. We shall restate them in symbolic language for all of the elements of R to show that R is a number field.

The Field Postulates

Given the set R and two binary operations called addition and multiplication.

P₁ For every $a, b \in R$, $a + b \in R$ and $ab \in R$. — The closure property for addition and multiplication

P₂ There is an element $0 \in R$ such that for every $a \in R$, $a + 0 = a$. — The additive identity

P₃ There is an element $1 \in R$, $1 \neq 0$ such that for every $a \in R$, $(a)(1) = a$. — The multiplicative identity

P₄ For every $a \in R$ there is an element $(-a) \in R$ such that $a + (-a) = 0$. — The additive inverse

P₅ For every $a, b \in R$ $a + b = b + a$ and $ab = ba$. — The commutative property for addition and multiplication

P₆ For every $a, b, c \in R$ $(a + b) + c = a + (b + c)$ and $(ab)(c) = (a)(bc)$. — The associative property for addition and multiplication

P₇ For every $a \in R$, except $a = 0$, there is an element $1/a \in R$ such that $(a)(1/a) = 1$. — The multiplicative inverse

P₈ For every $a, b, c \in R$ $a(b + c) = ab + ac$. — The distributive property of multiplication over addition

---

The elements of any set form a *field* if these eight postulates hold for the elements of the set under two binary operations.

A.4
The Construction of a Mathematical Proof

The algebra of real numbers is in large part the study of operations in a number field. The laws and theorems of algebra which we have assumed intuitively in this text all spring from the definitions, the field postulates, the equality axioms, the order postulates, and the substitution axiom. You will find that all of the operational results with these numbers that we assumed to be true can be *proved* to be true using only the definitions, postulates, and axioms.* While these postulates were stated only for the elements of R, any set of elements that form a field will automatically obey these same laws.

For example, consider the field postulate labeled P_4, which states that for every real number, a, there is a real number, $-a$, such that $a + (-a) = 0$. This could prompt the question: is there still another real number (besides $-a$) that one could add to a and get zero? Note that we have previously referred to $-a$ as *the* additive inverse of a, just as we refer to *the* Leaning Tower of Pisa, implying that there is only *one* of them.

Suppose now that we wish to *prove* that "the" additive inverse of a real number such as a is unique; i.e., we want to show that $-a$ is the only additive inverse of a. The statement we make of this fact which we wish to prove is called a *theorem*. By letting the symbol b represent *any* number which we can add to a and get zero, we can state the theorem at hand in symbolic language as follows:

Theorem: If $a, b \in R$ and $a + b = 0$, then $b = -a$

We then attempt to prove this theorem with an argument which leads to the desired conclusion; but each statement we make in our argument, before we proceed to the next statement, must be *justified* by the conditions given in our theorem (i.e. that a and $b \in R$ and that $a + b = 0$) or by postulates, axioms, or definitions. To emphasize the fact that each statement in a proof must be backed up with a justification, we shall number each statement and the corresponding reason alike in the following argument.

Theorem: If $a, b \in R$ and $a + b = 0$, then $b = -a$

Statement		Reason
1. $a, b \in R$	1.	Given
2. $a + b = 0$	2.	Given

Note that Statement 2 is an equation. Therefore, we may apply E_a and add the same number, $-a$, to both sides.

3. $(-a) + (a + b) = (-a) + 0$ 3. E_a

On the right of the equal mark in Statement 3 we have the sum, $(-a) + 0$. Since 0 is the additive identity, in Statement 4, we replace $(-a) + 0$ by its equal, $-a$. (The left side remains unchanged.)

4. $(-a) + (a + b) = -a$ 4. P_2 (additive identity)

---

*This statement applies to the laws of algebra considered in this text. The proofs of theorems dealing with such things as the sum of an infinite geometric series would also require the completeness axiom.

Now, applying the associative property to the terms on the left of the equal mark, we pair the first two terms rather than the last two.

5. $[(-a) + a] + b = -a$ 6. P_6 (associative property)

Applying P_4 to the two terms in brackets we see that $a + (-a) = 0$. Thus we have

6. $0 + b = -a$ 7. P_4 (additive inverse)

Postulate P_2 assures us that the sum on the left of the equal mark, $0 + b$, is equal to b. So we obtain

7. $b = -a$ 8. P_2 (additive identity)

In the seven statements and the corresponding reasons above, we have formulated an argument which shows that, if $a + b = 0$, then b must equal $-a$. Thus $-a$ is the only number we can add to a in order to get a sum equal to zero; i.e., we have proved that the additive inverse of a is unique.

The theorem above and the seven statements in the proof of it are repeated below without the intervening comments. Read the condensed proof from statement to statement, noting carefully in *each successive statement* what change has been made in the *preceding* statement and how that change was justified.

Theorem: If $a, b \in R$ and $a + b = 0$, then $b = -a$

1. $a,b \in R$ 1. Given
2. $a + b = 0$ 2. Given
3. $(-a) + (a + b) = (-a) + 0$ 3. E_a
4. $(-a) + (a + b) = -a$ 4. P_2 (additive identity)
5. $[(-a) + a] + b = -a$ 5. P_6 (associative property)
6. $0 + b = -a$ 6. P_4 (additive inverse)
7. $b = -a$ 7. P_2 (additive identity)

In learning to read and understand proofs of algebraic theorems, there is no great virtue in tackling at one swoop the entire sequence of statements in a completed proof. You can acquire facility at reading proofs just as easily by concentrating on a single statement and the justification of a deduction made from it. In fact, this is the only way to read a proof or any part of one with understanding. As an example, what postulate would justify the deduction made from the statement below?

If $b + 0 = a$ Given
Then $b = a$?

In Chapter 1, Section 1.9, by mentally adding specific quantities of zeros, we concluded that if any real number is multiplied by zero, the product is zero. While this was a good "educated guess" at the time, no guess is a proof. But if the sequence of boxed statements that follow is examined carefully, it should be clear that those statements, each justified by an axiom or a postulate, constitute a *proof* of the fact that the product of *any* real number and zero is equal to zero; i.e., if $a \in R$, then $a \cdot 0 = 0$.

Theorem 1: If $a \in R$, then $a \cdot 0 = 0$

1.	$a \cdot a \in R$	P_1
2.	$a + 0 = a$	P_2
3.	$a(a + 0) = a \cdot a$	E_m
4.	$a \cdot a + a \cdot 0 = a \cdot a$	P_8
5.	$a \cdot 0 + a \cdot a = a \cdot a$	P_5
6.	$(a \cdot 0 + a \cdot a) + [-(a \cdot a)] = a \cdot a + [-(a \cdot a)]$	E_a
7.	$(a \cdot 0 + a \cdot a) + [-(a \cdot a)] = 0$	P_4
8.	$a \cdot 0 + (a \cdot a + [-(a \cdot a)]) = 0$	P_6
9.	$a \cdot 0 + 0 = 0$	P_4
10.	$a \cdot 0 = 0$	P_2

In Theorem 1 we have proved that the product of the *additive identity*, 0, and any element of R is always equal to zero. Think that over for a bit and then compare the result of this theorem with field postulate P_3 which states the existence of the *multiplicative identity* in a number field.

P_3 There is an element $1 \in R$, $1 \neq 0$ such that for every $a \in R$, $(a)(1) = a$

Using these two statements, Theorem 1 and P_3, we can show that, unless we stipulate that the additive identity and the multiplicative identity are *different* elements (e.g., $1 \neq 0$), a field would become a triviality limited to one element only. To illustrate this consider the two statements

1. For every $a \in R$ $(1)(a) = a$
2. For every $a \in R$ $(0)(a) = 0$

Now if 1 and 0 *were* the same element, i.e. if $1 = 0$, we could use the substitution axiom to prove the following:

$$\text{For every } a \in R, \qquad a = (1)(a) = (0)(a) = 0$$

This would require that every element in the field be the same as the additive identity. Thus the field could contain only that one element, and this, in turn, would make Postulate 7 meaningless because the additive identity does not have a multiplicative inverse. Consequently, in stating the field postulates we stipulate that $1 \neq 0$, or, more generally, that the additive identity and the multiplicative identity are two different elements. For this reason, every field contains at least two elements.

In Section 1.4 we assumed that $-(-a) = a$. Using field postulates two (P_2), four (P_4), and six (P_6), and also the equality axioms for addition (E_a) and the symmetric property (E_s), we shall prove that $-(-a) = a$ for all real numbers. Study each step of the following proof carefully, and reread each postulate that is given as a reason.

Theorem 2: If $a \in R$, then $-(-a) = a$

1.	$a \in R$	Given
2.	$a + (-a) = 0$	P_4 (additive inverse)
3.	$[a + (-a)] + (-(-a)) = 0 + (-(-a))$	E_a
4.	$a + [(-a) + (-(-a))] = -(-a)$	P_6 (the associative property for addition) and P_2 (the additive identity)
	Note that $-(-a)$ is the *additive inverse* of $(-a)$.	
5.	$a + 0 = -(-a)$	P_4
6.	$a = -(-a)$	P_2
7.	$-(-a) = a$	E_s

In Section 1.7 we concluded that $(2)(-6)$ was equal to $-(2 \cdot 6)$ or -12 by adding two negative six's, and then we assumed, intuitively, that a positive number multiplied by a negative number always gave a negative product.

Consider the following theorem:

Theorem 3: If $a,b \in R$, then $(a)(-b) = -(ab)$

1.	$b + (-b) = 0$	P_4
2.	$a[b + (-b)] = a \cdot 0$	E_m
3.	$ab + (a)(-b) = 0$	P_8 and Theorem 1
4.	$(a)(-b) + ab = 0$	P_5
5.	$(a)(-b) + ab + [-(ab)] = 0 + [-(ab)]$	E_a
6.	$(a)(-b) + (ab + [-(ab)]) = -(ab)$	P_6 and P_2
7.	$(a)(-b) + 0 = -(ab)$	P_4
8.	$(a)(-b) = -(ab)$	P_2

Since this theorem is true for *all a* and $b \in R$, it would follow immediately that for any such specific case as $a = 2$ and $b = 6$ that $(2)(-6) = -(2 \cdot 6)$ or -12.

In all of the proofs given thus far, we have followed the direct pattern of statement and reason. It is frequently more convenient to use an indirect proof. We do this by assuming the *opposite* of that which we wish to prove and then show that this assumption must be false because it contradicts an established fact. Note the following examples of indirect proofs.

Theorem 4: If $a,b,c \in R$, $ab = c$ and $c \neq 0$, then $a \neq 0$ and $b \neq 0$

1. $ab = c \qquad c \neq 0$ — Given

2. Assume $a = 0$
   Then $(a)(b) = (0)(b) = 0$ — Substitution axiom and Theorem 1
   But $(a)(b) = c$ where $c \neq 0$ — Given
   Therefore $a \neq 0$

3. Assume $b = 0$
   Then $(a)(b) = (a)(0) = 0$ — Substitution axiom and Theorem 1
   But $(a)(b) = c$ where $c \neq 0$ — Given
   Therefore $b \neq 0$

Theorem 5: If $a,b \in R$, $a \neq 0$, $b \neq 0$, then $ab \neq 0$

1. $a \neq 0$ and $b \neq 0$ — Given
2. Assume $ab = 0$
3. $\frac{1}{a} \in R$, since $a \neq 0$ — P_7
4. $\frac{1}{a}(ab) = \frac{1}{a}(0)$ — E_m
5. $\left(\frac{1}{a}a\right)b = 0$ — P_6 and Theorem 1
6. $b = 0$ — P_7 and P_3
   But this is false because $b \neq 0$.
   Therefore $ab \neq 0$.

The following theorem establishes an important fact about products that are equal to zero:

Theorem 6: If $a,b \in R$ and $ab = 0$, then $a = 0$ or $b = 0$ or both

1. $ab = 0$ — Given
2. If $a \neq 0$, then $\frac{1}{a} \in R$ — P_7

3. $\dfrac{1}{a}(ab) = \dfrac{1}{a}(0)$ $\qquad\qquad\qquad\qquad$ E_m

4. $\left(\dfrac{1}{a}a\right)b = 0$ $\qquad\qquad\qquad\qquad$ P_6 and Theorem 1

5. $b = 0$ $\qquad\qquad\qquad\qquad\qquad\quad$ P_7 and P_3

   Similarly, if $b \neq 0$, we can show that $a = 0$.

Exercise A2:

Supply the missing justifications for each step in the proof of the following theorem:

Theorem: If $a, b \in R$ and $a \neq 0$, $b \neq 0$, then $\dfrac{1}{a} \cdot \dfrac{1}{b} = \dfrac{1}{ab}$

1. $a, b \in R$, so $ab \in R$ $\qquad\qquad\qquad$ 1. ?

2. $a \neq 0, b \neq 0$, so $ab \neq 0$ $\qquad\quad$ 2. Theorem 5

3. $ab \neq 0$, so $\dfrac{1}{ab} \in R$ $\qquad\qquad$ 3. ?

4. $a \cdot \dfrac{1}{a} = 1$ $\qquad\qquad\qquad\qquad$ 4. ?

5. $a \cdot \dfrac{1}{a} \cdot b \cdot \dfrac{1}{b} = 1 \cdot b \cdot \dfrac{1}{b}$ $\qquad\quad$ 5. ?

6. $a \cdot b \cdot \dfrac{1}{a} \dfrac{1}{b} = 1 \cdot 1$ $\qquad\qquad$ 6. ?? (Two statements needed)

7. $\dfrac{1}{ab}\left[ab \cdot \dfrac{1}{a} \dfrac{1}{b}\right] = \dfrac{1}{ab} \cdot 1$ $\qquad$ 7. ?? (Two statements needed)

8. $\left[\dfrac{1}{ab} \cdot ab\right]\dfrac{1}{a} \dfrac{1}{b} = \dfrac{1}{ab}$ $\qquad\quad$ 8. ?

9. $1 \cdot \dfrac{1}{a} \dfrac{1}{b} = \dfrac{1}{ab}$ $\qquad\qquad\quad$ 9. ?

10. $\dfrac{1}{a} \dfrac{1}{b} = \dfrac{1}{ab}$ $\qquad\qquad\qquad$ 10. ?

A.5
Proofs of Theorems about Inequalities

The theorems stated and illustrated in Chapter 7 may be proved by using the two order postulates, the definition of "less than," and the equality axioms. Some of these theorems are proved in the following discussion, and each one is identified by the same number, O_1, O_2, O_3, etc., that it had in Chapter 7.

Theorem O_1:   Let $a,b,c \in R$. If $a < b$ then $a + c < b + c$

(1) $a < b$ Given
(2) $a + p = b$ $p > 0$ Definition 1
(3) $a + p + c = b + c$ E_a
(4) $a + c + p = b + c$ Commutative Axiom
(5) $(a + c) + p = b + c$ Associative Axiom
(6) Then $a + c < b + c$ Definition 1

Theorem O_1 establishes the fact that the same number may be added to the members of an inequality without affecting the direction of the inequality sign. The following theorem, O_2, establishes the transitive property of inequalities:

Theorem O_2:   Let $a,b,c \in R$. If $a < b$ and $b < c$, then $a < c$

(1) $a < b$ and $b < c$ Given
(2) $a + p_1 = b$ $p_1 > 0$ Definition 1
(3) $a - b + p_1 = 0$ E_a
(4) $b + p_2 = c$ $p_2 > 0$ Definition 1
(5) Add the equations in steps (3) and (4)
$$a - b + p_1 = 0$$
$$\underline{\quad b + p_2 = c \quad}$$
$$a + p_1 + p_2 = c$$ E_a
(6) Let $p_1 + p_2 = p_3$, then $p_3 > 0$ O_c
(7) From (5)
$$a + p_3 = c$$ Substitution Axiom
(8) Then $a < c$ Definition 1

In the following theorem note carefully that c is a positive number:

Theorem O_3:   Let $a,b,c \in R$. If $c > 0$ and $a < b$, then $ac < bc$.

(1) $a < b$ and $c > 0$ Given
(2) $a + p = b$ $p > 0$ Definition 1
(3) $ac + pc = bc$ E_m
(4) $pc > 0$ O_c
(5) Since $ac + pc = bc$ Definition 1
Then $ac < bc$

Theorem O_4 states that if the members of an inequality are multiplied by a *negative* number, the direction of the inequality sign is reversed. The proof of this theorem may be given in a simplified manner by first establishing the truth of the following theorem, which we shall call Theorem O_0:

Theorem O_0:   Let $c \in R$. If c is positive, then $-c$ is negative

(1)	Since c is positive, then $0 < c$	Definition 3
(2)	$0 + p = c$ $p > 0$	Definition 1
(3)	$-c + 0 + p = -c + c$	E_a
	$-c + p = 0$	
(4)	$-c < 0$	Definition 1
(5)	Since $-c < 0$, then $-c$ is negative	Definition 2

Using Theorem O_0, we can state Theorem O_4 in the following manner:

Theorem O_4:   Let $a,b,c \in R$. If $c > 0$, and $a < b$, then $a(-c) > b(-c)$.

(1)	$a < b$ and $c > 0$	Given
(2)	$a + p = b$ $p > 0$	Definition 1
(3)	$a(-c) + p(-c) = b(-c)$	E_m
(4)	$c > 0$, then $-c < 0$	Theorem O_3
(5)	$p(-c) = -(pc)$	Theorem 3, Section A.4
(6)	$a(-c) + [-(pc)] = b(-c)$	Substitution Axiom
	$a(-c) - (pc) = b(-c)$	
(7)	$a(-c) = b(-c) + (pc)$	E_a
(8)	$b(-c) + pc = a(-c)$	E_s
(9)	$pc > 0$	O_c
(10)	Then $b(-c) < a(-c)$	Definition 1
	or $a(-c) > b(-c)$	

Of the theorems stated in Section 7.2, the four which have been proved are the most useful for our work with inequalities. The proofs of the remaining theorems, except Theorem O_8, require, in part, a more lengthy treatment than our purposes justify and will be omitted. The proof of Theorem O_8 is given as an exercise below in which you are asked to supply the reasons for the statements made.

Exercise A3:

Theorem O$_8$:   For all $a,b \in R$, if $a < b$, then $a < \dfrac{a+b}{2}$ and $\dfrac{a+b}{2} < b$

(1) $a < b$ Given

(2) $a + a < a + b$?
     $2a < a + b$

(3) $a < \dfrac{a+b}{2}$?

(4) $a < b$ Given

(5) $a + b < b + b$?
     $a + b < 2b$

(6) $\dfrac{a+b}{2} < b$?

Table 1
Power, Roots

n	n^2	n^3	$\sqrt{n}$	$\sqrt[3]{n}$	n	n^2	n^3	$\sqrt{n}$	$\sqrt[3]{n}$
1	1	1	1.000	1.000	51	2,601	132,651	7.141	3.708
2	4	8	1.414	1.260	52	2,704	140,608	7.211	3.733
3	9	27	1.732	1.442	53	2,809	148,877	7.280	3.756
4	16	64	2.000	1.587	54	2,916	157,464	7.348	3.780
5	25	125	2.236	1.710	55	3,025	166,375	7.416	3.803
6	36	216	2.449	1.817	56	3,136	175,616	7.483	3.826
7	49	343	2.646	1.913	57	3,249	185,193	7.550	3.849
8	64	512	2.828	2.000	58	3,364	195,112	7.616	3.871
9	81	729	3.000	2.080	59	3,481	205,379	7.681	3.893
10	100	1,000	3.162	2.154	60	3,600	216,000	7.746	3.915
11	121	1,331	3.317	2.224	61	3,721	226,981	7.810	3.936
12	144	1,728	3.464	2.289	62	3,844	238,328	7.874	3.958
13	169	2,197	3.606	2.351	63	3,969	250,047	7.937	3.979
14	196	2,744	3.742	2.410	64	4,096	262,144	8.000	4.000
15	225	3,375	3.873	2.466	65	4,225	274,625	8.062	4.021
16	256	4,096	4.000	2.520	66	4,356	287,496	8.124	4.041
17	289	4,913	4.123	2.571	67	4,489	300,763	8.185	4.062
18	324	5,832	4.243	2.621	68	4,624	314,432	8.246	4.082
19	361	6,859	4.359	2.668	69	4,761	328,509	8.307	4.102
20	400	8,000	4.472	2.714	70	4,900	343,000	8.367	4.121
21	441	9,261	4.583	2.759	71	5,041	357,911	8.426	4.141
22	484	10,648	4.690	2.802	72	5,184	373,248	8.485	4.160
23	529	12,167	4.796	2.844	73	5,329	389,017	8.544	4.179
24	576	13,824	4.899	2.884	74	5,476	405,224	8.602	4.198
25	625	15,625	5.000	2.924	75	5,625	421,875	8.660	4.217
26	676	17,576	5.099	2.962	76	5,776	438,976	8.718	4.236
27	729	19,683	5.196	3.000	77	5,929	456,533	8.775	4.254
28	784	21,952	5.292	3.037	78	6,084	474,552	8.832	4.273
29	841	24,389	5.385	3.072	79	6,241	493,039	8.888	4.291
30	900	27,000	5.477	3.107	80	6,400	512,000	8.944	4.309
31	961	29,791	5.568	3.141	81	6,561	531,441	9.000	4.327
32	1,024	32,768	5.657	3.175	82	6,724	551,368	9.055	4.344
33	1,089	35,937	5.745	3.208	83	6,889	571,787	9.110	4.362
34	1,156	39,304	5.831	3.240	84	7,056	592,704	9.165	4.380
35	1,225	42,875	5.916	3.271	85	7,225	614,125	9.220	4.397
36	1,296	46,656	6.000	3.302	86	7,396	636,056	9.274	4.414
37	1,369	50,653	6.083	3.332	87	7,569	658,503	9.327	4.431
38	1,444	54,872	6.164	3.362	88	7,744	681,472	9.381	4.448
39	1,521	59,319	6.245	3.391	89	7,921	704,969	9.434	4.465
40	1,600	64,000	6.325	3.420	90	8,100	729,000	9.487	4.481
41	1,681	68,921	6.403	3.448	91	8,281	753,571	9.539	4.498
42	1,764	74,088	6.481	3.476	92	8,464	778,688	9.592	4.514
43	1,849	79,507	6.557	3.503	93	8,649	804,357	9.644	4.531
44	1,936	85,184	6.633	3.530	94	8,836	830,584	9.695	4.547
45	2,025	91,125	6.708	3.557	95	9,025	857,375	9.747	4.563
46	2,116	97,336	6.782	3.583	96	9,216	884,736	9.798	4.579
47	2,209	103,823	6.856	3.609	97	9,409	912,673	9.849	4.595
48	2,304	110,592	6.928	3.634	98	9,604	941,192	9.899	4.610
49	2,401	117,649	7.000	3.659	99	9,801	970,299	9.950	4.626
50	2,500	125,000	7.071	3.684	100	10,000	1,000,000	10.000	4.642
n	n^2	n^3	$\sqrt{n}$	$\sqrt[3]{n}$	n	n^2	n^3	$\sqrt{n}$	$\sqrt[3]{n}$

Table II
Common Logarithms log t

t	0	1	2	3	4	5	6	7	8	9
10	0000	0043	0086	0128	0170	0212	0253	0294	0334	0374
11	0414	0453	0492	0531	0569	0607	0645	0682	0719	0755
12	0792	0828	0864	0899	0934	0969	1004	1038	1072	1106
13	1139	1173	1206	1239	1271	1303	1335	1367	1399	1430
14	1461	1492	1523	1553	1584	1614	1644	1673	1703	1732
15	1761	1790	1818	1847	1875	1903	1931	1959	1987	2014
16	2041	2068	2095	2122	2148	2175	2201	2227	2253	2279
17	2304	2330	2355	2380	2405	2430	2455	2480	2504	2529
18	2553	2577	2601	2625	2648	2672	2695	2718	2742	2765
19	2788	2810	2833	2856	2878	2900	2923	2945	2967	2989
20	3010	3032	3054	3075	3096	3118	3139	3160	3181	3201
21	3222	3243	3263	3284	3304	3324	3345	3365	3385	3404
22	3424	3444	3464	3483	3502	3522	3541	3560	3579	3598
23	3617	3636	3655	3674	3692	3711	3729	3747	3766	3784
24	3802	3820	3838	3856	3874	3892	3909	3927	3945	3962
25	3979	3997	4014	4031	4048	4065	4082	4099	4116	4133
26	4150	4166	4183	4200	4216	4232	4249	4265	4281	4298
27	4314	4330	4346	4362	4378	4393	4409	4425	4440	4456
28	4472	4487	4502	4518	4533	4548	4564	4579	4594	4609
29	4624	4639	4654	4669	4683	4698	4713	4728	4742	4757
30	4771	4786	4800	4814	4829	4843	4857	4871	4886	4900
31	4914	4928	4942	4955	4969	4983	4997	5011	5024	5038
32	5051	5065	5079	5092	5105	5119	5132	5145	5159	5172
33	5185	5198	5211	5224	5237	5250	5263	5276	5289	5302
34	5315	5328	5340	5353	5366	5378	5391	5403	5416	5428
35	5441	5453	5465	5478	5490	5502	5514	5527	5539	5551
36	5563	5575	5587	5599	5611	5623	5635	5647	5658	5670
37	5682	5694	5705	5717	5729	5740	5752	5763	5775	5786
38	5798	5809	5821	5832	5843	5855	5866	5877	5888	5899
39	5911	5922	5933	5944	5955	5966	5977	5988	5999	6010
40	6021	6031	6042	6053	6064	6075	6085	6096	6107	6117
41	6128	6138	6149	6160	6170	6180	6191	6201	6212	6222
42	6232	6243	6253	6263	6274	6284	6294	6304	6314	6325
43	6335	6345	6355	6365	6375	6385	6395	6405	6415	6425
44	6435	6444	6454	6464	6474	6484	6493	6503	6513	6522
45	6532	6542	6551	6561	6571	6580	6590	6599	6609	6618
46	6628	6637	6646	6656	6665	6675	6684	6693	6702	6712
47	6721	6730	6739	6749	6758	6767	6776	6785	6794	6803
48	6812	6821	6830	6839	6848	6857	6866	6875	6884	6893
49	6902	6911	6920	6928	6937	6946	6955	6964	6972	6981
50	6990	6998	7007	7016	7024	7053	7042	7050	7059	7067
51	7076	7084	7093	7101	7110	7118	7126	7135	7143	7152
52	7160	7168	7177	7185	7193	7202	7210	7218	7226	7235
53	7243	7251	7259	7267	7275	7284	7292	7300	7308	7316
54	7324	7332	7340	7348	7356	7364	7372	7380	7388	7396
t	0	1	2	3	4	5	6	7	8	9

t	0	1	2	3	4	5	6	7	8	9
55	7404	7412	7419	7427	7435	7443	7451	7459	7466	7474
56	7482	7490	7497	7505	7513	7520	7528	7536	7543	7551
57	7559	7566	7574	7582	7589	7597	7604	7612	7619	7627
58	7634	7642	7649	7657	7664	7672	7679	7686	7694	7701
59	7709	7716	7723	7731	7738	7745	7752	7760	7767	7774
60	7782	7789	7796	7803	7810	7818	7825	7832	7839	7846
61	7853	7860	7868	7875	7882	7889	7896	7903	7910	7917
62	7924	7931	7938	7945	7952	7959	7966	7973	7980	7987
63	7993	8000	8007	8014	8021	8028	8035	8041	8048	8055
64	8062	8069	8075	8082	8089	8096	8102	8109	8116	8122
65	8129	8136	8142	8149	8156	8162	8169	8176	8182	8189
66	8195	8202	8209	8215	8222	8228	8235	8241	8248	8254
67	8261	8267	8274	8280	8287	8293	8299	8306	8312	8319
68	8325	8331	8338	8344	8351	8357	8363	8370	8376	8382
69	8388	8395	8401	8407	8414	8420	8426	8432	8439	8445
70	8451	8457	8463	8470	8476	8482	8488	8494	8500	8506
71	8513	8519	8525	8531	8537	8543	8549	8555	8561	8567
72	8573	8579	8585	8591	8597	8603	8609	8615	8621	8627
73	8633	8639	8645	8651	8657	8663	8669	8675	8681	8686
74	8692	8698	8704	8710	8716	8722	8727	8733	8739	8745
75	8751	8756	8762	8768	8774	8779	8785	8791	8797	8802
76	8808	8814	8820	8825	8831	8837	8842	8848	8854	8859
77	8865	8871	8876	8882	8887	8893	8899	8904	8910	8915
78	8921	8927	8932	8938	8943	8949	8954	8960	8965	8971
79	8976	8982	8987	8993	8998	9004	9009	9015	9020	9025
80	9031	9036	9042	9047	9053	9058	9063	9069	9074	9079
81	9085	9090	9096	9101	9106	9112	9117	9122	9128	9133
82	9138	9143	9149	9154	9159	9165	9170	9175	9180	9186
83	9191	9196	9201	9206	9212	9217	9222	9227	9232	9238
84	9243	9248	9253	9258	9263	9269	9274	9279	9284	9289
85	9294	9299	9304	9309	9315	9320	9325	9330	9335	9340
86	9345	9350	9355	9360	9365	9370	9375	9380	9385	9390
87	9395	9400	9405	9410	9415	9420	9425	9430	9435	9440
88	9445	9450	9455	9460	9465	9469	9474	9479	9484	9489
89	9494	9499	9504	9509	9513	9518	9523	9528	9533	9538
90	9542	9547	9552	9557	9562	9566	9571	9576	9581	9586
91	9590	9595	9600	9605	9609	9614	9619	9624	9628	9633
92	9638	9643	9647	9652	9657	9661	9666	9671	9675	9680
93	9685	9689	9694	9699	9703	9708	9713	9717	9722	9727
94	9731	9736	9741	9745	9750	9754	9759	9763	9768	9773
95	9777	9782	9786	9791	9795	9800	9805	9809	9814	9818
96	9823	9827	9832	9836	9841	9845	9850	9854	9859	9863
97	9868	9872	9877	9881	9886	9890	9894	9899	9903	9908
98	9912	9917	9921	9926	9930	9934	9939	9943	9948	9952
99	9956	9961	9965	9969	9974	9978	9983	9987	9991	9996
t	0	1	2	3	4	5	6	7	8	9

ANSWERS TO EVEN-NUMBERED PROBLEMS

Chapter 1

Exercise 1

2. int.   4. int., n.n.   6. int., n.n.   8. neither   10. int.
12. neither

Exercise 2

2. 32   4. 7   6. 316   8. 25   10. 110   12. 20
14. 20   16. 7   18. 10   20. 5   22. 6

Exercise 3

2. -15 4. -15 6. -24 8. -23 10. -31
12. -22 14. -63 16. -44 18. -189 20. -1833

Exercise 4

2. -3 4. 2   6. 13   8. -12 10. 14   12. 4
14. -5 16. 15   18. 7   20. 579   22. -16 24. 39
26. 6   28. 6   30. -14 32. -18 34. 504

Exercise 5

2. 9   4. 6   6. 3   8. -13 10. 18   12. -15
14. 27   16. 2   18. -9 20. -24 22. 5   24. -43
26. 27   28. -40 30. 29   32. 627   34. -86

Exercise 6

2. 20	4. 90	6. -24	8. -20	10. -36	12. 36
14. -36	16. -36	18. 36	20. 36	22. -27	24. 4
26. -28	28. -1	30. 24			

Exercise 7

2. 2	4. -2	6. -9	8. 9	10. -8	12. 11
14. 7	16. -5	18. -4	20. 10	22. 25	24. -6
26. -2	28. -1	30. 1			

Exercise 8

2. -4	4. -6	6. $\dfrac{7}{8}$
8. 0	10. 0	12. undefined
14. 0	16. -18	18. undefined
20. undefined	22. -4	24. indeterminate
26. 0	28. -1	30. 3
32. 0	34. 75	36. 1

Chapter 2

Exercise 9

2. $-3xy^2$	4. $-3a^3b^2$	6. $-2x^4y^5$	8. $-5xy^2z^5$

Exercise 10

2. one term; monomial	4. 3 terms; multinomial	6. one term; monomial
8. one term; monomial	10. one term; monomial	12. 3 terms; multinomial

Exercise 11

2. $10x$	4. $-4x + 11xy$	6. $10a + 2ab$
8. $6a - 9b$	10. $10x$	12. $11x$
14. $6a^2 + ab + b^2$	16. $a^2 + 7a - 4b - b^2$	18. $12m^2 - 11n^2 - 6$
20. $9x^2yz - 6xy + 3y^2$	22. $8x^2y + 3x^2 - 6xy^2$	

Exercise 12

2. $7x^2 + 6y^2$	4. $4x^2y + 3xy - x^2 + x^2y^2 - 6$
6. $5a^2 - 4a - 2ab + 6$	8. 16

Exercise 13

2. $11x^2y^2 - 9xy + 5x^2 - 4y^2$ 4. 0
6. $4x^2y^2 - 14x^2 + 3xy$ 8. $-2x^2 + 6xy + 11y^2$
10. $a^3 - b^3$

Exercise 14

2. $6a^4b$ 4. $-12x^4y^2$ 6. $4x^5y$ 8. a^5b^3
10. $-30x^6y^2$ 12. $64x^3y^4z^3$ 14. $-12x^5y^3z^4$ 16. $-6x^5y^4$
18. $-10a^5b^3c^2$ 20. $16a^6$ 22. $9x^2$ 24. $-27x^3$
26. $-8a^9$ 28. $-27x^6y^3$ 30. $9a^6b^4$ 32. $16x^{12}$

Exercise 15

2. $-3x^3y + 6x^2y^2 - 3xy^3$ 4. $-4a^3 + 8a^2b + 4ab$
6. 20 8. 4
10. $12a^2 + 8ab$ 12. $12a^2 - 8ab$
14. $2x^4yz - 2x^2y^3z + 6x^2yz^3$ 16. $ab + ac + ad$
18. $mx + my$ 20. $6s^4 - 10s^3 + 12s^2$
22. $-6a^4 + 9a^2b^2 + 12a^3 - 6a^2b$ 24. $12m^5 - 4m^3n^2 + 8m^4n$
26. $3a^3b^2 - 4a^2b^3 + 2a^3b^3$

Exercise 16

2. $14x^2 - 26x - 4$ 4. $2a^2 + 15a + 28$
6. $4a^2 - 16$ 8. $a^2 - b^2$
10. $6x^3 + 18x^2 - 2x^2y - 9xy + y^2$ 12. $a^3 + b^3$
14. $m^2 + 2mn + n^2$ 16. $x^2 + y^2 + z^2 + 2xy + 2xz + 2yz$
18. $2a^4 - 3a^3 - 19a^2 - 6a + 8$ 20. $3x^4 - 16x^3 + 7x^2 + 18x - 8$
22. $9x^2 - 12xy + 4y^2$ 24. $49x^2 - 14x + 1$
26. $4x^2 - 4xy + 4xz - 2yz + y^2 + z^2$ 28. $16a^2 - 8ab + b^2 - 1$

Exercise 18

2. $r^3 + 6r^2s + 12rs^2 + 8s^3$ 4. $x^3 - 3x^2y + 3xy^2 - y^3$
6. $27a^3 - 27a^2 + 9a - 1$ 8. $125y^3 - 150y^2 + 60y - 8$

Exercise 19

2. $4a^2 + 4ab + b^2$ 4. $-27x^3y^3$ 6. $-10x^3 + 35x^2y^2$
8. $3x^3 - 17x^2 + 40x - 80$ 10. $2r^4s^4$
12. $27x^3 - 1$ 14. $20a^3b^3$ 16. $10a^2 + 7ab + b^2$
18. $-6x^4y - 2xy^3$ 20. $-28x^3y^3 + 4x^2y^4$ 22. $16x^2 - 25y^2$
24. $16a^2 - 24a + 9$ 26. $x^4 - y^4$

Exercise 20

2. 40

4. $2xy - 3x^2y$

6. $x^2 + 3x - 4$

8. $2a^2 - 3ab - 2b^2$

10. $6x^2 - 7x - 20$

12. $2x + 6$

14. $-16x$

16. $-2a^2 - 9a + 14$

18. $3x$

20. $8a + 2$

Exercise 21

2. no　　　4. polynomial　　　6. no　　　8. no　　　10. no

Exercise 22

2. $\dfrac{1}{a^2}$　　4. $-b^2$　　6. $\dfrac{1}{a^2}$　　8. $-\dfrac{5}{x}$　　10. $\dfrac{x}{y}$　　12. $\dfrac{1}{ab}$

14. $-7ac$　　16. $\dfrac{-2y}{x}$　　18. $\dfrac{4a^2}{b^3}$　　20. $\dfrac{5a^3}{2b^3c^2}$　　22. xy　　24. $-2ab$

26. $\dfrac{1}{16a^3b^2c^2}$

Exercise 23

2. $6 - \dfrac{x}{3}$　　4. $\dfrac{x}{5} + 1$　　6. $\dfrac{x}{2} + 1$　　8. $1 - \dfrac{b}{a}$

10. $1 - \dfrac{y}{x}$　　12. $2x - 4$　　14. $5b^2 - 1$　　16. $b^2 - \dfrac{2b}{a}$

18. $\dfrac{2}{b^2} - \dfrac{3a}{7b}$　　20. $\dfrac{3x^2}{y} - 2xy + \dfrac{y}{x}$　　22. $-2xy + \dfrac{3}{y} + 1$　　24. $-\dfrac{6a}{b^2} + \dfrac{2b}{a^2} - 1$

Exercise 24

2. $2y + 1$

4. $x^2 + 3x - 4$

6. $2x^2 - 5x - 3, R = 3$

8. $x + 4$

10. $y^2 + y + 1$

12. $3x^4 - 7x^3 + 18x^2 - 35x + 65, R = -130$

14. $3a^3 + a^2 + 2a + 1, R = 2$

16. $2a^4 - 2a^3 + a^2 - 6a - 2$

18. $a^4 - a^3 + a^2 - a + 1$

Exercise 25

2. $\dfrac{-3x^2z^4}{2}$　　4. $4x + 5y$　　6. $\dfrac{2cd^3}{3}$　　8. $x, R = -x$

10. $\dfrac{4ab^3c^6}{7}$　　12. $a - 2, R = 8$　　14. $1 - \dfrac{2b}{3a} + \dfrac{5c}{3a}$

16. $\dfrac{1}{6a^3} - \dfrac{1}{3a^4}$　　18. $x^2 + 3x + 9$　　20. $b^4 + 2b^3 + 4b^2 + 8b + 16$

Chapter 3

Exercise 26

2. $m^2 - 2mn + n^2$ 4. $4m^2 - 4mn + n^2$ 6. $x^2 - 2x + 1$
8. $4x^2 - 4xy + y^2$ 10. $16b^2 - 24b + 9$ 12. $25x^2 - 20xy + 4y^2$
14. $4a^4 + 20a^2 + 25$ 16. $9b^8 - 12b^4 + 4$ 18. 81
20. $9a^2x^2 - 24ax + 16$ 22. $x^4 + 10x^2 + 25$ 24. $x^6 - 6x^3y^4 + 9y^8$
26. $25x^{10} + 20x^5y^3 + 4y^6$

Exercise 27

2. no   4. $(y + 6)^2$ 6. no   8. $(3a - 4)^2$
10. $(2a - 3b)^2$ 12. $(3y - 2)^2$ 14. no   16. $(6a + 1)^2$
18. $(m^2 + 8)^2$ 20. $(2a^3 + 3)^2$ 22. $(5x^2 - 1)^2$ 24. $(7x + 5y)^2$

Exercise 28

2. $a^2 + 5a + 6$ 4. $b^2 + 9b + 20$ 6. $x^2 - x - 6$
8. $x^2 - 5x + 6$ 10. $2x^2 - 7x - 4$ 12. $2x^2 - 9x + 4$
14. $5y^2 + 4y - 12$ 16. $6x^2 + x - 35$ 18. $x^4 + 6x^2 + 8$
20. $6a^4 + 7a^2 - 3$ 22. $x^6 + 6x^3 + 5$ 24. $y^8 - 6y^4 - 16$
26. $20a^2 + 17ab - 10b^2$ 28. $72y^2 + 35y + 3$ 30. $14a^2 - 31ab + 15b^2$

Exercise 29

2. $(x + 2)(x + 1)$ 4. $(y + 4)(y + 1)$

Exercise 30

2. $(a - 5)(a - 4)$ 4. $(y - 5)(y - 3)$

Exercise 31

2. $(b - 4)(b + 2)$ 4. $(a + 5)(a - 2)$

Exercise 32

2. $(a + 3)(a + 1)$ 4. $(x - 8)(x - 3)$ 6. $(b - 6)(b - 1)$
8. $(x - 4)(x - 5)$ 10. $(x + 5)(x - 4)$ 12. $(x - 7)(x - 7)$
14. $(b - 8)(b + 7)$ 16. $(x - 5y)(x + 3y)$ 18. $(x + 10y)(x + 3y)$
20. $(a^2 - 6)(a^2 + 1)$ 22. $(a^2 + 3)(a^2 - 2)$ 24. $(y^3 - 7)(y^3 - 5)$
26. $(x^2 - 7y)(x^2 - 3y)$ 28. $(x^3 - 7y)(x^3 + 4y)$ 30. $(a^2 - 5b^2)(a^2 + 2b^2)$

Exercise 33

2. $(3a + 2)(2a + 1)$ 4. $(4x - 3)(x - 5)$ 6. $(5a - 4)(2a + 3)$
8. $(5a + 7)(a - 4)$ 10. $(3y - 2)(2y - 3)$ 12. $(5a - 3b)(a + 2b)$
14. $(3a + 5b)(2a + 3b)$ 16. $(2x^2 - 5)(3x^2 + 1)$ 18. $(3x^2 - 5)(2x^2 - 3)$
20. $(7x^3 + 3)(2x^3 + 1)$ 22. $(2b + 1)(8b + 3)$ 24. $(4a - 3b)(2a + 5b)$
26. $(6a + 5)(2a - 1)$ 28. $(5b^3 + 3)(b^3 - 2)$ 30. $(3a^4 + 8)(2a^4 + 3)$

Exercise 34

2. (a) $\{1, 2, 3, 6, 9, 18\}$ (b) $\{2, 4, 6, 8, \ldots\}$
   (c) $\{-2, -1, 0, 1, 2, 3, 4\}$ (d) $\{41, 43, 47, 53, 59\}$
4. (a) $\notin$ (b) $\in$ (c) $\in$ (d) $\subseteq$ (e) $\in$ (f) $\subseteq$ (g) $\notin$ (h) $\nsubseteq$
   (i) $\notin$ (j) $\in$ (k) $\nsubseteq$ (l) $\notin$

Exercise 35

2. 6   4. 3   6. 2   8. 1   10. 4

Exercise 36

2. $(y^2 - 8)(y^2 + 1)$ 4. $(7a^2 + 3)(a^2 + 4)$
6. $(4y^5 - 1)(2y^5 - 3)$ 8. $(3a - 2b)(2a + b)$

Exercise 37

24. $7xy^2(3xy - x^2 + 5)$ 26. $9rt(5r^2 - 2t^2)$
28. $2a(11a - 2b + 5c - 1)$ 30. $8x^3y(7y - 2x)$
32. $11m^2n(4m^2 - 11n^3)$ 34. $25xyz^2(3xy^2z^2 + yz - 2x^3)$
36. $17x^2y(2x^2 + 3xy^2 - 1)$ 38. $x^3y^2z^2(y - 4x^2z + 36xyz^2)$
40. $3a^2b^3(8a^3 - 5a^2b + 7ab^2 - 10b^2)$ 42. $3(b - 5)$ or $-3(-b + 5)$
44. $7a(a - 2)$ or $-7a(-a + 2)$ 46. $a^2(b - 1)$ or $-a^2(-b + 1)$
48. $2x^2(-1 + 2x - 4x^2)$ or   50. $9a^2(a^2 + 2a - 7)$ or
     $-2x^2(1 - 2x + 4x^2)$ $-9a^2(-a^2 - 2a + 7)$

Exercise 38

2. $(m + n)(c - d)$ 4. $(a + b)(2x - 5y)$ 6. $(3a - b)(4x^2 - 7y)$
8. $(3x + y)(a - 2b + 4c)$ 10. $(a + b - c)(x - y^2)$
12. $2(2x + y)(2a + b)$ 14. $(x - y)(5a + 2b)$ 16. $x(3x - 5)(4a + b)$
18. $4(3a + b)(x + 1)(2x - 3)$ 20. $(x + y)(1 + x - y)$
22. $x(3a + b)(5x^2 - 2x + 1)$

Exercise 39

2. $3x(x^2 + 9)(x^2 - 6)$
4. $2xy(3y + 4)(4y - 3)$
6. $4xy(x + 4)(x + 2)$
8. $ab(a - 12)(a - 2)$
10. $xy(x - 5y)(x + 2y)$
12. $3x(2x - 1)(3x + 1)$
14. $(2a + b)(x + 1)(x + 1)$

Exercise 40

2. 91
4. $c^2 - 2cd + d^2$
6. $4a^2 - b^2$
8. $9x^2 - 6xy + y^2$
10. $16x^2 + 24xy + 9y^2$
12. $4x^2 - \dfrac{1}{9}$
14. $9a^2 - 25b^2$
16. $81x^2 - 18x + 1$
18. $x^6 - 36$
20. $y^2 - 16y + 64$

Exercise 41

2. $(m + n)(m - n)$
4. $(a + 3b)(a - 3b)$
6. $(b + 1)(b - 1)$
8. $(b^3 + 6)(b^3 - 6)$
10. $(4a + 3b^2)(4a - 3b^2)$
12. $\left(\dfrac{2a}{5} + b\right)\left(\dfrac{2a}{5} - b\right)$
14. $(2x + 7y)(2x - 7y)$
16. $(x^2 - 5y)(x^2 + 5y)$
18. $(b^3 + 3c)(b^3 - 3c)$
20. $\left(a + \dfrac{1}{2}\right)\left(a - \dfrac{1}{2}\right)$

Exercise 42

2. $(6 - a - b)(6 + a + b)$
4. $(1 - 2a + b)(1 + 2a - b)$
6. $(7 - a + b)(7 + a - b)$
8. $3(3x + 1)(x + 1)$
10. $(6 - x + y)(6 + x - y)$
12. $(2a + b - x + y)(2a + b + x - y)$
14. $(2a + 3b + 2)(2a - 3b + 4)$

Exercise 43

2. $3(x + 3)(x - 3)$
4. $2x(x^2 + 2)(x^2 - 2)(x^4 + 4)$
6. $(a^2 + 4)(a + 5)(a - 5)$
8. $3a^2(a + 1)(a - 1)(a + 1)(a - 1)$
10. $x(2x + 1)(2x - 1)(2x^2 + 5)$
12. $a^3(a + 1)(a - 1)$
14. $3x(a + b - 3x)(a + b + 3x)$
16. $x^3(2a - b + x)(2a - b - x)$

Exercise 44

2. $(r^2 - rs + s^2)$
4. $(b^2 - 2b + 4)$
6. $(m^2 + mn + n^2)$
8. $(a^2 + a + 1)$
10. $(y^2 + 3y + 9)$

Exercise 45

2. $(a - 1)(a^2 + a + 1)$
4. $(x + 4)(x^2 - 4x + 16)$
6. $(x - y^2)(x^2 + xy^2 + y^4)$
8. $(x^2 + 1)(x^4 - x^2 + 1)$

10. $(a - 5)(a^2 + 5a + 25)$
14. $2(b + 2)(b^2 - 2b + 4)$
18. $xy(x - y)(x^2 + xy + y^2)$
22. $(x + 1)(x^2 - x + 1)(x^6 - x^3 + 1)$
26. $2b(b - 5)(b^2 + 5b + 25)$

12. $(x - y^3)(x^2 + xy^3 + y^6)$
16. $a(a^2 + 2)(a^4 - 2a^2 + 4)$
20. $3y(3x + 1)(9x^2 - 3x + 1)$
24. $a^2(a - 1)(a^2 + a + 1)$
28. $x^2(x - y)(x^2 + xy + y^2)$

Exercise 46

2. $(x - 1 - y)(x - 1 + y)$
6. $(a - 3 - b)(a - 3 + b)$
10. $(a + b + 2c)(3 + x)$
14. $(m + n)(1 - m - n)$
18. $(x + y)(2x - 2y - 3)$

4. $(2 - c)(a + 2b)$
8. $(x + y)(x - y - 1)$
12. $(x - y)(1 + x - y)$
16. $2(a + 2b)(x - y)$
20. $(x - 2y)(x + 2y - 1)$

Exercise 47

2. $9a^2 - 25$
6. $10a^2 - 29ab + 21b^2$
10. $x^2 - x + \dfrac{1}{4}$
14. $x^6 + 1$
18. $(4x + 1)(4x - 1)$
22. $\left(a + \dfrac{1}{2}\right)^2$
26. $(5x - 4)^2$

4. $9x^5 - 6x^4 + 15x^3$
8. $16x^2 - 24xy + 9y^2$
12. $x^4 - 1$
16. $(5x + 1)(2x - 3)$
20. $(6a + 1)(a - 4)$
24. $(8a + 1)(8a - 1)$
28. $\left(\dfrac{x}{3} + 5\right)\left(\dfrac{x}{3} - 5\right)$

30. $ab(a^2 + b^2)(a + b)(a - b)$
34. $2x(2x - 1)(4x^2 + 2x + 1)$
38. $(x - y)(2a + 3b)$
42. $(4 - 2a - b)(4 + 2a + b)$
46. $(a + b - c - d)(a + b + c + d)$
50. $(b + c)(a - x)$
54. $(a + b)(1 + a - b)$
58. $(x - 3 - y)(x - 3 + y)$

32. $2ab(2a + 5b)(2a - 5b)$
36. $(a^4 + 1)(a^2 + 1)(a + 1)(a - 1)$
40. $3a(2x + y)(a - 2b)$
44. $2a(3a + b)(a + b)$
48. $(a - b + c)(a + b - c)$
52. $(x + y)(1 - a)$
56. $(a - 5 - b)(a - 5 + b)$
60. $(a - b + 1)(a + b - 1)$

Chapter 4

Exercise 48

2. $\dfrac{2a}{a + b}$ or $\dfrac{-2a}{-a - b}$
6. $\dfrac{b - a}{b - a}$ or $\dfrac{a - b}{a - b}$ (both = 1)

4. $\dfrac{x}{x + y}$ or $\dfrac{-x}{-x - y}$
8. $\dfrac{2x + 5y}{2x + 5y}$ or $\dfrac{-2x - 5y}{-2x - 5y}$ (both = 1)

Exercise 49

2. $\dfrac{x}{y - 2x}$ or $-\dfrac{x}{2x - y}$

4. $\dfrac{x + 2y}{y}$ or $-\dfrac{x + 2y}{-y}$

6. $\dfrac{x + y}{x + y}$ or $-\dfrac{x + y}{-x - y}$ (both $= 1$)

8 $\dfrac{4a + b}{2a - 3b}$ or $-\dfrac{4a + b}{3b - 2a}$

Exercise 50

2.

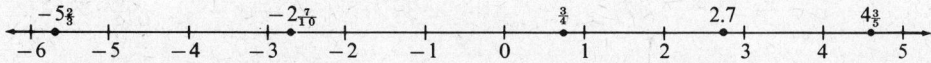

Exercise 51

2. $.333333\overline{3}\ldots$

4. $.285714285714\overline{285714}\ldots$

6. $8.0000\overline{0}\ldots$

8. $.515151\overline{51}\ldots$

Exercise 52

2. Q, R 4. J, Q, R 6. Q, R 8. Q, R 10. Q, R

Exercise 53

2. A number that can be written in the form $\dfrac{a}{b}$, where a is an integer and b is a nonzero integer.

Exercise 54

2. $\dfrac{15}{4}$

4. $\dfrac{21}{5}$

6. $-\dfrac{9}{4}$

8. $\dfrac{x^2}{y}$

10. $\dfrac{-acx}{bdy}$

12. $\dfrac{a^2c}{bd}$

14. $\dfrac{-40x^2}{3y^3z}$

16. $\dfrac{a^2 + 3ab + 2b^2}{a^2 - 4ab + 3b^2}$

18. $\dfrac{x^2 - 2xy + y^2}{x + y}$

20. $\dfrac{2x + 3}{x^2 + x - 6}$

Exercise 55

2. $\dfrac{2}{3}$

4. $\dfrac{1}{7}$

6. $-\dfrac{4}{3}$

8. $\dfrac{x}{2y}$

10. $-\dfrac{1}{2}$ 12. $-\dfrac{1}{2x}$ 14. $-4b$ 16. $\dfrac{1}{2y}$

18. $\dfrac{y}{x+y}$ 20. $x-2y$ 22. $\dfrac{1}{x-1}$ 24. $2b-4$

26. $\dfrac{x+2}{2}$ 28. $-(a+b)$ 30. $\dfrac{b(a+b)}{a-b}$ 32. $\dfrac{2x-3}{3x-2}$

34. $\dfrac{4x+1}{5x-4}$ 36. $\dfrac{x+y}{x^2+xy+y^2}$

Exercise 56

2. 18 4. $\dfrac{x^2+3}{a+1}$ 6. $\dfrac{4}{a-b}$

8. $\dfrac{2y}{x^2+xy+y^2}$ 10. $y-x$ 12. $2b+3a$

14. $\dfrac{a+5}{a-3}$ 16. $\dfrac{1}{a+1}$ 18. $\dfrac{x+1}{x}$

20. $\dfrac{(a-4)(a-2)(a+1)}{a}$ or $\dfrac{a^3-5a^2+2a+8}{a}$ 22. $\dfrac{1}{3(a+4)}$

Exercise 57

2. $\dfrac{3}{32}$ 4. 36 6. 21

8. 16 10. $\dfrac{a}{b}$ 12. $\dfrac{6x}{5}$

14. $\dfrac{x+2}{x+3}$ 16. $\dfrac{b^4}{a}$ 18. $\dfrac{y-5}{y+3}$

20. $\dfrac{a-3}{2(a+3)^2}$ 22. $\dfrac{(x-4)(x-1)}{2(3x-2)}$ 24. $\dfrac{1}{a-1}$

26. $\dfrac{4a(a-5)}{3}$

Exercise 58

2. $\dfrac{7}{10}$ 4. $\dfrac{7}{a}$ 6. 1

8. $\dfrac{a+b-c}{7}$ 10. $\dfrac{5}{x}$ 12. $\dfrac{a+b-c}{x}$

14. $\dfrac{4x-5}{x+1}$ 16. $\dfrac{4x-5}{(x+1)(x-4)}$ 18. $\dfrac{4x-5}{7}$

20. $\dfrac{3a+1}{a-1}$ 22. $\dfrac{-a+7}{a-1}$ 24. $\dfrac{7x}{x+1}$

26. $\dfrac{x+2}{x+1}$ 28. $\dfrac{x+5}{(x+2)(x-1)}$ 30. $\dfrac{6x-4}{(x+2)(x-1)}$

32. $\dfrac{2a+8}{(a+1)(a-2)}$

Exercise 59

2. $\dfrac{7}{5}$

4. 2

6. $\dfrac{1}{5x}$

8. $\dfrac{2}{5}$

10. 4

12. $\dfrac{b}{a+b}$

14. $\dfrac{x}{x+2}$

16. $\dfrac{1}{x+1}$

Exercise 60

2. $\dfrac{ad+bc}{bd}$

4. $\dfrac{108}{35}$

6. $\dfrac{3y^2-2xy+3x^2}{x^2y^2}$

8. $\dfrac{x^2+1}{x}$

10. $\dfrac{b-a}{ab}$

12. $\dfrac{y^2+2}{y}$

14. $\dfrac{14-3x}{x(x-4)}$

16. $\dfrac{a+7}{2(a+3)^2}$

18. $\dfrac{-4x+23}{3(x-3)(x-2)}$

20. $\dfrac{3x^2+11x+34}{(x-6)(x+2)(x+2)}$

22. $\dfrac{2x+2y+1}{x+y}$

24. $\dfrac{2+x^2-y^2}{x+y}$

26. $\dfrac{3a^3+a+2}{3(a+1)(a-1)}$

28. $\dfrac{18x^3-14x^2+6x+5}{2x(2x+1)(3x-2)}$

30. $\dfrac{2x^3-4x^2-25x+8}{(x-5)(x+3)}$

32. $\dfrac{2(x^2+x-4)}{(x+2)^2(x-2)}$

Exercise 61

2. $\dfrac{1}{6}$

4. $\dfrac{4}{3}$

6. $\dfrac{ad-bc}{bd}$

8. $\dfrac{ad}{bc}$

10. $\dfrac{2-xy}{x}$

12. $\dfrac{2}{xy}$

14. $\dfrac{9}{2y}$

16. $\dfrac{2x}{(3x-1)^2}$

18. $\dfrac{2x(x-3)}{(x-1)^3}$

20. $\dfrac{a+2}{a(a-1)}$

22. $\dfrac{x(x^2+y^3)}{2y^3}$

24. $\dfrac{2x+xy^2}{2y^2}$

Exercise 62

2. $\dfrac{8}{7}$

4. $\dfrac{b}{a}$

6. $\dfrac{y(2x+1)}{x(y-1)}$

8. $\dfrac{y-x}{y}$

10. $\dfrac{3x-3y}{2x+2y}$

12. $\dfrac{b+a}{b-a}$

14. $\dfrac{y^2(2x^2-1)}{x^2(2y^2+1)}$

16. $\dfrac{2x^2-2y^2+x-y}{3x^2-3y^2-x-y}$

18. $\dfrac{(a+5)(a-3)}{(3a+11)(a+3)}$

20. $\dfrac{2(a^2-b^2-2)}{(a-b)(a+b+2)}$

Chapter 5

Exercise 63

2. -6 4. 12 6. -3 8. $\dfrac{-7}{4}$ 10. $\dfrac{3}{5}$

12. 2 14. -1 16. -4 18. $-\dfrac{64}{3}$

Exercise 64

2. $\{y \mid -4y = 20\} = \{-5\}$ 4. $\{m \mid -5m = -45\} = \{9\}$

6. $\{y \mid 10y = -40\} = \{-4\}$ 8. $\left\{b \mid \dfrac{1}{3}b = 9\right\} = \{27\}$

Exercise 65

2. $\left\{\dfrac{4}{3}\right\}$ 4. $\{-5\}$ 6. $\{3\}$ 8. $\{-2\}$ 10. $\{-10\}$

12. $\{0\}$ 14. $\{2\}$ 16. $\{5\}$ 18. $\{0\}$ 20. $\{-4\}$

22. $\left\{\dfrac{19}{3}\right\}$ 24. $\{1\}$ 26. $\left\{\dfrac{-19}{4}\right\}$ 28. $\{3\}$ 30. $\{-23\}$

32. $\{3\}$ 34. $\{4\}$ 36. $\left\{-\dfrac{13}{2}\right\}$ 38. $\left\{\dfrac{p}{q}\right\}$ 40. $\left\{\dfrac{r}{m}\right\}$

Exercise 66

2. $\{6\}$ 4. $\{4\}$ 6. $\{-10\}$ 8. $\left\{\dfrac{11}{2}\right\}$ 10. $\left\{\dfrac{4}{9}\right\}$

12. $\varnothing$ 14. $\{-5\}$ 16. $\{0\}$ 18. $\{-9\}$ 20. $\varnothing$

22. $\{1\}$

Exercise 67

2. (a) $\left\{\dfrac{c-b}{a}\right\}$ (b) $\left\{\dfrac{d}{t}\right\}$ (c) $\left\{\dfrac{V}{lh}\right\}$

(d) $\left\{\dfrac{3V}{\pi r^2}\right\}$ (e) $\left\{\dfrac{c-1}{b}\right\}$ (f) $\left\{\dfrac{c-1}{a}\right\}$

(g) $\left\{\dfrac{s+bw}{a}\right\}$ (h) $\left\{\dfrac{at-s}{b}\right\}$ (i) $\left\{\dfrac{c}{a+b}\right\}$

(j) $\left\{\dfrac{e}{c+d}\right\}$ (k) $\left\{\dfrac{d}{b-c}\right\}$ (l) $\left\{\dfrac{t}{p-1}\right\}$

(m) $\{3 - 5y\}$ (n) $\left\{\dfrac{3 - x}{5}\right\}$ (o) $\left\{\dfrac{7 - 2y}{3}\right\}$

(p) $\left\{\dfrac{7 - 3x}{2}\right\}$ (q) $\left\{\dfrac{4 + 7b}{5}\right\}$ (r) $\left\{\dfrac{5a - 4}{7}\right\}$

(s) $\left\{\dfrac{8 + 5y}{4}\right\}$ (t) $\left\{\dfrac{4x - 8}{5}\right\}$ (u) $\left\{\dfrac{bc}{b - c}\right\}$

(v) $\left\{\dfrac{a}{ac - 1}\right\}$ (w) $\left\{\dfrac{A - p}{p}\right\}$ (x) $\left\{\dfrac{9c + 160}{5}\right\}$

Exercise 68

2. $x + 3$ 4. $x - 7$ 6. $\dfrac{x}{2} - 4$ 8. x^3

10. $(x + 4)^2$ or $x^2 + 8x + 16$ 12. $.65x$ 14. $x - .20x$

16. $x + .15x$ 18. $.08x$ liters   20. $\dfrac{20}{x}$ dollars   22. $x(x + 6)$ or $x^2 + 6x$ square meters

24. $\dfrac{200}{x}$ hours   26. $x - 30$ km/h   28. $\dfrac{800}{x - 40}$ hours   30. $\dfrac{1}{x}$

32. $\dfrac{1}{x - 2}$ 34. $\dfrac{75 + 86 + x}{3}$ 36. $x + 2x + x + 2x$ or $6x$ meters

Exercise 69

2. 9, 10, 11   4. 150   6. 5   8. 4, 6
10. 48   12. 6 meters, 20 meters   14. 6 centimeters, 6 centimeters, 2 centimeters

16. 40 meters, 80 meters   18. 5 centimeters   20. 50 meters

22. 3 meters, 4 meters, 5 meters   24. $1\dfrac{3}{22}$ hrs, $4\dfrac{6}{11}$ kilometers   26. $177\dfrac{3}{11}$ kilometers

28. 2 km/h   30. 160 km/h   32. Woochow reaches both at the same time; 2 seconds

34. 18 cm³   36. 60 liters   38. 35 30-cent cards and 15 50-cent cards

40. 80 cm³ cream, 80 cm³ skimmed milk   42. 50 cm³ of each solution   44. 83

46. 5 days   48. $3\dfrac{3}{7}$ hours   50. 20 cm³

52. 3 centimeters, 6 centimeters   54. $34,516.67

Exercise 70

2. $\dfrac{5}{9}$ 4. $\dfrac{8}{9}$ 6. $\dfrac{7}{10}$ 8. $\dfrac{9}{11}$ 10. $\dfrac{48}{11}$

12. $\dfrac{13}{11}$ 14. $\dfrac{79}{111}$ 16. $\dfrac{157}{111}$ 18. 8   20. $\dfrac{3}{4}$

22. $\dfrac{16}{125}$

Chapter 6

Exercise 71

2.

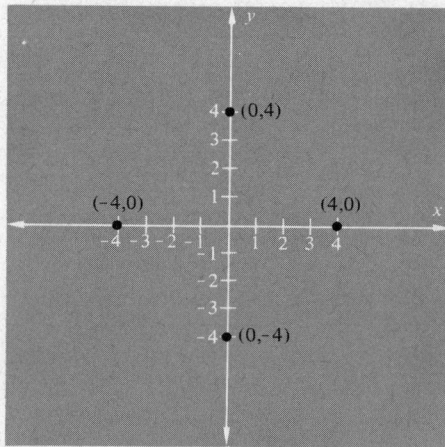

4.

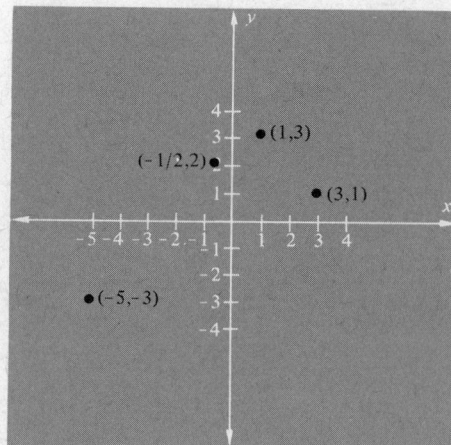

6. Q_1 8. 0   10. $(-4,-3)$

Exercise 72

2. $\{(0,-2),(1,-1),(2,0),(-1,-3),(-2,-4),\ldots\}$

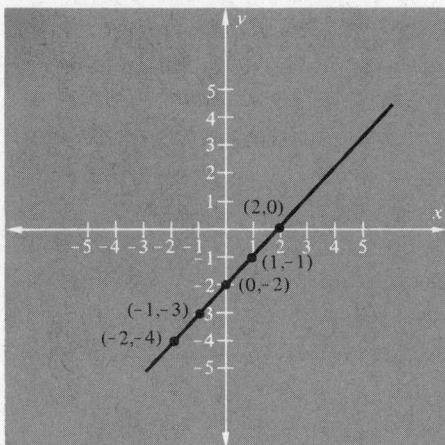

4. $\{(0,-2),(3,-1),(6,0),(-3,-3),(-6,-4),\ldots\}$

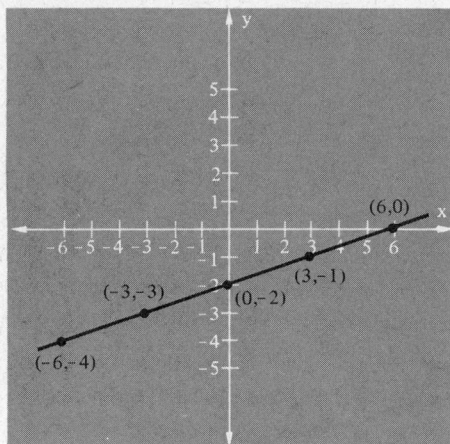

6. $\{(1,-5), (2,0), (3,5), (4,10), (5,15), \ldots\}$

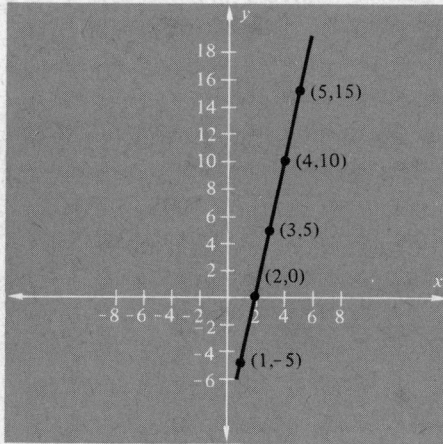

8. $\left\{(0,-1), \left(1,-\frac{2}{5}\right), \left(2,\frac{1}{5}\right), \left(-1,-\frac{8}{5}\right), \left(-2,-\frac{11}{5}\right), \ldots\right\}$

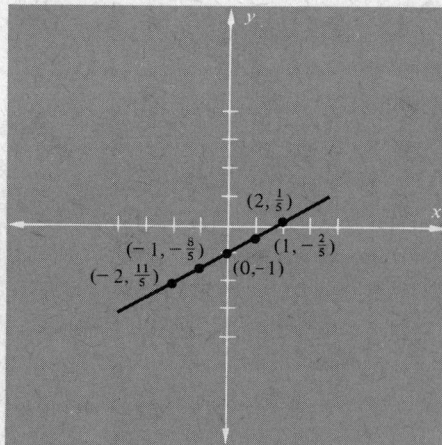

10. $\{(0,-5),(1,-5),(2,-5),(-1,-5),(-2,-5),\ldots\}$

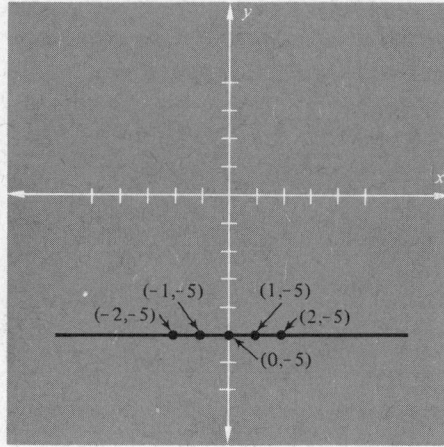

12. $\left\{(0,-4),\left(2,-\dfrac{8}{3}\right),\left(4,-\dfrac{4}{3}\right),\left(-1,-\dfrac{14}{3}\right),\left(-2,-\dfrac{16}{3}\right),\ldots\right\}$

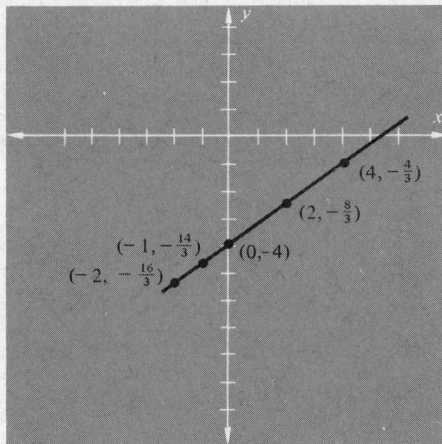

Exercise 73

2. a) $\{(2,-1), (-5,0), (8,2), (4,7), (2,8), (-3,5), (0,-5)\}$
   b) $\{(2,-1), (4,7)\}$
4. a) $A \cup B = \{-5, -1, 0, 2, 3, 5, 7\}$
   b) $A \cup C = \{-5, -3, -1, 0, 2, 3, 7\}$
   c) $B \cup C = \{-5, -3, -1, 0, 2, 5, 7\}$
   d) $A \cap B = \{-1, 0, 7\}$
   e) $A \cap C = \{-5, 0, 7\}$
   f) $B \cap C = \{0, 2, 7\}$

Exercise 74

2.

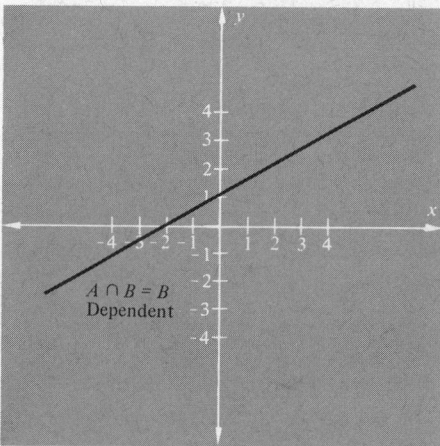

$A \cap B = B$
Dependent

4.

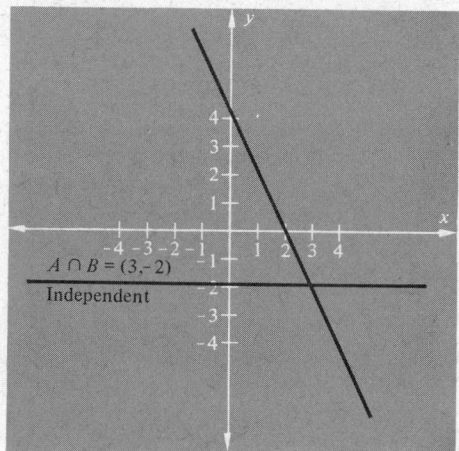

$A \cap B = (3,-2)$
Independent

6.

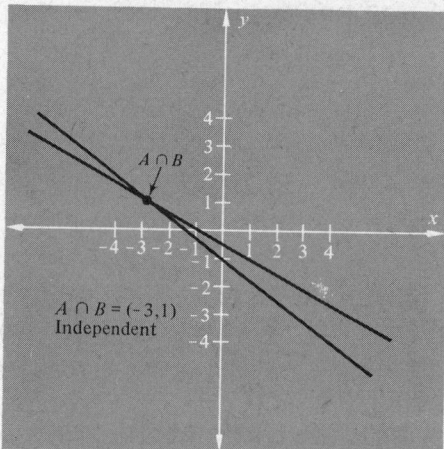

$A \cap B$

$A \cap B = (-3,1)$
Independent

8.

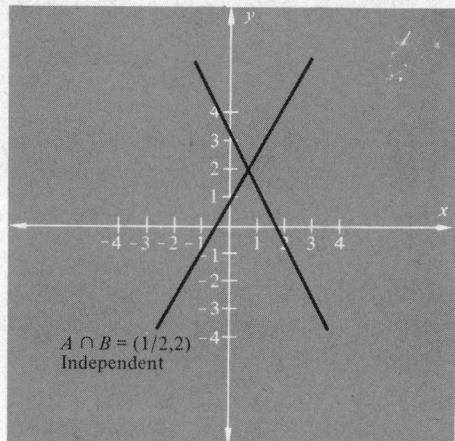

$A \cap B = (1/2,2)$
Independent

10.

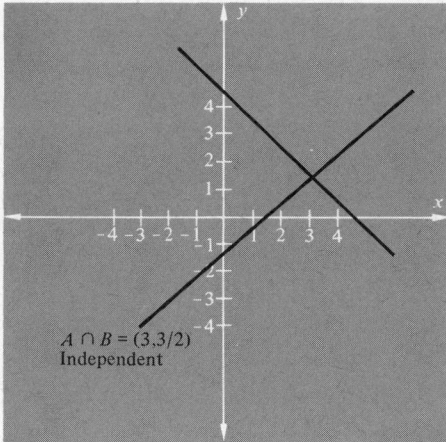

$A \cap B = (3,3/2)$
Independent

12.

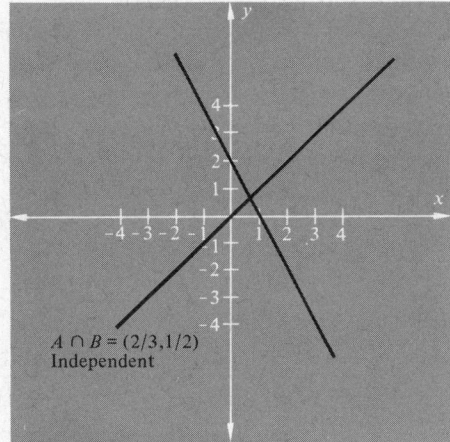

$A \cap B = (2/3,1/2)$
Independent

Exercise 75

2. $\{(4,1)\}$ 4. $\{(2,-4)\}$ 6. $\left\{\left(\frac{1}{3},2\right)\right\}$ 8. $\left\{\left(\frac{2}{3},\frac{1}{2}\right)\right\}$ 10. $\left\{\left(\frac{3}{5},\frac{1}{2}\right)\right\}$

Exercise 76

Same as 1–10, Exercise 75

Exercise 77

2. $x = 1, y = 0, z = -1$
4. $x = -1, y = 0, z = 0$
6. $x = -\frac{1}{2}, y = -3, z = \frac{1}{2}$

Exercise 78

2. 3,8
4. 84
6. 98
8. 4 meters, 7 meters
10. 50 meters, 36 meters
12. 40 meters, 30 meters
14. $6\frac{1}{2}$ hours from A to B;

$5\frac{1}{2}$ hours from B to A;

$357\frac{1}{2}$ kilometers distance
16. 40 kilometers, 50 km/h

18. 4 km/h
20. 85 liters skimmed milk, 15 liters cream
cream

22. 20 liters water, 80 liters brine
24. 3 fifty-cent flowers, 6 twenty-cent
flowers, and 6 thirty-five-cent flowers
26. $300 to college;
$400 to high school
28. $4000 at 6%, $8000 at $5\frac{1}{2}$%

30. 200 shares Stock A;
    125 shares Stock B

32. 30 radios, 20 record players

34. 896 kilometers

Exercise 79

4. $T = f(n)$ 6. $F = f(d)$ 8. $V = f(e)$ 10. $A = f(r)$

Exercise 80

2. $\{(-3,-5), (-2,0), (-1,3), (0,4), (1,3), (2,0), (3,-5), \ldots\}$

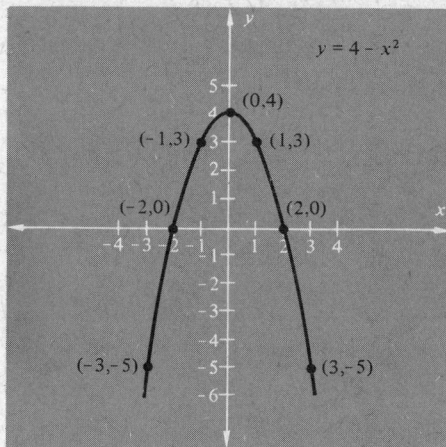

4. $\{(-6,-9), (-4,-8), (-2,-7), (0,-6), (2,-5), (4,-4), (6,-3)\}$

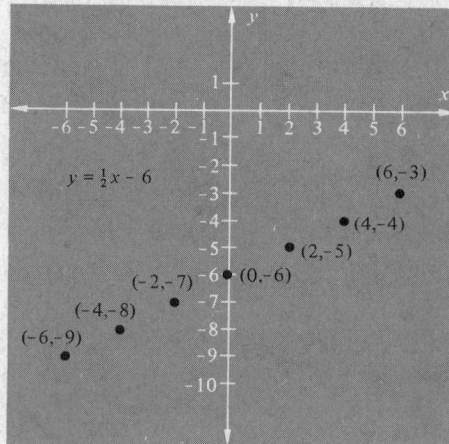

6. $\left\{ (0,0), \ldots, (x,g(x)), \ldots, \left(\frac{1}{2},\frac{7}{4}\right), (1,3), (2,4), \left(\frac{5}{2},\frac{15}{4}\right), (3,3), (4,0) \right\}$

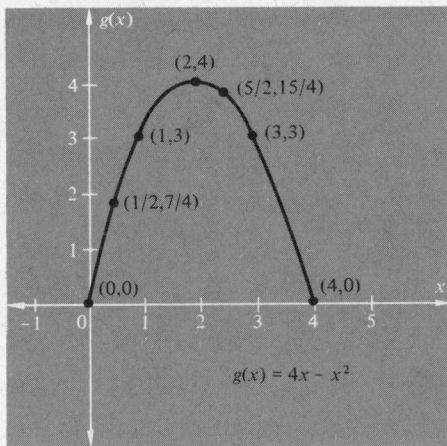

$g(x) = 4x - x^2$

8. $\{ (0,25), \ldots, (x,j(x)), \ldots, (3,16), (4,9), (5,0) \}$

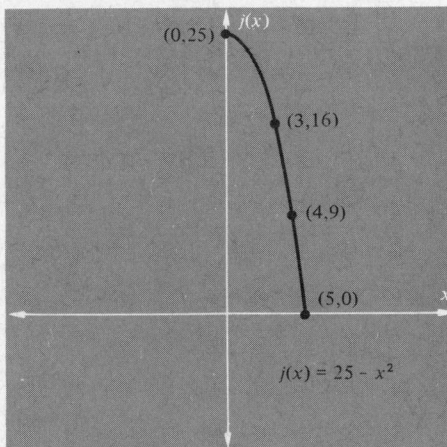

$j(x) = 25 - x^2$

10. $\{ (0,0), \ldots, (x,g(x)), \ldots, (1,1), (4,2), (9,3) \}$

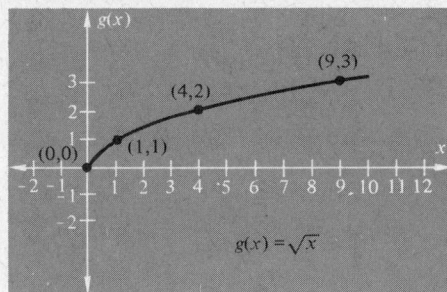

$g(x) = \sqrt{x}$

Exercise 81

(2) $f(-3) - g(5) = 10 - 2 = 8$

(4) $\dfrac{f(-1)}{g(0)} = \dfrac{2}{-3} = -\dfrac{2}{3}$

(6) $\dfrac{f(2)}{g(1)} = \dfrac{5}{-2} = -\dfrac{5}{2}$

(8) $\dfrac{f(5)}{g(3)} = \dfrac{26}{0}$ undefined

(10) $f(a)g(b) = (a^2 + 1)(b - 3) = a^2b - 3a^2 + b - 3$

(12) $[g(2)]^3 = (-1)^3 = -1$

Chapter 7

Exercise 82

2. $-8 < -5$

4. $-10 > -14$

6. $100 > 25$

8. $-\dfrac{1}{6} > -\dfrac{1}{3}$

10. $\dfrac{1}{4} > \dfrac{1}{5}$

12. $-6 > -24$

Exercise 83

2. $x > 11$

4. $x > -2$

6. $x > -4$

8. $x < -12$

10. $x < 3$

12. $x < -\dfrac{5}{2}$

14. $-10 < x < -6$

16. $x < 2$

18. $2 < x < 8$

Exercise 84

2. $\{x \mid x < -1\}$

4. $\{x \mid x > 3\}$

6. $\{x \mid x \geq 8\}$

8. $\{x \mid x > 4\}$

10. $\{x \mid x \leq 9\}$

12. $\{x \mid x \geq -2\}$

14. $\left\{x \mid x > \dfrac{15}{7}\right\}$

Exercise 85

2. $\{x \mid x \geq -8\} \cap \{x \mid x \leq 8\}$

4. $\{x \mid x > -2\} \cap \{x \mid x < 2\}$

6. $\{x \mid x > -4\} \cap \{x \mid x < 8\}$

8. $\left\{x \mid x > -\dfrac{7}{3}\right\} \cap \{x \mid x < 1\}$

10. $\{x \mid x > -1\} \cap \{x \mid x < 3\}$

12. $\{x \mid x \geq -7\} \cap \{x \mid x \leq 1\}$

14. $\{x \mid x > 0\} \cap \{x \mid x < 4\}$

16. $\{x \mid x > -5\} \cap \{x \mid x < 4\}$

Exercise 86

2. $\{x \mid x < -4\} \cup \{x \mid x > 4\}$

4. $\{x \mid x < -1\} \cup \{x \mid x > 9\}$

6. $\{x \mid x < -1\} \cup \{x \mid x > 5\}$

8. $\{x \mid x \leq -3\} \cup \{x \mid x \geq 9\}$

10. $\{x \mid x \leq -2\} \cup \{x \mid x \geq 7\}$

Exercise 87

2.

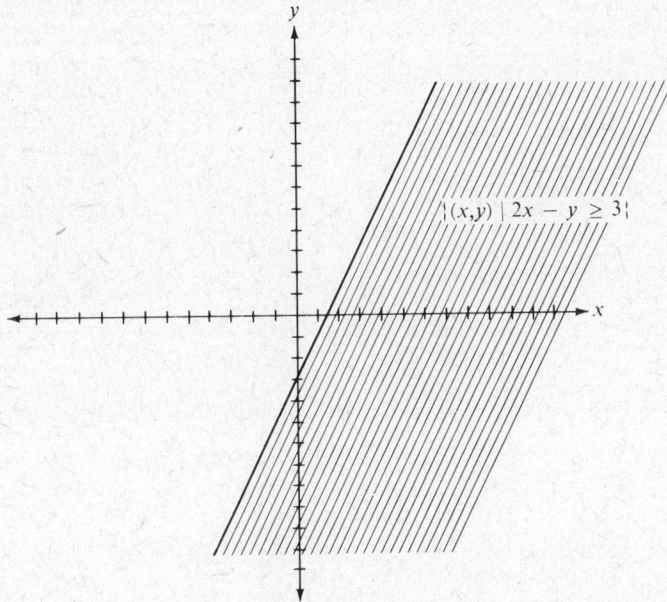

$\{(x,y) \mid 2x - y \geq 3\}$

4.

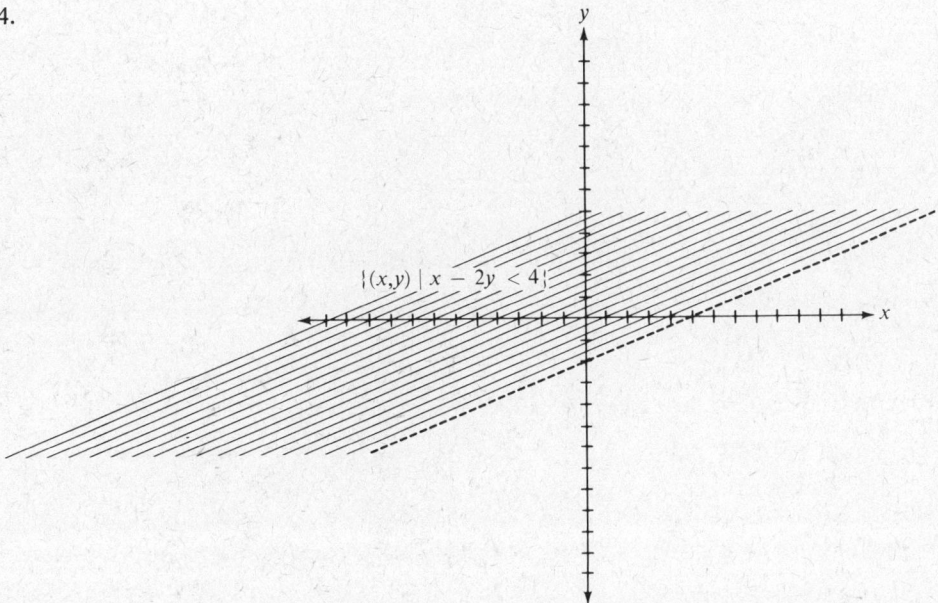

$\{(x,y) \mid x - 2y < 4\}$

6.

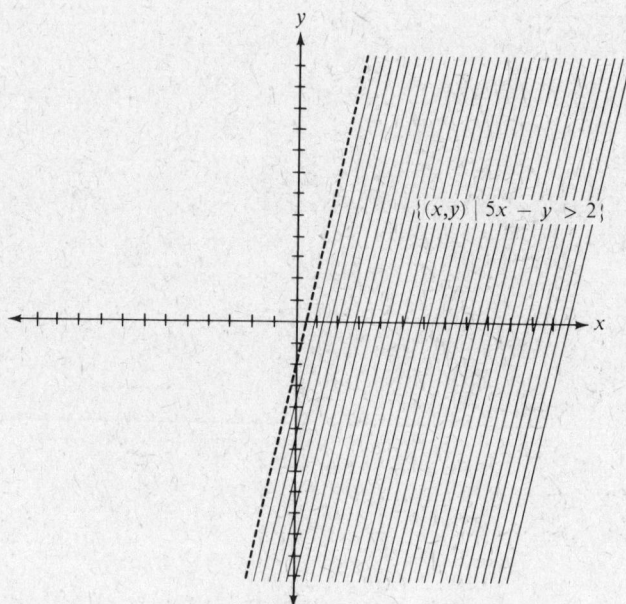

$\{(x,y) \mid 5x - y > 2\}$

8.

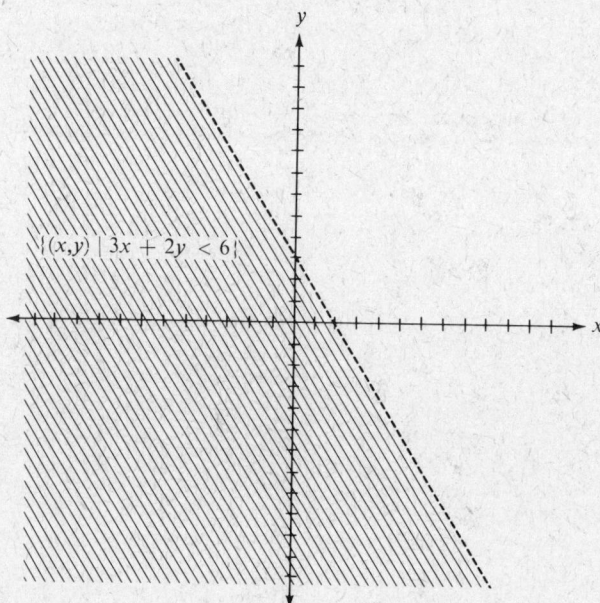

$\{(x,y) \mid 3x + 2y < 6\}$

Exercise 88

2.

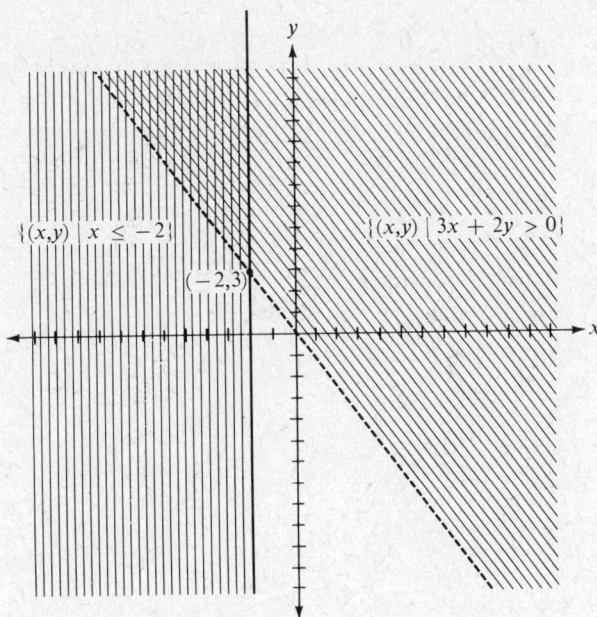

$\{(x,y) \mid x \leq -2\}$

$\{(x,y) \mid 3x + 2y > 0\}$

$(-2,3)$

4.

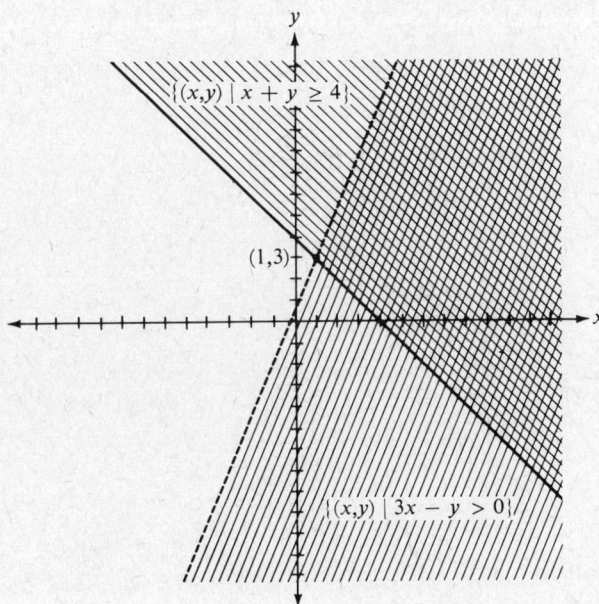

$\{(x,y) \mid x + y \geq 4\}$

$(1,3)$

$\{(x,y) \mid 3x - y > 0\}$

6.

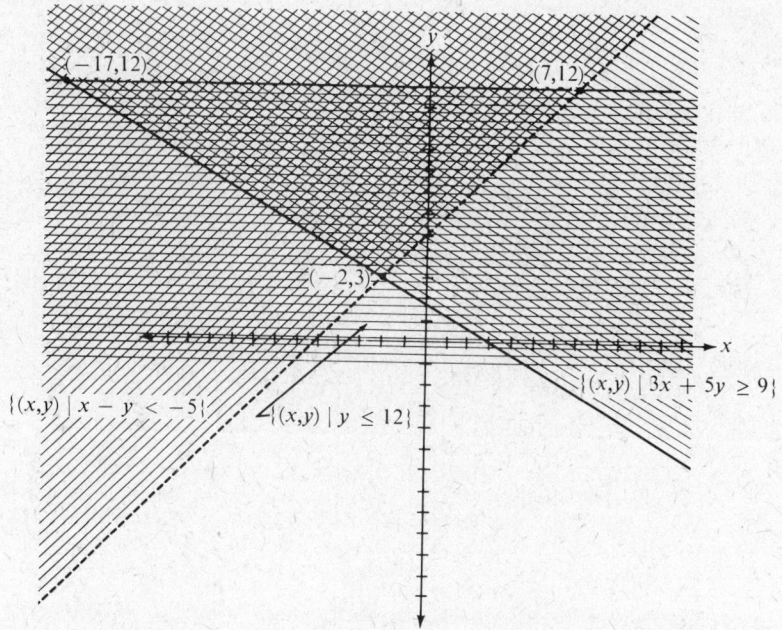

$\{(x,y) \mid x - y < -5\}$

$\{(x,y) \mid y \leq 12\}$

$\{(x,y) \mid 3x + 5y \geq 9\}$

(−17,12) (7,12) (−2,3)

8.

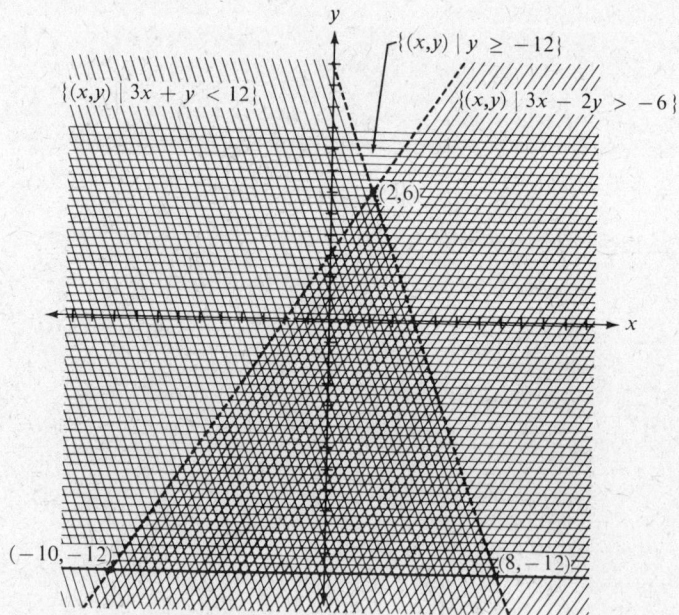

$\{(x,y) \mid y \geq -12\}$

$\{(x,y) \mid 3x + y < 12\}$

$\{(x,y) \mid 3x - 2y > -6\}$

(2,6)

(−10,−12) (8,−12)

Chapter 8

Exercise 89

2. (a) $\dfrac{1}{4}$ (b) $\dfrac{2}{3}$ (c) $\dfrac{3}{5}$ (d) $\dfrac{1}{2}$ (e) $\dfrac{13}{30}$

   (f) $\dfrac{1}{4}$ (g) $\dfrac{17}{800}$ (h) $\dfrac{3}{8}$ (i) $\dfrac{1}{5}$ (j) $\dfrac{11}{20}$

Exercise 90

2. 4 people per car
4. $\dfrac{10}{3}$ cents per gram

6. 47 km/h
8. 54 km/h
10. 25 cents per liter
12. 250 milligrams per tablet
14. 1.5 grams per cm³

Exercise 91

2. (a) 74.3 grams
   (b) 2.83 liters
   (c) 11,428 centigrams
   (d) 5.45 kilograms
   (e) 188.7 centiliters
   (f) 19.4 kilometers
   (g) 2.35 liters
   (h) 257 grams; .257 kilograms; 25,700 centigrams
   (i) 12.7 centimeters
   (j) 9.82 meters
   (k) 32.26 kilometers
   (l) 16.4 feet
   (m) 21.7 miles
   (n) .49 ounce

Exercise 92

2. 8
4. $\dfrac{3}{8}$
6. 32

8. 12 kilograms; $\dfrac{1}{6}$ (your weight)
10. 12.5 cups
12. $5939.76

14. 365 kilograms per square meter
16. $44\dfrac{4}{9}$ hours
18. 4

20. $112.50
22. 8 grams

Exercise 93

2. -12 4. $\dfrac{5}{6}$ 6. $\dfrac{40}{9}$

8. 100.5 centimeters   10. 100 newtons per square meter   12. 180 kilometers

14. 3.60 foot-candles

Chapter 9

Exercise 94

2. y^{13} 4. $2^7 = 128$ 6. $2^{y+2}y^b$

8. 4^{3+n} 10. $(-6)^7$ 12. y^{10}

14. $(-2)^5 = -32$ 16. $(-4)^5 = -1024$ 18. $(x - 2y)^5$

20. $(2x + 1)^{11}$ 22. $6y^{2b+5}$ 24. $(2a + 3)^6$

26. $(3y - x)^{2n}$ 28. $3x^{n+1}$ 30. $(-5)^{a+2}$

Exercise 95

2. 25 4. y^{2b-1} 6. $2x^2$ 8. $\dfrac{7}{2}x^4$

10. $4x^{n-1}$ 12. $(-2x)^a$ 14. $\dfrac{y^{2b}}{2}$ 16. $(2x - y)^{3a+b}$

18. $(2x - y)^{2n-1}$ 20. $\dfrac{5xy^2}{2}$ 22. a^2c^2 24. $x^{2a}y^{2a}$

Exercise 96

2. $81x^8$ 4. $125x^3y^9$ 6. $\dfrac{32c^5}{d^{25}}$ 8. $a^{3y}b^{xy}$

10. $-8a^{3n}b^9$ 12. $81a^{4m}b^{4n}$ 14. $\dfrac{3^a x^{2a} y^a}{z^{5a}}$ 16. $\dfrac{9x^8y^6}{4z^{14}}$

18. $\dfrac{625x^8y^{12}}{256z^{16}}$ 20. $\dfrac{x^{ac}y^{2c^2}}{z^{bc}}$ 22. $\dfrac{x^{3b^2}}{y^{2b^2}}$

Exercise 97

2. $\dfrac{1}{x^n}$ 4. $\dfrac{x^3}{y^2}$ 6. $\dfrac{y^3}{x^2}$ 8. $\dfrac{2}{x^2}$

10. $\dfrac{b}{a^2}$ 12. $\dfrac{6}{a^3}$ 14. y^3 16. a^3c^2

18. $\dfrac{6b^4y^2}{ax^3}$ 20. $\dfrac{1}{10,000\ x^8y^{16}}$ 22. $\dfrac{4a^3x^2}{7by^4}$ 24. $\dfrac{b^2 - a^3b}{a^3}$

26. $(a + b)^2$ 28. $\dfrac{y^2 + x^3}{x^3 y^2}$ 30. $\dfrac{y^2 + x}{y^2 - x}$ 32. $\dfrac{b^2 + a^2}{ab^2}$

34. $\dfrac{216 a^{12} y^3}{b^3 x^9}$

Exercise 98

2. $3x^2$ 4. -5 6. $5y^2$ 8. -3 10. y^4
12. -3 14. $9x^2 y$ 16. $8x^3 y^6$ 18. $4x^2 y^4$ 20. $11a^5 b^4$

Exercise 99

2. $y^{4/5}$ 4. $2x^{1/2} y^{3/4}$ 6. $a^{5/8}$ 8. $5x^{2/3} y^{1/3}$ 10. $-2x^{3/5} y^{1/5}$
12. $\dfrac{1}{11}$ 14. $\dfrac{1}{3}$ 16. $\dfrac{1}{6}$ 18. $4a^4 b^{10}$ 20. 64

Exercise 100

2. x^{n+1} 4. $y^{5/6}$ 6. $a + b$ 8. $18y^3$ 10. y^8
12. $x^{1/6}$ 14. $\dfrac{1}{x^{18}}$ 16. $3a^3 c^2$ 18. $\dfrac{1}{2500 x^{10} y}$ 20. $\dfrac{b^2 - a^2 b}{a^2}$
22. $6a^2 b^3$ 24. $3x^3 y^5$ 26. $\dfrac{x^2}{16}$ 28. $\dfrac{y^2}{2x}$ 30. $\dfrac{x^2 y^2 + y^2}{x}$

Chapter 10

Exercise 101

2. rational   4. rational   6. rational   8. irrational   10. irrational
12. rational   14. irrational   16. irrational   18. rational   20. rational

Exercise 102

2. $\sqrt{6ab}$ 4. 11 6. $7a$ 8. $3x$ 10. 30
12. $6\sqrt{10}$ 14. $4\sqrt{10}$ 16. $10\sqrt{21}$ 18. 24 20. 28
22. 216 24. $\sqrt[12]{a^7}$ 26. $2ab$ 28. $5xy$

Exercise 103

2. $2\sqrt[3]{4}$ 4. $-6\sqrt{3}$ 6. $-x\sqrt{x}$ 8. $4\sqrt{5}$
10. x^3 12. $-3x^4 \sqrt{3x}$ 14. $2x\sqrt{3}$ 16. $-10\sqrt[3]{2}$

18. $4y^3\sqrt[3]{2}$ 20. $10y\sqrt[3]{y^2}$ 22. $12x^2\sqrt{3y}$ 24. $2x^2\sqrt[3]{2}$

26. $2x^2\sqrt[3]{5x}$ 28. $a^2b^3c^4\sqrt{ab}$ 30. $2a\sqrt[4]{2ab^2}$ 32. $-6ab^2\sqrt{2ab}$

34. $2\sqrt{a^2-1}$ 36. $x\sqrt{y^2-1}$ 38. $(a-b)\sqrt{2}$ 40. $(x+1)\sqrt{x}$

Exercise 104

2. $\sqrt[4]{a^2b^3}$ 4. $\sqrt[4]{xy^3}$ 6. $\sqrt[7]{xy^3}$ 8. $2a^2b\sqrt{ab}$

Exercise 105

2. $\dfrac{\sqrt{10}}{4}$ 4. $\dfrac{\sqrt[3]{6}}{3}$ 6. $\dfrac{5\sqrt[3]{4}}{2}$ 8. $\dfrac{3\sqrt{x}}{x}$ 10. $\dfrac{2\sqrt[3]{x}}{x}$

12. $\dfrac{5\sqrt{7x}}{7x}$ 14. $\sqrt{2}$ 16. $\dfrac{\sqrt{15xy}}{5y}$ 18. $\dfrac{\sqrt[3]{12ab^2}}{2b}$ 20. $\dfrac{ab\sqrt{21bx}}{3x}$

22. $\dfrac{8x\sqrt{y}}{3y}$ 24. $\dfrac{2a^2\sqrt{6ab}}{3b^4}$ 26. $\dfrac{\sqrt[3]{6x^2y^2}}{2y}$ 28. $\dfrac{2a\sqrt[3]{b}}{3b}$ 30. $\sqrt[3]{3a}$

Exercise 106

2. $\sqrt{7}+3\sqrt{3}$ 4. $13x\sqrt{2x}$ 6. $\dfrac{10\sqrt{3}}{3}$ 8. $2a\sqrt{2a}$

10. $3\sqrt{3x}$ 12. $\dfrac{16\sqrt{3}}{3}$

14. $7\sqrt{2}+2\sqrt{3}$ 16. $11a^3\sqrt{2ab}$ 18. $\dfrac{3\sqrt{x}}{x}$

20. $\left(\dfrac{3b}{2}+\dfrac{a}{4}\right)\sqrt{2a}+\left(\dfrac{1}{a}-\dfrac{a^2}{3}\right)\sqrt{3b}$ 22. $\dfrac{9x\sqrt{2x}}{2y}$

Exercise 107

2. $11+6\sqrt{2}$ 4. $-4-2\sqrt{11}$ 6. 4 8. -12

10. $23+8\sqrt{7}$ 12. $76+10\sqrt{3}$ 14. $15+\sqrt{21}$ 16. -68

18. $-16-7\sqrt{14}$ 20. $a-b$ 22. $x-25$ 24. $9x^2-x$

26. $4x-y$ 28. $5x-9$

Exercise 108

2. $\sqrt{5}$ 4. $\sqrt{2}-1$

6. $3+2\sqrt{3}$ 8. $\sqrt{2}$

10. $\sqrt{2}$ 12. $\sqrt{a}$

14. $3-2\sqrt{2}$ 16. $\sqrt{a}$

18. $\dfrac{1}{3}$

20. $4 - 2\sqrt{3} + 2\sqrt{6} - 3\sqrt{2}$

22. $\dfrac{2\sqrt{5} + 2\sqrt{15} + \sqrt{3} + 3}{-2}$

24. $\dfrac{a^2 - 2a\sqrt{b} + b}{a^2 - b}$

26. $\dfrac{a\sqrt{a} - a\sqrt{b} + b\sqrt{a} - b\sqrt{b}}{a - b}$

Exercise 109

2. $\dfrac{2 + \sqrt{2}}{2}$ 4. $\dfrac{1 + \sqrt{15}}{4}$ 6. -1 8. $\dfrac{3}{4}$ 10. $3 - \sqrt{5}$

12. $1 + \sqrt{2}$ 14. $\dfrac{1 - 2\sqrt{2}}{2}$ or $\dfrac{1}{2} - \sqrt{2}$

Exercise 110

2. $10l$ 4. $i\sqrt{7}$ 6. $2i\sqrt{6}$ 8. $3ix$ 10. $9iab\sqrt{a}$

12. $\dfrac{ixy\sqrt{15x}}{5}$ 14. $8ai$ 16. $\dfrac{2iy\sqrt{15ax}}{3ab}$ 18. $1 - 3i$ 20. $a - 3i\sqrt{a}$

Exercise 111

4. R, C 6. C 8. C 10. C 12. R, C 14. C

Exercise 112

2. $-5 - i$ 4. $6 + 2i$ 6. $1 - 8i\sqrt{5}$ 8. $15 + 8i$

10. $17 + 20i\sqrt{2}$ 12. $15i$ 14. $6 + 13i\sqrt{2}$ 16. 25

18. $\dfrac{6 + 12i}{5}$ or $\dfrac{6}{5} + \left(\dfrac{12}{5}\right)i$

20. $\dfrac{-1 + 18i}{13}$ or $-\dfrac{1}{13} + \left(\dfrac{18}{13}\right)i$

22. $\dfrac{-7 - 6i}{10}$ or $-.7 - .6i$

24. $\dfrac{-1 + i\sqrt{5}}{2}$ or $-\dfrac{1}{2} + \dfrac{\sqrt{5}}{2}i$

26. $2 - i\sqrt{2}$

28. $\dfrac{-1 - 2i}{2}$ or $-\dfrac{1}{2} - i$

30. $\dfrac{1 + i\sqrt{5}}{2}$ or $\dfrac{1}{2} + \dfrac{\sqrt{5}}{2}i$

32. $(2 - 3i)^2 - 4(2 - 3i) + 13 = 0$
$4 - 12i + 9i^2 - 8 + 12i + 13 = 0$
$4 + 9(-1) - 8 + 13 = 0$
$4 - 9 - 8 + 13 = 0$
$17 - 17 = 0$
$0 = 0$

Chapter 11

Exercise 113

2. $2, -3, 8$ 4. $4, -7, 0$ 6. $8, -2, 3$ 8. $7, 0, -4$

10. $7, 1, 0$ 12. $6, 2, 11$ 14. $3, 5, -16$ 16. $5, 1, 2$

Exercise 114

2. $\{-2\sqrt{2}, 2\sqrt{2}\}$ 4. $\{-i\sqrt{3}, i\sqrt{3}\}$ 6. $\{-2\sqrt{6}, 2\sqrt{6}\}$

8. $\{-2i\sqrt{10}, 2i\sqrt{10}\}$ 10. $\{-4 + \sqrt{7}, -4 - \sqrt{7}\}$

12. $\{-1 - i\sqrt{2}, -1 + i\sqrt{2}\}$ 14. $\left\{-\dfrac{3}{5}, \dfrac{1}{5}\right\}$

16. $\left\{\dfrac{1 - i\sqrt{2}}{3}, \dfrac{1 + i\sqrt{2}}{3}\right\}$ 18. $\left\{\dfrac{1 - i\sqrt{3}}{2}, \dfrac{1 + i\sqrt{3}}{2}\right\}$

Exercise 115

2. $\left\{3, \dfrac{1}{2}\right\}$ 4. $\left\{\dfrac{1}{3}, -\dfrac{3}{2}\right\}$

6. $\{7, -5\}$ 8. $\left\{\dfrac{1 + \sqrt{3}}{2}, \dfrac{1 - \sqrt{3}}{2}\right\}$

10. $\{1 + \sqrt{5}, 1 - \sqrt{5}\}$ 12. $\left\{\dfrac{1 + \sqrt{3}}{4}, \dfrac{1 - \sqrt{3}}{4}\right\}$

14. $\{4 + 3i, 4 - 3i\}$ 16. $\left\{\dfrac{4 + i}{3}, \dfrac{4 - i}{3}\right\}$

18. $\{2 + i\sqrt{3}, 2 - i\sqrt{3}\}$ 20. $\left\{\dfrac{2 + i\sqrt{5}}{3}, \dfrac{2 - i\sqrt{5}}{3}\right\}$

22. $\{-3 + 2\sqrt{3}, -3 - 2\sqrt{3}\}$ 24. $\left\{\dfrac{1}{2}, -1\right\}$

26. $\left\{\dfrac{1 + \sqrt{51}}{10}, \dfrac{1 - \sqrt{51}}{10}\right\}$ 28. $\left\{\dfrac{1 + \sqrt{3}}{2}, \dfrac{1 - \sqrt{3}}{2}\right\}$

30. $\left\{0, \dfrac{9}{5}\right\}$ 32. $\left\{\dfrac{2 + \sqrt{3}}{3}, \dfrac{2 - \sqrt{3}}{3}\right\}$

Exercise 116

2. $\{3, -4\}$ 4. $\{\sqrt{7}, -\sqrt{7}\}$

6. $\{4\}$ 8. $\left\{0, \dfrac{5}{3}\right\}$

10. $\left\{\dfrac{3}{2}\right\}$

12. $\left\{\dfrac{1 + \sqrt{17}}{4}, \dfrac{1 - \sqrt{17}}{4}\right\}$

14. $\left\{\dfrac{4}{3}\right\}$

16. $\{4, -4\}$

18. $\left\{\dfrac{1 + i}{5}, \dfrac{1 - i}{5}\right\}$

20. $\{2i\sqrt{3}, -2i\sqrt{3}\}$

22. $\left\{3, \dfrac{1}{2}\right\}$

24. $\left\{\dfrac{-7 + \sqrt{133}}{14}, \dfrac{-7 - \sqrt{133}}{14}\right\}$

26. $\{7, -5\}$

28. $\left\{\dfrac{1 + \sqrt{3}}{2}, \dfrac{1 - \sqrt{3}}{2}\right\}$

30. $\{1 + \sqrt{5}, 1 - \sqrt{5}\}$

32. $\left\{\dfrac{1 + \sqrt{3}}{4}, \dfrac{1 - \sqrt{3}}{4}\right\}$

34. $\{4 + 3i, 4 - 3i\}$

36. $\left\{\dfrac{4 + i}{3}, \dfrac{4 - i}{3}\right\}$

38. $\{2 + i\sqrt{3}, 2 - i\sqrt{3}\}$

40. $\left\{\dfrac{2 + i\sqrt{5}}{3}, \dfrac{2 - i\sqrt{5}}{3}\right\}$

42. $\left\{\dfrac{1 + \sqrt{51}}{10}, \dfrac{1 - \sqrt{51}}{10}\right\}$

44. $\left\{\dfrac{1 + \sqrt{3}}{2}, \dfrac{1 - \sqrt{3}}{2}\right\}$

46. $\left\{0, \dfrac{9}{5}\right\}$

48. $\left\{\dfrac{2 + \sqrt{3}}{3}, \dfrac{2 - \sqrt{3}}{3}\right\}$

50. $\left\{\dfrac{1}{2}, -1\right\}$

Exercise 117

2. $\{7, -7\}$

4. $\left\{\dfrac{5}{2}, -1\right\}$

6. $\{-2\}$

8. $\left\{-\dfrac{1}{2}, 5\right\}$

10. $\left\{0, \dfrac{1}{2}\right\}$

12. $\left\{\dfrac{3}{4}, \dfrac{5}{2}\right\}$

14. $\left\{-\dfrac{1}{3}\right\}$

16. $\left\{\dfrac{5}{3}, 3\right\}$

18. $\left\{\dfrac{1}{4}, -\dfrac{1}{4}\right\}$

20. $\left\{0, \dfrac{1}{3}\right\}$

22. $\left\{\dfrac{3}{4}, -\dfrac{5}{2}\right\}$

24. $\left\{-\dfrac{3}{4}, -\dfrac{1}{2}\right\}$

26. $\left\{0, \dfrac{3}{4}\right\}$

28. $\left\{\dfrac{3}{4}, -\dfrac{1}{3}\right\}$

30. $\left\{\dfrac{5}{6}, -8\right\}$

32. $\left\{\dfrac{3}{2}, \dfrac{10}{3}\right\}$

34. $\left\{\dfrac{5}{6}, -\dfrac{5}{6}\right\}$

36. $\{2, -2\}$

38. $\left\{-\dfrac{13}{2}, 3\right\}$

40. $\left\{\dfrac{4}{3}, -\dfrac{3}{4}\right\}$

Exercise 118

2. $\left\{\dfrac{2}{3}, -\dfrac{1}{2}\right\}$

4. $\{-3, 1\}$

6. $\{3\}$

8. $\{4, -1\}$

10. $\{2\}$

12. $\left\{\dfrac{1}{2}\right\}$

14. $\{-2\}$

Exercise 119

2. $a^2 + 2ab + b^2$

4. $x + 4\sqrt{x} + 4$

6. $5 - 6\sqrt{x - 4} + x$

8. $4x + 12\sqrt{x + 1} + 13$

10. $5x + 2\sqrt{2x - 1}\sqrt{3x + 1}$

12. $11x - 6\sqrt{x + 3}\sqrt{2x - 1} + 26$

Exercise 120

2. $\{3\}$
4. $\{3\}$
6. $\{5, 1\}$
8. $\varnothing$
10. $\{2, -2\}$

12. $\{3\}$
14. $\left\{\dfrac{1}{4}\right\}$
16. $\{1\}$
18. $\{5\}$
20. $\{7\}$

22. $\{6\}$
24. $\{4\}$
26. $\{5, 1\}$

Exercise 121

2. $4, 5$

4. 3

6. $\left\{\dfrac{5 + \sqrt{37}}{2}, \dfrac{5 - \sqrt{37}}{2}\right\}$

8. 1 meter

10. 8 meters, 16 meters

12. 6 centimeters, 8 centimeters

14. 6 meters

16. 5 centimeters

18. 8 meters, 6 meters

20. 5.4%

22. 60 km/h, 80 km/h

24. 10 hours

26. 6 km/h

28. 26 meters, 52 meters

30. 56 km/h and 76 km/h

Chapter 12

Exercise 122

2. $\log_2 32 = 5$

4. $\log_8 64 = 2$

6. $\log_{64} 8 = \dfrac{1}{2}$

8. $\log_{27} 3 = \dfrac{1}{3}$

10. $\log_{23} 23 = 1$

12. $\log_5 5 = 1$

14. $\log_{2/3} 1 = 0$

16. $\log_8 \dfrac{1}{2} = -\dfrac{1}{3}$

18. $\log_{64} \dfrac{1}{4} = -\dfrac{1}{3}$

20. $\log_{100} \dfrac{1}{10} = -\dfrac{1}{2}$

Exercise 123

2. $7^1 = 7$

4. $b^1 = b$

6. $25^0 = 1$

8. $\left(\dfrac{5}{3}\right)^0 = 1$

10. $3^4 = 81$

12. $10^3 = 1000$

14. $10^{-1} = \dfrac{1}{10}$

16. $36^{-1/2} = \dfrac{1}{6}$

18. $81^{-1/4} = \dfrac{1}{3}$

20. $10^{-2} = \dfrac{1}{100}$

22. $2^6 = 64$

24. $4^3 = 64$

Exercise 124

2. 125	4. $\frac{1}{3}$	6. 1000	8. 2	10. 9
12. 10	14. 10	16. 3	18. 1	20. 1
22. 0	24. 0	26. $-\frac{1}{3}$		

Exercise 125

2. 13	4. 7	6. 3	8. 4	10. 6	12. 118
14. 6	16. 12	18. 17	20. 12	22. 13	24. 56

Exercise 126

2. 3	4. 5	6. 0	8. 1	10. 0	12. 0
14. −3	16. 1	18. 2	20. −1	22. −3	24. 0

Exercise 127

2. 7.9872 − 10	4. 5.8357	6. 9.9542 − 10	8. 1.8797
10. 8.9474 − 10	12. 3.8837	14. 6.5899 − 10	16. 4.3729
18. 8.6749 − 10	20. 7.8561 − 10	22. 4.9217 − 10	24. 2.5340

Exercise 128

2. .277	4. 499	6. 184	8. .0269	10. 430
12. 7.84	14. 224	16. 529000	18. .250	

Exercise 129

2. 1.6887	4. 7.0192 − 10	6. 9.3018 − 10	8. 0.8027
10. 3.6705			

Exercise 130

2. .03961	4. 64.92	6. .001884	8. 1962	10. 5.392

Exercise 131

2. 262,100	4. 3.732	6. .4446	8. .6324	10. .7883

Exercise 132

2. 3.585   4. 4.930   6. .6126   8. 2.919
10. 4.308   12. 1.176

Exercise 133

2. 4.3219   4. 2.5789   6. 1.8679   8. 3.6275   10. .7245

Exercise 134

2. 6.307   4. 52.21   6. 2549   8. 492,400   10. 416.1
12. .6625

Appendix

Exercise A1

2. (a) 8   (b) 8   (c) 28   (d) 1   (e) 46   (f) 0   (g) no

Exercise A2

1. P_1 3. P_7 4. P_7 5. E_m 6. P_5, P_7
7. E_m, P_3 8. P_6 9. P_7 10. P_3

Exercise A3

2. Theorem O_1 3. E_m 5. Theorem O_1 6. E_m

INDEX